# BIOCHEMICAL ANALYSIS OF MEMBRANES

# Biochemical Analysis of Membranes

EDITED BY

A. H. MADDY

*Department of Zoology*
*University of Edinburgh*

LONDON
CHAPMAN AND HALL
A Halsted Press Book
JOHN WILEY & SONS, INC., NEW YORK

*First published 1976*
*by Chapman and Hall Ltd*
*11 New Fetter Lane, London EC4P 4EE*

*Typeset by Preface Ltd, Salisbury, Wilts*
*and printed in Great Britain by*
*William Clowes & Sons Ltd,*
*London, Colchester and Beccles*

ISBN 0 412 12440 8



Distributed in the U.S.A. by Halsted Press,
a Division of John Wiley & Sons, Inc., New York

**Library of Congress Cataloging in Publication Data**
Main entry under title:

Biochemical analysis of membranes.

1. Membranes (Biology) 2. Biological chemistry.
I. Maddy, A. H. [DNLM: 1. Cell membrane – Analysis.
QH601.B611]
QH601.B47 1976 574.8′75 75-40188
ISBN 0-470-14988-4

# Contents

# Contributors


P. S. Agutter — Department of Physiology, University of St. Andrew's, St. Andrew's, Fife, Scotland.

O. J. Bjerrum — Protein Laboratory, University of Copenhagen, Sigurdsgade 34, Copenhagen, DK-2200, Denmark.

T. C. Bøg-Hansen — Protein Laboratory, University of Copenhagen, Sigurdsgade 34, Copenhagen, DK-2200, Denmark.

R. M. Broekhuyse — Department of Biochemistry and Institute of Ophthalmology, University of Nijmegen, Nijmegen, The Netherlands.

Z. A. Cohn — The Rockefeller Institute, New York, New York 10021, U.S.A.

G. M. K. Cook — Strangeways Research Laboratory, Wort's Causeway, Cambridge, CB1 4RN, England.

G. Dallner — Department of Pathology, Sabbatsberg Hospital, Karolinska Institute, Stockholm, Sweden.

J. DePierre — Department of Biochemistry, Arrhenius Laboratory, University of Stockholm, Stockholm, Sweden.

M. J. Dunn — Department of Zoology, University of Edinburgh, West Mains Road, Edinburgh, EH9 3JT, Scotland.

J. R. Harris — Department of Physiology, University of St. Andrew's, St. Andrew's, Fife, Scotland.

A. L. Hubbard — Yale University School of Medicine, New Haven, Connecticut 06510, U.S.A.

A. H. Maddy — Department of Zoology, University of Edinburgh, West Mains Road, Edinburgh, EH9 3JT, Scotland.

| | |
|---|---|
| D. M. Neville | Laboratory of Neurochemistry, National Institute of Mental Health, 9000 Rockville Pike, Bethesda, M.D. 20014, U.S.A. |
| S. Razin | Department of Clinical Microbiology, The Hebrew University, Hadassah Medical School, Jerusalem, Israel. |
| B. Roelofsen | Rijksuniversiteit Te Utrecht, Department of Biochemistry, Transitorium B, Universiteitcentrum 'De Uithof', Padualaan 8, Utrecht, The Netherlands. |
| S. Rottem | Department of Clinical Microbiology, The Hebrew University, Hadassah Medical School, Jerusalem, Israel. |
| G. L. Sottocasa | Instituto di Chimica Biologica, Villa Sevastopulo, Via Favio Severo 150, 34127 Trieste, Italy. |
| J. H. Veerkamp | Department of Biochemistry and Institute of Ophthalmology, University of Nijmegen, Nijmegen, The Netherlands. |
| R. F. A. Zwaal | Rijksuniversiteit Te Utrecht, Department of Biochemistry, Transitorium B, Universiteitcentrum 'De Uithof', Padualaan 8, Utrecht, The Netherlands. |

# Preface

In natural membranes biochemists have, perhaps for the first time, attempted the analysis of a dynamic supramolecular structure. This study requires the application of the most sophisticated physico-chemical techniques, yet the interpretation of the data obtained by these techniques is ultimately limited by our knowledge of the molecular components of membranes, and the provision of information on these building units falls largely within the province of the biochemist. It is hoped that this book will assist biochemists, and biologists without a formal biochemical training, involved in this quest in two ways; firstly, by providing practical guidance on the most important current techniques, and secondly by providing sufficient theoretical background to these techniques to enable workers to modify the details to fit their own precise requirements.

The book falls into two sections, the first of which deals with methods for the isolation of various types of membrane, (the absence of a chapter on the erythrocyte membrane is not due to an oversight but due to the fact that several excellent modern reviews on this subject already exist, e.g. G. Schwach and M. Passow (1973), *Molec. Cell. Biochem.*, 2, 197; thus making a further one superfluous), whilst the second is devoted to biochemical methods for the study of the isolated membranes. In this latter section the first four chapters relate to the isolation and chemical analysis of the membrane proteins, lipids and carbohydrates, while later chapters are concerned with the spatial distribution of lipids and proteins within membranes, and a review of the application of modern immunological methods to membrane study. Each chapter ends with a special section devoted to experimental procedures.

*August 1975* A. H. Maddy

# SECTION ONE

# 1 Techniques for the manipulation of mycoplasma membranes

SHMUEL RAZIN and SHLOMO ROTTEM



## 1.1 Introduction

Mycoplasmas are minute prokaryotic micro-organisms, varying in shape from spherical structures with diameters of 0.3 to 1.0 μm to slender branched filaments of uniform diameter (0.2–0.3 μm) ranging in length from a few μm to 150 μm. Hence, the name *Mycoplasma*, fungus form (Fig. 1.1). Electron microscopy of thin sections of mycoplasmas reveals an extremely simple ultrastructure, supporting the idea that these may be the simplest and most primitive organisms capable of autonomous growth. The mycoplasma cell is built of only three organelles: the cell membrane, the ribosomes, and the prokaryotic nucleoid (Fig. 1.2). Unlike all other prokaryotes, mycoplasmas have no cell walls or intracellular membranous structures. The absence of a cell wall in mycoplasmas is a characteristic of outstanding importance, to which the mycoplasmas owe many of their peculiarities, for example, their morphological instability, osmotic sensitivity, tendency to penetrate and grow in depth of solid media, resistance to antibiotics which interfere with cell wall synthesis, susceptibility to lysis by detergents, alcohols, specific antibody and complement etc. [1,2].

The fact that mycoplasma cells contain only one membrane type, the plasma membrane, constitutes one of their most useful properties for membrane studies since it is certain that the plasma membrane, once isolated, is uncontaminated with other membrane types. This is also true for erythrocytes. However, working

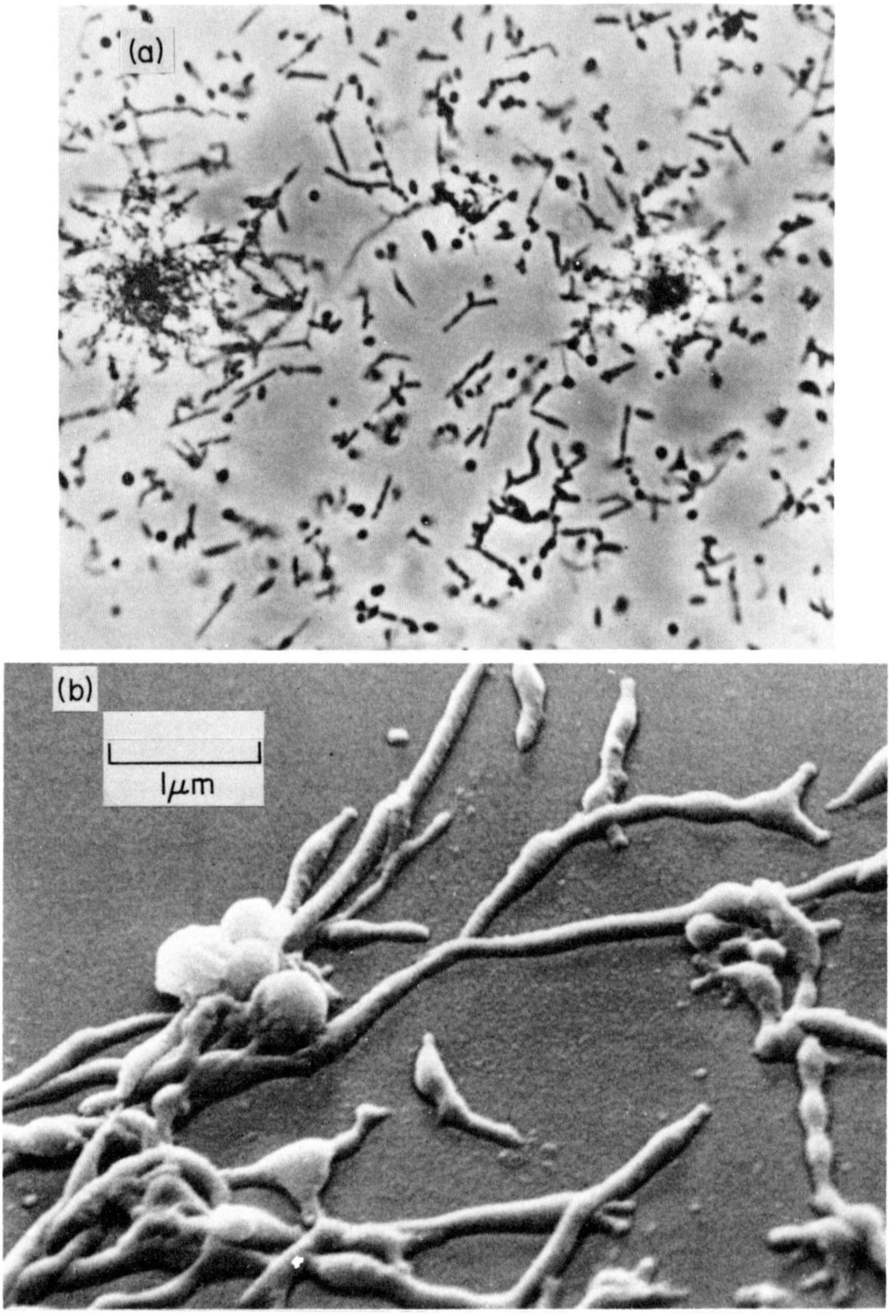


**Fig. 1.1** (a) Phase-contrast micrograph of a 24 hour broth culture of *Acholeplasma laidlawii* (x1250). From Razin, S., Cosenza, B. J. and Tourtellotte, M. E. (1967), *Ann. N.Y. Acad. Sci.*, **143**, 66. (b) Scanning electron micrograph of a 6 day culture of *Mycoplasma pneumoniae*. From Biberfeld, G. and Biberfeld, P. (1970), *J. Bact.*, **102**, 855. The micrographs show the characteristic morphological elements that can be seen in mycoplasma cultures. These include branched filaments, chains of cocci and single spherical cells.

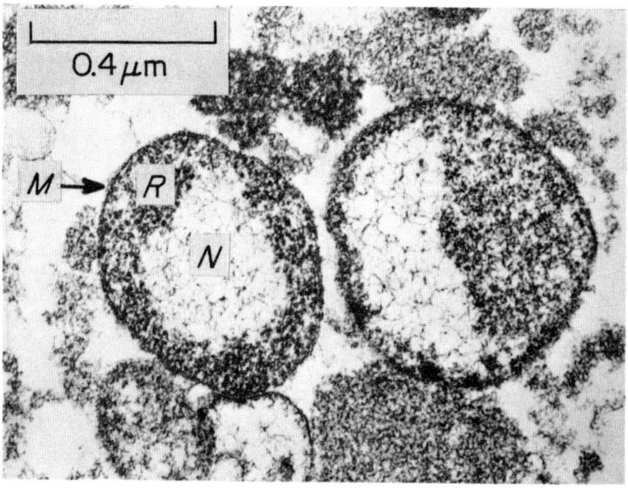


**Fig. 1.2** Thin sections of *Mycoplasma hominis* cells showing the trilaminar shape of the plasma membrane (*M*), ribosomes (*R*), and the fibrillar nucleoid region (*N*). From Anderson, D. R. and Barile, M. F. (1965), *J. Bact.*, **90**, 180.

with mycoplasmas has the added advantage that the mycoplasma cell is fully capable of growth and reproduction. This can be exploited to introduce variations in membrane composition in ways not at all possible with erythrocyte membranes. The minute mycoplasmas, which are all parasites, lack many of the biosynthetic pathways found in the more familiar bacteria, such as *E. coli*, and therefore need complex growth media to provide the numerous essential nutrients. These include membrane lipid components or their precursors. This dependence on an external supply of lipids provides a powerful tool for the introduction of controlled changes in membrane lipids, so that the effects of specific lipid components on membrane structure and function can be analysed as will be described in detail in a later section. For extensive recent reviews on the physiology of mycoplasmas and their membrane biochemistry the reader is referred to Maniloff and Morowitz [3] and to Razin [2,4].

The number of established species of mycoplasmas has been increasing most markedly during the past few years. Some mention should, therefore, be made of the taxonomy and nomenclature of these organisms. All the mycoplasmas are members of the class Mollicutes (soft-skin organisms) established in 1967 to include all the prokaryotes without cell walls. This class so far contains only one order, Mycoplasmatales, and two families, *Mycoplasmataceae* and *Acholeplasmataceae*. The *Mycoplasmataceae* is divided into two genera: *Mycoplasma*, which includes the cholesterol-requiring species, and *Ureaplasma*, which are also capable of hydrolysing urea (T-mycoplasmas), while the *Acholeplasmataceae*, which do not require cholesterol for growth, contains so far a single genus, *Acholeplasma* [5–7]. The trivial term 'mycoplasmas' is generally used to include all the organisms classified in both families of the order

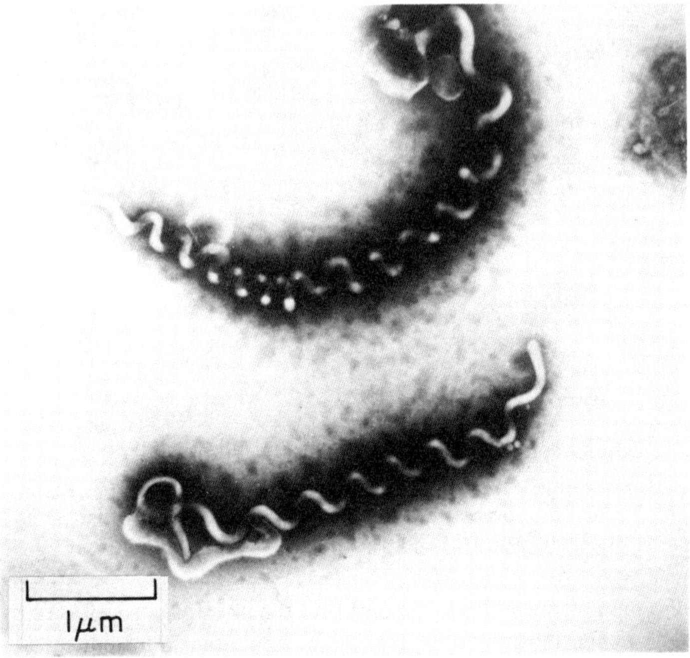


**Fig. 1.3** A negatively-stained 2 day culture of *Spiroplasma citri* showing the characteristic helical filaments. From Cole, R. M., Tully, J. G., Popkin, T. J. and Bové, J. M. (1973), *J. Bact.*, **115**, 367.

Mycoplasmatales. A group of organisms resembling the mycoplasmas in most characteristics, but differing in that they have a helical shape (Fig. 1.3), has recently been described and cultured from diseased plants and insects and named *Spiroplasma* [8,9]. Another group of micro-organisms which seem to be less closely related to mycoplasmas though they, too, lack a cell wall are thermophilic organisms isolated from coal refuse piles [10]. These organisms grow best at 59°C and at pH 2.0 and so were named *Thermoplasma acidophilum*. Their remarkable stability to lysis by osmotic shock, by detergents, and by heat [11] suggests an unusual membrane structure. Recent analysis of their membrane components indeed showed some unusual features such as the presence of long-chain isopranols instead of fatty acids [12], a remarkably low number of free amino and carboxyl groups in membrane proteins [13] and 'highly rigid' lipid regions [14].

## 1.2 Cultivation of mycoplasmas

### *1.2.1 Growth media*

Mycoplasmas are usually cultivated in complex media containing serum or serum components, heart infusion, peptone and yeast extract. Only a few species have so far been cultivated in defined or semi-defined media [15]. Most of the *Mycoplasma* species and all the *Acholeplasma* species can be grown in the modified Edward medium [16] consisting (per litre) of:

| | |
|---|---|
| Bacto-heart infusion broth, 13 g | $K_2HPO_4$, 2.4 g |
| Bacto-peptone, 5 g | Sodium deoxyribonucleate, 0.02 g |
| Bacto-yeast extract, 7 g | Bacto-PPLO serum fraction, 20 ml |
| Glucose, 5 g | Penicillin G, 300 000 units |
| NaCl, 2.5 g | Thallium acetate, 0.25 g. |

A basal medium containing the heart infusion broth, peptone, yeast extract, sodium deoxyribonucleate and NaCl is prepared and its pH is adjusted to 8.0. The basal medium is autoclaved and may be kept at room temperature. Stock solutions of glucose (500 mg ml$^{-1}$, w/v), $K_2HPO_4$ (240 mg ml$^{-1}$, w/v) and thallium acetate (25 mg ml$^{-1}$, w/v) are prepared, autoclaved and kept at 4°C. For compounding the complete medium, 10 ml of each of the glucose, $K_2HPO_4$ and thallium acetate solutions, 20 ml of the PPLO serum fraction and a solution containing 300 000 units of penicillin G are added to each litre of the basal medium. The complete medium should be stored in the cold and used within a period of 1–2 weeks. Commercial lots of PPLO serum fraction may differ in their growth-promoting activity [17]. For growth of most of the laboratory-adapted strains, 3–5% (v/v) of horse serum can replace the more expensive and often less effective PPLO serum fraction.

To cultivate the more fastidious mycoplasmas, in particular for primary isolation, the PPLO serum fraction and the Bacto-yeast extract are replaced by 20% (v/v) horse serum and 10% (v/v) yeast extract prepared from dry yeast according to the procedure of Hayflick [18] and commercially available from several sources (cf. Microbiological Associates, Bethesda, Md.). The addition of 20 mM L-arginine to the medium greatly improves the growth of the non-fermentative mycoplasmas that possess the arginine dihydrolase pathway [19]. Growth of *Ureaplasma* (T-mycoplasmas) is most markedly improved by the addition of 0.05% urea to the medium [20]. *Spiroplasma citri* will grow in the modified Edward medium supplemented with sorbitol (7%, w/v), tryptone (Difco) (1%, w/v), sucrose (1%, w/v), fructose (0.1%, w/v), and glucose (0.1%, w/v) according to Saglio *et al.* [8]. The concentration of horse serum in the culture medium may be reduced to 5% (v/v) without greatly reducing the growth of laboratory-adapted *S. citri* strains (Razin, unpublished observation).

Since mycoplasmas incorporate exogenous fatty acids into their polar lipids, the membrane phospholipids and glycolipids of mycoplasmas can be specifically labelled by adding a radioactive fatty acid to the growth medium. Effective labelling of membrane lipids of all the mycoplasmas tested so far has been achieved by adding 25–50 $\mu$Ci of $^3$H-oleic acid or $^3$H-palmitic acid per litre of the growth medium. Membrane proteins may by labelled by adding 100–200 $\mu$Ci of $^{14}$C-L-phenylalanine to the medium [21].

### *1.2.2 Growth conditions*

Although most mycoplasma species are facultative anaerobes, they grow better aerobically. However, primary isolation of many strains requires an atmosphere of 95% $N_2$ and 5% $CO_2$. The optimal pH for mycoplasma growth varies between 6.0 and 8.5. The fermentative strains, such as *A. laidlawii*, usually produce higher yields when the pH of the medium is adjusted to 8.0–8.5, while the arginine-hydrolysing strains, such as *M. hominis*, and the urea-splitting *Ureaplasma* strains produce higher yields when the initial pH of the medium is adjusted to 6.5. These strains will stop growing and eventually lyse once the pH of the growth medium reaches 8.0–8.5 due to the ammonia liberated from the arginine or urea hydrolysed [7,22].

Growth of most laboratory-adapted mycoplasmas can be followed by pH changes in the medium and may be assessed by turbidity measurements at 640 nm [23]. With the fast-growing mycoplasmas, yields of up to 100–150 mg cell protein per litre of medium may be expected. For successful membrane isolation the organisms must be harvested before the end of the logarithmic phase of growth, when turbidity is still on the rise. *A. laidlawii* cultures may reach this point after 16 to 24 hours of incubation, depending on medium batch and inoculum size. The turbidity of the culture at this point is about 0.4 at 640 nm and the pH of the medium drops from the initial value of about 8.5 to about 7.5.

### *1.2.3 Criteria for culture purity*

Since the media used for growing mycoplasmas are very rich, even a few contaminating bacteria may overgrow the slower-growing mycoplasmas. The incorporation of high concentrations of penicillin and thallium acetate into the mycoplasma media usually prevents growth of air-borne contaminants but may fail to prevent the growth of contaminants originating from water, in particular pseudomonads. A rapid and simple check for bacterial contamination can be made by examining a drop of the culture by phase-contrast microscopy. The contaminating bacteria can easily be distinguished from mycoplasmas by their much larger dimensions, different shape, and frequently by their motility, as is the case with pseudomonads. A Gram-stained preparation might also be useful for detecting contamination. Mycoplasmas are Gram-negative and stain rather poorly.

On the whole, it is relatively easy to detect the contamination of a mycoplasma culture by other bacteria, yeasts or fungi. A heavier and coarser turbidity are hints of possible contamination, which should be verified by phase-contrast microscopy. It is much more difficult to detect cross-contamination of a mycoplasma culture with another mycoplasma species, an unfortunate event which may occur in laboratories working on different mycoplasma species simultaneously. In this case turbidity changes, phase

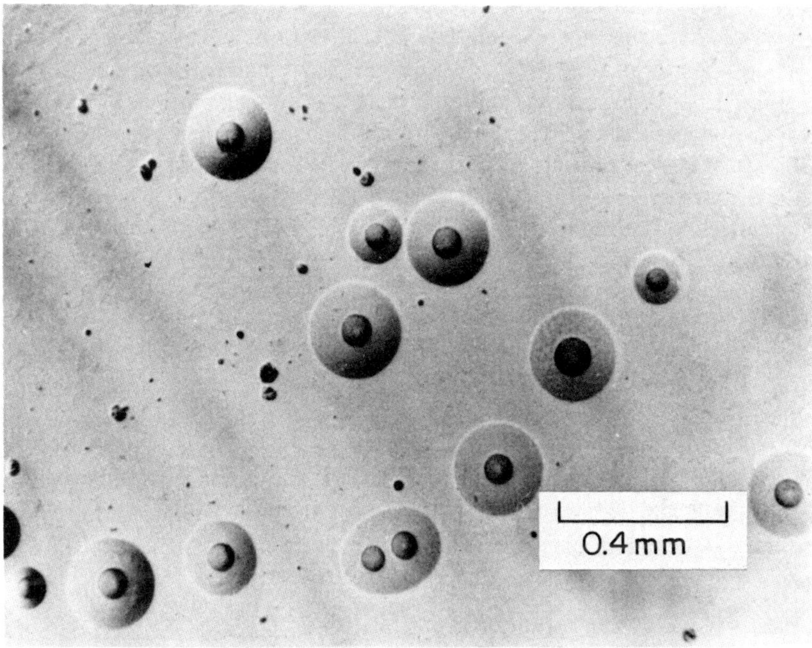


**Fig. 1.4** A 4 day culture of *Acholeplasma laidlawii* on solid medium showing the characteristic 'fried-egg' colonies. From Razin, S. and Oliver, O. (1961), *J. gen. Microbiol*, **24**, 225.

contrast microscopy of culture and even colony morphology have little or no value. The different mycoplasmas have an essentially similar colony shape resembling a fried egg (Fig. 1.4). The colony consists of a central region embedded in the agar and a peripheral zone spread on the agar surface [24]. The best, and frequently the only, way to distinguish a mixed mycoplasma culture is by use of specific antisera. Antisera conjugated with fluorescein *iso*thiocyanate may be used for the direct identification of mycoplasma colonies on the plates [25]. Isolated colonies from the suspect culture may be cloned [26] and their identity verified by the growth inhibition test using standard antisera [27]. Reference antisera and seed are available from the FAO/WHO International Reference Centre for Animal Mycoplasmas, Institute of Medical Microbiology, University of Aarhus, DK 8000 Aarhus C, Denmark, or from Research Resources Branch, National Institute of Allergy and Infectious Diseases, National Institutes of Health, Bethesda, Md. 20014, U.S.A.

### *1.2.4 Storage of mycoplasma cultures*

Some of the mycoplasma broth cultures may be kept in the cold for at least several weeks without total loss of viability. However, it is a routine practice in many laboratories to subculture mycoplasma strains once or twice a week. The best way to keep the titre of viable cells in broth cultures relatively constant for prolonged periods of time with no need of subculturing is by storage at –70°C. For this purpose a logarithmic culture is divided into small portions (1 ml) which are quickly frozen in a mixture of dry ice and alcohol and then transferred to a –70°C freezer. A significant drop in viable counts occurs during the freezing process. However, those cells which survive are kept alive at –70°C for months and even years. The addition of dimethylsulfoxide (DMSO) or glycerol to the broth culture to a final concentration of 1.5 M reduces the killing of cells during freezing to a minimum (M. Raccach and S. Razin, unpublished observation). By using samples from the same frozen batch of organisms a standard inoculum can be assured. Freeze drying is another effective method for keeping a mycoplasma culture alive. Properly prepared freeze-dried ampoules were shown to contain viable organisms after more than 20 years of storage in the cold [28]. Recommended procedures for freeze-drying of mycoplasmas have just been published by the FAO/WHO Programme on Comparative Mycoplasmology and can be obtained by request from Dr. K. Bögel (Veterinary Public Health, Division of Communicable Diseases, WHO, 1211 Geneva 27, Switzerland).

### 1.3 Harvest and washing of organisms

The lack of a protective cell wall renders the mycoplasmas much more fragile than ordinary bacteria. The mycoplasmas are thus easily damaged or even lysed during their harvest and washing. With some, such as *Spiroplasma citri*, significant cell lysis occurs on centrifugation of the culture, as evidenced by the highly viscous pellet containing highly-polymerized DNA released from the cells [29]. The great susceptibility of *S. citri* to lysis by centrifugation creates a dilemma since reduction of the centrifugal force reduces the number of organisms collected, while increasing the force results in increased cell lysis. Fortunately, most other mycoplasmas are less problematical. The pellet of sedimented organisms obtained by centrifugation of the culture for 15 min at 12 000 x *g* can easily be resuspended in the wash fluid with no evidence of any significant release of cellular DNA. The fluid most frequently used in our laboratory for washing mycoplasmas is a 0.25 M NaCl solution. For membrane isolation, one wash of sedimented organisms with a large volume of 0.25 M NaCl will suffice.

The tendency of washed mycoplasmas to lose intracellular components and lyse on prolonged incubation in non-nutrient media at 37°C [30,31] poses a serious problem in studies on the disposition of proteins and lipids in mycoplasma membranes [32]. The current procedures used in these studies are based on comparison of labelling or susceptibility to digestion of membrane components in intact cells and isolated membranes, employing agents which cannot penetrate the permeability barrier of the cell [33,34]. Hence, in order to obtain meaningful results, the cells must remain intact throughout the labelling procedure. Leakage of small solutes and macromolecules from mycoplasma cells suspended in different non-nutrient media has, therefore, been studied by Amar *et al.* [32]. Release of hexokinase, a cytoplasmic enzyme in fermentative mycoplasmas, is a convenient indicator for the leakage of macromolecules through the cell membrane, while release of thymidine-labelled compounds, from cells grown with radioactive thymidine, can serve as an indicator for the leakage of nucleotides and DNA. Hexokinase activity can be tested in 1 ml of the following reaction mixture: 8 $\mu$mol $MgCl_2$, 16 $\mu$mol KF, 4 $\mu$mol ATP, 80 $\mu$mol $K_2HPO_4$ at pH 8.0, and 0.6 ml of the preparation to be tested. The assay is begun at 37°C by the addition of 1 $\mu$mol of $^{14}$C-labelled glucose (0.05 $\mu$Ci $\mu$mol$^{-1}$) and is terminated after 60 min of incubation by the addition of 5 ml of ice-cold deionized water. The phosphorylated sugar is separated from the unphosphorylated sugar by transfer through a Bio-Rad AG 1–X2 (50–100 mesh) anion exchange resin column. The unphosphorylated sugar is washed from the column with 20 ml of deionized water and the phosphorylated sugar is eluted with 6 ml of 1.0 M LiCl. The eluate, collected directly in scintillation-counting vials, is counted in a Tri-Carb liquid scintillation spectrometer after the addition of 15 ml of a mixture containing 333 ml of Triton X–100 (Spectro-quality, Packard Instrument Co., Inc., Downer's Grove, Ill.), 666 ml of toluene, 5.5 g of 2,5-diphenyloxazole and 125 mg of dimethyl-1,4-bis-2-(5-phenyloxazolyl)-benzene.

## 1.4 Membrane isolation

### *1.4.1 Osmotic lysis*

Mycoplasmas resemble bacterial protoplasts and erythrocytes in their sensitivity to osmotic shock due to their lack of protective cell walls. Osmotic lysis is certainly the simplest and gentlest method to isolate mycoplasma cell membranes. The osmotic sensitivity of mycoplasmas decreases most markedly with the aging of the culture [35,36] and the presence even of traces of divalent cations in the suspension medium may provide complete protection against lysis, thus interfering with membrane isolation [36]. Hence, osmotic lysis is most effective when the organisms are harvested before the end of the logarithmic phase of growth, with the divalent cation concentration in the lysis medium kept to a minimum. Perfect control of the growth rate is essential and can only be gained when the strain is well-adapted to the growth medium. This is relatively easy to achieve with the fast-growing mycoplasmas such as *A. laidlawii*, *M. mycoides* var. *capri* or *M. hominis*, but can prove extremely difficult with the more exacting, slow-growing mycoplasmas, such as *M. pneumoniae* [37].

The preliminary assessment of the osmotic fragility of the cells helps to determine whether the regular osmotic lysis procedure will be effective for the isolation of membranes from a specific batch of washed mycoplasma cells. The osmotic fragility can be roughly estimated by adding 0.05 ml samples of the washed cell suspension (containing 5–10 mg cell protein $ml^{-1}$) in 0.25 M NaCl to 2 ml volumes of deionized water and 0.25 M NaCl. Rapid clearing of the suspension in deionized water, but not in 0.25 M NaCl, indicates sensitivity to osmotic shock. The same procedure can be utilized to compare the osmotic fragility of different batches of washed organisms. For this purpose, 0.05 ml of the cell suspension is added to 2 ml volumes of 0.25 M NaCl and 0.05 M NaCl. The turbidity of the suspensions is measured at 500 nm after 15 min incubation at 37°C. Results are expressed as per cent lysis in 0.05 M NaCl according to the formula:

$$\frac{A_{500} \text{ in } 0.25 \text{ M NaCl} - A_{500} \text{ in } 0.05 \text{ M NaCl}}{A_{500} \text{ in } 0.25 \text{ M NaCl}} \times 100$$

Batches of washed *A. laidlawii* cells showing osmotic fragility values of 40% and above can be used for membrane isolation by the regular osmotic lysis procedure.

This procedure consists of transferring one volume of the washed cell suspension in 0.25 M NaCl (10–20 mg cell protein $ml^{-1}$) to 50–100 volumes of deionized or double-distilled water pre-heated to 37°C. The membranes are collected after 15 min of incubation by centrifugation at 48 000 x *g* for 30 min. The membranes are washed consecutively in deionized water, 0.05 M NaCl in 0.01 M phosphate buffer, pH 7.5, and deionized water. They are then resuspended in β-buffer (0.15 M NaCl; 0.05 M Tris; 0.01M 2-mercaptoethanol, in deionized water adjusted to pH 7.4 with HCl) diluted 1:20 in deionized water (dilute β-buffer). The membrane suspensions should preferably be kept at –70°C to prevent degradation of membrane phospholipids and proteins during prolonged storage periods. Membrane-associated endogenous phospholipases and peptidases have been demonstrated in mycoplasma membranes [38–41]. Slow activity of the endogenous phospholipase of *M. hominis* membranes could be detected even at –20°C, but not at –70°C [41].

Pre-loading with glycerol may increase the sensitivity to osmotic lysis of strains like *M. gallisepticum* which are relatively resistant to lysis by the regular procedure [42]. Glycerol, which penetrates freely through the cell membrane of mycoplasmas [43], intensifies the osmotic shock by increasing the internal osmotic pressure of the cells. The washed cell pellet is resuspended in 5 ml of a 2 M glycerol solution and incubated at 37°C for 10 min. The suspension is then rapidly injected into 250 ml of deionized water and incubated at 37°C for 15 min. The membranes are collected as described above.

### *1.4.2 Digitonin-induced lysis*

Mycoplasmas, in particular the sterol-requiring species, contain considerable quantities of cholesterol in their cell membranes and are therefore very sensitive to lysis by digitonin [22,44]. For lysis by digitonin, washed mycoplasmas

(20–50 mg of cell protein) are added to 200 ml of 0.25 M NaCl containing 25 μg digitonin per ml and incubated at 37°C for 15 min. The membranes are then collected and washed as described above. In some cases treatment of the membranes with deoxyribonuclease (page 15) is essential for their effective washing. One of the advantages of digitonin lysis over osmotic lysis is that it is less dependent on the age of the culture, making it a useful technique for the isolation of membranes from slow-growing mycoplasmas, where the difficulty of determining the right harvesting time constitutes a handicap to osmotic lysis. Another advantage of digitonin lysis is that it can take place in the presence of divalent cations [22]. Elimination of the need to reduce the ionic strength and divalent cation concentration of the lysis medium may be of importance in view of the finding that some loosely-bound membrane proteins may become dissociated and lost under those conditions [45,46]. On the other hand, a possible drawback with the membranes obtained by digitonin lysis is that they seem to retain the cholesterol-digitonide complex, and are thus somewhat altered [22].

### *1.4.3 Other techniques*

Alternate freezing and thawing [47,48], gas cavitation [48] or ultrasonic treatment [48–50] have been utilized to lyse mycoplasmas. Freezing and thawing most frequently fails to lyse the bulk of the cells [48]. Gas cavitation, likewise, proved unsatisfactory [48]. The use of sonic or ultrasonic oscillators and mechanical presses rupture mycoplasma cells most effectively. Their use is not recommended, however, since disintegration of the membranes into minute particles invariably results, and some of these cannot be sedimented even at very high gravitational forces [50,51].

## 1.5 Criteria for membrane purity

It is the general consensus that our criteria for membrane purity are arbitrary or operational rather than absolute. Some proteins are bound so loosely to the membrane that even slight changes in the ionic strength of the solution used for cell lysis and membrane washing will detach them (e.g. see [52]). We should therefore allow for the possibility that some loosely-bound proteins and ions may be lost from mycoplasma membranes during the osmotic lysis of the cells in deionized water and subsequent washings [45,46,53]. Furthermore, the recent demonstration of endogenous peptidases [38,39] and phospholipases [40,41] in mycoplasma membranes indicates that some degree of hydrolysis of membrane components is inevitable during membrane isolation and purification and so some modification in membrane composition must be expected. To minimize this kind of degradation, washing of membranes should be carried out at a low temperature and the membranes should be stored at –70°C as soon as possible after their isolation.

### 1.5.1 Chemical composition

Gross chemical analysis of isolated mycoplasma membranes (Table 1.1) shows that they, like other biological membranes, consist mainly of proteins and lipids. The protein comprises, in rough terms, about two-thirds of the mass of the mycoplasma membrane, the balance being mostly lipid. Since protein can easily be determined by the Lowry technique [57] the content of other membrane constituents is most frequently expressed in relation to mg membrane protein rather than in percentage of total dry weight [22,29]. Polyacrylamide-gel electrophoresis of mycoplasma membranes in the presence of acetic acid and urea [58] or sodium dodecyl sulphate [32] reveals 15 to 40 different polypeptide bands (Fig. 1.5). The electrophoretic patterns are species-specific and, since they are affected only slightly by variations in growth conditions, they can be used for the identification and classification of mycoplasmas [58].

Polar lipids (phospholipids and glycolipids) are the major lipid components in mycoplasma membranes as in other biomembranes. Cholesterol is also present in large quantities in mycoplasma membranes. A detailed description of the composition, distribution and biosynthesis of the mycoplasma membrane lipids can be found elsewhere [1,2].

As with other microbial membranes, ribosomes and DNA may stick to mycoplasma membranes. Alternate washings of the membranes in deionized water and 0.05 M NaCl in 0.01 M phosphate buffer, pH 7.5, proved to be quite an effective method for removing them [42]. They may also be removed by treating the membranes with ribonuclease and deoxyribonuclease. The addition of nucleolytic enzymes to *A. laidlawii* membranes is usually unnecessary, because the potent endogenous nucleases of this organism, both cytoplasmic and membrane-associated, remove the ribosomes and DNA sufficiently when cell lysis is carried out at 37°C [46,59]. With several other mycoplasmas, however,

TABLE 1.1
*Gross chemical composition of mycoplasma membranes*

| Organism | Protein | Lipid | Carbohydrate | References |
|---|---|---|---|---|
| *Acholeplasma laidlawii* | 57 | 32 | 0.5 | [54] |
| *Mycoplasma bovigenitalium* | 59 | 37 | 2.2 | [54] |
| *Mycoplasma mycoides* var. *capri* | 50 | 40 | 2.0 | [54] |
| *Mycoplasma mycoides* var. *mycoides* | 51 | 39 | 3.2 | [55] |
| *Mycoplasma hominis* | 57 | 41 | 0.8 | [22] |
| *Mycoplasma gallisepticum* | 80 | 19 | 0.2 | [44] |
| *Mycoplasma fermentans* | 60 | 29 | 9.0[a] | [56] |
| *Spiroplasma citri* | 57 | 34 | 2.2[b] | [29] |
| *Thermoplasma acidophilum* | 60 | 25 | 10.0 | [13] |

[a] Not clear whether the determination was carried out before or after removal of glycolipids and nucleic acids.
[b] Including hexosamines.

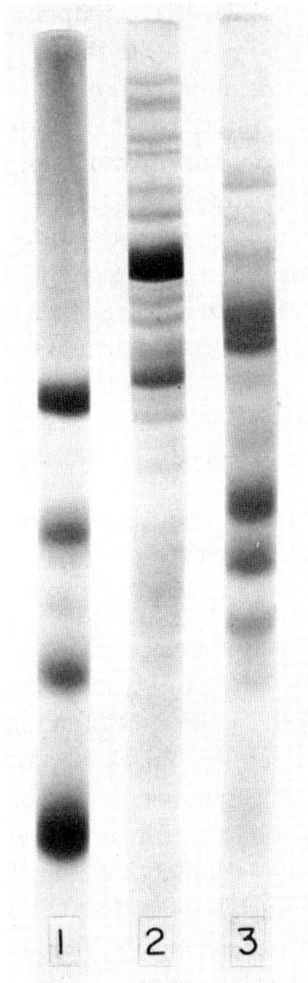


**Fig. 1.5** Electrophoretic patterns of membrane proteins of *Mycoplasma hominis* (2), *Mycoplasma mycoides* var. *capri* (3), and protein standards (1), in polyacrylamide gels containing 0.1% sodium dodecyl sulphate. The protein standards were (from top to bottom): bovine serum albumin (mol. wt. 67 000), ovalbumin (mol. wt. 45 000), chymotrypsinogen A (mol. wt. 25 000) and lysozyme (mol. wt. 14 400).

treatment with deoxyribonuclease is an essential step in the isolation and purification of their membranes, particularly when large quantities of cells are lysed in relatively small volumes of medium. This was shown to be the case with *Spiroplasma citri* lysed by osmotic shock [29] and with *M. gallisepticum* lysed by digitonin [44]. The viscous sediment obtained after centrifugation of the lysed organisms is suspended in a small volume (about 50 ml) of 0.02 M $MgCl_2$ containing 10 $\mu$g of purified pancreatic deoxyribonuclease per ml. After 15 min of incubation at 37°C with shaking the membranes are collected by centrifugation and washed alternately with 0.05 M NaCl in 0.01 M phosphate buffer, pH 7.5, and deionized water.

Total carbohydrate values for mycoplasma membranes are rather low (Table 1.1), and may in part represent contamination with nucleic acids. Yet, recent studies indicate that carbohydrate residues are present on the membrane surface of many *Mycoplasma, Acholeplasma* and *Spiroplasma* species, as evidenced by the ability of the cells to bind [60] and be agglutinated [61] by lectins. The reactive carbohydrate groups may be part of glycolipid molecules [61], lipopolysaccharide-like compounds [62] or glycoproteins [63]. Carbohydrates in the form of free polymers are rarely found as components of biomembranes; this is also the case with mycoplasma membranes. Nevertheless two, rather unusual, carbohydrate polymers can be found associated with mycoplasma membranes. One of these, the galactan of *M. mycoides* var. *mycoides*, covers the exterior cell surface forming a slime layer [64]. The other, a polymer consisting of *N*-acetylgalactosamine and *N*-acetylglucosamine residues, was found in *A. laidlawii* [65] and seems to be even more closely associated with the membrane than the galactan since a considerable part of it remains attached to the membrane even after extensive washing.

Detailed methods for the chemical analysis of mycoplasma membranes are not provided here since the methods are basically the same as those employed in the chemical analysis of other membrane types. (See Chapters ???).

### *1.5.2 Density-gradient analysis*

Another way to check the purity of a membrane preparation is to determine its density. Density-gradient centrifugation has also been found useful in the separation of mycoplasma membranes from unlysed cells and from precipitates of the growth medium ingredients [66]. Most frequently a precipitate of unidentified black material sediments together with the cells and accompanies the membranes throughout the washing procedure. When the yield of membranes is high it is easy to separate a large part of them from the black material simply by collecting only the periphery of the pellet. The black material is much heavier than the membranes and concentrates at the centre of the pellet. Obviously, this procedure leads to a considerable loss of membrane material. Hence, when membrane yields are low another method must be used [67]. Portions of a thick membrane suspension are layered on top of a 45% sucrose solution distributed in Corex centrifuge tubes. Centrifugation at 27 000 x *g* for 30 min results in the sedimentation of the black material, while the much lighter membranes float on top of the sucrose solution.

The density of purified mycoplasma membranes can be determined by sucrose density gradient centrifugation [42,68]. The technique used in our laboratory is as follows: samples (0.15 ml) of membrane suspensions (1 mg protein $ml^{-1}$) are layered over 11.5 ml of a linear sucrose gradient (30–60%). The gradients are centrifuged at 40 000 rev/min in an SW 41 rotor of a Spinco ultracentrifuge for 2 h at 4°C. The density of the membrane band is then determined according to its position on the gradient. The density of purified mycoplasma membranes was found to vary between 1.140 g $cm^{-3}$ to 1.20 g $cm^{-3}$, depending on the mycoplasma species and on growth conditions [22,42,44], reflecting mostly differences in the ratio of membrane protein to lipids. The density of membranes of several *Mycoplasma* and *Acholeplasma* species has been shown to increase with the age of the culture, reflecting a significant increase in the ratio of membrane protein to lipid occurring during aging [69,70]. Similar changes were observed when membranes were isolated from cultures of *A. laidlawii* grown at different pH values, a factor which affects the growth rate of the organisms [68]. Hence, membrane density values cannot be regarded as constant for a given species.

### *1.5.3 Electron microscopy*

Electron microscopy can also be used to determine the purity of a membrane preparation. Electron micrographs of thin-sectioned material may show the presence of unlysed cells in the membrane preparation and the presence of cytoplasmic contaminants, such as ribosomes, adhering to the membranes. Thin-sectioned mycoplasma membranes fixed with osmium tetroxide show the characteristic triple-layered shape, with a total thickness of about 10 nm (Fig. 1.6).

## 1.6 Enzymic activities

Marker enzymes are used as an aid in distinguishing different membrane types isolated from the same cell. Obviously, this has no application to mycoplasma

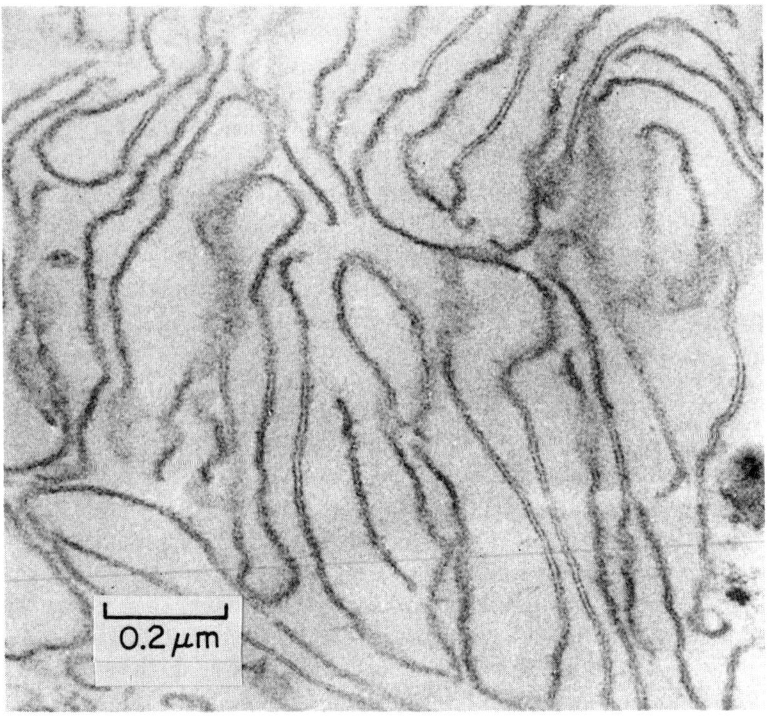


**Fig. 1.6** Thin sections of isolated *Acholeplasma laidlawii* membranes.

cells with their single membrane type. However, checking the activity of cytoplasmic and membrane-associated enzymes can aid in determining the purity of mycoplasma membrane preparations. Although enzyme localization studies have not yet covered all mycoplasmas, they have invariably shown the ATPase activity (EC 3.6.1.3) of the organisms to be exclusively associated with the membrane [59,71,72], whereas, the hexokinase activity of fermentative mycoplasmas is restricted to the cytoplasm [32]. Hence, hexokinase activity in membrane preparations indicates their contamination with cytoplasmic material, while ATPase activity in cytoplasmic fluid fraction indicates its contamination with membrane material. The NADH oxidase activity (EC 1.6.99.3) is found in the soluble cytoplasmic fraction of all the mycoplasmas tested apart from *A. laidlawii*, where it is membrane-bound [15,37,59,73]. The NADH oxidase activity of *A. laidlawii* membranes decreases most markedly on storage at temperatures of $-20°C$ and higher; it can be reactivated by the addition of high concentrations (up to 20 mM) of SH-containing compounds such as $\beta$-mercaptoethanol, glutathione and cysteine [74].

ATPase activity decreases very markedly on aging of the culture and with prolonged storage of the membrane preparations. The enzyme is not activated by sodium and potassium nor is it inhibited by oubain, resembling the ATPase activity of other microbial plasma membranes. The enzyme is dependent, however, on $Mg^{2+}$ ions that can be replaced by $Mn^{2+}$ or $Co^{2+}$ but not by $Ca^{2+}$ [71]. *A. laidlawii* and *M. mycoides* var. *capri* are comparable with eukaryotes, in so far as their plasma membrane ATPases are also integral lipid-dependent membrane proteins. Mycoplasma ATPases differ in this respect from the ATPases characterized to date in other microbial plasma membranes [2,4]. The mycoplasma ATPase cannot be released from the membrane by decreasing the ionic strength of the medium and the addition of EDTA [46]. Moreover, the physical state of membrane lipids have a determinative influence on its activity. Thus, the Arrhenius plots of the ATPase activity of *A. laidlawii* and *M. mycoides* var. *capri* membranes showed breaks at temperatures corresponding to the lipid phase transition in these membranes [72,75].

## 1.7 Alteration of membrane composition

One of the greatest advantages in taking mycoplasmas as models for membrane studies stems from the fact that their membrane lipid composition can be altered in a controlled manner. This results from the total or partial inability of the organisms to synthesize long-chain fatty acids and cholesterol, making them dependent on the supply of these from the growth medium. The ability to introduce controlled alterations in the fatty acid composition and cholesterol content of mycoplasma membranes has been utilized to study the effects of these alterations on the biophysical and biochemical properties of the membranes (reviewed in [2 and 4]).

### *1.7.1 Alteration of fatty acid composition*

None of the *Mycoplasma* species tested so far is able to synthesize long-chain fatty acids, or interconvert any preformed fatty acid added to the growth medium [1,76–79]. *A. laidlawii*, on the other hand, is capable of synthesizing the saturated fatty acids only, so that unsaturated fatty acids have to be provided exogenously [76,80]. Moreover, *A. laidlawii* was found capable of elongating unsaturated fatty acids supplied in the growth medium [81–83]. Obviously, the most dramatic alterations in fatty acid composition can be achieved with *Mycoplasma* strains, which are totally unable to synthesize or transform any fatty acids. Rodwell and Peterson [78] made up a defined medium which facilitated the precise determination of the growth response of a goat *Mycoplasma* species (strain Y) to fatty acids. None of the straight-chain saturated fatty acids could support growth of this organism when tested alone, but good growth was obtained with branched-chain acids such as *iso*palmitate, *iso*stearate, or *anteiso*heptadecanate. No *cis*-monoenoic fatty acid supported growth when tested alone but good growth was obtained with *trans*-monoenoic fatty acids, such as elaidate and *trans*-vaccenate. Good growth was also obtained with a combination of a *cis*-monoenoic fatty acid and a straight-chain fatty acid, but the chain length of the monoenoic acid and the position of the double bond had a marked effect on the range of saturated fatty acids with which it could be successfully paired [84]. Thus, with myristoleate, long-chain saturated acids of chain length C20 and C22 gave the best growth, while with oleate the range extended from C12 to C20. The growth-promoting fatty acids were incorporated unchanged into the polar lipid fraction of the organism and in cases where the organism grew well when a single fatty acid was supplied (branched-chain or *trans*-monoenoic acid) this was the *sole* fatty acid found.

Although the alterations in fatty acid composition of *Mycoplasma* sp. strain Y are well controlled and rather dramatic, little use has been made so far of this organism outside of Rodwell's laboratory, probably due to the need of compounding a complex defined growth medium for demonstrating the variations in fatty acid composition. *A. laidlawii* has been much more extensively used for this purpose, in spite of the fact that the possible variations in its fatty acid composition are less dramatic than with *Mycoplasma* strain Y due to the ability of *A. laidlawii* to synthesize saturated fatty acids [85,86]. Nevertheless, as Table 1.2 shows, the alterations in the fatty acid composition of *A. laidlawii* membrane lipids are still very pronounced. The exogenous fatty acid is incorporated in large amounts, reaching 60–90 mol % of the total fatty acids. Lauric, myristic, palmitic and stearic acids are present throughout as a biosynthetic background upon which exogenous fatty-acid incorporation is superimposed. The growth requirement for monoenoic acids can be partially met by a cyclopropane-containing fatty acid, such as lactobacillic acid, or by a branched-chain fatty acid [87]. The range of acceptable fatty acids does, however, have its limits. Increasing the number of double bonds or methyl

TABLE 1.2
*Fatty acid composition of the total membrane lipids of* A. laidlawii *B grown with various fatty acids at 35°C*

| Fatty acids found | None | 18:0* | 16:0 | $18:0^i$ | $18:1_t$ | $17:0^a$ | $18:1_c$ | $18:2_{c,c}$ |
|---|---|---|---|---|---|---|---|---|
| | Fatty acid added to the growth medium | | | | | | | |
| 12:0 | 5.0 | 8.7 | 8.5 | 4.3 | 2.9 | 1.3 | 2.2 | 4.3 |
| 13:0 | 3.4 | trace | trace | trace | trace | trace | trace | trace |
| 14:0 | 30.8 | 12.7 | 6.9 | 8.9 | 4.2 | 2.6 | 6.6 | 6.9 |
| 15:0 | 6.6 | trace | trace | trace | trace | trace | trace | trace |
| 16:0 | 40.1 | 11.6 | 80.4 | 6.8 | 7.2 | 2.2 | 22.1 | 31.2 |
| 17:0 | 3.2 | trace | trace | trace | trace | trace | trace | trace |
| $17:0^a$ | – | – | – | – | – | 90.7 | – | – |
| 18:0 | 7.7 | 58.4 | 1.9 | trace | trace | 1.0 | 4.5 | 2.6 |
| $18:0^i$ | – | – | – | 76.4 | – | – | – | – |
| 18:1 | 1.8 | 4.9 | 1.6 | 1.5 | 85.4 | 1.5 | 64.4 | 3.4 |
| 18:2 | 1.2 | 3.7 | 0.6 | 1.9 | 0.3 | 0.7 | trace | 51.3 |

*Fatty acids are designated by the number of carbon atoms followed by the number of double bonds, if any, present in the molecule; the subscripts *c* and *t* denote the *cis* or *trans* configurations, respectively, of these double bonds. The superscripts i and a indicate a methyl group attached to the penultimate carbon atom (an isobranched fatty acid) and the antepenultimate carbon atom (an anteisobranched acid), respectively. The fatty acid composition values are expressed as mol %. From McElhaney [93].

branches on a fatty acid lowers and eventually eliminates its incorporability into membrane lipids. Thus, arachidonic acid is barely incorporated. Similarly, long-chain saturated fatty acids (22 carbons and higher) are not incorporated in significant quantities [83,86].

For altering the fatty acid composition of *A. laidlawii* membrane lipids, a lipid-poor or lipid-depleted growth medium to which known fatty acids are added is used. Good results are obtained with Tryptose medium or with a modified Edward medium. The Tryptose medium consists of 2% (w/v) lipid-extracted tryptose (Difco Laboratories Inc., Detroit, Mich.), 0.5% (w/v) glucose, 0.5% (w/v) sodium acetate, 0.5% (w/v) charcoal-treated bovine serum albumin, 10–25 $\mu$g ml$^{-1}$ of a known fatty acid added as an ethanolic solution and 500 units ml$^{-1}$ of penicillin G [87]. The pH of the medium is adjusted to 8.0. Edward medium is the same as described in the section on Cultivation of Mycoplasmas (Section 1.2.1) but with 0.5% of charcoal-treated bovine serum albumin instead of PPLO serum fraction. For the extraction of tryptose, 500 g of dry tryptose are transferred to a 6 l beaker to which 3 l of chloroform-methanol (2:1) is added. After stirring the mixture for 2 h at room temperature, the tryptose granules are allowed to sediment to the bottom of the beaker and the chloroform-methanol is removed. Extraction is repeated twice more and the extracted tryptose is freeze dried. Fatty acids which contaminate the bovine serum albumin are removed by charcoal-treatment. Activated charcoal (Darco or Norit A) is added to a 10% albumin solution to a final concentration of 5%. The pH of the mixture is brought down to 3.0 with a solution of 1 N HCl and the mixture is incubated at 4°C for 1 h with constant stirring. The mixture is then centrifuged at 35 000 x *g* for 30 min to remove the charcoal particles and the pH is adjusted to 7.0 with a solution of 1 N NaOH. The solution is then filtered through a Seitz filter for sterilization and removal of minute non-sedimentable charcoal particles. The charcoal-treated albumin solution is kept at –20°C until used.

### *1.7.2 Alteration of cholesterol content*

The cell membranes of *Mycoplasmataceae* are unique among those of other prokaryotes in their high cholesterol content. In this respect they resemble the plasma membranes of eukaryotes. Moreover, cholesterol is an essential membrane component of the mycoplasmas belonging to this family and since none of them is capable of synthesizing it, cholesterol must be provided exogenously. The *Acholeplasmataceae*, which do not require cholesterol for growth, can also incorporate exogenously-provided cholesterol; however, the highest amounts recorded in membranes of growing *A. laidlawii* do not exceed 10–12% of the total membrane lipids [88], whereas in the sterol-requiring *Mycoplasma* species cholesterol may account for up to 40% of the total membrane lipid [89]. As with fatty acids, the dependence of mycoplasmas on an external supply of cholesterol may be utilized to change the amount of cholesterol in the membrane, within certain limits, enabling an analysis to be made of the effects of variations in the cholesterol content on membrane properties [2,4].

The cholesterol content of *A. laidlawii* can be varied from 0 to about 10% of the total membrane lipids by growing the organisms in the lipid-poor Tryptose or Edward medium described above, containing 0.5% of charcoal-treated bovine serum albumin. Growth of the organisms in these lipid-depleted media enables the simultaneous alteration of the cholesterol content and fatty acid composition of membrane lipids. The fatty acids and cholesterol are added to the growth medium as ethanolic solutions, provided the final ethanol concentration in the medium does not exceed 0.5%; at 1.0% ethanol, growth of *A. laidlawii* is already inhibited to some extent. The maximal concentration of cholesterol that can be achieved in this way is less than 10 μg cholesterol per ml of medium. At higher concentrations part of the cholesterol crystallizes, forming an opalescent turbidity in the medium. Cholesterol crystals will in this case co-sediment with the cells upon centrifugation and will contaminate the membrane preparation [90,91]. In order to raise the concentration of cholesterol in the growth medium without forming crystals, de Kruyff *et al.* [88] proposed to add it as part of liposomes prepared with the polar lipids of *A. laidlawii*. The liposomes are formed in the following way: 90 mg of total *A. laidlawii* lipids (isolated from cells grown on 0.03 mM palmitic and 0.03 mM oleic acids) and 45 mg cholesterol are mixed in chloroform, evaporated to dryness and resuspended in 120 ml of the complete, fatty acid-poor growth medium. The suspension is treated with a Branson ultrasonic oscillator at maximum power for 15 min at room temperature. The resulting liposome suspension is centrifuged at 0°C for 15 min at 37 000 x *g* to remove the large liposomes. The supernatant fluid, containing approximately 30 mg of cholesterol, is added to 880 ml of growth medium. If the liposome suspension is added to a growth medium which already contains the maximal level of cholesterol obtained with an ethanolic solution, the final concentration of cholesterol will reach a value of over 40 $\mu$g ml$^{-1}$ with no crystallization.

When only the cholesterol content of the membrane is to be controlled, it is more convenient to add the cholesterol as a solution in Tween 80 [90]. Tween 80 serves in this case both as an emulsifier of cholesterol and as a source for the required oleic acid [85]. A stock solution of cholesterol (20 mg ml$^{-1}$) is prepared in ethanol. Since cholesterol crystallizes from this solution in the cold, it has to be warmed before use. A stock solution of Tween 80 (10%, v/v) is prepared in deionized water and sterilized at 121°C for 15 min. Different volumes of up to 1 ml of the cholesterol solution are added to 1 ml volumes of the Tween 80 solution. Ethanol is used to bring the final volume of all mixtures to 2 ml. The turbid mixtures are then carefully warmed until they are cleared. Portions (0.2 ml) of these solutions are immediately transferred to flasks containing 100 ml of the lipid-poor growth medium. A maximal level of 20 $\mu$g of cholesterol per ml of medium can be obtained by this method. However, part of the cholesterol at this level crystallizes immediately upon its introduction to the growth medium. Pre-incubation of the medium at 37°C and heating of the pipette used to transfer the concentrated cholesterol solution helps to decrease or even prevent crystallization. The final concentration of Tween 80 in the growth medium is 0.01%, a concentration which does not affect *A. laidlawii* growth, but may affect growth of other mycoplasmas [90,91].

Another, perhaps even better, model than *A. laidlawii* for studying cholesterol effects has recently been developed by Rottem *et al.* [23], who succeeded in adapting the sterol-requiring *Mycoplasma mycoides* var. *capri* to grow with very little cholesterol. The medium used for adaptation was the modified Edward medium containing 0.5% of charcoal-treated bovine serum albumin, 5 $\mu$g ml$^{-1}$ of oleate, 5 $\mu$g ml$^{-1}$ of palmitate and decreasing concentrations of cholesterol added as ethanolic solutions. Twenty-five serial transfers of the organism in this medium with decreasing concentrations of cholesterol sufficed to adapt it to grow with no added cholesterol. Nevertheless, it is unclear as yet whether the adapted strain can grow in the total absence of cholesterol. The available evidence rather suggests that very small amounts of cholesterol are still required for growth and that these are apparently provided by traces of cholesterol present in the growth medium. A drop below a certain critical level in the cholesterol of the growth medium is probably responsible for the frequent failure to obtain satisfactory growth of the adapted strain in certain batches of the medium unless these are supplemented with 0.1 $\mu$g cholesterol per ml [23]. These observations are in accord with those of Rodwell *et al.* [92] who failed to adapt the closely related goat mycoplasma strain Y to grow in a defined medium with no cholesterol. The addition of 0.5 to 1.0 $\mu$M (0.19–0.38 $\mu$g ml$^{-1}$) cholesterol was found to be essential for growth in this case.

The cholesterol content of membranes of the adapted *M. mycoides* var. *capri* strain amounted to less than 3% of the total membrane lipids as against 22%–25% in membranes of the parent strain. The most remarkable difference between membranes of the two strains was the demonstration of a thermal phase transition in the cholesterol-poor membranes. The high cholesterol content in membranes of the native strain apparently eliminates the phase transition [72]. It is harder to demonstrate the effects of cholesterol on the phase transition of *A. laidlawii* membrane lipids since the maximum amounts of cholesterol that can be incorporated into the membranes of this species during growth are too low to eliminate the phase transition, but a reduction in the energy content of the phase transition could still be observed [75].

### *1.7.3 Alteration of the protein content*

As mentioned in the section on density-gradient analysis (p. 16), the ratio of membrane protein to lipid markedly increases on aging of the culture. Similar changes can be brought about by chloramphenicol. The addition of 20 to 100 $\mu$g chloramphenicol per ml (depending on the mycoplasma strain and on growth conditions) to a young logarithmic culture of mycoplasmas arrests membrane protein synthesis, while membrane lipid synthesis continues for several hours, as measured by the incorporation of radioactive oleic or palmitic acid. The membranes of the chloramphenicol-treated cells are consequently richer in lipid and have a lower buoyant density [21,69]. The possibility of introducing significant changes in the protein to lipid ratio of mycoplasma membranes has been utilized to study the effects of the membrane protein component on the physical state of membrane lipids [70] and their cholesterol content [69].

### References

1. Smith, P. F. (1971), *The Biology of Mycoplasmas*, Academic Press, New York and London.
2. Razin, S. (1973), In *Advances in Microbial Physiology*, (Rose, A. H. and Tempest, D. W., eds), Vol. 10, pp. 1–80, Academic Press, New York and London.
3. Maniloff, J. and Morowitz, H. J. (1972), *Bact. Rev.*, **36**, 263.
4. Razin, S. (1974), In *Progress in Surface and Membrane Science*, (Cadenhead, D. A., Danielli, J. F. and Rosenberg, M. D., eds), Vol. 9, pp. 257–312, Academic Press, New York and London.
5. Freundt, E. A. (1974), In *Bergey's Manual of Determinative Bacteriology*, (Buchanan, R. E. and Gibbons, N. E. eds), 8th edition, Williams and Wilkins Co., Baltimore.

6. Razin, S. (1973), In *Handbook of Microbiology*, (Laskin, A. I. and Lechevalier, H. A. eds), Vol. 1, pp. 105–129. CRC Press, Cleveland.
7. Shepard, M. C., Lunceford, C. D., Ford, D. K., Purcell, R. H., Taylor-Robinson, D., Razin, S. and Black, F. T. (1974), *Int. J. syst. Bact.*, **24**, 160.
8. Saglio, P., L'hospital, M., Laflèche, D., Dupont, G., Bové, J. M., Tully, J. G. and Freundt, E. A. (1973), *Int. J. syst. Bact.*, **23**, 191.
9. Davis, R. E. and Worley, J. F. (1973), *Phytopathology*, **63**, 403.
10. Darland, G., Brock, T. D., Samsonoff, W. and Conti, S. F. (1970), *Science*, **170**, 1416.
11. Belly, R. T. and Brock, T. D. (1972), *J. gen. Microbiol.*, **73**, 465.
12. Langworthy, T. A., Smith, P. F. and Mayberry, W. R. (1972), *J. Bact.*, **112**, 1193.
13. Smith, P. F., Langworthy, T. A., Mayberry, W. R. and Hougland, A. E. (1973), *J. Bact.*, **116**, 1019.
14. Smith, G. G., Ruwart, M. J. and Haug, A. (1974), *FEBS Letters*, **45**, 96.
15. Rodwell, A. W. (1969), In *The Mycoplasmatales and the L-Phase of Bacteria*, (Hayflick, L. ed.), pp. 413–449, Appleton-Century-Crofts (Prentice-Hall), New York.
16. Edward, D. G. ff. (1947), *J. gen. Microbiol.*, **1**, 238.
17. Hughes, J. H., Thomas, D. C., Hamparian, V. V. and Somerson, N. L. (1974), *J. med. Microbiol.*, **7**, 35.
18. Hayflick, L. (1965), *Tex. Rep. Biol. Med.*, **23**, suppl. 1, 285.
19. Schimke, R. T., Berlin, C. M., Sweeney, E. W. and Carroll, W. R. (1966), *J. biol. Chem.*, **241**, 2228.
20. Shepard, M. C. and Lunceford, C. D. (1970), *Appl. Microbiol.*, **20**, 539.
21. Kahane, I. and Razin, S. (1969), *Biochim. biophys. Acta*, **183**, 79.
22. Rottem, S. and Razin, S. (1972), *J. Bact.*, **110**, 699.
23. Rottem, S., Yashouv, J., Ne'eman, Z. and Razin, S. (1973), *Biochim. biophys. Acta*, **323**, 495.
24. Razin, S. and Oliver, O. (1961), *J. gen. Microbiol.*, **24**, 225.
25. Del Giudice, R. A., Robillard, N. F. and Carski, T. R. (1967), *J. Bact.*, **93**, 1205.
26. Fox, H., Purcell, R. H. and Chanock, R. M. (1969), *J. Bact.*, **98**, 36.
27. Clyde, W. A. Jr. (1964), *J. Immun.*, **92**, 958.
28. Tully, J. G. and Ruchman, I. (1964), *Proc. Soc. exp. Biol. Med.*, **115**, 554.
29. Razin, S., Hasin, M., Ne'eman, Z. and Rottem, S. (1973), *J. Bact.*, **116**, 1421.
30. Smith, P. F. and Sasaki, S. (1958), *Appl. Microbiol.*, **6**, 184.
31. Butler, M. and Knight, B. C. J. G. (1960), *J. gen. Microbiol.*, **22**, 470.
32. Amar. A., Rottem, S. and Razin, S. (1974), *Biochim. Biophys. Acta*, **352**, 228.
33. Wallach, D. F. H. (1972), *Biochim. biophys. Acta*, **265**, 61.
34. Bretscher, M. S. (1973), *Science*, **181**, 622.
35. Razin, S. (1963), *J. gen. Microbiol.*, **33**, 471.
36. Razin, S. (1964), *J. gen. Microbiol.*, **36**, 451.
37. Pollack, J. D., Somerson, N. L. and Senterfit, B. (1970), *Infect. Immunity*, **2**, 326.
38. Choules, G. L. and Gray, W. R. (1971), *Biochem. biophys. Res. Commun.*, **45**, 849.
39. Pecht, M., Giberman, E., Keysary, A., Yariv, J. and Katchalski, E. (1972), *Biochim. biophys. Acta*, **290**, 267.

40. van Golde, L. M. G., McElhaney, R. N. and van Deenen, L. L. M. (1971), *Biochim. biophys. Acta,* **231**, 245.
41. Rottem, S., Hasin, M. and Razin, S. (1973), *Biochim. biophys. Acta,* **323**, 520.
42 Rottem, S., Stein, O. and Razin, S. (1968), *Archs. Biochem. Biophys.* **125**, 46.
43. McElhaney, R. N., de Gier, J. and van Deenen, L. L. M. (1970), *Biochim. biophys. Acta,* **219**, 245.
44. Levisohn, S., and Razin, S. (1973), *J. Hyg., Cambridge,* **71**, 725.
45. Ne'eman, Z., Kahane, I. and Razin, S. (1971), *Biochim. Biophys. Acta,* **249**, 169.
46. Ne'eman, Z, and Razin, S. (1975), *Biochim. Biophys. Acta,* **375**, 54.
47. Williams, M. H. and Taylor-Robinson, D. (1967), *Nature,* **215**, 973.
48. Hollingdale, M. R. and Lemcke, R. M. (1969), *J. Hyg. Cambridge,* **67**, 585.
49. Argaman, M. and Razin, S. (1969), *J. gen Microbiol.,* **55**, 45.
50. Kahane, I. and Razin, S. (1969), *J. Bact.,* **100**, 187.
51. Pollack, J. D., Razin, S., Pollack, M. E. and Cleverdon, R. C. (1965), *Life Sci.,* **4**, 973.
52. Hoogeveen, J. T., Juliano, R., Coleman, J. and Rothstein, A. (1970), *J. Memb. Biol.,* **3**, 156.
53. Kahane, I., Ne'eman, Z. and Razin, S. (1973), *J. Bact.,* **113**, 666.
54. Razin, S. (1967), *Ann. N.Y. Acad, Sci.,* **143**, 115.
55. Razin, S., Ne'eman, Z. and Ohad, I. (1969), *Biochim. biophys. Acta,* **193**, 277.
56. Gabridge, M. G. and Murphy, W. H. (1971), *Infect. Immunity,* **4**, 678.
57. Lowry, O. H., Rosebrough, N. J., Farr, A. L. and Randall, R. J. (1951) *J. biol. Chem.,* **193**, 265.
58. Rottem, S. and Razin, S. (1967), *J. Bact.,* **94**, 359.
59. Pollack, J. D., Razin, S. and Cleverdon, R. C. (1965), *J. Bact.,* **90**, 617.
60. Kahane, I. and Tully, J. G. (1974), *Proc. int. Congress on Mycoplasmas of Man, Animals, Plants and Insects*, pp. 23, Bordeaux.
61. Schiefer, H. G., Gerhardt, U., Brunner, H. and Krüpe, M. (1974), *J. Bact.,* **120**, 81.
62. Smith, P. F., Mayberry-Carson, K. J., Langworthy, T. A., Mayberry, W. R. and Sugiyama, T. (1974), In *Les Mycoplasmes,* (Bore, J. M. and Duplan, J. F. eds), Vol. 33, pp. 63–67, Colloques INSERM, Paris.
63. Kahane, I. and Marchesi, V. T. (1973), *Ann. N.Y. Acad. Sci.,* **225**, 38.
64. Gourlay, R. N. and Thrower, K. J. (1968), *J. gen. Microbiol.,* **54**, 155.
65. Gilliam, J. M. and Morowitz, H. J. (1972), *Biochim. biophys. Acta,* **274**, 353.
66. Engelman, D. M., Terry, T. M. and Morowitz, H. J. (1968), *Biochim. biophys. Acta,* **135**, 381.
67. Razin, S., Wormser, M. and Gershfeld, N. L. (1974), *Biochim. biophys. Acta,* **352**, 385.
68. Kahane, I. and Razin, S. (1970), *FEBS Letters,* **10**, 261.
69. Razin, S. (1974), *FEBS Letters,* **47**, 81.
70. Rottem, S. and Greenberg, A. (1975), *J. Bact.,* **121**, 631.
71. Rottem, S., and Razin, S. (1966), *J. Bact.,* **92**, 714.
72. Rottem, S., Cirillo, V. P., de Kruyff, B., Shinitzky, M. and Razin, S. (1973), *Biochim. biophys. Acta,* **323**, 509.
73. Low, I. E. and Zimkus, S. M. (1973), *J. Bact.,* **116**, 346.

74. Ne'eman, Z. (1974), Ph.D. Thesis, The Hebrew University, Jerusalem.
75. de Kruyff, B., van Dijck, P. W . M., Goldbach, R. W., Demel, R. A. and van Deenen, L. L. M. (1973), *Biochim. biophys. Acta,* **330**, 269.
76. Rottem, S. and Razin, S. (1967), *J. gen. Microbiol.,* **48**, 53.
77. Rodwell, A. W. (1968), *Science,* **160**, 1350.
78. Rodwell, A. W. and Peterson, J. E. (1971), *J. gen. Microbiol.,* **68**, 173.
79. Herring, P. K. and Pollack, J. D. (1974), *Int. J. Syst. Bact.,* **24**, 73.
80. Pollack, J. D. and Tourtellotte, M. E. (1967), *J. Bact.,* **93**, 636.
81. Panos, C. and Henrikson, C. V. (1969), *Biochemistry,* **8**, 652.
82. Panos, C. and Rottem, S. (1970), *Biochemistry,* **9**, 407.
83. Romijn, J. C., van Golde, L. M. G., McElhaney, R. N. and van Deenen, L. L. M. (1972), *Biochim. biophys. Acta,* **280**, 22.
84. Rodwell, A. W. (1967), *Ann. N.Y. Acad. Sci.,* **143**, 88.
85. Razin, S., Tourtellotte, M. E., McElhaney, R. N. and Pollack, J. D. (1966), *J. Bact.,* **91**, 609.
86. McElhaney, R. N. and Tourtellotte, M. E. (1969), *Science,* **164**, 433.
87. Rottem, S. and Panos, C. (1969), *J. gen. Microbiol.,* **59**, 317.
88. de Kruyff, B., Demel, R. A. and van Deenen, L. L. M. (1972), *Biochim. biophys. Acta,* **255**, 331.
89. Rottem, S. and Razin, S. (1973), *J. Bact.,* **113**, 565.
90. Razin, S. and Tully, J. G. (1970), *J. Bact.,* **102**, 306.
91. Rottem, S., Pfendt, E. A. and Hayflick, L. (1971), *J. Bact.,* **105**, 323.
92 Rodwell, A. W., Peterson, J. E. and Rodwell, E. S. (1972), *Pathogenic Mycoplasmas Ciba Found. Symp.,* pp. 121–139.
93. McElhaney, R. N. (1974), *J. molec. Biol.,* **84**, 145.

# 2 The preparation of cell surface membrane enriched fractions

DAVID M. NEVILLE, JR



## 2.1 Introduction

The cell surface membrane or plasma membrane is the physical boundary which separates the internal environment of the cell from the external environment. This crucial location involves the surface membrane in a wide variety of cellular functions. All stimuli from the environment which the cell utilizes as information must cross or interact with the boundary. Likewise, all modifications of the external environment by the cell must cross or occur at the boundary and the narrow range of chemical, electrostatic, and mechanical potentials required for life must be maintained across this boundary. The importance and diversity of these functions, and their biochemical correlates, accounts for the present widespread interest in surface membrane research. It is the purpose of this chapter to provide investigators who may have little background in this field with the means of preparing and evaluating surface membrane enriched fractions from any required tissue or cell line.

## 2.2 The present state of surface membrane isolation methodology

### *2.2.1 The general problem*

Today, procedures for obtaining surface membrane enriched fractions are cited in the literature for a wide variety of tissues and cell types. These are listed in Table 2.1 with the average enrichment of a putative surface membrane

TABLE 2.1
*Published plasma membrane enriched fractionation procedures, 1960–1974*

| Tissue or cell | Species | Morphology | PM marker enrichment | References |
|---|---|---|---|---|
| *Large solid organs* | | | | |
| Liver | Rodent, bovine, porcine | Sheets, vesicles | 25x | 1–24 |
| Kidney | Rodent, rabbit | Brush border, vesicles | 15x | 25–33 |
| Pancreas | Rodent | Sheets | 5–19x | 34 |
| Skeletal muscle | Rodent, frog | Vesicles | 8x | 35–40 |
| Smooth muscle | Rodent | Vesicles | 20x | 41–44 |
| Cardiac muscle | Frog | Vesicles | 6–16x | 45–46 |
| Fat | Rodent | Ghosts, vesicles | 14x | 47–51 |
| *Epithelial surfaces* | | | | |
| Intestine | Rodent | Brush borders, vesicles | 20x | 52–56 |
| Urinary Bladder | Rodent | Sheets | 4.5x | 57–58 |
| Taste papillae | Bovine | Vesicles | 12x | 59 |
| Olfactory epithelium | Bovine | Vesicles | 47x | 60 |
| Epidermis | Bovine | Desmosomes | – | 61 |
| *Neural Tissues* | | | | |
| Brain | Rodent | Vesicles | 9x | 62–69 |
| Retina | Chicken | Vesicles | 13x | 70 |
| *Endocrine and reproductive organs* | | | | |
| Thyroid | Bovine | Sheets, vesicles | 10–40x | 71–74 |
| Adrenal cortex | Bovine | Vesicles | – | 75 |
| Adrenal medulla | Bovine | Vesicles | 14x | 76 |

| | | | | |
|---|---|---|---|---|
| Corpus luteum | | Vesicles | 20× | 77 |
| Anterior pituitary | Bovine | Vesicles | 28× | 78 |
| Mammary gland | Rabbit, bovine | Vesicles | 15× | 79–80– |
| *Blood forming cells and Reticulo Endothelial system* | | | | |
| Lymphocytes | Human, porcine | Vesicles | 14× | 81–82 |
| Thymocytes | Rodent, avian | Vesicles | – | 83–85 |
| Macrophages | Rodent | Sheets | – | 86 |
| Leukemic cells | Murine | Sheets | 20× | 87–88 |
| *Solid tumors* | | | | |
| Hepatoma | Rodent | | – | 89–90 |
| Fibrosarcoma | Avain | | 8× | 91 |
| *Tissue cultured lines* | | | | |
| Fibroblasts | Rodent, primate, avian | Sheets, vesicles | 4–10× | 92–101 |
| HeLa | Human | Ghosts, vesicles | 20× | 102–105 |
| Ehrlich ascites | Rodent | Ghosts | 15× | 106–108 |
| Krebs ascites | Rodent | Vesicles | – | 109 |
| Muscle | Rodent | Vesicles | – | 110 |
| Liver | Rodent | Vesicles | 10× | 93, 111 |
| Lymphocyte | Human | Sheets | 5× | 112 |
| KB | Human | Sheets, vesicles | 25× | 113 |
| Neuroblastoma C1300 | Murine | Sheets | – | 114 |
| Sarcoma 180 | | | 13× | 115 |
| *Embryonic cells* | | | | |
| Chick embryo cells | Avian | Vesicles | 23× | 116 |

This table is generally limited to mammalian tissues and cell lines.

biochemical marker referenced to the original homogenate [1–116]. Many of the methods are modifications of this author's original procedure for liver plasma membranes [1] and utilize a combination of isopycnic density gradient and rate sedimentation centrifugations. Twenty-one tissues and eight different types of tissue cultured cell lines are listed. The enrichments vary from 5–50 fold. While these enrichments cannot be used generally to document purity, they do provide information bearing on the utility of a given preparation for a given need [117]. Since these methods have proved successful for such a variety of tissues and cell lines, it is likely that they will provide surface membrane enrichments for any tissue or cell line if properly modified and if a suitable surface membrane marker can be found.

Although surface membrane enriched fractions are easily available, surface membrane fractions of documented purity are not as common. Purity can be assessed by morphologic criteria when a distinctive surface membrane morphology exists such as a brush border or a large sheet-like membrane. Quantitative documentation of purity usually requires specific activities in both the initial homogenate and final membrane fraction of markers associated with all intracellular organelles (see Table 2.3). Smooth vesicles derived from endoplasmic reticulum (ER) provide the most difficulty. These organelles are the most frequent contaminant and their marker enzymes are not constant from tissue to tissue and often lack specificity. Of the thirty types of preparations in Table 2.1 only the first nine listed in Table 2.2 provide validation of the ER marker and sufficient data to calculate the ER contamination of the surface membrane preparation. Most of these preparations contain less than 20% of their membraneous material derived from endoplasmic reticulum, mitochondrial fragments or lysosomes. However, this fact does not permit the conclusion that these preparations consist of 80% surface membranes. Appreciable contamination with vesicles derived from endocytotic vesicles or Golgi apparatus can occur. In addition, surface membrane heterogeneity arising from multiple cell types and specialized membrane features often render a simple percentage purity statement ambiguous. The problems of heterogeneity purity requirements, and documentation will be considered in further detail.

### *2.2.2 Surface membrane heterogeneity*

The cell surface membrane or plasma membrane can be defined as the most peripheral membraneous portion of a cell. This is a morphological definition which excludes cytoplasmic vesicles which may have been derived from the surface membrane by pinocytosis or other forms of membrane recycling. The reason for this distinction is that in many investigations we are interested in quantifying membrane components at the crucial cell environment interface. The surface membrane thus defined is structurally heterogenous for many cells. The hepatocyte surface membrane, for example, exists in three forms: a single membrane exhibiting small protrusions and invaginations bordering on the sinusoid known as the blood front; a membrane closely opposed to the membrane of the neighbouring liver cells exhibiting numerous desmosomes known as the cell front; and a rich protrusion of long villous processes forming the bile canaliculus known as the bile front. Evidence for biochemical heterogeneity paralleling the structural heterogeneity exists for liver [15] but is best documented in intestinal epithelium [54] and in renal proximal convoluted tubule epithelium [27,29,32]. The brush border membranes of these cells are enriched 15-fold in alkaline phosphatase and maltase while enrichments for $Na^{+}$-$K^{+}$-ATPase and adenyl cyclase are low. Vesicular preparations from these tissues free of brush border and presumably derived from the blood front or cell front show a reversal of the enrichment patterns listed above. Thus, one sees that surface membrane marker enzymes are sometimes valid for only a portion of the specialized membrane.

Cells growing in tissue culture also have three interfaces; one with the dish below, one with the neighbouring cell and one with the media. The biochemical specialization of these regions has not been investigated.

An additional source of surface membrane heterogeneity must be considered when membranes are isolated from solid organs. All organs contain multiple cell types and it is often not possible, especially when dealing with vesicular fractions, to know the cellular origin of the surface membrane. It has been estimated that the littoral cells lining the liver sinusoids could contribute 40% of the total organ cell surface membrane [4]. Similar considerations would apply to any highly vascularized organ.

As a final source of membrane heterogeneity the sideness of the membrane must be considered. Recently, evidence has accumulated that the protein, lipid, and carbohydrate composition of the internal surface interface is quite different from the external surface interface. For example, 5′-nucleotidase activity [118] and insulin receptors [119] can only be demonstrated at the external interface. Most of the glycoprotein carbohydrate also appears at this interface [120]. Cell rupture and surface membrane vesiculation may leave the surfaces facing the same way, or the membranes may invert and reseal, thus presenting the original external interface to the inside of the vesicle. Enzyme markers localized to the original external interface might be difficult to detect if their substrates are impermeable [121].

The topological origin of the plasma membranes listed in the Table 2.1 preparations is generally unknown. Membranes derived from suspensions of tissue cultured cell lines isolated by techniques which give large ghosts or sheets may be considered to contain the entire surface membrane of that cell type. The last four entries in Table 2.2 fit this criteria. When mixed cell types are present specific morphological features, when present, can indicate topological source. Intestinal and renal brush borders are an example. For sheet-like liver membrane preparations bile canaliculi and desmosomal junctions indicate that most of the material is derived from the bile and cell front of the plasma membrane. The exact quantitation is, however, difficult. The source of the vesicular material seen in these preparations cannot be identified and could represent disrupted bile canaliculi or blood front membranes.

Vesicular surface membrane preparations contain no morphological markers, and their topological source cannot be identified except by unique biochemical markers which have been related to the membrane topology by some type of histochemical technique. In general when this has been claimed, the markers have not been demonstrated to be unique for that and only that specific topological feature. Fixation artifacts and the difficulty of relating quantitative histochemical data back to whole homogenate data complicate these studies. However, it is this author's opinion that there are unique markers such as protein and peptide receptors for each different cell type and topological feature, and in the future it will be possible to identify the origin of vesicular surface membrane preparations.

### *2.2.3 Fractionation techniques*

Techniques for the isolation of plasma membranes are based on general concepts and methodology of tissue organelle fractionation pioneered by Claude and largely developed by Palade, Porter, Hogeboom, Schneider, Siekevitz, and de Duve and co-workers [122]. These workers have shown that the particles released on cell rupture can be divided into distinct groups on the basis of their biochemical activities and morphology, and that these groups also have unique size and density distributions which make possible their fractionation by centrifugal fields with and without density gradients. The major groups or organelles that are now generally recognized are listed in Table 2.3 along with their sedimentation properties, density distributions in sucrose, biochemical markers, and the percentage of protein which they contribute to the homogenate. The data is largely from rat liver [123].

Much of this knowledge has been acquired throught quantitative fractionation techniques. As practiced by de Duve and co-workers [124]; Wattiaux-deConinck and co-workers [125] and Beaufay and co-workers [123] the homogenate is divided into three operational fractions on the basis of rate sedimentation: a crude nuclear fraction, 600 *g*, 15 min; a crude mitochondrial fraction, 10 000 *g*, 25 min; and a crude microsomal fraction consisting of the particulate material left in the post-mitochondrial supernatant. These three fractions are then subfractionated according to their sedimentation rates and densities by overlaying them on a density gradient and centrifuging for varying amounts of time. A large number of enzymes are monitored. Groups of enzymes can be observed to move together at a given rate and to stabilize at definite isopycnic points, thus establishing a class of organelle defined by a density, a sedimentation coefficient, and an enzyme profile. The morphology of each fraction can then be assessed by electron microscopy. The percentage of each organelle class within the three fractions and the original homogenate can then be calculated. Thus, a complete quantitative picture of the organelle composition of a tissue emerges. This technique has been applied to liver [123,124], kidney [126], rabbit aorta smooth muscle [43] and tissue cultured rat embryo fibroblasts [127], and should be applied to many of the issues listed in Table 2.1 for which only meagre knowledge of organelle composition and effective markers is now available. Beaufay has developed zonal rotors which require as little as 1 mg of tissue protein to perform complete quantitative fractionations [43].

The structural fate of the membrane and the crude fraction in which it will be concentrated is dependent on the method of cell rupture, the composition of the media and the cell type in question. If the cytoplasmic contents are extruded through one opening, an empty sac or ghost may result. This ghost may vesiculate to smaller vesicles which reseal leaving some trapped particles inside. Vesicles larger than 4 $\mu$m in diameter are distinguishable from the smaller intracytoplasmic vesicles and can provide positive morphologic identification of the surface membrane. When rupture results in vesiculation of the membrane to vesicles 2 $\mu$m and below, positive morphologic identification is lost and biochemical markers must be used to identify the surface membranes. Portions of a cell surface may separate as in intact sheet as is the case with brush borders and the cell and bile front of hepatocytes. In the latter case the membranes of previously adjacent cells remain adherent by desmosomal linkages. These sheets are larger than intracytoplasmic structures and provide positive identification under phase microscopy. Under electron microscopy desmosomal structures, bile canaliculi and brush borders all provide positive morphologic identification of surface membrane [1,52].

The ease of rupture of cells and the final morphological state of the membrane is often dependent on the divalent cation concentration. The divalent cation concentration cannot be increased beyond a critical point however, because irreversible agglutination of all the organelles takes place rendering fractionation impossible. Cells from different tissues, species, or after transformation vary in their sensitivity to divalent cations.

More effective than divalent cations in preventing vesiculation are agents which complex membrane SH groups. Warren and co workers [97] have been able to isolate a variety of surface membranes as sheets using these reagents.

Membrane sheets, ghosts, and brush borders sediment in low centrifugal fields (1500 x *g*, 15 min). Large vesicles require doubling or quadrupling these fields. Small vesicles are lost to the crude microsomal fraction (10 000 x *g*, 25 min supernatant) or, if large enough, to the crude mitochondrial fraction. Large membrane fragments can also be separated from small vesicles by two phase systems [98,13]. Large ghosts, sheets, and brush borders have isopycnic points in sucrose in the range of 1.185–1.16 g cc$^{-1}$. Vescular membrane fractions band between 1.16 and 1.08 g cc$^{-1}$. Most surface membrane isolation procedures use isopycnic density gradient centrifugation in either sucrose [1] or dextran and sucrose [108] as a second step. If the membranes are in the form of sheets, ghosts, or brush borders a low *g* pellet enriched in membranes is applied to the density gradient. If the membranes are vesiculated, a high *g* pellet is applied. The application pellets are also usually enriched in one or more of the major intracellular organelles, (see Table 2.3), and any organelle unique to a particular cell type. The degree of enrichment depends on the homogenizing conditions used; shear, ionic strength, pH, concentration of divalent cations, and the *g* force integrated over time for the application pellet and the conditions of any prior centrifugations.

### *2.2.4 Organelle quantitation*

After the density gradient centrifugation the position of the surface membranes is identified by phase microscopy if the surface membranes are in the form of ghosts or sheets. If they are in vesicular form the gradient is assayed with respect to a marker believed to be specific for the surface membrane, and enrichment is calculated with respect to the initial homogenate. Commonly used surface membrane markers are listed in Table 2.3. None of these markers is invariably present strictly localized to the plasma membrane. For example, 5′-nucleotidase, the most widely used plasma membrane marker, shows low enrichments in preparations from several tissue cultured cell lines [92,115]. EDTA used to dissociate cells from the petri dish surface could be artifactually involved. In HeLa cells when care is taken not to inactivate the enzyme in the homogenate with EDTA the 5′-nucleotidase enrichment is only one-third of the $Na^+$-$K^+$-ATPase enrichment [105]. 5′-Nucleotidase has also been found to be enriched in Golgi apparatus [128] and endoplasmic reticulum [118]. Insulin receptors and adenylate cyclase have been found to be enriched in Golgi apparatus [129,130]. This organelle also shares the high cholesterol to phospholipid ratio found in plasma membranes [123]. $Na^+$-$K^+$-ATPase (oubain sensitive) is the next widely used plasma membrane marker This enzyme is believed to be present in the surface membrane of every animal cell. Its use as a marker is often negated by the high activities of oubain insensitive ATPase, an enzyme which is not specific for surface membrane. In addition, as mentioned previously, the localization of $Na^+$-$K^+$-ATPase on intestinal epithelium [54] renal proximal convoluted tubule [27,29,32], and urinary bladder [58], shows polarity, the enzyme being deficient at the luminal surfaces.

Contaminating vesicles of cytoplasmic organelles are similarly identified by their specific markers. Often a second gradient or rate sedimentation step must be used to reduce the intracellular membrane contamination. The contamination of the final fraction is estimated by multiplying the purification ratio of the organelle marker in question in the surface membrane fraction by the percent of organelle protein in the initial homogenate. These calculations require total fractionation data on the cell type or tissue in question.

Of the intracellular organelles, mitochondria and nuclei give the least problem. The isopycnic points of mitochondria are relatively constant from tissue to tissue [127] and generally above those for cell surface membranes.

Large membrane fragments have higher sedimentation rates than mitochondria while vesicular membranes have lower sedimentation rates. Mitochondrial marker enzymes are constant from tissue to tissue [127], and those listed in Table 2.3 are unique for mitochondria. In general, the comments applicable to mitochondria are also applicable to lysosomes and peroxisomes.

Nuclei are easily identified by phase microscopy or DNA analysis and are generally eliminated in very low speed pellets, by use of hypotonic media which lead to solubilization, or are pelleted in the gradient due to their high density. However, care must be taken not to confuse nuclear membranes with cell surface membranes. The former sometimes share ER markers [131].

Endoplasmic reticulum is heterogeneous with respect to density and these densities, especially for smooth ER, overlap the surface membrane densities, rendering complete separation by density alone impossible. When the surface membrane exists as large fragments, separation by rate sedimentation is easily accomplished. As membrane fragments become smaller, approaching the microsomal vesicle size, complete separation becomes difficult to impossible.

In liver, ER markers include glucose-6-phosphatase, a number of electron transport enzymes, esterases, and a variety of drug metabolizing enzymes [123]. In other tissues, with the exception of kidney, glucose-6-phosphatase is present in low amounts and validity of this marker is questionable [132]. Assays should always be run in the presence of acid and alkaline phosphatase inhibitors (fluoride and EDTA). Evidence for the biochemical heterogeneity of endoplasmic reticulum exists [123,133]. The complete complement of the electron transport chain is not invariably present in the ER of all tissues. NADPH cytochrome *C* reductase and NADH cytochrome *C* reductase are present in low amounts in an induced hepatoma (D23) and cytochrome $b_5$ and cytochrome P450 were undetectable [90]. In rabbit aorta smooth muscle [43] and in cultured rat embryo fibroblasts [127] investigators were unable to find a satisfactory microsomal marker. It is necessary then, when dealing with tissues other than liver and kidney, to establish the validity of a marker for endoplasmic reticulum. Since this organelle contributes less than 25% of the homogenate protein in liver (an ER rich tissue by electron microscopy) a 4-fold marker enrichment in a microsomal fraction would appear to be a minimum requirement for a specific marker. This is often not observed and may reflect smooth membranes of extra ER origin in the microsomal fraction or localization of the markers in non-microsomal membranes. Both cases have been documented [123,127] and total quantitative fractionation as practiced by de Duve [124] and Beaufay [123] is the best approach to validating ER markers.

Smooth vesicles of a variety of origins (see Table 2.3) are likely to be the major contaminant of any cell membrane preparation which isolates the membrane in vesicular form. For this reason the microsomal fraction of a tissue or cell type used for cell surface membrane preparation should be well characterized. The very complete quantitative characterization of the liver microsomal fraction recently reported by Beaufay and co-workers [123] is recommended reading for any investigator isolating a vesicular plasma membrane preparation from a post mitochondrial supernatant. Besides ER derived vesicles, Golgi and endocytotic vesicles may be major contaminants. In liver [123] and pituitary [130], UDP-galactosyl transferase is a unique marker for Golgi. If this is true for other tissues this contamination can be quantifyed. Endocytotic vesicles are likely to be very similar in composition to the plasma membrane. There could however be important differences rendering the distinction worthwhile. The ratio of endocytotic vesicles to surface membrane is generally unknown but Steinman and co-workers report that for cultured fibroblasts 25% of the surface membrane is endocytosed per hour [134]. The presence of large quantities of endocytotic vesicles carrying surface membrane markers may account for the high yields of 'plasma membranes' achieved with some vesicular preparations [18]. These yields are larger than expected on the basis of morphological considerations and account for the wide range in the estimate of surface membrane contribution to homogenate protein [117].

'Non-penetrating' reagents designed to covalently label the cell surface membrane of intact cells specifically have been used as surface membrane markers [112]. These methods attempt to circumvent the problems posed by lack of uniqueness of biochemical surface membrane markers and the problems in quantitating total surface membrane protein. The most popular method is iodination utilizing $^{125}$Iodide and lactoperoxidase. Unfortunately, this method has not been assessed in parallel with conventional markers on any of the preparations listed in Table 2.2 where good quantitative data is available. However, a detailed study of the iodination method of Phillips and Morrison [135] has been performed by Tsai and co-workers [136]. They conclude that the depth of iodination in the membrane and even into the cytoplasm is a variable and is dependent on the conditions of iodination and the cell size. Conditions must be optimized for each cell type. Another major problem with surface labelling is how to keep cell breakage to a low level. If the surface membrane contributes 1% to the homogenate protein, 1% cell breakage could in theory produce 50% errors in surface membrane purification and contamination estimates.

The individual plasma membrane isolation procedures [1–116] will not be reviewed in detail because of limitations of space, redundency, and this author's patience. Some of the more generally used procedures have been reviewed in detail [4,12, ,137,138]. However, it should be emphasized that each cell type, tissue, species, and even age group has its own special characteristics of cell fragility and ease of membrane vesiculation which dramatically effects plasma membrane sedimentation properties and renders the conditions of homogenization a major variable. Conditions worked out for one tissue or cell type cannot be blindly applied to another. Rather, membrane purification must be followed by either phase microscopy or a suitable marker for each new tissue, age group or pathological state of the cell or tissue being used.

TABLE 2.2
*Plasma membrane fractionation procedures which include a valid endoplasmic reticulum contamination evaluation*

| Tissue or cell type | Endoplasmic reticulum marker | Unique PM morphological marker | References |
|---|---|---|---|
| Liver | Glucose-6-Pase<br>NADPH cyto *c* red | + | 1–24 |
| Kidney brush border | Glucose-6-Pase | + | 25–33 |
| Fat | NADH oxidase | – | 49 |
| Pancreas | NADPH cyto *c* red | + | 34 |
| Intestinal brush border | NADPH cyto *c* red | + | 55 |
| Urinary bladder epithelium | NADH cyto *c* red | + | 58 |
| Taste papillae | NADH cyto *c* red | – | 59 |
| Synaptic PM | NADH cyto *c* red | – | 64, 65 |
| Chick embryo fibroblast | Antimycin insensitive<br>NADH cyto *c* red | – | 92 |
| KB cell | – | + | 113 |
| Neuroblastoma | – | + | 114 |
| HeLa cell | – | + | 102, 104, 105 |
| Fibroblast | – | + | 97, 98, 99 |

### 2.2.5 *Purity requirements*

A variety of biochemical studies have been performed utilizing the plasma membrane fractions listed in Tables 2.1 and 2.2. Many studies concern the localization of certain enzymes to specific types of surface membranes. Others deal with the detailed properties of surface membrane bound enzymes. Studies on adenylate cyclase are an example [139]. A large literature exists on plasma membrane receptors for various polypeptide hormones [140,141], peptide releasing factors [142], drugs [143–145], neurotransmitters [146,147], and prostaglandins [148,149]. The distribution of these receptors with respect to cell type and the receptor properties in relation to sequence analogues and stereoisomers has been detailed [140,147]. Increasingly, surface membrane receptors for large molecules, not generally considered as informational carriers, are being discovered. The interaction of these components results in changes in the cells internal or external environment. The clearing of desialated glycoprotein from the circulation by specific liver receptors is one such example [150]. The diminution of protein synthesis resulting from membrane binding of a variety of bacterial and seed toxins is another [151,152]. As predicted [1], the availability of surface membrane fractions has helped to unravel some of the interrelationships between cellular functions and the surface membrane.

Most types of study cited above did not require precise knowledge of the quantity nor identity of the organelle contamination of the membrane fractions.

Recently, however, investigators have begun to accurately quantify surface membrane receptor concentration in an attempt to relate changes in receptor concentration to altered physiological and pathological states. Kahn and co-workers [153,154] have been able to correlate the *in vivo* insulin resistance induced by genetic, metabolic and hormonal manipulations with a decreased surface membrane insulin receptor concentration, and conclude that these phenomena are causally related. This type of study consists of comparing the number of receptors in surface membrane fractions derived from a normal and pathological tissue, and has been previously reviewed [4,155]. Two distinct problems arise. First, do both fractions contain the same quantity of surface membrane derived from the same topological site? Second, if the cell size or surface membrane surface area or topology is different in the two cases (i.e., the cell diameter is increased in one state) to what type of membrane component and units should the receptors be referenced? In regard to the second question it is probably best to reference receptor concentration in a variety of ways, i.e., per cell, per unit cell surface area (geometrical calculations or stereological technique [156]) per unit isolated membrane protein, per unit other receptor or membrane marker, and then look for a relevant correlation [153]. With regard to the first question, it is necessary when doing a comparative study to demonstrate that the isolation procedure is not differentially selecting between the two different starting materials [4,155]. This can be done by performing a total quantitative fractionation for membrane markers as well as contaminating organelles. Failure to do this results in misleading or uninterpretable data [155]. For example, a receptor may be found to disappear from a cell surface membrane fraction following viral oncogenic transformation. In actual fact, the vesicular size of the transformed membrane following homogenization changed from 3 $\mu$m to 0.05 $\mu$m moving the surface membrane derived vesicles from the 3000 $g$ x 15 min gradient application pellet to the post mitochondrial supernatant. This change would be most difficult to detect using phase microscopy and would only be found by keeping a total balance sheet of a surface membrane marker. These problems are formidable with vesicular membrane preparations because of the difficulties with ER markers discussed previously. They are less of a problem with large membrane fragments. An alternative to membrane isolation is to assay the receptor directly on the surface of the cell in a collagenase prepared cell suspension [153] or directly in a petri dish for tissue-cultured cell lines. In this manner Goldstein and Brown [157] were able to demonstrate that tissue-cultured fibroblasts derived from a patient with homozygous familial hypercholesterolemia did not exhibit high affinity receptors for low density lipoprotein. These mutant cells fail to control cholesterol synthesis or to degrade low density lipoprotein and the disease process appears to be a consequence of the receptor loss.

In summary, comparative studies using vesicular preparations should be undertaken with caution and avoided if alternatives are possible. If they must be

TABLE 2.3
*Physical and biochemical characteristics of cellular organelles*

| Organelle | % of homogenate protein (liver) | Particle dimension, $\mu m$ | Sedimenting centrifugal field | Density in sucrose, $g\ cc^{-1}$ | Biochemical markers |
|---|---|---|---|---|---|
| Plasma membrane (large) | 0.4–2.5? | 3–20 | 1500 *g*, 15 min | 1.19–1.15 | 5′-nucleotidase, $Na^{+}$-$K^{+}$-ATPase |
| (vesicle) | 0.4–2.5? | 0.05–3 | 145 000 *g*, 40 min | 1.19–1.07 | $Mg^{2+}$-ATPase, Adenylate cyclase, Alkaline Pase, Alkaline Phosphodiesterase, Polypeptide hormone receptors |
| Nuclei | 13 | 3–12 | 600 *g*, 15 min | > 1.30 | DNA |
| Nuclear membrane | – | 3–12 | 1500 *g*, 15 min | 1.22–1.18 | NAD diaphorase |
| Golgi apparatus (large) | 1.0 | 1 | 2000 *g*, 20 min | 1.16–1.12 | UDP-galactosyl transferase |
| (vesicle) | 1.0 | 0.05–0.5 | 145 000 *g*, 15 min | 1.16–1.12 | UDP-galactosyl transferase |
| Mitochondria | 16 | 0.5–2 | 10 000 *g*, 25 min | 1.21–1.17 | Monoamine oxidase, cytochrome oxidase, Succinate cyto *c* red |
| Mitochondrial outer membrane | – | – | 145 000 *g*, 40 min | 1.14 | Monoamine oxidase |
| Lysosomes | 2.0 | 0.5–0.8 | 10 000 *g*, 25 min | 1.216–1.19 | Acid phosphatase Acid hydrolases |
| Peroxisomes | 3 | 0.5–0.8 | 10 000 *g*, 25 min | 1.23–1.18 | Catalase |
| Endoplasmic reticulum | 24 | 0.05–0.3 | 145 000 *g*, 40 min | Rough 1.26–1.18 Smooth 1.23–1.06 | Glucose-6-Pase NADH and NADPH, Cytochrome *c* reductase, Cytochrome *b*5, aminopyrine demethylase, Esterase |
| Free and polyribosomes | – | 0.025–0.6 | 145 000 *g*, 40 min | > 1.26 | – |
| Endocytotic vesicles | – | 0.1–0.5 | 145 000 *g*, 40 min | – | – |
| Secretory vacuoles | – | 0.1–0.5 | 145 000 *g*, 40 min | – | – |
| Fat droplets | – | 0.5–5 | – | – | – |
| Glycogen | – | – | 145 000 *g*, 40 min | 1.6 | – |

The range of particle dimensions (longest axis), sedimenting centrifugal fields, plasma membrane densities, as well as biochemical markers, is a compendium of many tissues and cell types [1–117]. The remaining data is for liver and sources can be found in [124] and [128].

done, total quantitative fraction data on the normal and pathological tissue should be obtained.

## 2.3 Practical procedures for the isolation of plasma membranes

### *2.3.1 Comparison of plasma membranes isolated from ob/ob mice and their thin litter mates*

In this section methods for isolating mouse liver plasma membranes are presented. My laboratory, in conjunction with Drs Kahn and Roth of the National Institute of Arthritis, Metabolism and Digestive Diseases have found these methods useful in performing comparative studies of surface membrane insulin receptor concentration in insulin resistant states [153,154]. These studies are an outgrowth of work originally begun with Dr Pierre Freychet* [165] and recently continued with Dr Andrew Soll† [161,166].

The method used is presented in Section 2.3.3 and yields sheet-like membrane fragments The mouse strain used was C57BL/6J. Mice with the obese hyperglycemic syndrome carried a double dose of the recessive mutation *ob*. It was found that essentially the same method used for rat liver was

TABLE 2.4
*Comparison of plasma membrane purification of* ob/ob *mice and their thin litter mates. (All results are the average of 2 or 3 determinations.)*

| | *ob/ob* | | | Thin | | |
|---|---|---|---|---|---|---|
| Purification fraction (step in Section 2.3.3) | Total protein (mg) | 5′-Nucleotidase ($\mu$mol $mg^{-1}$ $h^{-1}$) | Relative specific activity | Total protein[a] | 5′-Nucleotidase ($\mu$mol $mg^{-1}$ $h^{-1}$)[b] | Relative specific activity |
| Dilute homogenate [2] | 1450 | 0.30 | 1.0 | 1517 | 0.35 | 1.0 |
| First sediment [3] | 110 | 1.8 | 6.3 | 101 | 1.9 | 5.4 |
| Floated particles [11] | 21 | 4.3 | 14 | 24 | 5.0 | 14 |
| Purified membranes [15] | 5.3 | 6.8 | 23 | 6.9 | 6.8 | 19 |

[a]The total Lowry protein is expressed as yield in mg per 10 g wet weight of liver (albumin standard) [163].
[b]The 5′-nucleotidase activity was determined at 37°C as described [164] with the addition of a deproteinization step.

From Neville and Kahn [4].

*Permanent address: Institut National de la Santé et de la Recherche Médicale, 06034 Nice Cedex, France.
†Present address: Veterans Administration Center, Wadsworth Hospital, Los Angeles, California 90073.

TABLE 2.4 continued

| Purification fraction (step in Section 2.3.3) | *ob/ob* Succinate cytochrome *c* reductase[c] (μmol mg$^{-1}$ min$^{-1}$) | *ob/ob* Relative specific activity | *ob/ob* Percentage mitochondrial protein in fraction | Thin Succinate cytochrome *c* reductase[c] (μmol mg$^{-1}$ min$^{-1}$) | Thin Relative specific activity | Thin Percentage mitochondrial protein in fraction |
|---|---|---|---|---|---|---|
| Dilute homogenate [2] | 0.027 | 1.0 | 15 | 0.035 | 1.0 | 15 |
| First sediment [3] | 0.043 | 1.6 | 24 | 0.061 | 1.7 | 26 |
| Floated particles [11] | 0.036 | 1.3 | 20 | 0.058 | 1.7 | 26 |
| Purified membranes [15] | 0.005 | 0.19 | 2.9 | 0.006 | 0.17 | 2.6 |

Homogenate mitochondrial and microsomal percentage protein is assumed equal to that of the rat and is taken from de Duve *et al.* [124].

| Purification fraction (step in Section 2.3.3) | *ob/ob* Glucose-6-phosphatase[d] (μmol mg$^{-1}$ h$^{-1}$) | *ob/ob* Relative specific activity | *ob/ob* Percentage microsomal protein in fraction | Thin Glucose-6-phosphatase[d] (μmol mg$^{-1}$ h$^{-1}$) | Thin Relative specific activity | Thin Percentage microsomal protein in fraction |
|---|---|---|---|---|---|---|
| Dilute homogenate [2] | 4.9 | 1.0 | 25 | 5.2 | 1.0 | 25 |
| First sediment [3] | 7.0 | 1.4 | 35 | 5.4 | 1.0 | 25 |
| Floated particles [11] | 0.84 | 0.17 | 4.3 | 1.1 | 0.21 | 5.3 |
| Purified membranes [15] | 0.42 | 0.09 | 2.3 | 0.54 | 0.10 | 2.5 |

[c]The succinate cytochrome *c* reductase activity was measured at 25° [158].

[d]The glucose-6-phosphatase activity was measured at 37°C in the presence of EDTA and KF as described [159].

applicable to mouse liver [3]. By using 10 g of tissue from each group, a 4-place rotor for the first sedimentation, and a 3-place rotor for the flotation and zonal centrifugation, the control group is centrifuged with the experimental group, thus minimizing procedural differences.

Examination of Table 2.4 shows that the yield of membrane protein, the enrichment of the membrane marker 5′-nucleotidase, and the contamination rates of the major liver organelles are comparable for the obese mouse and its

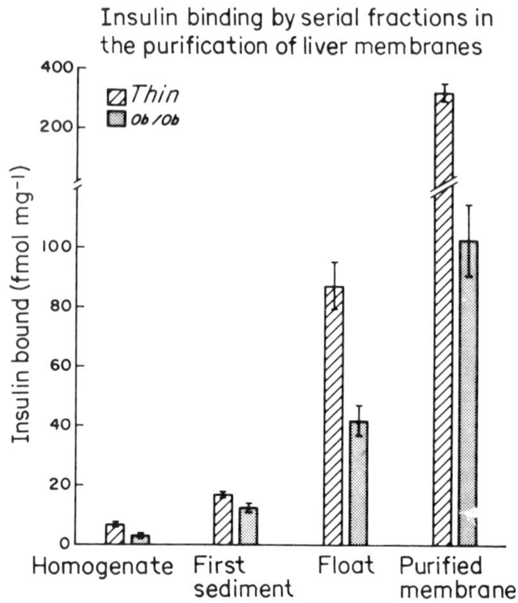


**Fig. 2.1** Insulin binding with purification of mouse liver plasma membranes isolated from C57BL/6J mice (thin) and a mutant *ob/ob*. Specific binding was measured with ($^{125}$I)-insulin at $5 \times 10^{-10}$ M with cellular fractions at concentrations of 0.2 to 7 mg protein ml$^{-1}$. The cellular fractions referred to are steps 2, 3, 11 and 15 in the purification. (From Kahn *et al.* [153].)

thin litter mates. This is in spite of a gross difference in the starting material. It is also apparent that at each step in the purification process the yields of protein, plasma membrane, microsomes, and mitochondria are comparable for the two groups. Likewise, as shown in Fig. 2.1, insulin receptor concentration is increased at each step of the procedure for both groups of animals. These findings indicate that the fractionation procedure is not selecting differentially with respect to the thin and fatty livers, at least by these criteria. In addition, as seen in Fig. 2.2, the purified membranes of each group are indistinguishable by phase microscopy.

The problem of selection of various parts of the plasma membrane is more difficult to approach. The same specific activity of the membrane marker 5′-nucleotidase in both groups gives confidence that extensive selection is not taking place since Evans [15] has shown that shear-induced subfractions of liver-plasma membranes have different specific activities of this enzyme. The subfractions described by Evans [15] had different protein subunit compositions, as revealed by acrylamide gel electrophoresis in the presence of sodium dodecyl sulphate. The gel electrophoretic patterns for both groups of purified membranes are shown in Fig. 2.3 and are remarkably similar.

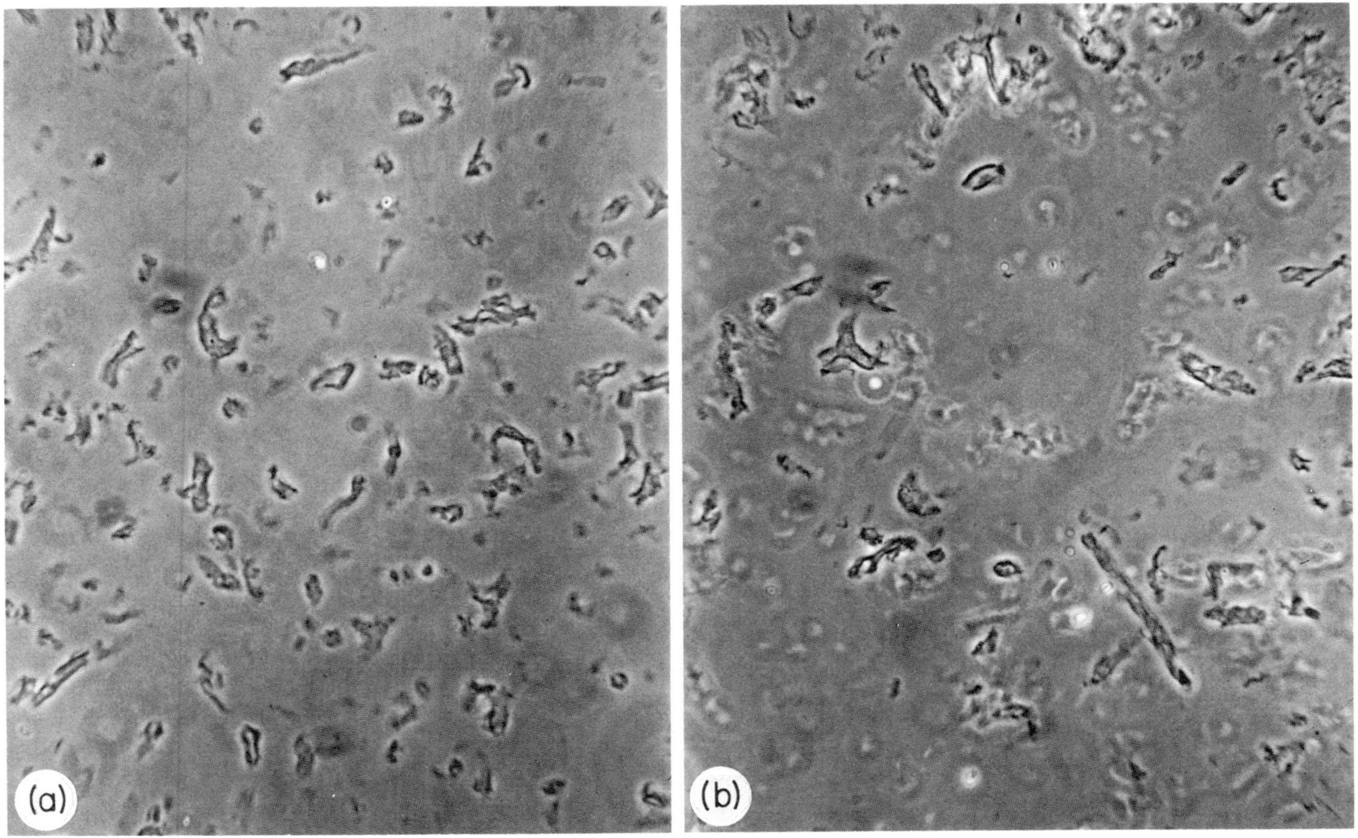


**Fig. 2.2** Appearance of the purified membranes using phase microscopy. Membranes from the obese mouse are on the right (b), from the thin mouse on the left (a). Base magnification ×250. Reprinted from Kahn *et al.* [153].

Another way to minimize the risk of differential plasma membrane selection between two groups is to study another receptor or component which is most similar to the receptor in question. In the present case we were interested in assessing the role of surface-membrane insulin-receptor concentration in the insulin-resistant state characteristic of the hyperglycemic mouse [153]. We chose the glucagon receptor as a reference receptor. Although glucagon has opposite effects from those of insulin on the liver, it, like insulin, is a small polypeptide hormone which is secreted by the pancreas directly into the portal blood circulation. In Fig 2.4 it can be seen that although the concentration of

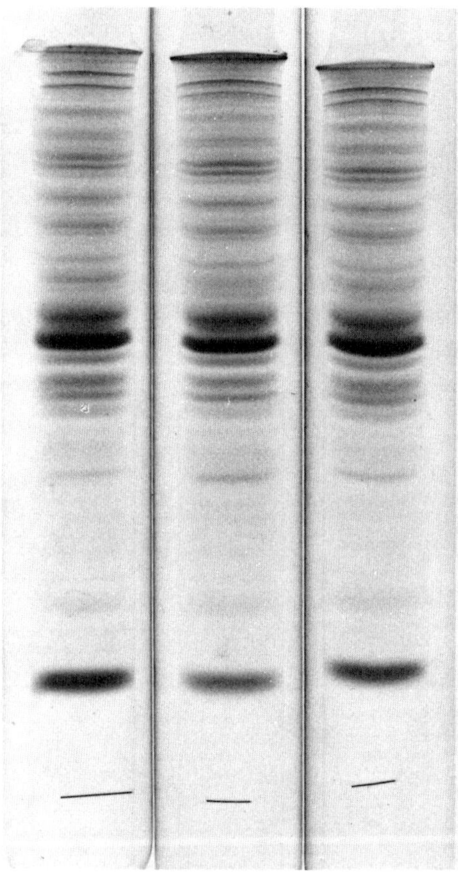

Fig. 2.3 Polyacrylamide gel electrophoresis in SDS. From left to right the gels are from purified plasma membranes of the obese, the thin, and a mixture of both thin and obese. Each gel exhibits 40 resolvable bands. There are no bands observed in one membrane which are not observed in the others, and 2 of the 40 differ slightly in intensity. The gels contain 11% acrylamide and are run with a discontinuous buffer system at pH 9.5 as described [162]. From Kahn *et al.* [153].

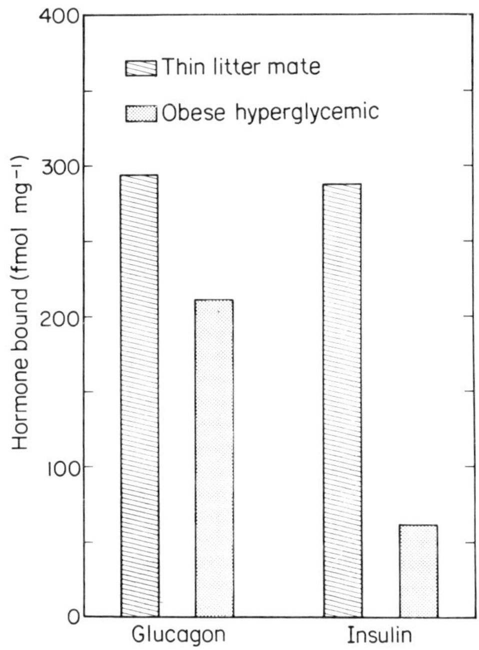


**Fig. 2.4** Comparison of binding of [$^{125}$I]-insulin and [$^{125}$I]-glucagon to liver plasma membranes of obese hyperglycemic mice and their thin litter mates. The concentration of insulin and glucagon was $5 \times 10^{-10}$ M and the concentration of membrane protein was 0.2 mg ml$^{-1}$ for all experiments. From Kahn *et al.* [153].

glucagon receptors per mg of membrane protein falls by about 20% in the *ob/ob* membranes, the fall in insulin receptors is 75%.

The amount of insulin bound to saturable receptors per 0.2 mg of membrane protein over a range of free insulin concentrations is shown for normal mice and for the *ob/ob* mutant by the shaded areas in Fig. 2.5. At all insulin concentrations, the binding by *ob/ob* membranes was decreased 3- to 4-fold. The binding defect was seen, whether the binding data were plotted per mg of membrane protein, per membrane unit of 5′-nucleotidase activity, per membrane unit of adenyl cyclase activity, per number of membrane glucagon receptors, per hepatocyte, or per unit of cell surface area [153]. In addition Fig. 2.5 shows binding data for another insulin resistant mouse of genetic origin *Db/Db,* for insulin resistance induced by gold thioglucose, and for the effects of fasting, dieting, and fasting plus insulin treatment on *ob/ob* mice.

The same type of data was obtained when binding was preformed directly to liver cell suspensions isolated with collagenase [153]. However, the isolated membranes could be stored frozen and when assayed on subsequent dates gave highly reproducible data.

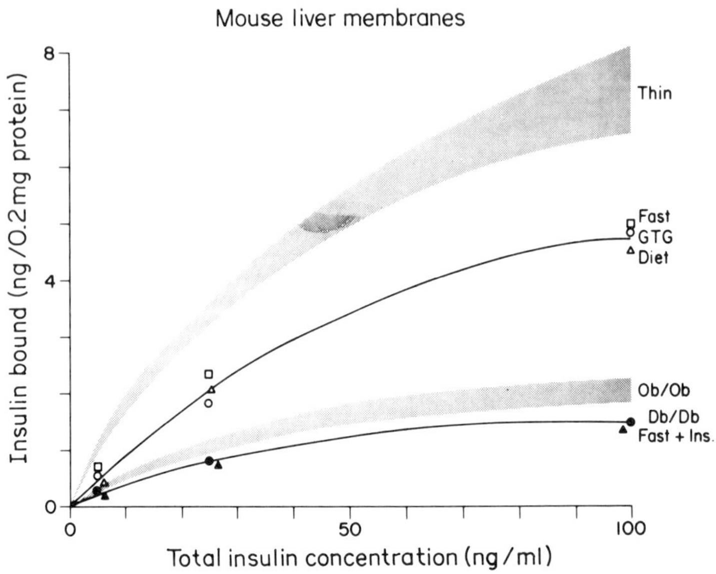


**Fig. 2.5** Specific binding curves at steady state for C57BL/6J *ob/ob* membranes and C57BL/6J +/+ controls. Shaded area, ± SEM 3 binding studies on 3 different membrane preparations. Each 150λ assay, incubated at 20°C for 6 h prior to rapid centrifugation and counting of the pellet, contained 30 μg of mouse PM protein, 15 pg [$^{125}$I]-insulin, unlabelled insulin from 0–750 ng in Krebs-Ringer buffer pH 7.4 + 1.5% serum albumin. Other membrane preparations are Δ, *ob/ob* dieted 10 days; 0, ++ gold thioglucose treated; □, *ob/ob* 20-h fast; ▲, *ob/ob* 20-h fast insulin treated; ●, C57BL/KsJ *db/db*. All points are averages of two preparations and two or more binding studies. (from Neville *et al.* [160] and Soll *et al.* [161].

### *2.3.2 Comments on the method*

The hypotonic media devoid of divalent cations allow for cell breakage to be achieved with a minimum of shear, thereby minimizing vesiculation of membranes which would be lost to the microsomal fraction. Homogenization should achieve at least 99% cell rupture; otherwise, cells will be floated and cytoplasmic contaminants will be carried through to the final step. Cell breakage is monitored by phase microscopy. Pestles of Dounce-type homogenizers often vary in diameter and should be standardized by timing the free descent of the pestle in the water-filled homogenizer. The dilution with media (step 2 of Section 2.3.3) aids in dispersion of clumped cytoplasm, dissolution of nuclei and nuclear membranes, and the subsequent sedimentation of membranes.

The centrifugal field (step 3 of Section 2.3.3) is chosen to maximize the sedimentation of the large membranes and minimize the sedimentation of mitochondria and clumps of endoplasmic reticulum. Angle-head rotors and plastic bottles cannot be used for this step, owing to inadequate packing of the pellet.

The choice of density for flotation (steps 6–8 of Section 2.3.3) is such as to maximize flotation of membranes and minimize flotation of mitochondria and microsomes. The density chosen is often quite critical because, although mitochondria, for example, have a higher average density than plasma membranes, their density distribution does overlap, and there are 1000 mitochondria for each plasma membrane. Flotation through the top layer of sucrose minimizes the flotation of microsomes and mitochondria. This step is an equilibrium flotation with respect to the large membranes, but the smaller particles have insufficient time to reach equilibrium. Reproducibility is ensured by the use of a refractometer. The density may need to be varied to suit species, nutrition, and other differences. Eighty percent of the membranes should float to form a skin at the liquid-air interface. If membranes do not lie evenly on the cushion, the linearity of the gradient should be checked.

### *2.3.3 Experimental details*

Mouse (C57BL/6J) liver-cell-plasma membranes were isolated using a modification of the author's original technique [1,3]. The method is described below in detail because attention to these details is necessary to ensure reproducibility. In particular, the sucrose concentrations are quite critical. These are given here in percent (w/w) at 20°C and are checked by an Abbe refractometer. All steps are carried out between 0° and 4°C. Medium refers to 0.001 M $NaHCO_3$. Three 10 g lots of liver can be processed together.

(1) Decapitate enough mice to provide 10 g of liver. Trim livers free of the gall bladder (without rupture) and connective tissue, place in an iced beaker, and mince with scissors.

(2) Place 10 g of minced liver in a large Dounce homogenizer (available from Blaessig Glass Company, Rochester, N.Y.). Add 25 ml medium. Homogenize at 4°C with 7 vigorous strokes of the loose pestle. Add 250 ml medium (4°C) and stir for 2 min, then filter first through 2 layers of cheesecloth, then through 4 layers.

(3) Distribute the filtered homogenate equally between two 250 ml glass centrifuge bottles (do not prechill the bottles) and spin 1500 x *g* max for 10 min using a swinging bucket rotor. Use a speed of 2800 revs/min for a Head #947 IEC B–20 centrifuge or Sorvall RC–2 using head #HS–4, or 2700 revs/min for IEC PR–2 centrifuge using head #284. To accommodate HS–4 or 947 heads, the tops of 250 ml Kimax bottles (Kimble #14700) are sawn off leaving a cup which stands 11 cm high.

(4) Carefully pour off the supernatant and, while the tube is upside down, insert absorbent paper into the neck to remove excess supernatant. Now pour off pellets into a large Dounce homogenizer.

(5) Homogenize pellets with three gentle strokes of the loose pestle.

(6) Adjust 69% stock sucrose solution to 69 ± 0.1% using a refractometer. Place 11 ml into a 50 ml cylinder and cool in an ice bucket. Pour in the homogenate from (5). Add $H_2O$ (4°C) to make 20 ml. Mix vigorously with a rod until no Schlerin patterns are evident. Check the mixture in the refractometer. Adjust with $H_2O$ or 69% sucrose until the homogenate sucrose concentration reads 44.0 + 0.1%.

(7) Pour 20 ml into an S–25.1 tube (Beckman Instruments, Inc).

(8) Carefully overlay this amount with 10 ml of 42.3 ± 0.1% sucrose (checked by the refractometer).

(9) Balance all three tubes with ± 0.005 g by adding 42.3% sucrose.

(10) Load the tubes into a prechilled swinging bucket rotor (SW 25.1) and spin 25 000 revs/min (90 000 x *g* max) for 2 h, with the brake on. Handle the tubes carefully to preserve the density interface. Make certain that the caps are well greased to keep the tubes at 1 atm. (See step 13 now.)

(11) Remove the float with a spatula. Add 8 ml medium. Spin, to pack the sediment, 25 000 x *g* max for 10 min.

(12) Add 1 ml medium and resuspend the pellet by squirting once through a 4-in No. 22 needle.

(13) Into the bottom of a siliconized S–25.1 tube (Siliclad, Clay Adams, Parsippany, N.J., 07054), place a 'cushion' of 4.1 ml of 50 ± 1% sucrose. Over each cusion form a linear sucrose gradient from 27–3%. For a Beckman gradient former, use the following solutions and settings: heavy 30%, light 0%, 50 ml syringes containing 20 ml, hold up –8, start 0, end 62. Run on slow.

(14) Overlay the gradient with the resuspended homogenate ( ~ 1.3 ml) and spin 2000 revs/min (550 x *g* max) for 1 h with the brake off.

(15) Using a syringe and long blunt No. 20 needle, remove the material at the cushion interface. Dilute with the medium to make 9.2 ml. Place 1.5 ml into 6 small conical sealable plastic centrifuge tubes (Eppendorf) and centrifuge at 15 000 x *g* for 3 min. Discard the supernatant and freeze the pellet at –70°C. Examine the remainder of the diluted membranes by phase microscopy.

## References


1. Neville, D. M., Jr. (1960), *J. biophys. biochem. Cytol.*, **8**, 413–422.
2. Emmelot, P. C., Bos, C. J., Benedetti, E. L. and Rumke, P. (1964), *Biochim. biophys. Acta*, **90**, 126–145.
3. Neville, D. M., Jr. (1968), *Biochim. biophys. Acta*, **154**, 540–552.
4. Neville, D. M., Jr. and Kahn, C. R. (1974), *Methods in Molecular Biology*, Vol. 5, (Laskin, A. I. and Last, J. A. eds), pp. 57–88, Dekker, New York.
5. Coleman, R., Michell, R. H., Finean, J. B. and Hawthorne, J. N. (1967), *Biochim. biophys. Acta*, **135**, 573–579.
6. House, P. D. R. and Weidman, M. J. (1970), *Biochem. biophys. Res. Commun.*, **41**, 541–548.
7. Takeuchi, M. and Terayama, M. (1965), *Exp. Cell Res.*, **40**, 32–44.
8. Stein, Y., Widnell, C. and Stein, O. (1968), *J. Cell Biol.*, **39**, 185–192.

9. Berman, H. M., Gram, W. and Spirtes, M. A. (1969), *Biochim. biophys. Acta,* **183**, 10–18.
10. Ray, T. K. (1970), *Biochim. biophys. Acta,* **196**, 1–9.
11. Nigam, V. N., Morais, R. M. and Karasaki, S. (1971), *Biochim. biophys. Acta,* **249**, 34–40.
12. Lauter, C. J., Soyom, A. and Trams, E. G. (1972), *Biochim. biophys. Acta,* **266**, 511–523.
13. Lesko, L., Donlon, M., Marinetti, G. V. and Horn, J. D. (1973), *Biochim. biophys. Acta,* **311**, 173–179.
14. Dorling, P. R. and LePage, R. N. (1973), *Biochim. biophys. Acta,* **318**, 33–40.
15. Evans, W. H. (1970), *Biochem. J.,* **116**, 833–842.
16. Hinton, R. H., Dobrota, M., Fitzsimons, J. R. T. and Reid, E. (1970), *Eur. J. Biochem.,* **12**, 349–361.
17. El-Aaser, A. A., Fitzsimons, J. T. R., Hinton, R. H., Reid, E., Klucis, E. and Alexander, P. (1966), *Biochim. biophys. Acta,* **127**, 553–556.
18. Touster, O., Aronson, N. N., Jr., Dulaney, J. T. and Hendrickson, H. (1970), *J. Cell Biol.,* **47**, 604–618.
19. Weaver, R. A. and Boyle, W. (1969), *Biochim. biophys. Acta,* **173**, 377–388.
20. Herzenberg, L. A. and Herzenberg, L. A. (1961), *Proc. natn. Acad. Sci. U.S.A.,* **47**, 762–767.
21. Song, C. S., Rubin, W., Rifking, A. B. and Kappas, A. (1969), *J. Cell Biol.,* **41**, 124–132.
22. Henning, R., Kaulen, H. D. and Stoffel, W. (1970), *Hoppe-Seyler's Z. physiol. Chem.,* **351**, 1191–1199.
23. Fleischer, B. and Fleischer, S. (1969), *Biochim. biophys. Acta,* **183**, 265–275.
24. Lutz, F. and Frimmer, M. (1970), *Hoppe-Seyler's Z. physiol. Chem.,* **351**, 1429–1434.
25. Thuneberg, L. and Rostgaard, J. (1968), *Exp. Cell Res.,* **51**, 123–140.
26. Kinne, R. and Kinne-Saffran, E. (1969), *Pfluegers Arch. Eur. J.,* **308**, 1–15.
27. Wilfong, R. F. and Neville, D. M., Jr. (1970), *J. biol. Chem.,* **245**, 6106–6112.
28. Berger, S. J. and Sacktor, B. (1970), *J. Cell Biol.,* **47**, 637–645.
29. Heidrich, H. G., Kinne, R., Kinne-Saffran, E. and Hannig, K. (1972), *J. Cell Biol.,* **54**, 232–245.
30. Binkley, F., King, N., Milikin, E., Wright, R. K. O'Neal, C. H. and Wundram, I. J. (1968), *Science,* **162**, 1009–1011.
31. Fitzpatrick, D. F., Davenport, G. R., Forte, L. and Landon, E. J. (1969), *J. biol. Chem.,* **244**, 3561–3569.
32. Marx, S. J., Fedak, S. A. and Aurbach, G. D. (1973), *J. biol. Chem.,* **247**, 6913–6918.
33. Coleman, R., Michell, R. H., Finean, J. B. and Hawthorne, J. N. (1967), *Biochim. biophys. Acta,* **135**, 573–579.
34. Meldolesi, J., Jamieson, J. D. and Palade, G. E. (1971), *J. Cell Biol.,* **49**, 150–158.
35. Schapira, G., Dobocz, I., Piau, J. P. and Delain, E. (1974), *Biochim. biophys. Acta,* **345**, 348–358.
36. Kono, T. and Colowick, S. P. (1961), *Archs. Biochem. Biophys.,* **93**, 520–533.

37. McCollester, D. L. (1962), *Biochim. biophys. Acta,* **57**, 427–437.
38. Rosenthal, S. L., Edelman, P. M. and Schwartz, I. L. (1965), *Biochim. biophys. Acta,* **109**, 512–517.
39. Sulakhe, P. V., Fedelesova, M., McNamara, D. B. and Dhalla, N. S. (1971), *Biochem. biophys. Res. Commun.,* **42**, 793–800.
40. Andrew, C. G. and Appel, S. H. (1973), *J. biol. Chem.,* **248**, 5156–5163.
41. Carroll, P. M. and Sereda, D. D. (1968), *Nature,* **217**, 666–667.
42. Devynck, M. A., Pernollet, M. G., Meyer, P., Permandjian, S. and Fromageot, P. (1973), *Nature New Biology.,* **245**, 55–57.
43. Peters, T. J., Muller, M. and de Duve, C. (1972), *J. exp. Med.,* **136**, 1117–1139.
44. Kidwai, A. M., Radcliffe, M. A. and Daniel, E. E., (1971), *Biochim. biophys. Acta,* **233**, 538–549.
45. Barr, C., Connor, J. A., Dewey, M. M., Aprille, J. and Johnston, P. V. (1974), *Biochim. biophys. Acta,* **345**, 336–347.
46. Kidwai, A. M., Radcliffe, M. A., Duchon, G. and Daniel, E. E. (1971), *Biochem. biophys. Res. Commun.* **45**, 901–910.
47. Rodbell, M. (1967), *J. biol. Chem.,* **242**, 5744–5750.
48. McKeel, D. W. and Jarett, L (1970), *J. Cell Biol.,* **44**, 417–432.
49. Avruch, J. and Wallach, D. F. H. (1971), *Biochim. biophys. Acta,* **233**, 334–347.
50. Laudat, M. H., Pairault, J., Bayer, P., Martin, M. and Laudat, P. (1972), *Biochim. biophys. Acta,* **255**, 1005–1008.
51. Giacobino, J. P. and Perrelet, A. (1971), *Experientia,* **27**, 259–261.
52. Miller, D. and Crane, R. K. (1961), *Biochim. biophys. Acta,* **52**, 293–298.
53. Forstner, G. G., Sabesin, S. M. and Isselbacher, K. J. (1968), *Biochem. J.,* **106**, 381–390.
54. Fujita, M., Ohta, H., Kawai, K., Matsui, H. and Nakao, M. (1972), *Biochim. biophys. Acta,* **274**, 336–347.
55. Douglas, A. P., Kerley, R. and Isselbacher, K. J. (1972), *Biochem. J.,* **128**, 1329–1338.
56. Porteous, J. W. (1972), *Subcellular Components* (Birnie, G. D., ed.), pp. 157–183, Butterworths, London, University Park Press, Baltimore.
57. Hicks, R. M . and Ketterer, B. (1970), *J. Cell Biol.,* **45**, 542–553.
58. Vergara, J., Zambrano, F., Robertson, J. D. and Elrod, H. (1974), *J. Cell Biol.,* **61**, 83–94.
59. Lo, C. -H. (1973), *Biochim. biophys. Acta,* **291**, 650–661.
60. Koyama, N., Sawada, K. and Kurihra, K. (1971), *Biochim. biophys. Acta,* **241**, 42–48.
61. Skerrow, C. J. and Matoltsy, A. G. (1974), *J. Cell Biol.,* **63**, 515–523.
62. Whittaker, V. P. (1968), *Biochem. J.,* **106**, 412–417.
63. De Robertis, E., Azcurra, J. M. and Fiszer, S. (1966), *Brain Res.,* **5**, 45–56.
64. Cotman, C. W. and Matthews, D. A. (1971), *Biochim. biophys. Acta,* **249**, 380–394.
65. Levitan, I. B., Mushynski, W. E. and Ramirez, G. (1972), *J. biol. Chem.,* **247**, 5376–5381.
66. Morgan, I. G., Wolfe, L. S., Mandel, P. and Gambos, G. (1971), *Biochim. biophys. Acta,* **241**, 737–751.
67. Hemminki, K. and Suovaniemi, O. (1973), *Biochim. biophys. Acta,* **298**, 75–83.

68. Henn, F. A., Hansson, H. -A. and Hamberger, A. (1972), *J. Cell Biol.*, **53**, 654–661.
69. Bretz, U., Baggiolini, M., Hauser, R. and Hodel, C. (1974), *J. Cell Biol.*, **61**, 466–480.
70. Merrell, R. and Glaser, L. (1973), *Proc. natn. Acad. Sci. U.S.A.*, **70**, 2794–2798.
71. Yamashita, K. and Field, J. B. (1970), *Biochem. biophys. Res. Commun.*, **40**, 171–178.
72. Wolff, J. and Jones, A. B. (1971), *J. biol. Chem.*, **246**, 3939–3947.
73. Amir, S. M., Carraway, T. F., Jr., Kohn, I. D. and Winnand, R. J. (1973), *J. Biol. Chem.*, **248**, 4092–4100.
74. Turkington, R. W. (1962), *Biochim. biophys. Acta*, **65**, 386–388.
75. Finn, F. M., Widnell, C. C. and Hofmann, K. (1972), *J. biol. Chem.*, **247**, 5695–5702.
76. Nijjar, M. S. and Hawthorne, J. N. (1974), *Biochim. biophys. Acta*, **367**, 190–201.
77. Gospodarowicz, D. (1973), *J. biol. Chem.*, **248**, 5050–5056.
78. Poirier, G., Labrle, F., Barden, N. and Lemaire, S. (1972), *F.E.B.S. Letters*, **20**, 283–286.
79. Shiu, R. P. C. and Friesen, H. G. (1974), *Biochem. J.*, **140**, 301–311.
80. Keeran, T. W., Morre, D. J., Olson, D. E., Yunghans, W. N. and Patton, S. (1970), *J. Cell Biol.*, **44**, 80–93.
81. Allan, D. and Crumpton, M. J. (1970), *Biochem. J.*, **120**, 133–143.
82. Demus, H. (1973), *Biochim. biophys. Acta*, **291**, 93–106.
83. Ragland, W. L., Pace, J. L. and Doak, R. L. (1973), *Biochem. biophys. Res. Commun.*, **50**, 118–126.
84. Allan, D. and Crumpton, M. J. (1972), *Biochim. biophys. Acta*, **274**, 22–27.
85. Ladoulis, C. T., Misra, D. N., Estes, L. W. and Gill, T. J. (1974), *Biochim. biophys. Acta*, **356**, 27–35.
86. Ryan, J. W. and Smith, U. (1971), *Biochim. biophys. Acta*, **249**, 177–180.
87. Dods, R. F., Essner, E. and Barclay, M. (1972), *Biochem. biophys. Res. Commun.*, **46**, 1074–1081.
88. Hourani, B. T., Chace, N. M. and Pincus, J. H. (1973), *Biochim. biophys. Acta*, **328**, 520–532.
89. Emmelot, P. and Bos, C. J. (1969), *Int. J. Cancer*, **4**, 705–722.
90. Raftell, M. and Blomberg, F. (1973), *Biochim. biophys. Acta*, **291**, 421–430.
91. Perdue, J. F., Warner, D. and Miller, K. (1973), *Biochim. biophys. Acta*, **298**, 817–826.
92. Perdue, J. F. and Sneider, J. (1970), *Biochim. biophys. Acta*, **196**, 125–140.
93. Perdue, J. F., Kletzien, R. and Miller, K. (1971), *Biochim. biophys. Acta*, **249**, 419–439.
94. Bingham, R. W. and Burke, D. C. (1972), *Biochim. biophys. Acta*, **274**, 348–352.
95. Russell, T. and Pastan, I. (1973), *J. biol. Chem.*, **248**, 5835–5840.
96. Israel, A., Verjus, M. -A. and Semmel, M. (1973), *Biochim. biophys. Acta*, **318**, 155–166.
97. Warren, L., Glick, M. C. and Nass, M. K. (1967), *J. Cell Physiol.*, **68**, 269–287.

98. Brunette, D. M. and Till, J. E. (1971), *J. Memb. Biol.,* **5**, 215–224.
99. Klenk, H. -D. and Choppin, P. W. (1969), *Virology,* **38**, 255–268.
100. Renkonen, O., Gahmberg, C. G., Simons, K. and Kääriäinen, L. (1970), *Acta chem. scand.,* **24**, 733–735.
101. Klenk, H. -D. and Choppin, P. W. (1969), *Virology,* **38**, 255–268.
102. Atkinson, P. H. and Summers, D. F. (1971), *J. biol. Chem.,* **246**, 5162–5175.
103. Boone, C. W., Ford, L. E., Bond, H. E., Stuart, D. C. and Lorenz, D. (1969), *J. Cell Biol.,* **41**, 378–392.
104. Bosmann, H. B., Hagopian, A. and Eylar, E. H. (1968), *Archs. Biochem. Biophys.,* **128**, 51–69.
105. Johnsen, S., Stokke, T. and Prydz, H. (1974), *J. Cell Biol.,* **63**, 357–363.
106. Molnar, J., Markovic, G. and Chao, H. (1969), *Archs. Biochem. Biophys.,* **134**, 524–532.
107. Forte, J. G., Forte, T. M. and Heinz, E. (1973), *Biochim. biophys. Acta,* **298**, 827–841.
108. Kamat, V. B. and Wallach, D. F. H. (1965), *Science,* **148**, 1343–1345.
109. Stonehill, E. H. and Huppert, J. (1968), *Biochim. biophys. Acta* **155**, 353–359.
110. Reporter, M. and Raveed, D. (1973), *Science,* **18**, , 863–865.
111. Perdue, J. F., Kletzien, R., Miller, K., Pridmore, G. and Wray, V. L. (1971), *Biochim. biophys. Acta,* **249**, 435–453.
112. Kennel, S. J. and Lerner, R. A. (1973), *J. molec. Biol.,* **76**, 485–502.
113. Charalampous, F. C., Gonatas, N. K. and Melbourne, A. D. (1973), *J. Cell Biol.,* **59**, 421–435.
114. Truding, R., Shelanski, M. L., Daniels, P. M. and Morell, P. (1974), *J. biol. Chem.,* **249**, 3973–3982.
115. Shin, B. C. and Carraway, K. C. (1973), *Biochim. biophys. Acta,* **330**, 254–268.
116. Krizanova, O. and Ciampor, F. (1971), *Acta Virol.,* **15**, 19–27.
117. Neville, D. M., Jr. (1975), *Methods in Membrane Biology*, Vol. 3 (Korn, E., ed.), p. 1, Plenum Press, New York.
118. Widnell, C. C. (1972), *J. Cell Biol.,* **52**, 542–558.
119. Jarett, L. and Smith, R. M. (1974), *J. biol. Chem.,* **249**, 7024–7031.
120. Steck, T. L. and Dawson, G. (1974), *J. biol. Chem.,* **249**, 2135–2142.
121. DePierre, J. W. and Karnovsky, M. L. (1973), *J. Cell Biol.,* **56**, 275–303.
122. de Duve, C. (1971), *J. Cell Biol.,* **50**, 20D–55D.
123. Beaufay, H., Amar-Costesec, A., Thinès-Sempoux, D., Wibo, M., Robbi, M. and Berthet, J. (1974), *J. Cell Biol.,* **61**, 213–231.
124. de Duve, C., Pressman, B. C., Gianetto, R., Wattiaux, R. and Appelmans, F. (1955), *Biochem. J.,* **60**, 604–617.
125. Wattiaux-deConinck, S. and Wattiaux, R. (1969), *Biochim. biophys. Acta,* **183**, 118–128.
126. Wattiaux-deConinck, S., Rutgeerts, M. J. and Wattiaux, R. (1965), *Biochim. biophys. Acta,* **105**, 446–459.
127. Tulkens, P., Beaufay, H. and Trouet, A. (1974), *J. Cell Biol.,* **63**, 383–401.
128. Farquhar, M. G., Bergeron, J. J. M. and Palade, G. E. (1974), *J. Cell Biol.,* **60**, 8–25.
129. Bergeron, J. J. M., Evans, W. H. and Geschwind, (1973), *J. Cell Biol.,* **59**, 771–776.
130. McKeel, D. W., Jr. and Jarett, L. (1974), *J. Cell Biol.,* **62**, 231–236.
131. Kasper, C. B. (1971), *J. biol. Chem.,* **246**, 577–581.

132. Weber, G. and Cantero, A. (1955), *Cancer Res.*, **15**, 105–108.
133. Glaumann, H. and Dallner, G. (1970), *J. Cell. Boil.*, **47**, 34–48.
134. Steinmen, R. M., Silver, J. M. and Cohn, Z. A. (1974), *J. Cell Biol.*, **63**, 949–969.
135. Phillips, D. R. and Morrison, M. (1971), *Biochemistry,* **10**, 1766–1771.
136. Tsai, C., Huang, C. and Canellakis, E. S. (1973), *Biochim. biophys. Acta,* **332**, 47–58.
137. Korn, E. D. (1974), *The Inflammatory Process*, Vol. 1, 2nd ed. (Zweiarch, B. F., Grant, L. and McCluskey, R. T., eds), pp. 52–113, Academic Press, New York.
138. Hinton, R. H. (1972), *Subcellular Components*, (Birnie, G. D., ed), pp. 119–155, Butterworths, London, University Park Press, Baltimore.
139. Pohl, S. L., Birnbaumer, L. and Rodbell, M. (1971), *J. Biol. Chem.*, **246**, 1849–1856.
140. Kahn, C. R. (1975), *Methods in Membrane Biology,* Vol. 3 (Korn, E., ed), pp. 81–146, Plenum Press, New York.
141. Roth, J. (1973), *Metabolism,* **22**, 1059–1073.
142. Labrie, F., Barden, N., Poirier, G. and DeLean, A. (1972), *Proc. natn. Acad. Sci. U.S.A.*, **69**, 283–287.
143. Silverman, M. (1974), *Biochim. biophys. Acta,* **339**, 92–102.
144. Glossmann, H. and Neville, D. M., Jr. (1972), *J. biol. Chem.*, **247**, 7779–7789.
145. Klee, W. A. and Nirenberg, M. (1974), *Proc. natn. Acad. Sci., U.S.A.*, **71**, 3474–3477.
146. Festoff, B. W. and Engel, W. K. (1974), *Proc. natn. Acad. Sci., U.S.A.*, **71**, 2435–2439.
147. Cuatrecasas, P. (1974) *Annual Review of Biochemistry*, Vol. 43 (Snell, E. E., ed), pp. 169–214, Annual Reviews Inc., Palo Alto, California.
148. Smigel, M. and Fleischer, S. (1974), *Biochim. biophys. Acta,* **332**, 358–373.
149. Rao, C. V. (1974), *J. biol. Chem.*, **249**, 7203–7209.
150. Pricer, W. E. and Ashwell, G. (1971), *J. biol. Chem.*, **246**, 4825–4833.
151. Uchida, T., Pappenheimer, A. M., Jr. and Harper, A. A. (1973), *J. biol. Chem.*, **248**, 3845–3850.
152. Nicolson, G. L. (1974), *Nature,* **251**, 628–630.
153. Kahn, C. R., Neville, D. M., Jr. and Roth, J. (1973), *J. biol. Chem.*, **248**, 244–250.
154. Neville, D. M., Jr. (1974), *In Vitro,* **9**, 445–454.
155. Neville, D. M., Jr. (1974), *Proceedings of the Fogarty Conference on Cell Surfaces and Malignancy*, Sept. 11–13, 1947, Bethesda, Maryland.
156. Weibel, E. R., Stäubli, W., Gnägi, H. R. and Hess, F. A. (1969), *J. Cell Biol.*, **42**, 68–91.
157. Goldstein, J. L. and Brown, M. S. (1974), *J. biol. Chem.*, **249**, 5153–5162.
158. Tisdale, H. D. (1967), *Meth. Enzym.* **10**, 213.
159. Hubscher, G. and West, G. R. (1965) , *Nature,* **205**, 799–800.
160. Neville, D. M., Jr., Kahn, C. R., Soll, A. and Roth, J. (1973), *Protides of the Biological Fluids* XXI Colloquium May 1973, Bruges (Peeters, H., ed), pp. 269–273, Pergamon Press, New York.
161. Soll, A. H., Kahn, C. R., Neville, D. M., Jr. and Roth, J. (1975), *J. clin. Invest.*, **56**, 769–780.
162. Neville, D. M., Jr. and Glossmann, H. (1971), *J. biol. Chem.*, **246**, 6335–6338.

163. Lowry, O. H., Rosebrough, N. J., Farr, A. L. and Randall, R. J. (1951), *J. biol. Chem.*, **193**, 265–275.
164. Widnell, C. C. and Unkeless, J. C. (1968), *Proc. natn. Acad. Sci. U.S.A.*, **61**, 1050–1057.
165. Freychet, P., Roth, J. and Neville, D. M., Jr. (1971), *Proc. natn. Acad. Sci. U.S.A.*, **68**, 1833–1837.
166. Soll, A. H., Kahn, C. R. and Neville, D. M., Jr. (1974), *J. biol. Chem.*, **250**, 4702–4707.

# 3 The isolation of mitochondria and their membranes

G. L. SOTTOCASA



## 3.1 Introduction

Mitochondria are intracellular organelles present in all eukaryotic cells. Their description at the light microscopic level can be traced to the second half of the past century. The historical background of early research on mitochondria was reviewed in 1953 by Cowdry [1]. The fine structure of the organelle at the electron-microscopic level has been fully described by Palade [2,3]. *In situ*, in tissue thin sections, mitochondria appear as vescicular structures of more or less elongated appearance. The profiles are 0.3–5 μm long and 0.2–0.5 μm wide. The variability in dimensions reflects not only the inclination of the section but also the flexibility of the structure as revealed by direct phase-contrast microscopic observation of living cells. Despite the early interpretation by Sjöstrand [4] and more recent alternative models proposed by Green's group [5–7], it is generally held that mitochondria consist of two concentric distinct vescicular membranes the inner of which is invaginated to form the '*cristae mitochondriales*'. Two spaces are therefore delimited inside the mitochondrion: one between the two membranes, the intermembrane space, and another one surrounded by the inner membrane. The latter compartment contains material more or less finely dispersed and has been named the mitochondrial matrix [2]. For this reason the compartment is also referred to as matrix space. Due to the presence of the invaginated cristae, the intermembrane space is continuous with the space within the single cristae, a space which can be called intracristael space; the matrix space, on the other hand, is somewhat concamerated by the protruding cristae and in certain cases, when the cristae are particularly numerous, it may be appropriate to call this compartment intercristael space. Once isolated, mitochondria retain more or less the described structure, with a certain tendency to approach a spherical appearance. Characteristically however, mitochondria isolated in 0.25 M sucrose show a much higher electron density in the matrix space. The degree of this phenomenon may vary, from one organelle to another in the same specimen and may be so intense as to obscure the cristae profiles. Reversible transitions from this 'condensed' state to the normal 'orthodox' state have been described in detail by Hackenbrock [8] in intact mitochondria and by Greenawalt [9] in the same organelles after removal of the outer membrane. The transitions are correlated with the metabolic state of the mitochondrion and have been considered indicative of the existence of different conformational states of the inner membrane related to the energy-conserving mechanism in oxidative phosphorylation [7,10,11]. In contrast to tissue slices, isolated mitochondria can also be studied morphologically using negative staining. By means of this technique a few characteristic features have been identified. The outer mitochondrial membrane appears as smooth sheets easily recognizable from its spatial relation to the inner membrane. Small projecting units have been seen on its surface [12–14] but these particles can be easily distinguished from the inner membrane projecting spheres first described by Fernandez-Morán [15]. Subsequent study of these particles [16,17] has shown that they line the matrix surface of the inner mitochondrial membrane and can therefore be seen only under conditions where the contrasting medium has access to this compartment. The particles have now been [18] identified with the coupling factor $F_1$ first described by Pullman *et al.* [19] and Penefski *et al.* [20].

The morphological features so far described refer mainly to liver organelles but are qualitatively recognized in all mitochondria irrespective of their origin. Quantitative differences are found, however, especially concerning the number and extent of the cristael system. Mitochondria from heart, [21–25] skeletal muscle [3,26] and kidney [3,27] are particularly rich in cristae, which appear as transverse septa packed together leaving little space for the matrix. Adrenal cortex mitochondria often show tubular or vescicular cristae [28,29]. Peculiar organization of the cristae have been described in very specialized tissues such as the cricothyroid muscle of bat [30].

By now it is well established that mitochondria are complex organelles capable of carrying out independent integrated functions including electron transport from metabolites to oxygen, electron transport-coupled ATP biosynthesis, oxidation of Krebs cycle intermediates and fatty acids, fatty acid elongation, partial reactions of the urea cycle and synthesis and oxidation of ketone bodies. Pioneering work on mitochondrial functions was carried out by Warburg who, in 1913, [31] discovered that oxygen-consuming reactions in anaerobic cells were confined to particulate material; by Keilin who, in the late 1920's [32,33], provided evidence for the existence of electron-transport mediated by particle-bound cytochromes; by Green [34,35] who proposed that not only electron transport but also citric acid cycle oxidations, fatty acid oxidation and oxidative phosphorylation were carried out by an integrated enzyme system called 'cyclophorase'; and, eventually, by Schneider and Potter [36] and Lehninger and Kennedy [37] who could show that isolated mitochondria can indeed perform all the stated functions. A number of review articles on the subject appeared around 1950 [38–46].

Later another important function was discovered in mitochondria, namely independent protein synthesis [47–50]. A prerequisite to this function is the existence of mitochondrial DNA [51–54], DNA synthesis [55–58], ribosome-like RNA particles [59–61], and transfer RNA [62–63]. The rapidly expanding field of mitochondrial nucleic acids and protein biosynthesis has been recently reviewed [64] and is the subject of specialized meetings [65–66].

Another essential function carried out by isolated mitochondria is energy-linked translocation of ions, both of positive and negative charge. The subject has been studied intensely in many laboratories since 1953, and is covered by numerous reviews [66–110].

Nowadays isolated mitochondria have become routine preparations in many laboratories. In the present chapter the general principles of the procedure will be outlined and its limitations discussed. Mention will be made of the criteria which can be used to assess the quality of the material in relation to the experimental goal.

## 3.2 Isolation procedure

The isolation of mitochondria is a special case of the more general problem of cell fractionation. Pioneering work in this field was carried out in the early 1940's by Porter, Claude and Fullman [111–113] and later by Hogeboom and Schneider [40,41,114]. The general principles have been clearly expressed and discussed by De Duve [115] in his opening lecture at the Biochemical Society Symposium held in Louvain in 1962. A very clear review of the subject appeared in 1964 [116].

Mitochondria have been isolated from a variety of tissues both of animal and plant origin (for detailed techniques see [117]). In addition, the size of the preparation ranges, according to different experimental designs, from a few milligrams to many grams total protein. Although the general principles for the isolation are obviously the same, small-scale and large-scale preparations deserve separate descriptions. Four steps can be recognized in the procedure: (1) homogenization; (2) separation by differential centrifugation of the mitochondrial fraction; (3) purification of the fraction by repeated washings and (4) assay of purity and quality of the preparation.

### *3.2.1 Homogenization*

The aim of this step is to disaggregate the tissue and break the plasma membrane in order to obtain a suspension of single subcellular organelles. Two points are to be considered: (a) the homogenizers and (b) the dispersing media used. The type of homogenizer should be chosen in relation to the type of tissue to be processed: the harder the tissue the more mechanically efficient the device. Liver, kidney, adrenal cortex and brain among the animal tissues may be easily homogenized by means of a motor-driven Potter homogenizer. The device consists essentially of a glass-tube with a close fitting glass or teflon pestle. Before homogenization tissues are usually cut into small pieces by means of a pair of scissors and the pieces are washed in aliquots of medium to remove blood, if present. Scissors may be used successfully to reduce skeletal muscle to a pulp which can be processed in a Potter homogenizer. The operation should be carried out directly on the fresh tissue [118]. The same homogenizer has been used on comparatively hard tissues, such as heart, following pretreatment with proteolytic enzymes [119,120].

Other means of homogenization have been used, including mortar and pestle in combination with sand or glass beads. In large-scale preparations of mitochondria from slaughter-house organs the most frequently used homogenizers are Waring blendors, 'Ultra turrax' and 'Polytron' tissue processors. Usually the tissue is first ground by means of a commercial meat grinder.

Many different suspending media have been used in the past, including salts, sugars and high molecular weight compounds in different concentrations, and/or combinations. The simplest and most commonly used solution is 0.25 M sucrose. This medium has been employed largely for the preparation of liver mitochondria from laboratory animals. Some advantages have been found with the use of mannitol, at the same or slightly higher concentrations, or by combinations of sucrose and mannitol. The use of mannitol has been extended to the preparation of mitochondria from a number of tissues including kidney, thyroid, thymus, ascites tumour, cells, lymphocytes as well as plant mitochondria [117]. Salt solutions, such as 0.15 M KCl, may also be used as media for isolation of mitochondria from certain tissues. Electrolyte-containing media are particularly suited for the preparation of mitochondria from skeletal muscle where myofibrils would otherwise complicate the separation of organelles by differential centrifugation. The medium of choice for this tissue also includes a buffer (0.05 M Tris-HCl) pH 7.4, $Mg_2SO_4$ (0.005 M), ATP (0.001 M) and EDTA (0.001 M) [121]. EDTA and buffers are frequent additions where calcium (and other metal ions detrimental to mitochondrial activity) are present in the preparation, and if glycolysis is particularly efficient in the tissue so that pH has a tendency to drop to low values.

In certain cases bovine serum albumin is included in the medium. The addition is aimed to adsorb free fatty acids which have uncoupling effects [122–127] and exerts a rather general beneficial effect on mitochondrial functions. Brown adipose tissue mitochondria, for example, are completely uncoupled when prepared in the absence of bovine serum albumin [128–132].

Aggregation phenomena interfere severely in the separation of brain mitochondria: this problem may be overcome by the addition of plyanions such as heparin or polyethylene sulphonate to the medium [133–134]. In the preparation of these mitochondria the density of the medium may be increased by the addition of Ficol, a high molecular weight polysaccharide.

This first step in the isolation procedure deserves some comments. First, the time between sacrifice of the animal and use of the tissue should be as short as possible. When the material has to be transferred from slaughter-house to the laboratory the operation should be carried out as soon as possible at 0°C. Chilling of the organs is best obtained by immersion in the pre-cooled suspending medium. It is always better to have a number of small pieces than a whole organ which obviously takes longer to cool. A second point concerns the temperature during homogenization. While room temperature should be avoided during transport and standing of whole organs, it is even more essential to avoid exposure to temperatures above 0–4°C once the tissue has been ground or homogenized. Grinding or homogenizing produces, in fact, partial unmasking of lysosomal enzymes and lowering of the pH. Both factors are harmful to mitochondria. For these reasons it is always advisable to work in a cold room, possibly using ice-cooled containers, avoiding heating the homogenizer by friction as much as possible. For the same reasons large volumes of homogenate which cannot be immediately processed must be avoided.

Another important point concerns the time and speed of homogenization. Excess homogenization can damage the organelles, inadequate homogenization causes incomplete distruption of the cells which, in turn reflects on the yield.

### *3.2.2 Separation of the fraction by differential centrifugation*

Differential centrifugation is a technique based on the sequential removal from the suspension of different populations of particles by sedimentation in a centrifugal field. At constant centrifugal field, the sedimentation velocity is directly proportional to the volume of the particle and to the density difference between particle and medium. It is, on the other hand, inversely proportional to the viscosity of the medium [135]. For these reasons the separation of nuclear and mitochondrial fractions are achieved in different laboratories and with different tissues at different speeds and for variable times.

In general the nuclear fraction is first removed from the homogenate by centrifugation at 600–1600 x $g$ for 10–20 min and mitochondria are sedimented from the supernatant by centrifugation at 4000–12 000 x $g$ for 5–20 min. The centrifuge to be used may be any commercially available apparatus compatible with the size of the homogenate to be processed, provided the rotor is maintained at low temperature. Even rather inexpensive centrifuges, such as portable ones, operated in the cold room give excellent results. The separation of the mitochondrial fraction from the remaining supernatant should be as quick as possible to minimize contact of the particles with the potentially harmful soluble enzymes present in the homogenate. It is preferable therefore to reduce the time required to obtain the mitochondrial fraction to the minimum. Many conditions may be manipulated to this aim, such as volume of the homogenate and its concentration in relation to the size of the centrifuge, and time and speed of the two centrifugation steps. Obviously the volume of the homogenate depends essentially on the experiment planned; its concentration, on the other hand, cannot be freely changed in view of the fact that when the homogenate is too thick the resolution of the technique drops and heavily cross-contaminated fractions are obtained. The same limitations apply to changes in time and speed of the centrifugation.

### *3.2.3 Washings*

The mitochondrial sediment usually appears as a more or less brown pellet. However, the colour is seldom evenly distributed. This phenomenon reflects the presence in the population of mitochondria of different sizes and often of particles other than mitochondria.

Erythrocytes are common contaminants in the first mitochondrial pellet and can be easily seen at the bottom of the centrifuge tube as a red spot. Care should be taken when resuspensing the sediment to leave them adhering to the container. In tissues where the endoplasmic reticulum is particularly abundant, microsomes may also be a quantitatively important contaminant. When present, they appear as a superficial pinkish layer which can be easily removed by gently shaking the tube before pouring out the last third of the supernatant. The washings are usually repeated two or three times so decreasing the level of contamination.

Two points are to be considered in choosing the conditions at this stage (a) resuspension and (b) volume and conditions of centrifugation. The choice of the technique for (a) depends on the size of the pellet. On a small scale a glass stirring rod is perfectly adequate to obtain a virtually homogeneous suspension in most cases provided the medium is added in very small aliquots. The operation may be completed by means of a pipette through which the suspension is aspirated and blown out several times. In large scale preparations the use of motor-driven Potter homogenizers may be more appropriate. The use of such homogenizers with too tight a pestle is to be avoided in view of the possible damage caused to the particles by the operation. As for (b) the pellet may be taken up in the same volume as the original homogenate or, more frequently, in a smaller one. A rather commonly used procedure is to reduce the volume to ½ at each washing. This reduces the number of tubes at each step thereby saving time.

During washing, mitochondria are suspended in the pure medium, without all the soluble protein present in the original homogenate. The viscosity of the system is therefore lower and the sedimentation velocity higher. The centrifugation time is therefore shorter during washings and the speed the same or slightly higher.

### *3.2.4 Test of purity and quality*

Obviously purity of the mitochondrial fraction and quality of the organelles present in the fraction are two different concepts and their relative importance depends on the purpose of the experiment. It may not be important that mitochondrial fractions are relatively contaminated for the study of oxidative phosphorylation, provided that the mitochondria are tightly coupled. On the contrary, cytochemical studies, aimed to assess enzyme localization require special care in obtaining fractions of the highest degree of purity, or, at least, the level of contaminants must be known. The level of contamination by other organelles is best tested by the use of marker enzymes. An enzyme may be used as marker of a certain cellular structure when (1) it is firmly bound to it, (2) it is neither activated nor inhibited during the separation procedure and (3) it is localized exclusively in that structure.

The possible contaminants of mitochondria fractions include: nuclei, microsomes, plasma membranes, lysosomes and related organelles, submitochondrial particles and bacteria. Table 3.1 lists the corresponding marker enzymes. Lysosomal enzymes show the well known phenomenon of latency, i.e. display fully their activity only when tested in the presence of detergents such as Triton X–100, or generally when the fraction has been previously subjected to membrane-disrupting treatments. The assessment of lysosomal contamination of a mitochondrial fraction is difficult not only for this reason but, even more, because a variable percentage of these organelles are disrupted during homogenization and the remaining membranes may co-sediment with mitochondria. The membranes in question cannot be monitored by any known enzyme activity. Fortunately lysosomes on a protein basis always represent a negligible proportion of a mitochondrial fraction. The presence of submitochondrial particles (essentially inner membrane fragments) cannot

TABLE 3.1
*Marker enzymes for possible contaminants of the mitochondrial fraction*

| Contaminant | Enzyme | Reference |
|---|---|---|
| Nuclei | NAD-pyrophosphorylase | [136, 137] |
| Microsomes | e.g. Glucose-6-phosphatase, NADPH-cytochrome *c* reductase | [138, 139] |
| Plasma membrane | 5′Nucleotidase | [140–142] |
| Lysosomes | e.g. β-glucuronidase, acid phosphatase | [143, 157] |
| Uricosomes- | urate oxidase, D-amino acid oxidase | |
| Peroxisomes | catalase | [144, 157] |

obviously be determined on the basis of the presence of a given marker enzyme. Their presence may be detected, however, taking advantage of the fact that, in contrast to submitochondrial particles, intact mitochondria do not oxidize aerobically exogenous NADH [145–147] ; also their maximal rate of substrate oxidation is independent of added cytochrome *c* and they show virtually no succinate-ferricyanide reductase activity in the presence of Antimycin A [148–152]. These phenomena may provide a more qualitative estimate than a quantitative one. Bacterial contamination, which is particularly important in protein synthesis studies, can be assessed by microbiological means [153]. The use of electron microscopy to investigate the degree of purity of subcellular fractions is definitely a less quantitative technique than enzymatic analysis, due to the size of the sample analysed. In order to obtain quantitatively significant results it would be necessary to examine a very large number of microscopic fields, a procedure definitely too long to be compatible with most experimental designs. It may be added that the technique is less specific, especially when dealing with disrupted membranes whose origin is sometimes difficult to establish.

To test the quality of mitochondrial preparations, classically, two parameters are taken into consideration: (a) phosphate and phosphate-acceptor control and (b) P/O ratio. Intact tightly coupled mitochondria are known to respire at low rate when $P_i$ or ADP are made limiting. The ratio between respiratory rate in the presence of ADP and $P_i$ and that in their absence is usually referred to as respiratory control ratio (RCR) and provides a measure of the tightness of coupling of respiration to phosphorylation [154–155]. The most commonly used method for the measure of the respiration rate is direct polarography with a Clark electrode [156]. Values of RCR between 5 and 10 are common figures in the literature, occasionally higher values may be obtained. P/O ratio, i.e. the efficiency of the system, may be measured by classical respirometry concomitant with the determination of inorganic phosphate disappearance. The technique is described in great detail by Slater [158]. Alternatively ADP/O ratios (equivalent to P/O ratios) [154–155] may be measured polarographically. The measure of partial reactions of oxidative phosphorylations may also provide useful information on the efficiency of the system [151–152].

## 3.3 Experimental procedures

### *3.3.1 Preparation of ox adrenal cortex mitochondria*

Adrenals are obtained from the animals as soon as possible and transferred immediately from slaughter-house to the laboratory immersed in ice-cold 0.25 M sucrose. The organs are decorticated by many tangential cuts and the small pieces of tissue are collected in a 50 ml-beaker containing the medium consisting of: 0.25 M sucrose, 1 mM ATP and 0.1% fatty-acid free [159] bovine serum albumin. The pH of the medium is adjusted to 7.5. After weighing, the size of tissue slices is further reduced by means of a pair of scissors. This operation is tedious and requires very sharp scissors but it is very important because the smaller the pieces the shorter the homogenization required and the better the condition of the mitochondria. During the operation the suspending medium is changed two or three times in order to remove blood. Aliquots of 5 g of the tissue are homogenized in 50 ml of medium with a Potter homogenizer driven by a motor operated at 1250 revs/min. The operation is continued until no visible tissue fragments are present in the suspension. The nuclear fraction is sedimented at 600 x *g* for 10 min. Mitochondria are obtained from the supernatant by centrifugation for 20 min at 1500 x *g*. Fluffy layers are removed in the usual way and the sediment resuspended in half the volume of the medium and spun down at 3000 x *g* for 10 min. The operation is repeated with a quarter of the original volume and the last sediment resuspended in 0.25 M sucrose at 20–40 mg protein $ml^{-1}$.

Similarly to liver, ox adrenal cortex mitochondria can perform the oxidation of pyruvate and Krebs cycle intermediates; in contrast they do not oxidize β-hydroxy butyrate and only very slowly oxidize glutamate. RCR values usually obtained range between 3 and 7 with different substrates [160]. In order to test the degree of contamination by microsomes, glucose-6-phosphatase has been measured in the mitochondrial suspension and compared with that of microsomes isolated from the same homogenate by centrifugation of the post-mitochondrial supernatant at 105 000 x *g* for 1 h. Results of a typical experiment are presented in Table 3.2. This table offers an example of the use of marker enzymes. Dividing the specific activity of glucose-6-phosphatase found in mitochondria by that of microsomes, one finds that the mitochondrial fraction contains 5.7% microsomal protein. On the other hand, dividing the microsomal succinate oxidase by the mitochondrial activity one reaches the conclusion that the microsomal fraction does not contain more than 1.2% mitochondrial protein. As with liver, adrenal cortex offers an example of bi-location of rotenone-insensitive-NADH cytochrome *c* reductase, which is present in the two fractions with virtually the same specific activity. The importance of this finding will be further discussed below.

TABLE 3.2
*Comparison of some enzymatic parameters of mitochondria and microsomes from ox adrenal cortex*

| | Glucose-6-phosphatase (μmol $P_i$ 20 $min^{-1}$ $mg^{-1}$ protein) | Rotenone-insensitive NADH-cytochrome *c* reductase (μmol cytochrome *c* red $min^{-1}$ $mg^{-1}$ protein) | *Succinate oxidase* μ $AO_2$ $min^{-1}$ $mg^{-1}$ protein | P/O | RCR |
|---|---|---|---|---|---|
| Mitochondria | 0.019 | 0.473 | 0.247 | 1.88 | 6.3 |
| Microsomes | 0.330 | 0.478 | 0.003 | | |

The quality of the mitochondrial preparation is defined by the high values of P/O found and, perhaps more important, by a good level of respiratory control. The degree of contamination by microsomes, in the vicinity of 5%, is also satisfactory. In tissues containing a well developed endoplasmic reticulum only seldom does one obtain lower levels. The conditions needed to achieve this result depend essentially: (a) on the careful removal of fluffy layers after the first sedimentation of mitochondria and (b) on the dilution of the homogenate. The more concentrated the homogenate *during* the homogenization the higher the contamination. 10% homogenate represent the optimum from this point of view, further dilution does not considerably improve the result. The yield of this technique is around 10 mg mitochondrial protein $g^{-1}$ fresh tissue.

### *3.3.2 Large-scale preparation of mitochondria from ox liver*

The organ, obtained within 10 min of sacrifice, is cut in pieces of approximately 200 g weight and rapidly transferred to the laboratory immersed in ice-cold 0.25 M sucrose. The tissue is ground in a stainless-steel meat-grinder, after removal of connective tissue, large blood and biliary ducts. The mince is washed on cheese-cloth with aliquots of cold 0.25 M sucrose. 180 g of the washed mince is added to 1620 ml cold solution containing 0.25 M sucrose, 0.1 mM EDTA (pH 7.5) and homogenized by means of an 'Ultraturrax' (Tübingen – Germany) operated at full speed (20 000 revs/min) for 20 s. The operation is carried out in a 2000 ml plastic cylinder with nitrogen bubbling. The pH of the suspension is maintained at 7.4 during homogenization by addition of Tris base. After the treatment the homogenate still contains a number of visible tissue fragments which are removed by filtration on nylon net or cheese cloth. Homogenization for longer times may improve the yield but is detrimental to the quality of the organelle. Up to 1 kg of tissue may be so processed. The homogenate, after filtration, is passed through a turbine-moved Sharples centrifuge equipped with n.8–R 34 rotor operated at 18 000 revs/min, with a flux of 250 ml $min^{-1}$. To avoid clogging of the inlet, the O-ring situated at the bottom of the rotor is removed. The inner surface of the rotor is coated with a sheet of Teflon to allow a rapid and easy removal of the sediment. The supernatant is collected in ice. After removal of the sediment (mainly erythrocytes, unbroken cells and nuclei) the rotor is again equipped with the O-ring and the supernatant is centrifuged at 50 000 revs/min, with a flux of 360 ml $min^{-1}$.

The sediment collected on the Teflon sheet is clearly divided into two distinct zones. A brown one in the two lower thirds of the rotor and a transparent, pinkish one occupying the remaining surface. The sediment has been divided in three portions by means of a large spatula and suspended in 0.25 M sucrose. The three suspensions have been analysed for glucose-6-phosphatase activity and the data compared with those of microsomes obtained from the same homogenate in the usual way. The results of this experiment are reported in Table 3.3.

TABLE 3.3
*Glucose 6-phosphatase activity present in different portions of the mitochondrial fraction sedimented on the inner surface of Sharples Rotor 8-R 34.*

| | Specific activity ($\mu$mol $P_i$ 20 $min^{-1}$ $mg^{-1}$ protein) |
|---|---|
| Lower portion | 0.74 |
| Middle portion | 1.22 |
| Upper portion | 2.20 |
| Microsomes | 10.60 |

From the table it appears that contaminating microsomes are unevenly distributed along the inner surface of the rotor, and as expected, are much more concentrated in the upper portion of it. It may be noticed that in the lower two thirds taken together the microsomal contamination is less than 10%. This level has been considered satisfactory for most experimental purposes based on large scale preparations. Without any further washing the mitochondria sedimented in the lower portion of the rotor show respiratory control by ADP and phosphate and high P/O ratios. Further washing of the resuspended mitochondria by means of the same centrifuge notably decreases the quality of the organelle. This effect is probably caused by the very high speed of the rotor which may damage the particles mechanically. For preparative purposes we decided to use the mitochondrial fraction as it is sedimented. For other purposes, when a higher degree of purity is required, the fraction may be washed once or twice, as already described, using a conventional refrigerated centrifuge.

### *3.3.3 Separation of inner and outer mitochondrial membranes*

Around 1965 in a number of laboratories attempts were made to separate the outer envelope of mitochondria from the inner compartment. Four different approaches were undertaken: (a) selective solubilization of the outer membrane by digitonin: (b) osmotic disruption of the outer membrane: (c) disruption of the particles by ultrasound and (d) disruption of mitochondria by organic solvents, detergents, phospholipase, freezing and thawing, fatty acids and diethyl stilbestrol. A procedure based on approach (a) was first worked out by Levy *et al.* [161–163] and subsequently elaborated further by Schnaitman *et al.* [164]. It is based on the selective action of digitonin which, at low concentrations, causes the outer mitochondrial membrane to disaggregate into small fragments leaving well preserved inner membrane and matrix particles, still capable of most mitochondrial functions [165–168]. By differential centrifugation, inner membrane and matrix particles may be separated from outer membrane fragments. Approach (b) was developed first by Parsons *et al.* [169–171] who, on the basis of electron microscopy, could show that mitochondria exposed to hypotonic media undergo osmotic swelling with consequent rupture of the outer membrane. By discontinuous density gradient centrifugation a fraction containing essentially pure outer membrane could be recovered.

A technique based on approach (c) was developed by Sottocasa *et al.* [139, 172–178]. The authors found that sonication resulted in the detachment from mitochondria of a particulate fraction very poor in respiratory-chain linked activity, enriched, however, in rotenone-insensitive NADH-cytochrome *c* reductase. This fraction could be separated from the bulk of other mitochondrial protein by continuous density gradient centrifugation. This finding suggested that the outer mitochondrial membrane could be monitored by following rotenone-insensitive NADH-cytochrome *c* reductase as a marker enzyme. The activity was in fact the reflection of a system, similar to that present in microsomes consisting of NADH-cytochrome $b_5$ reductase and cytochrome $b_5$. Parsons *et al.* simultaneously and independently confirmed and extended these findings [71]. At the time another enzyme had been discovered by the group working in Baltimore [164,168] namely monoamino oxidase. These two enzyme activities proved very useful in the refinement of the different procedures for membrane separation from mitochondria.

Approaches listed under (d) have been followed by Green's group [179–186]. The authors based their work essentially on the assumption that all enzyme activities found in solution after the treatment were originally located in the outer mitochondrial membrane. This working hypothesis was later disproved by experimental findings from many laboratories and the conclusions arrived at in those papers should be considered invalid (for a complete discussion on the subject see [187]).

The three main separation procedures have lead to the same conclusions regarding intramitochondrial localization of enzymes. It has been possible to establish, for example, that the matrix space contains all the enzymes of the citric acid cycle (with the exception of succinate dehydrogenase), pyruvate oxidation, transamination, fatty acid activation and oxidation, glutamate oxidation, and mitochondrial protein synthesis. On the other hand, in addition to respiratory chain components, the enzymes involved in ATP synthesis, a number of respiratory-chain linked dehydrogenases, and the enzymes involved in heme synthesis are firmly bound to the inner membrane. Fatty acid elongation [188–189] and amino acid incorporation into protein [190–192] have also been reported to occur at the level of the inner mitochondrial membrane. A number of kinases catalysing the interconversion of nucleotides have been located in the intermembrane space. In particular, adenylate kinase is a good marker of this compartment. The outer membrane has been shown to contain, in addition to the enzymes already mentioned, ATP-dependent acyl CoA synthetase and a number of enzymes involved in phospholipid metabolism. The intramitochondrial location of enzymes was reviewed in 1969 and 1970 by Ernster and Kuylenstierna [187–193] and by Smoly *et al.* [194].

The following section will be devoted to the detailed description of inner and outer membranes separation from ox adrenal cortex and from liver mitochondria. These two techniques, developed in our laboratory, will give an example of mitochondrial subfractionation on a small and large scale respectively.

### *3.3.4 Subfractionation of adrenal cortex mitochondria*

The subfractionation of adrenal cortex mitochondria is based upon previous experience with rat liver mitochondria. A few differences have been encountered however, and will be underlined and discussed. In this case the principle of the technique was also to cause the outer membrane to swell by hypotonic treatment. After contraction of the inner compartment, induced by hypertonic sucrose in the presence of ATP and $Mg^{2+}$, the suspension was subjected to sonic oscillation to facilitate the detachment of the fragments of the outer membrane from the inner compartment. The separation of the two populations of particles was achieved by continuous density gradient centrifugation.

#### *(i) Swelling and contraction*

Mitochondria are taken up, at 20 mg protein $ml^{-1}$, in 10 mM Tris-$H_3PO_4$ buffer (pH 7.8) containing 0.1 mM EDTA. The resuspension is carried out by the aid of a hand-moved Potter homogenizer. After standing in ice for 15 min, an equal volume of 1.8 M sucrose containing 2 mM ATP and 2 mM $Mg^{2+}$ is added. It may be noticed that, in contrast to liver mitochondria, EDTA is included in the swelling medium, and that the hypertonic solution is twice as concentrated. The reason for the former modification is based on the results obtained by Pfaff and Schwalback [195], whereas the latter modification has been introduced on account of the sonication step (see below) which in the case of adrenal cortex is prolonged with the production of smaller particles. Since the inner membrane is impermeable to sucrose, once loaded with this compound by sonication, the inner membrane particles have a higher density and they should reach the equilibrium density in a reasonably short time, despite their unfavourable size.

#### *(ii) Sonication*

This step is carried out using a Branson 'Sonifier' equipped with a micro-tip operated at 3.5 A. Aliquots of 3.5 ml of the mitochondrial suspension are subjected to sonic irradiation in a system consisting of a centrifuge tube (SW–39 Spinco Rotor) inserted through a rubber stopper in a cooling jacket. The system is perfused with ice-cold water during the operation. 40 s is the optimal time of sonication in the presence of 0.9 M sucrose. This result suggests that adrenal cortex mitochondria are much more resistent to sonic oscillation than liver mitochondria, which under comparable conditions are treated for 10–15 s.

### *(iii) Density-gradient centrifugation*

The density gradient is prepared by carefully pipetting 11 layers, 2 ml each, of sucrose solutions of density decreasing linearly from 1.205 to 1.135. After standing overnight at 0°C, a linear continuous gradient is formed by diffusion. The tubes are loaded with 5 ml each of the mitochondrial suspension. Centrifugation is performed using an SW-25 Spinco Rotor, operated at full speed for 5 h.

### *(iv) Separation and analysis of the fractions*

The gradient at the end of the centrifugation is fractionated by inserting a hypodermic needle into the bottom of the centrifuge tube, through a rubber septum. The needle is connected via a thin 'tygon' tubing with an optical unit (Uvicord–LKB–Sweden) and a fraction collector. Constant flow is ensured by a peristaltic pump. Seventeen fractions per gradient are collected. The fractions are

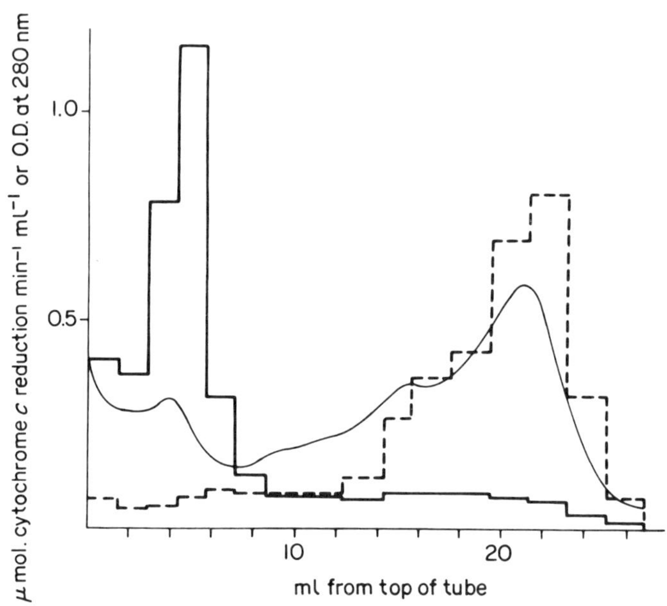


**Fig. 3.1** Distribution of rotenone-insensitive NADH-cytochrome *c* reductase, succinate-cytochrome *c* reductase and protein in subfractions obtained from *ox* adrenal cortex mitochondria. Conditions described in the text: optical density recording at 280 nm; solid histogram: rotenone-insensitive NADH-cytochrome *c* reductase; dashed histogram: succinate-cytochrome *c* reductase.

analysed for rotenone-insensitive NADH-cytochrome *c* reductase and succinate-cytochrome *c* reductase activities (as already described [139]) as markers of outer and inner membranes respectively.

Results of a typical experiment are presented in Fig. 3.1. Rotenone-insensitive NADH-cytochrome *c* reductase shows a maximum in a fraction at very low density, in the vicinity of 1.135; the fractions collected at higher density values are virtually devoid of this enzyme activity. Succinate-cytochrome *c* reductase is practically confined to a broad band at much higher density, around 1.19. The degree of overlaping of the two enzyme activities is very low. Obviously this technique allows one to discriminate without ambiguity between enzymes localized in the inner and in the outer membrane. By means of this procedure it has been possible to show that the mitochondrial cytochrome $P_{450}$ is located in the inner membrane [152], a result confirmed by other methods [196].

This method and similar ones applied to other mitochondria [139,175–197] are particularly suited for cytochemical studies aimed to investigate intramitochondrial location of enzymes. The size of the preparation is at the analytical level and not more than 150 mg mitochondrial protein can be processed, with a maximal yield of outer mitochondrial membrane of the order of 10 mg. We have recently developed a preparative technique by which gram quantities of mitochondria may be treated, for preparative purposes. An example is given in the next paragraph.

### *3.3.5 Large-scale subfractionation of mitochondria from ox liver*

Once again the general principle of the technique is to disrupt the outer mitochondrial membrane by hypotonic swelling, to contract the inner compartment by hypertonic sucrose in the presence of ATP and $Mg^{2+}$ and to facilitate the detachment of outer membrane fragments by sonication. The separation is in this case carried out in three stages: (a) Centrifugation under density conditions which do not allow the sedimentation of outer membrane. Two fractions are obtained: a pellet consisting of inner membrane and part of the matrix protein and a supernatant consisting of outer membrane vescicles and soluble material: (b) Purification of inner membrane by sonication and centrifugation: (c) Purification of the outer membrane by discontinuous density-gradient centrifugation.

#### *(i) Swelling-contraction and sonic treatment*

2.0 g of ox-liver mitochondria prepared as described above are resuspended at 20 mg ml$^{-1}$ in a medium containing 10 mM Triethanolamine-HCl (pH 7.8), 1 mM EDTA. After standing in ice for 20 min with occasional stirring, an equal volume of 1.8 M sucrose, 4 mM ATP, 4 mM $Mg^{2+}$ and 1 mM EDTA is added. Sonication is performed on aliquots of 50 ml by means of a Branson 'Sonifier' equipped with a macro-tip, operated at 2.8 A. The sonication vessel is cooled by circulating ice-cold water and each aliquot is irradiated three times for 30 s each, with 30 s intervals. The sonicated suspension is then spun down at 15 000 x *g* for 15 min using a No. 30 Spinco Rotor. A supernatant $SN_1$ and a pellet $P_1$ are obtained.

*(ii) Purification of the inner membrane by sonication and centrifugation*

The pellet $P_1$ is resuspended in 0.25 M sucrose containing 10 mM Triethanolamine-HCl buffer (pH 7.8) and 1 mM EDTA. The suspension is again subjected to sonic oscillation as described above. The purpose of this treatment is to solubilize the matrix protein still contained within the inner membrane vescicles. The suspension is then spun down at 105 000 x *g* in No. 40 Spinco Rotor for 1 h. A supernatant $SN_2$ and a Pellet $P_2$ are obtained.

*(iii) Purifiation of the outer membrane by discontinuous density-gradient centrifugation*

The supernatant $SN_1$, which contains in suspension the fragments of the outer mitochondrial membrane, is brought to 1.2 M sucrose by additions of crystals of the solute. The density of the suspension is now higher than that of outer membrane vesicles so that they float in a centrifugal field. 15 ml of the suspension are pipetted into No. 30 Rotor Spinco tubes. On top of the sample two layers are carefully pipetted consisting of 5 ml 1.12 M and 0.45 M sucrose respectively. The centrifugation is carried out at 78 000 x *g* for 15 h. Three fractions are clearly distinguishable: a light fraction (L) at the interface between 0.45 M and 1.12 M sucrose solutions; an intermediate fraction (IF) spread in the 1.12 M and upper part of 1.2 M sucrose layers and a heavy fraction ($P_3$) at the bottom of the tube.

*(iv) Distribution of enzyme markers in the mitochondrial subfractions*

The fractions obtained as described above have been analysed for protein, rotenone-insensitive NADH-cytochrome *c* reductase, succinate-cytochrome *c* reductase and malic dehydrogenase as markers of outer, inner membrane and matrix enzymes respectively. Data are graphically presented in Fig. 3.2 where the relative specific activities (specific activity of the fraction divided by specific activity of the starting mitochondrial suspension) are plotted versus the percentage protein recovered in that fraction. Clearly, the first treatment (middle portion of the figure) results in the separation of a sediment $P_1$, collecting some 30% of the original mitochondrial protein, in which the bulk of succinate-cytochrome *c* reductase is sedimented. Already at this stage the relative specific activity is twice that of the starting material. The contamination by outer membranes and soluble matrix is is very low and further reduced by the second treatment (left-hand side, Fig. 3.2). As a result the succinate-cytochrome *c* reductase activity is increased almost five times with respect to the starting mitochondrial suspension. The procedure allows the isolation of some 330 mg purified inner membrane protein. The supernatant $SN_1$, on the other hand, contains some 70% of the protein, the bulk of rotenone-insensitive NADH-cytochrome *c* reductase and some 75% of malic dehydrogenase. About 20% of inner membrane particles are still present in suspension, on the basis of succinate-cytochrome *c* reductase activity. Discontinuous density gradient centrifugation yields a light fraction in which the marker enzyme of the outer membrane is concentrated 13–14 times. The same fraction is virtually devoid of

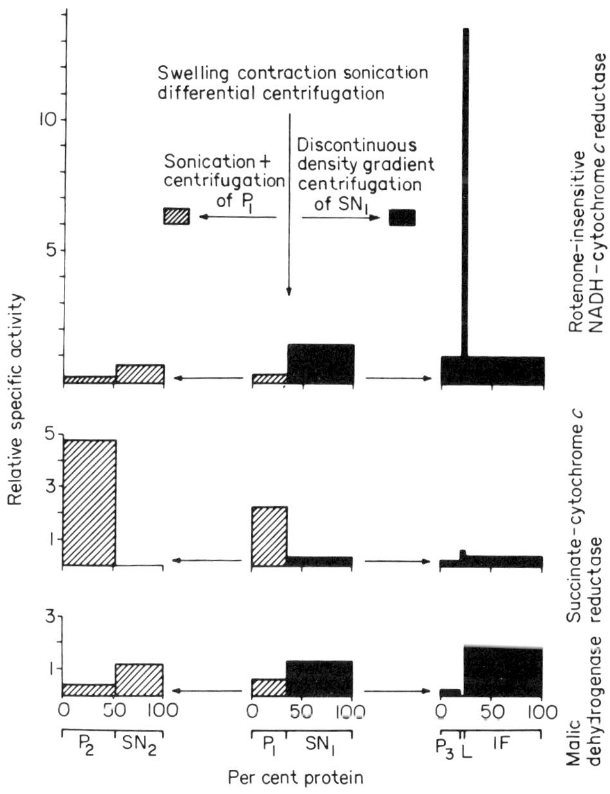


**Fig. 3.2** Subfractionation of ox-liver mitochondria: distribution of some enzymic parameters. Conditions as described in the text. P = pellet: SN = supernatant: L = light fraction: IF = intermediate fraction. Column Λ = enzyme distributions after intial centrifugation, $A_1$ = distributions after sonication and centrifugation of $P_1$,$A_2$ = distributions after discontinuous gradient centrifugation of $SN_1$.

matrix protein and contains very little inner membrane fragments. The total yield in outer mitochondrial membrane of the procedure was 37 mg protein. Considering that the outer membrane represents only 4–5% of the total mitochondrial protein, the yield of the procedure and the degree of purity of the fraction are highly satisfactory.

### 3.4 General considerations

The techniques described may be compared with others and the characteristics of the subfractions obtained may be discussed. The method described by Parsons [169–171] is particularly suitable for obtaining very small samples of very high degree of purity, necessary for morphological studies of isolated membranes. The procedure published by Schnaitman and Greenawalt [168] allows the separation of very high amounts of pure inner and outer membranes as well as inner membrane plus matrix particles still capable of most mitochondrial functions. The only limitation of the technique seems to be the presence of detergents in the isolated fractions. It is, in fact, difficult to imagine that digitonin which interacts with the outer membrane is not, at least in part, retained by the membrane after washing. The same line of thinking can also be applied to inner membrane plus matrix vesicles which show a pseudopoydal morphology suggesting that digitonin is also interacting with the inner membrane and indicating a chemical alteration. The same criticism applies even more to the use of lubrol in solubilizing matrix protein. The very clear results presented by the authors, however, are all consistent with the view that the chemical alteration introduced by the detergent has very little relevance at the enzymic, and even at much higher, levels of organization, in view of the substantial intactness of integrated mitochondrial functions, such as respiratory control and oxidative phosphorylation, measured in inner membrane plus matrix particles.

The techniques I have here described in detail and which are derived from that originally reported for rat-liver mitochondria [139,174–178] are based only on physical methods which should minimize reorganization of membranes during the procedure. The close agreement of findings obtained by different investigators with different approaches makes us confident in stating that the choice of the technique need only reflect the requirements of the experimental design planned.

### References

1. Cowdry, E. V. (1953), *J. Histochem. Cytochem.*, **1**, 183.
2. Palade, G. E. (1952), *Anat. Rec.*, **114**, 427.
3. Palade, G. E. (1953), *J. Histochem. Cytochem.*, **1**, 187.
4. Sjöstrand, F. S. (1953), *Nature*, **171**, 30.
5. Green , D. E. (1965), in *Oxidases and Related Redox Systems*, (King,

T. E., Mason, H. S. and Morrison, M., eds.), Vol II, p. 1061, John Wiley, New York.
6. Green, D. E. (1966), in *Comprehensive Biochemistry*, (Florkin, M. and Stotz, E., eds.), Vol. 14, p. 309, Elsevier, Amsterdam.
7. Penniston, J. T., Harris, R. A., Asai, J. and Green, D. E. (1968), *Proc. natn. Acad. Sci. U.S.A.*, **59**, 624.
8. Hackenbrock, C. R. (1966), *J. Cell. Biol.*, **30**, 269.
9. Greenawalt, W. J. (1969), *Fedn. Proc. Fedn. Am. Socs. exp. Biol.*, **28**, 663.
10. Harris, R. A., Penniston, J. T., Asai, J. and Green, D. E. (1968), *Proc. natn. Acad, Sci. U.S.A.*, **59**, 830.
11. Green, D. E., Asai, J., Harris, R. A. and Penniston, J. T. (1968), *Archs. Biochem. Biophys.*, **125**, 684.
12. Smith, D. S. (1963), *J. Cell. Biol.*, **19**, 115.
13. Wlodawer, P., Parsons, D. F., Williams, G. R. and Wojtczak, L. (1966), *Biochim. biophys. Acta.*, **128**, 34.
14. Parsons, D. F. (1965), *Int. Rev. exp. Path.*, **1**, 1.
15. Fernandez-Moràn, H. (1962), *Circulation*, **26**, 1039.
16. Stoeckenius, W. (1963), *J. Cell. Biol.*, **17**, 443.
17. Fernandez-Moràn, H., Oda, T., Blair, P. V. and Green, D. E. (1964), *J. Cell Biol.*, **22**, 63
18. Kagawa, Y. and Racker, E. (1966), *J. biol. Chem.*, **241**, 2475.
19. Pullman, M. E., Penefsky, H. S., Datta, A. and Racker, E. (1960), *J. biol. Chem.*, **235**, 3322.
20. Penefsky, H. S., Pullman, M. E., Datta, A. and Racker, E. (1960), *J. biol. Chem.*, **235**, 3330.
21. Moore, H. D. and Ruska, H. (1957), *J. biophys. biochem. Cytol.*, **3**, 261.
22. Porter, K. R. and Palade, G. E. (1957), *J. biophys. biochem. Cytol.*, **3**, 269.
23. Simpson, F. O. and Oertelis, S. J. (1962), *J. Cell. Biol.*, **12**, 91.
24. Stenger, R. J. and Spiro, D. (1961), *J. biophys. biochem. Cytol.*, **9**, 325.
25. Slautterback, D. B. (1965), *J. Cell. Biol.*, **24**, 1.
26. Luft, R., Ikkos, D., Palmieri, G., Ernster, L. and Afzelius, B. (1962), *J. clin. Invest.*, **41**, 1776.
27. Sjöstrand, F. S. (1963), *J. Ultrastruct. Res.*, **9**, 340.
28. Luft, J. and Hechter, O. (1957), *J. biophys. biochem. Cytol.*, **3**, 615.
29. Idelman, S. (1966), *Annls. Sci. nat. (Zool.)*, **8**, 205.
30. Revel, J. P., Fawcett, D. W. and Philpott, C. W. (1963), *J. Cell Biol.*, **16**, 192.
31. Warburg, O. (1913), *Pflüger's Arch. ges. Physiol.*, **154**, 599.
32. Keilin, D. (1929), *Proc. R. Soc. Ser. B.*, **104**, 206.
33. Keilin, D. (1933), *Ergebn. Enzymforsch.*, **2**, 139.
34. Green, D. E., Loomis, W. F. and Auerbach, V. H. (1948), *J. biol. Chem.*, **172**, 389.
35. Green, D. E. (1951), *Biol. Rev.*, **26**, 410.
36. Schneider, W. C. and Potter, V. R. (1949), *J. biol. Chem.*, **177**, 893.
37. Lehninger, A. L. and Kennedy, E. P. (1949), *J. biol. Chem.*, **179**, 957.
38. Dounce, A. L. (1950) in *The Enzymes*, (Sumner, J. B. and Myrbäck, K., eds.), Vol. I, p. 187, Academic Press, New York.
39. Dounce, A. L. (1952), *J.C.C.P.*, **39**, 43.
40. Hogeboom, G. H. (1951), *Fedn. Proc. Fedn. Am. Socs. exp. Biol.*, **10**, 640.
41. Schneider, W. C. and Hogeboom, G. H. (1951), *Cancer Res.*, **11**, 1.
42. Green, D. E. (1951), *Biol. Rev.*, **260**, 410.

43. Green, D. E. (1952), *J.C.C.P.*, **39**, suppl. 2, 75.
44. DeDuve, C. (1952), *Exposes Annuels de Biochimie Medical,* **14**, 47.
45. Holter, H. (1952), *Adv. Enzym.*, **13**, 1.
46. Lindberg, O. and Ernster L. (1954), *Protoplasmatologia,* **3**, A4.
47. Sisakyan, N. M. and Filippovich, I. I. (1957), *Biokhimiya,* **22**, 375.
48. MacLean, J. R., Cohn, G. L., Brandt, J. K. and Simpson, M. V. (1958), *J. biol. Chem.*, **233**, 657.
49. Fraser, M. J. and Gutfreund, H. (1958), *Proc. R. Soc. Ser. B.*, **149**, 392.
50. Roodyn, D. B., Reis, P. J. and Work, T. S. (1961), *Biochem. J.*, **80**, 9.
51. Steinert, M. (1960), *J. biophys. biochem. Cytol.*, **8**, 452.
52. Nass, M. K. and Nass, S. (1962), *Exptl. Cell. Res.*, **26**, 424.
53. Nass, M. K. and Nass, S. (1963), *J. Cell. Biol.*, **19**, 593.
54. Nass, M. K., Nass, S. and Afzelius, B. A. (1965), *Exptl. Cell. Res.*, **37**, 516.
55. Moses, M. (1964), *J. Histochem. Cytochem.*, **12**, 115.
56. Swift, H. (1965), *Ann. Nat.*, **99**, 201.
57. Scherbaum, O. (1960), *Ann. N.Y. Acad. Sci.*, **90**, 655.
58. Parsons, J. (1965), *J. Cell. Biol.*, **25**, 641.
59. Watson, M. and Aldridge, W. (1964), *J. Histochem, Cytochem.*, **12**, 96.
60. Swift, H., Adams, B. and Larsen, K. (1964), *Jl R. microsc. Soc.*, **83**, 161.
61. André, J. and Marinozzi, V. (1965), *J. Microscopie,* **4**, 615.
62. Ashwell, M. and Work, T. S. (1970), *Ann. Rev. Biochem.*, **39**, 251.
63. Rabinowitz, M. and Swift, H. (1970), *Phys. Rev.*, **50**, 376.
64. Borst, P. (1972), *Ann. Rev. Biochem.*, **41**, 333.
65. Slater, E. C., Tager, J. M., Papa, S. and Quagliariello, E. (1968), (eds.), *Biochemical Aspects of the Biogenesis of Mitochondria*, Adriatica Editrice, Bari.
66 Kroon, A. M. and Saccone, C. (1974), *The Biogenesis of Mitochondria* Academic Press, New York.
67. MacFarlane, M. G. and Spencer, A. G. (1953), *Biochem. J.*, **54**, 569.
68. Bartley, W. and Davies, R. E. (1954), *Biochem. J.*, **57**, 39.
69. Bartley, W. and Amoore, J. E. (1958), *Biochem. J.*, **70**, 718.
70. Brierley, G. P., Bachmann, E. and Green, D. E. (1962), *Proc. natn. Acad. Sci. U.S.A.*, **48**, 1928.
71. Gamble, J. L. Jr. (1957), *J. biol. Chem.*, **228**, 955.
72. Vasington, F. D. and Murphy, J. V. (1962), *J. biol. Chem.*, **237**, 2670.
73. Lehninger, A. L., Rossi, C. S. and Greenawalt, J. W. (1963), *Biochem. biophys. Res. Commun.*, **10**, 444.
74. Vasington, F. D. (1963), *J. biol. Chem.*, **238**, 1841.
75. Saris, N. E. (1959), *Finska Kemistanfundets Medd.*, **68**, 98.
76. Saris, N. E. (1963), *Soc. Sci. Fennica Commentationes Phys. Math.*, **28**, 11.
77. Carafoli, E. (1965), *Biochim. biophys. Acta.*, **97**, 99.
78. Carafoli, E., Weiland, S. and Lehninger, A. L. (1965), *Biochim. biophys. Acta.*, **97**, 88.
79. Bartley, W. and Amoore, J. E. (1958), *Biochem. J.*, **69**, 348.
80. Chappell, J. B., Greville, G. D. and Bicknell, K. E. (1962), *Biochem. J.*, **84**, 61.
81. Chappell, J. B., Cohn, M. and Greville, G. D. (1963), in *Energy linked Functions of Mitochondria*, (Chance, B., ed.), p. 219, Academic Press, New York.
82. Chappell, J. B. and Crofts, A. R. (1966), in *Regulation of Metabolic*

*Processes in Mitochondria* (Tager, J. M., Papa, S., Quagliariello, E. and Slater, E. C., eds.), Vol. 7, p. 293, Biochimica et Biophysica Acta Library.

83. Chappell, J. B. and Haarhoff, K. N. (1967), in *Biochemistry of Mitochondria* (Slater, E. C., Kaniuga, Z. and Wojtczak, L., eds.), p. 75, Academic Press, London.
84. Fonyo, A. (1968), *Biochem. biophys. Res. Commun.*, **32**, 624.
85. Palmieri, F. and Klingenberg, M. (1967), *Eur. J. Biochem.*, **1**, 439.
86. Fritz, I. B. and Delisle, G. (1967), *Proc. nat. Acad. Sci. U.S.A.*, **58**, 790.
87. Tubbs. P. K. and Garland, P. B. (1968), *Br. med. Bull.*, **24**, 158.
88. Robinson, B. H. and Chappell, J. B. (1967), *Biochim. biophys. Res. Commun.*, **28**, 249.
89. Azzi, A., Chappell, J. B. and Robinson, B. H. (1968), *Biochem. biophys. Res. Commun.*, **32**, 624.
90. Heldt, H. W., Jacobs, H. and Klingenberg, M. (1965), *Biochem. biophys. Res. Commun.*, **18**, 174.
91. Chance, B. (1965), *J. biol. Chem.*, **240**, 2729.
92. Lehninger, A. L., Carafoli, E. and Rossi, C. S. (1967), *Adv. Enzym.*, **29**, 259.
93. Lehninger, A. L. (1970), *Biochem. J.*, **119**, 129.
94. Carafoli, E. and Lehninger, A. L. (1971), *Biochem. J.*, **122**, 681.
95. Borle, A. B. (1973), *Fedn. Proc. Fedn. Am. Soc. exp. Biol.*, **32**, 1944.
96. Lehninger, A. L. (1972), in *The Molecular Basis of Electron Transport*, (Schultz, J. and Cameron, B. F., eds.), Vol. 4, p. 133, Academic Press, New York.
97. Chappell, J. B. (1968), *Br. med. Bull.*, **24**, 150.
98. Pressman, B. C. (1970), in *Membranes of Mitochondria and Chloroplasts*, (Racker, E., ed.), p. 213, Van Nostrand Reinhold, New York.
99. Ernster, L. and Drahota, Z. (1969), (eds.), *Mitochondria Structure and Function*, Academic Press, New York.
100. Lehninger, A. L. (1971), in *The Dynamic Structure of Cell Membranes*, (Hölzl Wallach, D. F. and Fischer, H., eds.), p. 199, Springer-Verlag, Berlin.
101. Ovchinnikov, Ju.A., Ivanov, V. T. and Shkrob, A. M. (1974), in *Membrane-Active Complexones*, p. 331, Elsevier, Amsterdam.
102. Azzone, G. F., Massari, S., Colonna, R., Dell'Antone, P., Frigeri, L. and Beltrame, M. (1970), in *Dynamics of Energy-Transducing Membranes*, (Ernster, L. Estabrook, R. W. and Slater, E. C., eds.), p. 405, Elsevier, Amsterdam.
103. Papa, S., Guerrieri, F. and Lorusso, M. (1974), *ibid, idem.*, p. 417.
104. Van Dam, K. and Postma, P. W. (1974), *ibid, idem*, p. 433.
105. Reynefarje, B. and Lehninger, A. L. (1974), *ibid, idem*, p. 447.
106. Carafoli, E. and Sottocasa, G. L. (1974), *ibid, idem*, p. 455.
107. Scarpa, A. (1974), *ibid, idem*, p. 471.
108. Quagliariello, E., Passarella, S. and Palmieri, F. (1974), *ibid, idem*, p. 483.
109. Williams, G. R. and Orr, J. L. (1974), *ibid, idem*, p. 497.
110. Klingenberg, M. (1974), *ibid, idem*, p. 511.
111. Porter, K. R., Claude, A. and Fullman, E. F. (1945), *J. exp. Med.*, **81**, 233.
112. Claude, A. (1946), *J. exp. Med.*, **84**, 51,61
113. Claude, A. (1950), *Harvey Lectures Ser.*, **43**, 121.
114. Hogeboom, G. H., Schneider, W. C. and Striebich, M. J. (1952), *J. biol. Chem.*, **196**, 111.
115. De Duve, C. (1963), in *Methods of Separation of Subcellular Structural Components*, p. 1, Cambridge University Press, Cambridge.

116. De Duve, C. (1964), *Harvey Lectures Ser.*, **59**, 49.
117. Estabrook, R. W. and Pullman, M. E. (1967), (eds.), *Meth. Enzym.*, **10**.
118. Ernster, L. and Nordenbrand, K. (1967), *Meth. Enzym.*, **10**, 86.
119. Chance, B. and Hagihara, B. (1961), *Proc. 5th int. Congr. Biochem.*, Moscow, Vol. 5, p. 3, Pergamon Press, London.
120. Tyler, D. D. and Gonze, J. (1967), *Meth. Enzym.*, **10**, 75.
121. Chappel, J. B. and Perry, S. V. (1953), *Biochem. J.*, **55**, 586.
122. Pressman, B. C. and Lardy, H. A. (1956), *Biochim. biophys. Acta.*, **21**, 458.
123. Scholefield, P. G. (1956), *Can. J. Biochem. Physiol.*, **34**, 1227.
124. Lehninger, A. L. and Remmert, L. F. (1959), *J. biol. Chem.*, **234**, 2459.
125. Hülsmann, W. C., Elliott, W. B. and Slater, E. C. (1960), *Biochim. biophys. Acta.*, **39**, 267.
126. Wojtczak, L. and Wojtczak, A. B. (1960), *Biochim. biophys. Acta.*, **39**, 277.
127. Borst, P., Loos, J. A., Christ, E. J. and Slater, E. C. (1962), *Biochim. biophys. Acta.*, **62**, 509.
128. Lindberg, O., De Pierre, J., Rylander, E. and Afzelius, B. A. (1967), *J. Cell. Biol.*, **34**, 293.
129. Smith, R. E., Roberts, J. C. and Hittelman, K. J. (1966), *Science*, **154**, 653.
130. Pedersen, J. I., Christiansen, E. N. and Grav, H. J. (1968), *Biochem. biophys. Res. Commun.*, **32**, 492.
131. Hittelman, K. J., Lindberg, O. and Cannon, B. (1969), *Eur. J. Biochem.*, **11**, 183.
132. Nicholls, D. G., Cannon, B., Grav, H. J. and Lindberg, O. (1974), in *Dynamics of Energy-Transducing Membranes*, (Ernster, L., Estabrook, R. W. and Slater E. C., eds.), p. 529, Elsevier, Amsterdam.
133. Stahl, W. L., Smith, J. C., Napolitano, L. M. and Basford, R. E. (1961), *J. Cell. Biol.*, **19**, 293.
134. Basford, R. E. (1967), *Meth. Enzym.*, **10**, 96.
135. De Duve, C., Berthet, J. and Beaufay, H. (1959), *Progr. Biophys. biophys. Chem.*, **9**, 325.
136. Hogeboom, G. H. and Schneider, W. C. (1952), *J. biol. Chem.*, **197**, 611.
137. Roodyn, D. B. (1963), in *Methods of Separation of Subcellular Structural Components*, (Grant, J. K., ed.), p. 20, Cambridge University Press, Cambridge.
138. Hers, H. G. and De Duve, C. (1950), *Bull. Soc. chim. Biol.*, **32**, 20.
139. Sottocasa, G. L., Kuylenstierna, B., Bergstrand, A. and Ernster, L. (1967), *J. Cell. Biol.*, **32**, 415.
140. Novikoff, A. B., Essner, E., Goldfischer, S. and Heus, M. (1962), in *The Interpretation of Ultrastructure*, (Harris, J. C., ed.), p. 149, Academic Press, New York.
141. El-Aaser, A. A. and Reid, E. (1955), *Abst. fed. Eur. Biochem. Soc. Meeting, Vienna*, p. 41.
142. Coleman, R., Michell, R. H., Finean, J. B. and Hawthorne, J. N. (1967), *Biochim. biophys. Acta.*, **135**, 573.
143. Beaufay, H., Bendall, D. S., Baudhuin, P., Wattiaux, R. and De Duve, C. (1959), *Biochem. J.*, **73**, 628.
144. Beaufay, J., Jacques, P., Baudhuin, P., Sellinger, O. Z., Berthet, J. and De Duve, C. (1964), *Biochem. J.*, **92**, 184.
145. Lehninger, A. L. (1951), *J. biol. Chem.*, **190**, 345.

146. Lehninger, A. L. (1951), *Phosphorus Metabolism,* **1**, 344.
147. Lehninger, A. L. (1955), *Harvey Lectures Ser.,* **49**, 176.
148. Copenhaver, J. H. Jr. and Lardy, A. H. (1952), *J. biol. Chem.,* **195**, 225.
149. Pressman, B. C. (1955), *Biochim. biophys. Acta.,* **17**, 273.
150. Estabrook, R. W. (1961), *J. biol. Chem.,* **236**, 3051.
151. Lee, C. P., Sottocasa, G. L. and Ernster, L. (1967), *Meth. Enzym.,* **10**, 33.
152. Sottocasa, G. L. and Sandri, G. (1970), *Biochim. biophys. Acta.,* **223**, 36.
153. Kroon, A. M., Botman, M. J. and Saccone, C. (1968), *in Biochemical Aspects of the Biogenesis of Mitochondria,* (Slater, E. C., Tager, J. M., Papa, S. and Quagliariello, E., eds.), p. 439, Adriatica Editrice, Bari.
154. Chance, B. and Williams, R. G. (1955), *J. biol. Chem.,* **217**, 383.
155. Chance, B. and Williams, R. G. (1956), *Adv. Enzym.,* **17**, 65.
156. Estabrook, R. W. (1967), *Meth. Enzym.,* **10**, 41.
157. De Duve, C. (1967), *Meth. Enzym.,* **10**, 7.
158. Slater, E. C. (1967), *Meth. Enzym.,* **10**, 19.
159. Guilloroy, R. J. and Racker, E. (1968), *Biochim. biophys. Acta.,* **153**, 490.
160. Sottocasa, G. L. and Sandri, G. (1970), *Biochem. J.,* **116**, p. 16.
161. Lévy, M., Toury, R. and André, J. (1966), *C.r. hebd. Séanc. Acad. Sci., Paris. Ser. D.,* **262**, 1593.
162. Lévy, M., Toury, R. and André, J. (1966), *C.r. hebd. Séanc. Acad. Acad. Sci., Paris. Ser. D.,* **263**, 1766.
163. Lévy, M., Toury, R. and André, J. (1967), *Biochim. biophys. Acta.,* **135**, 599.
164. Schnaitman, C., Erwin, V. G. and Greenawalt, J. W. (1967), *J. Cell. Biol.,* **32**, 719.
165. Hoppel, C. and Cooper, C. (1968), *Biochem. J.,* **107**, 367.
166. Morton, D. J., Hoppel, C. and Cooper, C. (1968), *idem.,* **107**, 377.
167. Newman, H. A. I., Gordesky, S. E., Hoppel, C. and Cooper, C. (1968), *idem.,* **107**, 381.
168. Schnaitman, C. and Greenawalt, J. W. (1968), *J. Cell Biol.,* **38**, 158.
169. Parsons, D. F., Williams, G. R. and Chance, B. (1966), *Ann. N.Y. Acad. Sci.,* **137**, 643.
170. Parsons, D. F. and Williams, G. R. (1967), *Meth. Enzym.,* **10**, 443.
171. Parsons, D. F., Williams, G. R., Thompson, W., Wilson, D. and Chance, B. (1967), in *Round Table Discussion on Mitochondrial Structure and Compartmentation,* (Quagliariello, E., Papa, S., Slater, E. C. and Tager, J. M., eds.), p. 29, Adriatica Editrice, Bari.
172. Sottocasa, G. L. and Ernster, L. (1965), *Proc. 2nd Meeting Fed. Eur. Biochem. Soc., Vienna,* p. 112.
173. Sottocasa, G. L. and Kuylenstierna, B. (1966), *Proc. 3rd Meeting Fed. Eur. Biochem. Soc., Warsaw,* p. 118
174. Sottocasa, G. L., Ernster, L., Kuylenstierna, B. and Bergstrand, A. (1967), in *Round Table Discussion on Mitochondrial Structure and Compartmentation,* (Quagliariello, E., Papa, S., Slater, E. C. and Tager, J. M., eds.), p. 74.
175. Sottocasa, G. L., Kuylenstierna, B., Ernster, L. and Bergstrand, A. (1967), *Meth. Enzym.,* **10**, 448.
176. Sottocasa, G. L. and Sandri, G. (1968), *Italian J. Biochem.,* **17**, 17.
177. Wojtczak, L. and Sottocasa, G. L. (1972), *J. memb. Biol.,* **7**, 313.
178. Wojtczak, L. and Załuska, H. (1969), *Biochim. biophys. Acta.,* **193**, 64.
179. Bachmann, E., Allmann, D. W. and Green, D. E. (1966), *Archs. Biochem. Biophys.,* **115**, 153.

180. Allmann, D. W. and Bachmann, E. (1967), *Meth. Enzym.*, **10**, 438.
181. Allmann, D. W., Galzigna, L., McCaman, R. E. and Green, D. E. (1966), *Archs. Biochem. Biophys.*, **117**, 413.
182. Green, D. E., Bachmann, E., Allmann, D. W. and Perdue, J. F. (1966), *Archs. Biochem. Biophys.*, **115**, 172.
183. Allmann, D. W., Bachmann, E., Orme-Johnson, N., Tan, W. C. and Green, D. E. (1968), *Archs. Biochem. Biophys.*, **125**, 981.
184. Byington, K. H., Morey, A. V. and Smoly, J. (1968), *Fedn. Proc. Fedn. Am. Socs. exp. Biol.*, **27**, 461.
185. Byington, K. H., Smoly, J. M., Morey, H. V. and Green, D. E. (1968), *Archs. Biochem. Biophys.*, **128**, 762.
186. Smoly, J. M., Byington, K. H., Tan, C. W. and Green, D. E. (1968), *Archs. Biochem. Biophys.*, **128**, 774.
187. Ernster, L. and Kuylenstierna, B. (1970), in *Membranes of Mitochondria and Chloroplasts*, (Racker, E., ed.), p. 172, Van Nostrand Reinhold, New York.
188. Colli, W., Hinkle, P. and Pullman, M. E. (1968), *Fedn. Proc. Fedn. Am. Socs. exp. Biol.*, **27**, 648.
189. Quagliariello, E., Landriscina, C. and Coratelli, P. (1968), *Biochim. biophys. Acta.*, **164**, 12.
190. Beattie, D. S., Basford, R. E. and Koritz, S. B. (1967), *Biochemistry*, **6**, 3099.
191. Neupert, W., Brdiczka, D. and Bücher, T. (1967), *Biochem. biophys. Res. Commun.*, **27**, 488.
192. Work, T. S. (1968), in *Biochemical Aspects of the Biogenesis of Mitochondria*, (Slater, E. C., Tager, J. M., Papa, S. and Quagliariello, E., eds.), p. 367, Adriatica Editrice, Bari.
193. Ernster, L. and Kuylenstierna, B. (1969), *Mitochondria Structure and Function*, (Ernster, L. and Drahota, Z., eds.), p. 5, Academic Press, London.
194. Smoly, J. M., Kuylenstierna, B. and Ernster, L. (1970), *Proc. natn. Acad. Sci. U.S.A.*, **66**, 125.
195. Pfaff, E. and Schwalbach, K. (1967), in *Mitochondrial Structure and Compartmentation*, (Quagliariello, E., Papa, S., Slater, E. C. and Tager, J. M., eds.), p. 346, Adriatica Editrice, Bari.
196. Satre, M., Vignais, P. V. and Idelman, S. (1969), *FEBS Letters*, **5**, 135.
197. Sandri, G., Sottocasa, G. L., Panfili, E., Soranzo, M. R., Traniello, S. and Grazi, E. (1974), *Italian J. Biochem.*, **23**, 165.

# 4 Isolation, subfractionation and characterization of the endoplasmic reticulum

JOSEPH DEPIERRE and GUSTAV DALLNER



## 4.1 Introduction

The endoplasmic reticulum (ER) of the liver cell represents a complex membrane network on which are localized a large number of enzymes and enzyme systems. Among the functions associated with this organelle are the metabolism of xenobiotics; synthesis and transport of a number of proteins, including glycoproteins and lipoproteins; synthesis of cholesterol, phospholipids, and triglycerides; and participation in glycogen breakdown.

Quantitatively, the ER is a major cell component, containing a large portion of the cellular protein and phospholipid. Electron micrographs reveal the presence of this membrane network throughout the cytoplasm. The system is composed of tubules, vesicles, and lamellae having a total surface area as determined by morphometric studies on the order of 11 $m^2$ $ml^{-1}$ of liver tissue [1]. Approximately 60% of this surface is covered with ribosomes, amounting to some 13 million bound ribosomes per hepatocyte [1]. The areas with attached ribosomes are called rough ER, while the remaining areas are referred to as the smooth ER.

The ER is well known to respond to phenobarbitone, carcinogens, and other xenobiotics by increasing its volume and surface area (two-fold in the case of phenobarbitone) together with a substantial increase (five-fold in the case of phenobarbitone) in the specific activity of the cytochrome $P_{450}$–linked drug metabolizing system [2,3]. The adaptability and functional importance of this organelle have resulted in it being the object of numerous isolations, subfractionations, and other investigations. Here we will first discuss the principles and physicochemical properties involved in the isolation of microsomes, as well as review the approaches used and results obtained in various laboratories. Then we will describe in detail experimental procedures for isolation and subfractionation of microsomal fractions. Since most studies along these lines, including our own, have been with rat liver and since the most clear-cut results have been achieved with this organ, consideration will be limited almost exclusively to investigations on rat liver microsomes.

### *4.1.1 Aims of subcellular fractionation*

The aims involved in isolation of a subcellular particle are three-fold: first, the morphological, chemical, and enzymatic properties of the organelle should be altered as little as possible during the isolation procedure. Secondly, recovery of the organelle should be as high as possible in order to best insure obtaining a representative sample. Finally, the isolated fraction should be as free as possible from contamination by other organelles.

It is, of course, impossible in practice to achieve all of these goals; so that isolation procedures, especially those designed to isolate fragments of the ER, must represent a compromise between them. In the simplest terms the investigator must often choose between high recovery and lower purity or lower recovery and higher purity. The goals of most studies dictate a choice for higher purity.

## 4.2 Use of liver tissue in preparing microsomes

Because of its ready availability and relative homogeneity, liver is by far the most frequently used tissue for the preparation of microsomes. The liver of adult animals contains only two major cell types, parenchymal cells (hepatocytes) and reticuloendothelial (Kupffer) cells, in approximately equal numbers [4]. Two factors favour the recovery of ER chiefly from hepatocytes: first, the Kupffer cells are relatively small and contain only a limited amount of ER; and secondly, the non-hepatocytes are relatively resistant to mechanical shearing forces and thus are not disrupted by normal homogenization procedures [5]. These factors also make it possible to isolate ER chiefly from the hepatocytes of prenatal rats, even though almost 40% of the volume of the liver in these animals is composed of hematopoietic cells [4,5].

A problem encountered when using liver as starting material for the isolation of microsomes is the presence of large quantities of blood in this organ. Erythrocytes are disrupted by normal homogenization procedures, resulting in the release of haemoglobin and other proteins. Fragments of the endoplasmic reticulum, which are highly negatively charged [6], readily adsorb soluble constituents of broken erythrocytes [7]. Haemoglobin contamination may be especially troublesome in studies on the microsomal cytochromes. Serum components may also be adsorbed to microsomes and interfere with various investigations. For example, adsorption of serum glycoproteins can give a value for protein-bound microsomal sialic acid which is twice the true value [8]. Fortunately, these problems, caused by the presence of blood, can be avoided either by perfusing the liver before homogenization or by washing the microsomes with an appropriate medium (see below).

### *4.2.1 Homogenization*

#### *(i) Preliminary disruption of the liver*

The usual first step in the isolation of microsomes is to cut the liver into small pieces with a pair of scissors before homogenization. In some cases the liver has been further disrupted by forcing these pieces through a tissue press, but this procedure is to be avoided, since it damages organelles such as mitochondria and peroxisomes [9].

#### *(ii) Speed of the pestle and number of passes during homogenization*

The liver is then generally homogenized in 0.25 M sucrose using a Potter-Elvehjem homogenizer with a Teflon pestle. The clearance between the pestle and the vessel is not critical, but should not be too tight. The details concerning the speed at which the pestle is turning and the number of passes performed during homogenization are of great importance, but are often omitted in published reports. For maximal recovery of microsomes extensive homogenization at high speed is required to disrupt as many of the hepatocytes as possible; but such homogenization also results in increased damage to, and fragmentation of, cellular organelles and non-hepatocytes. Increased disruption of the plasma membrane [10], Golgi cisternae [11], connective tissue cells and cells of erythropoesis [5,12], release of outer mitochondrial membranes [13] and rupture of lysosomes [14] can all cause a serious increase in the contamination of the microsomal fraction.

TABLE 4.1
*Comparison of the homogenization procedures used by different laboratories*

| | *Laboratory* Adelman *et al.* [15, 16] | Amar-Costesec *et al.* [17] | DePierre and Dallner |
|---|---|---|---|
| Speed of pestle revolution during homogenization (revs/min) | 1000–2000 | 1300 | 400 |
| Number of passes | 8–10 | 1 x 3[a] (1 x 6)[b] | 4 |
| Properties of the microsomal fraction | | | |
| protein (mg $g^{-1}$ liver, wet wt.) | 22 | 41 | 20–25 |
| phospholipid (mg $g^{-1}$ liver, wet wt.) | 15 | 23 | 6–7 |
| recovery of microsomal markers (%) | | | |
| glucose-6-phosphatase | – | 76 | 55 |
| NADPH-cytochrome *c* reductase | – | 66 | 60 |
| nucleoside diphosphatase | – | 66 | – |
| recovery of other markers (%) | | | |
| mitochondrial | | | |
| cytochrome oxidase | 10 | – | – |
| monoamine oxidase | – | 20 | – |
| succinate-cytochrome *c* reductase | – | – | 0.7 |
| lysosmal | | | |
| acid phosphatase | 30 | 17 | 11 |
| 'plasma membrane' | | | |
| 5′-nucleotidase | – | 50 | 29 |
| peroxisomal | | | |
| catalase | 9 | – | – |

[a]See text for explanation

[b]The authors do not describe in detail how they wash the mitochondrial pellet, which they do 3 times. They probably resuspend the pellet by rehomogenization in this procedure.

This can be seen in Table 4.1, which compares the results from our laboratory and two others. In the laboratory of Adelman and co-workers [15,16] homogenization at 1000–2000 revs/min with 8–10 passes results in a microsomal fraction containing 10% of the total cytochrome oxidase (a marker for the inner mitochondrial membrane) and 30% of the acid phosphatase (a lysosomal marker). On the other hand, when we homogenize at 400 revs/min with 4 passes, the microsomes contain only 0.7% of the succinate-cytochrome *c* reductase (also a marker for the inner mitochondrial membrane) and 11% of the acid phosphatase. The Louvain group (Amar-Costesec *et al.* in Table 4.1) [17] accomplishes extensive disruption by first homogenizing the liver with a single pass, then sedimenting and rehomogenizing the 'nuclear' pellet, and once again collecting and homogenizing the 'nuclear' pellet. These homogenizations are performed at 1300 revs/min and a relatively high recovery of microsomal marker enzymes is achieved from the combined supernatants. However, the price paid for this higher recovery is that the total microsomal fraction contains twice as much protein and phospholipid as those isolated in our laboratory and Adelman *et al.'s*, but only about 10% more of the microsomal marker enzymes. This would seem to indicate that the microsomes isolated in Louvain are more heavily contaminated with membranes derived from organelles other than the ER.

It can be argued that the extra protein and phospholipid recovered by the Louvain group represents fragments of the ER that are poor in the traditional markers and which are lost by other laboratories in the 'nuclear' pellet. We tested this hypothesis by gently resuspending our 'nuclear' pellet without homogenization and then collecting a second 'crop' of microsomes by the standard scheme of differential centrifugation. This second 'crop' possessed NADPH-cytochrome *c* reductase activity which had the same specific activity as that of the original microsomes. Thus, it seems likely to us that the Louvain group recovers an additional 10% of the microsomal markers by randomly disrupting more hepatocytes and that the homogenization force required for this additional breakage of cells disrupts organelles and results in a more heavily contaminated microsomal fraction.

In addition to increasing the contamination of the microsomal fraction prolonged and/or vigorous homogenization may release bound ribosomes from the ER, modify vesicle structure, and affect enzyme activity. Therefore, we feel that while our homogenization procedure is certainly not ideal, that one would prefer to achieve a higher recovery of microsomes, this procedure nonetheless offers the best available compromise resulting in relatively good recovery, relatively low contamination, and a representative sample.

*(iii) The medium used for homogenization*

Almost all procedures for the isolation of microsomes utilize the same medium during homogenization, namely, 0.25 M sucrose. There is something to be said for using a higher concentration of sucrose in this medium. Siekevitz and Palade [18] obtained better morphological preservation of the elongated, tube-like structure of the rough endoplasmic reticulum by homogenizing the liver in 1.46 M sucrose and by carrying out centrifugations in 0.88 M sucrose. In addition, higher concentrations of sucrose reduce microsomal aggregation (see Table 4.2) [19,20].

However, it would seem desirable to use lower concentrations of sucrose whenever possible because of the possible inhibition of various enzymes by this solute. For example, the ATPase activity of rough microsomes is increasingly inhibited by increasing concentrations of sucrose, until at 1.2 M sucrose this enzyme is 75% inhibited (see Fig. 4.1) [21]. It seems advisable to routinely test for such enzyme inhibition by sucrose, especially after centrifugation in sucrose

TABLE 4.2
*Stability of rat liver smooth microsomes*

| Experiment | Microsomal concentration (g ml$^{-1}$) | Sucrose concentration (M) | % aggregated |
|---|---|---|---|
| 1. Smooth microsomes | 0.05 | 0.15 | 81 |
| | 0.1 | 0.15 | 65 |
| | 0.2 | 0.15 | 36 |
| | 0.4 | 0.15 | 22 |
| 2. Smooth microsomes | 0.05 | 0.88 | 36 |
| | 0.1 | 0.88 | 30 |
| | 0.2 | 0.88 | 25 |
| | 0.4 | 0.88 | 19 |
| 3. Smooth microsomes and 0.3% albumin | 0.05 | 0.15 | 40 |
| 4. Smooth microsomes, pelleted and resuspended | 0.4 | 0.88 | 80 |

The degree of aggregation was determined by filtration of microsomes through a 0.45 μm millipore filter. Data taken from [19] and [20];

gradients. Another disadvantage of using high sucrose concentrations is the resulting high viscosity, a viscosity which is extremely dependent on temperature. Much longer centrifugation times are required to harvest microsomes from solutions of high viscosity.

But why is sucrose so widely used in homogenization and centrifugation media instead of another solute? The original choice may have been based on such non-scientific considerations as the fact that sucrose is inexpensive. (This is no small consideration when one realizes that sucrose is used in kilogram quantities for making media.) In any case the choice of sucrose was fortuitous

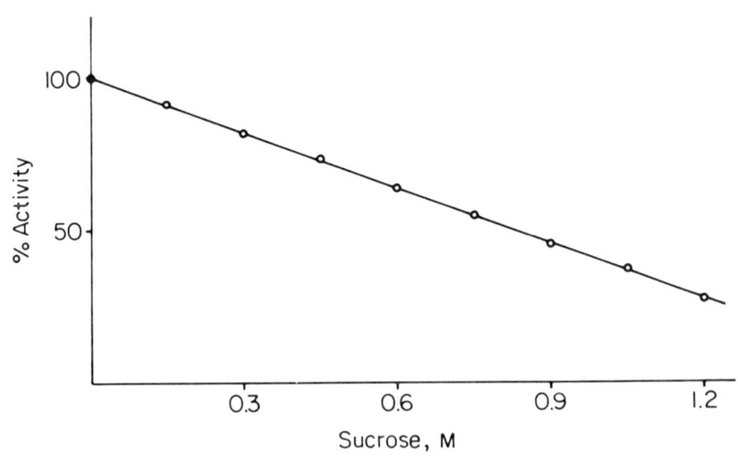


**Fig. 4.1** Inhibition of ATPase activity of rough microsomes by sucrose. Data taken from [21].

TABLE 4.3
*Apparent penetration of various substances through the microsomal membranes*

| Substance | Extramicrosomal $H_2O$ space | Intramicrosomal $H_2O$ space | Microsomal substance space | (Substance space / Intra-microsomal space) x 100 |
|---|---|---|---|---|
| | μl mg⁻¹ dry weight | | | |
| – | 1.88 | 1.48 | – | – |
| Glycerol | 1.80 | 1.41 | 1.09 | 77 |
| Glucose | 1.83 | 1.38 | 1.10 | 80 |
| Sucrose | 1.85 | 1.38 | 1.06 | 77 |
| Maltotriose | 1.93 | 1.41 | 0.96 | 68 |
| Ouabain | 1.78 | 1.43 | 1.22 | 85 |
| Inulin | 1.91 | 1.51 | 0.05 | 3 |
| Acetate | 1.75 | 1.19 | −0.04 | −3 |
| Mevalonate | 1.76 | 1.23 | 0.04 | 3 |
| G6P | 1.71 | 1.20 | 0–0.36 | 0–30 |

Microsomes were mixed with the radioactively-labelled substance and centrifuged for 2 h at 105 000 *g*. Radioactivity was then determined in the pellet. The Dextran (mol. wt. 80 000) space was considered to represent the extramicrosomal water. The total water was measured by weighing the sample before and after drying over sulphuric acid at 80°C. Intramicrosomal water was calculated as the difference between total and extramicrosomal water. 14% of the intramicrosomal water space determined in this fashion represented water of hydration (measured with $D_2O$). For more details, see [22]. G6P = glucose-6-phosphate.

for the isolation of microsomes from rat liver. Under conditions of ultracentrifugation liver microsomal vesicles are completely permeable to sucrose (and other small uncharged molecules; see Table 4.3) [22], a property which they seem to share with peroxisomes alone among the other closed organelles [23]. Thanks to this fact, sucrose causes little if any osmotic damage to microsomal vesicles. In addition, penetration by sucrose means that the equilibrium density of the vesicles is provided mainly by the membrane material and its water of hydration. In this manner a high equilibrium density advantageous in separation procedures is obtained.

High molecular weight polymers such as dextran, glycogen, and Ficoll which do not penetrate the microsomal membrane can also be used in the isolation and subfractionation of microsomes (e.g. [24]). One would expect that use of a non-penetrating solute would result in such a low equilibrium density of both rough and smooth microsomes that it would be considerably more difficult to separate these subfractions in a density gradient (see below).

*(iv) Ions in the homogenization medium and aggregation*

When considering the use of ions in the homogenization medium, it is important to remember that microsomes are very easily aggregated. The microsomal surface is highly negatively charged, rough microsomes more so than smooth [6]. This charge results in extensive adsorption of basic cytoplasmic proteins and may tend to prevent aggregation of microsomal vesicles in the absence of ions. When the negative surface charge is neutralized by cations, adsorbed protein is released but the microsomal vesicles have a markedly increased tendency to aggregate.

Microsomal aggregation is not necessarily deleterious to microsomal function and may be acceptable if no subfractionation of the total microsomes is planned. If one wishes to use the microsomes to study protein synthesis, $Mg^{2+}$ should be present in the medium in the final stages of preparation, despite the aggregation this causes (e.g. [16]). But if subfractionation is to be carried out, aggregation must be avoided. For this reason, buffers and other ions are generally not used during the preparation of microsomes. Because of the buffering capacity of proteins, the pH of a rat liver homogenate in sucrose is usually close to 7 anyway. However, in cases where the water used contains large amounts of dissolved $CO_2$, the pH may be too low. If this problem is encountered, the water may be boiled and stored in closed containers.

It is interesting to note that certain types of microsomal aggregation are reversible, whereas others are not. Spontaneous aggregation – favoured by low concentrations of sucrose, low concentrations of microsomal protein, and pelleting of the microsomes (see Table 4.2) – cannot be reversed. Thus, it is advisable to avoid concentrations of microsomal protein lower than 5 mg $ml^{-1}$ and sucrose concentrations lower than 0.25 M. If it is difficult to achieve this high a concentration of particle protein, protection against aggregation can also be achieved with colloids such as albumin [19], spermine or spermidine [25] or with higher sucrose concentrations. Other types of aggregation, such as the specific aggregation of rough microsomes by low concentrations of monovalent cations, are reversible [19].

If, because of these considerations, one wishes to achieve an ion-free medium for the homogenization procedure, great care must be taken with distilled water, glassware, and the preparation of sucrose used. In order to obtain the low concentrations of ions required for successful subfractionation of microsomes, a laboratory must have experience and special equipment, such as a glass distilling apparatus and a very effective system for washing glassware.

Of course, a totally ion-free homogenate can never be obtained because of the substantial amounts of ions which the liver itself contains. It has been shown that one-third of the $Mg^{2+}$ present in the liver is associated with the microsomal fraction [26], and this can result in an aggregation of the microsomes even if the medium used in homogenization is itself ion-free. Thus, care must be taken with the animals used in the preparation of microsomes. The water, food, and air of the animal room must be controlled to insure suitable ion concentrations in the liver.

Another observation also indicates that it is important to control such seemingly small details. It is our experience that various drugs that bind to the

ER affect the subfractionation of microsomes. For example, microsomes from rats pretreated with methylcholanthrene aggregate violently at the concentration of $Cs^+$ normally used to separate rough and smooth microsomes, and lower concentrations of $Cs^+$ must be used. In addition, proliferation of the ER of rats is often induced by the various disinfectants used in keeping animal rooms clean; and ingestion of material used to cover the bottom of the cages, such as sawdust from impregnated wood, can also cause such induction. Inductions of this sort result in microsomes with an increased tendency to aggregate.

There seems little doubt that many of the discrepancies in results reported by different laboratories arise from the use of microsomal preparations which are aggregated to different extents. Microsomal aggregation can be tested for simply by filtering a microsomal suspension through a millipore filter with holes of appropriate size and then determining protein or NADPH–cytochrome *c* reductase in the filtrate. In good non-aggregated microsomal preparations about 90% of the membranes should pass through a filter with 0.45 $\mu$m pores [19].

### *4.2.2 Centrifugation*

#### *(i) Removal of unbroken cells and large particles*

Unbroken cells and large particles are routinely removed from the homogenate by differential centrifugation for 10 or 20 minutes (depending on whether a 10% or 20% homogenate, respectively, is used) at 10 000 x *g*. This step removes unbroken cells, nuclei, mitochondria, lysosomes, and peroxisomes, as well as large fragments of the plasma membrane and of the cisternal Golgi. However, this centrifugation also sediments a significant portion of the fragments of the ER.

If fed animals are used, a considerable portion of this loss of microsomes in the 10 000 *g* pellet is caused by the presence of glycogen in the liver homogenate. The majority of this glycogen is associated with the ER both in electron micrographs of the hepatocyte [27] and physically in the homogenate [28]. As a result, 30–40% more microsomes are lost in the low-speed pellet when fed instead of starved rats are used. In addition, association of glycogen with the microsomes presents other difficulties, for example making it extremely difficult to solubilize NADPH–cytochrome *c* reductase from these membranes.

For these reasons the animals are generally starved for 20 h before sacrifice, a procedure which lowers the liver content of glycogen to approximately 1 mg $g^{-1}$ wet weight. After prolonged starvation (2 or 3 days) glycogen begins to be built up in the liver again as a result of increased gluconeogenesis. It is very important to remember that starvation itself may cause changes in the biochemistry of the rats, in particular changes in the endoplasmic reticulum. Thus, starvation is known to increase the glucose-6-phosphatase activity of microsomes [29].

### (ii) Preparation of the total microsomal fraction

The usual procedure for obtaining a total microsomal fraction from the 10 000 x $g$ supernatant is to centrifuge at 105 000 x $g$ for 60 minutes. When this centrifugation is performed in 0.25 M sucrose in an angle-head rotor, a loose fluffy layer (shown electron microscopically to contain small smooth vesicles and free ribosomes) can be seen at the top of the pellet. In addition, a small amount of microsomal vesicles remain behind in the supernatant after this procedure. Thus, if one wishes to obtain a post-microsomal supernatant free of membrane fragments, centrifugation should be carried out for a longer time, not less than two hours, and the supernatant should be drawn off, not decanted.

Because of the small size of microsomal vesicles, it is very tempting to use higher centrifugal forces to sediment them in order to decrease the time of centrifugation required. However, this temptation should be resisted. Centrifugation of microsomes in 0.25 M sucrose in an angle-head rotor at forces greater than 105 000 $g$ damages the vesicles; their permeability is increased and they are seen in electron micrographs to be oval in shape rather than round.

Recently, a method for the rapid isolation of the total rat liver microsomal fraction by low speed centrifugation has appeared [30]. In this procedure the 10 000 $g$ supernatant is adjusted to contain approximately 8 mM $CaCl_2$ and 5 mM $MgCl_2$, and the aggregated microsomes are collected by centrifuging at 1500 x $g_{max}$ for 10 minutes. For some purposes such a preparation may be quite adequate. For instance, microsomes prepared in this manner exhibit normal levels of drug metabolizing activity and of cytochromes $P_{450}$ and $b_5$ [31]. However, these microsomes must certainly be too highly aggregated for further subfractionation. In addition, it should be remembered that microsomes contain endogenous phospholipases [32,33] which may be activated by $Mg^{2+}$ and/or $Ca^{2+}$ and damage the vesicle membrane. These divalent cations may affect the properties of the membrane directly as well.

A method for isolating microsomes from the 10 000 $g$ supernatant by passage through a Sepharose 2B column rather than by centrifugation has also been developed [34]. We have no experience with this method, but it may prove to be useful in situations where an ultracentrifuge is not available or where pelleting of the total microsomal fraction must be avoided.

### (iii) Washing the microsomes

In the total microsomal pellet normally obtained, approximately 50% of the water phase is extra-microsomal and will therefore contain elements of the hepatocyte cytoplasm, the blood serum, and the erythrocyte cytoplasm (including haemoglobin). In addition, proteins – and especially hepatocyte cytoplasmic proteins, which are overwhelmingly basic – tend to adsorb strongly to the negatively charged surface of endoplasmic reticulum fragments.

Fortunately, these soluble proteins included in and adsorbed to the microsomal pellet can be removed simply by washing. By washing with sucrose one can remove the extra-microsomal water, but sucrose seems to act almost as a 'glue', promoting the adsorption of proteins to the microsomal surface. Thus, the total microsomal pellet is virtually never washed with sucrose. A relatively effective means of removing non-microsomal protein from the 105 000 *g* pellet is to wash with a salt solution containing a monovalent cation, e.g., with 0.15 M KCl or, even better, 0.5 M KCl. Care must be taken to remove as much sucrose solution from the pellet as conveniently possible before washing with salt; otherwise, the wash will not be effective. More effective washing, especially with respect to removal of haemoglobin, can be achieved with an alkaline buffer such as 0.15 M Tris at pH 8.0. In this case not only can the Tris cation compete with soluble proteins for binding to negative charges on the membrane surface, the alkaline pH also lowers the positive charge of the cytoplasmic proteins and decreases their ability to bind. Washing of the microsomal pellet with a buffer of acid pH only partially removes adsorbed proteins.

Washing of microsomes with KCl or Tris has been found to reduce their protein content by 30–35% without removing typical microsomal markers such as glucose-6-phosphatase and NADPH-cytochrome *c* reductase. It is possible that some of the protein removed by this procedure is peripheral protein which actually belongs to the microsomal membrane and not adsorbed protein. In cases where the function of a protein is known, this distinction is often easily made; thus, haemoglobin is an adsorbed and not a peripheral protein. However, in other cases it may be extremely difficult to determine whether a protein removed by washing actually belongs to the membrane or not.

In addition to possibly removing peripheral proteins, repeated washing of the total microsomal pellet will cause ribosomes to come off. It has been shown that about 50% of the ribosomes located on the ER are bound to the membrane by nascent polypeptide chains, while the remaining ribosomes are bound mainly through electrostatic forces involving $Mg^{2+}$ [35,36]. Furthermore, the small ribosomal subunit is also bound to the large subunit by $Mg^{2+}$ [35]. Repeated washing removes $Mg^{2+}$ from the microsomes and results in loss of ribosomal subunits and ribosomes.

### *(iv) Removing the contents of microsomal vesicles*

The washing procedure described above eliminates non-membrane proteins from the outside of microsomes, but leaves a more serious problem unsolved. Homogenization does not disrupt the endoplasmic reticulum randomly. Instead, this organelle seems to 'actively' pinch off at various points to form closed, semipermeable vesicles with the same inside-outside orientation as *in vivo* [18]. As a result, ribosomes are found on the outside of microsomal vesicles and the vesicles contain that which was present in the lumen of the intact endoplasmic reticulum. In addition, this pinching off process is such that if a solute to which isolated microsomes are impermeable – e.g., dextran – is included in the homogenization medium, none of this solute will be found inside the isolated microsomes.

Consequently, microsomal vesicles contain secretory proteins, other, as yet unidentified, proteins including lipoproteins, and lipids such as cholesterol and cholesterol esters. This content seriously interferes with qualitative and quantitative studies on the proteins and lipids of microsomal membranes. For instance, when the proteins of microsomes still containing luminal proteins are characterized by SDS disc gel electrophoresis, it is not possible to distinguish true membrane proteins from the trapped 'contaminant' proteins.

To date, no one has really tackled the problem of removing luminal lipids from microsomal vesicles. This may prove to be very difficult, since one would have to wash out trapped lipids – for instance, with detergent solutions – without seriously disrupting the phospholipid bilayer of the membrane.

The situation is considerably better in the case of luminal proteins. There are two major types of secretory proteins found within microsomal vesicles. The first type are proteins which upon completion are released from the ribosomes into the lumen and which are transported in free, non-membrane-bound form through the endoplasmic reticulum and the Golgi finally to be secreted from the cell [37,38]. Serum albumin belongs to this group, and rat liver microsomes contain approximately 1 mg of albumin per 20–25 mg of microsomal protein. Methods are available for removing such proteins from the microsomes (see below).

The second type of enclosed secretory proteins are glycoproteins which are transported along the ER toward the Golgi [39] in a membrane-bound form [40]. The functional explanation for this finding is that sugar moieties are added stepwise to the glycoproteins during their passage through the ER by transferases present in the membrane; and since a substrate-enzyme-acceptor complex is necessary for this addition of sugar moieties, the reaction is facilitated by having the substrate-glycoprotein bound to the membrane as well. Certain types of experiments are seriously complicated by the fact that a significant amount of secretory glycoprotein may be bound to the microsomal membranes and there is at present no means of selectively removing such glycoproteins.

Three approaches to the removal of secretory proteins such as albumin from microsomal vesicles have been applied. The first approach utilizes mechanical forces to temporarily disrupt the microsomal membrane and allow escape of the contents before the vesicle reseals. Sonication, Ultra Turrax, and French press treatment have all been used in such studies. In general, this approach is relatively ineffective in releasing the microsomal contents and may have deleterious effects on the microsomes. For instance, disruption of microsomes with mechanical forces can release loosely bound peripheral membrane proteins such as nucleoside diphosphatase [41].

In an investigation in our laboratory [42] rough microsomes were sonicated in 15 s intervals, with pauses in between for cooling, until the microsomes were reduced to 1/100 of their original mass. This procedure released no more than 3% of the total microsomal protein. At the same time, some loss of cytochromes $b_5$ and $P_{450}$ was seen and the ribosomes came off the membranes. Additional sonication caused loss of NADH- and NADPH-cytochrome *c* reductase activities and additional damage to the microsomal vesicles. Another danger inherent in sonicating microsomes is that one may hybridize vesicles and thus obscure any heterogeneities which may be present in the vesicle population, but appropriate controls eliminated this possibility in our case [43].

The effects of subjecting rough microsomes to French press treatment and sonication in the presence of both high and low concentrations of salt has also been studied [44]. These treatments preferentially released up to 50% of the secretory proteins from microsomal vesicles and pronounced changes in the sedimentation properties of the microsomes and partial ribosome detachment were seen. In addition, release of microsomal contents by these procedures was not as effective as the release achieved with low concentrations of detergent (see below).

A second method for removing the contents of microsomal vesicles has been investigated in our laboratory [45]. Microsomes were first washed with 0.15 M Tris, pH 8.0, and then resuspended in distilled water. Incubation of this suspension was carried out for 15 minutes at 30°C and then the microsomes were cooled in ice, resulting in cold-induced rupture of the vesicles, a treatment similar to that employed by Schramm *et al.* [46] for zymogen granules. The microsomes were then recollected by centrifugation and examined. It was found that this procedure removes 40–50% of the total protein present in the original microsomal pellet, including approximately 30–40% from the outer surface of the membrane and 10% from the lumen. About 14% of the original phospholipid and most of the bound ribosomes are also removed. The morphology, cytochromes, and enzymatic activities of the microsomal membranes are not significantly altered. However, permeability of the microsomal vesicles to ions is irreversibly increased by this treatment [22].

Recently, a detailed investigation on the use of low concentrations of detergents to release the microsomal content without disassembling the membrane has been carried out [44,47]. Rough microsomes were used. The most extensive studies were carried out with sodium deoxycholate, but Brij 35, Triton X–100, sodium dodecyl sulphate, and hexadecyltrimethylammonium bromide can all be used. The non-polar detergents are said to be most effective. Short-term labelling with radioactive leucine and glucosamine were used to follow the release of secretory proteins and glycoproteins, respectively.

When the microsomes were incubated for 30 minutes at 0°C with low concentrations of deoxycholate (0.025–0.05%), release of the microsomal contents was achieved. Microsomal β-glucuronidase and nucleoside diphosphatase are also quantitatively released (O. Nilsson, G. Dallner, unpublished observations), but other components of the membrane were not removed. Thus, about 50% of the rapidly labelled protein, 15% of the total protein, 3% of the phospholipids, 5% of the NADPH-cytochrome *c* reductase, 6% of the NADH-cytochrome *c* reductase, and 2% of the cytochrome $b_5$ were removed by this treatment. Slowly labelled proteins, which presumably belong to the membrane, were released to about the same extent as the phospholipids and microsomal markers. Rapidly labelled glycoproteins, presumably secretory, were partially removed in such a way as to suggest that some of these glycoproteins are incorporated into or more tightly associated with the microsomal membrane than the bulk of secretory proteins, a finding similar to that of Redman and Cherian [40].

Electron microscopy of these detergent-treated microsomes revealed that the average vesicle size was decreased about 30%, but the vesicles looked intact and the ribosomes were still bound to the membrane. Centrifugation in sucrose gradients showed no change in the sedimentation properties of these microsomes. However, sedimentation in density gradients of dextran or Ficoll led to the conclusion that the detergent-treated microsomes were permeable to these macromolecules.

The concentration of detergent required for release of the microsomal content without serious disruption of the membrane was found to be highly dependent on the ionic strength of the medium and on the amount of microsomal phospholipid present. Increases in the ionic strength decreased the concentration of detergent required. Thus, in a medium containing 0.5 M KCl 0.025% deoxycholate was as effective as was 0.05% in a medium of much lower ionic strength. In addition, it was found that lysis of the membrane began when microsomes had bound more than 1 molecule of deoxycholate per 40 molecules of phospholipid. In media of low ionic strength breakdown of the membrane began at 0.098% detergent and complete solubilization of the rough microsomes was achieved with concentrations of deoxycholate greater than 0.392%.

Sabatini and his co-workers [44,47] suggested on the basis of these experiments that low concentrations of detergent lead to the formation of pores or openings in the microsomal membrane that are too small to be detected electron microscopically. Furthermore, they found that the formation of these holes was reversible upon removal of detergent. This procedure seems to be a convenient and effective way to remove the contents of microsomal vesicles, although it is not clear exactly how much of these contents can not be released in this manner. Also, the possibility of getting macromolecules into microsomes offers new approaches to the study of the architecture of the microsomal membrane. For example, one can now use trypsin to digest proteins localized on the inner surface of the microsomal membrane or one can use lactoperoxidase, $H_2O_2$, and iodide to label these proteins [48]. However, it should not be forgotten that treatment of microsomes with enough detergent to punch holes in the membrane may also cause serious disruption of certain aspects of the membrane architecture too subtle to be detected by the approaches used by Sabatini and his co-workers in their study.

In conclusion, it can be seen that all of the presently available methods for removing the content of microsomal vesicles without seriously damaging the membrane have their limitations. However, the luminal proteins greatly interfere with certain types of investigations and must be removed as best one can.

### *(v) Purity of the total microsomal fraction*

Historically, because the microsomal fraction was being isolated and studied before it was known from which morphological entities in the cell it originates, the definition of this fraction has been an operational one. Whereas the mitochondrial fraction consists chiefly of mitochondria and the nuclear fraction consists chiefly of nuclei, the microsomal fraction simply consists of those elements of a tissue homogenate which can be sedimented by ultracentrifugation from the mitochondrial (10 000 *g*) supernatant. Thus, since virtually all of the membrane elements present in the mitochondrial supernatant are pelleted by ultracentrifugation, the purity of the microsomal fraction is determined by the purity of the 10 000 *g* supernatant.

As discussed above, homogenization of rat liver by an appropriate procedure results in a microsomal fraction that consists chiefly of fragments of the ER. However, this is not necessarily the case for other tissues. Of course, rough microsomes will originate from the ribosome-coated sections of the ER in all tissues, but smooth microsomes from pancreas [49] and testis [50] originate mainly from the Golgi apparatus; the smooth microsomes from kidney [51], Ehrlich ascites tumour cells [52], and fibrocytes [53] contain parts of the plasma membranes; and this fraction from other tissues may contain still other elements.

Contamination of the total microsomal fraction from rat liver by nuclei is thought to be negligible. Intact nuclei are much too large to be found in the mitochondrial supernatant. Pieces of the nuclear membrane, for which no good marker is known, may be present in this supernatant and thus may contaminate the microsomes. However, it has been observed that if nuclei are broken, DNA is released into the homogenization medium and causes formation of a viscous gel from which it is impossible to isolate microsomes. For this reason it seems unlikely that the microsomal fraction is seriously contaminated by nuclear membranes.

Nor are mitochondria a serious contaminant of the microsomal fraction from rat liver. Our preparations contain no more than 0.7% of the total cytochrome oxidase activity (a marker for intact mitochondria and inner mitochondrial membranes) of the homogenate. However, the use of a tissue press and/or vigorous homogenization to disrupt the liver can strip the outer membrane from mitochondria [9,13], and these outer membranes can end up in the microsomal fraction. Even when reasonable homogenization procedures are used, the microsomes can be significantly contaminated with mitochondrial outer membranes. Thus, Amar-Costesec and his co-workers [17] found 20% of the total monoamine oxidase activity (a marker for these membranes) of the homogenate in their microsomal fraction. It would seem advisable to measure markers for both the outer and inner mitochondrial membranes in one's microsomal preparation, even though monoamine oxidase activity is technically difficult to determine and is located to a certain extent on the ER itself. It is reassuring that electron micrographs of rat liver fractions reveal only few mitochondria that have been stripped of their outer membranes and support the conclusion that microsomes are contaminated with these outer membranes to only a limited degree.

Rat liver microsomes seem to be significantly contaminated by lysosomes and peroxisomes. Of the total acid phosphatase activity (a lysosomal hydrolase) of the homogenate, we found 11% in our microsomes and Amar-Costesec *et al.* found 17% [17]. Adelman *et al.* [15,16] found 9% of the total homogenate activity of the peroxisomal enzyme catalase in their rough microsomes. Certain observations make it difficult to assess the reliability of these figures. Many hydrolases may dephosphorylate $\beta$-glycerophosphate, the substrate used for measuring acid phosphatase activity, at a significant rate even at pH values around 5. As a result, acid phosphatase activity inhibitable by tartrate (which is thought to be a specific inhibitor of the lysosomal enzyme) [54] is used as a marker for lysosomes. On the other hand, catalase activity seems to be localized in the soluble cytoplasm as well as in the peroxisomes [55]. The most serious difficulty is that when lysosomes and peroxisomes are damaged, they can lose their contents; and since no markers are known for the membranes of these organelles, these membranes may contaminate the microsomal fraction undetected. Fortunately, as judged by electron microscopy and lysosomal latency, mild homogenization of liver does not seem to extensively disrupt lysosomes and peroxisomes [56].

It is also impossible at present to determine exactly the contamination of microsomes by vesicles originating from the plasma membrane; but it is clear once again in this case that the extent of this contamination depends very much on the homogenization procedure used. The regions of the hepatocyte plasma membrane in the vicinity of the bile canaliculi are apparently strengthened by the specialized areas involved in intercellular junctions. These regions are not easily fragmented and after mild homogenization they are found in the nuclear pellet. However, other regions of the plasma membrane are probably broken into relatively small vesicles even by mild homogenization, and vigorous homogenization can also break up the areas around the bile canaliculi (see [57] for a review). Unfortunately, there do not seem to be any good markers for the hepatocyte plasma membrane available. In fact, it is possible that no single enzymatic activity or component will prove to be a good marker for all areas of this membrane, since the different specialized regions of the hepatocyte plasma membrane are apparently different biochemically [57–59]. It is at least clear that the most widely used marker for the plasma membrane, 5′-nucleotidase, is not an ideal one, because this enzyme is also known to be present on the

ER [60]. The Louvain group [17] finds 50% of the total 5′-nucleotidase of the homogenate in their microsomal fraction, and we find 29%. Cholesterol which is sometimes used as a marker for the plasma membrane [57,61,62], is also present in the ER. 15–30% of the total phospholipids of the plasma membrane fraction from rat liver is sphingomyelin [63]; whereas the corresponding figure for the microsomal fraction is about 4% [64]. In addition, the specific content of gangliosides is 12 times lower in the microsomes than in the plasma membrane fraction [65]. This last observation suggests that contamination of the microsomal fraction by fragments of the plasma membrane is minimal.

The situation is somewhat different with regards to contamination of microsomes by the Golgi apparatus. Morphologically, this organelle consists not only of large cisternal elements but also of small peripheral vesicles whose size and density closely resemble those of fragments from the smooth ER [66]. In addition, the large cisternal components are maintained intact only by very gentle homogenization with a pestle speed of 100 revs/min [11]; so that these structures are fragmented by the homogenization procedures generally used in the isolation of microsomes. As a result of these considerations, total microsomal preparations contain 40–50% of the Golgi fragments present in the original homogenate. At present the only means of obtaining microsomes reasonably free of Golgi is to prepare the microsomal fraction from the liver of rats that have been subjected to acute alcohol intoxication [66]. The Golgi vesicles in this fraction are filled with very low density lipoproteins, so that they can be flotated away from the microsomal vesicles in a sucrose density gradient. This procedure removes 80–90% of the UDP-galactosyl-transferase activity, a Golgi marker, from the microsomal fraction. The possible deleterious effects on the microsomal membranes of subjecting rats to such high doses of alcohol have not yet been investigated.

Free ribosomes are also present in the total microsomal fraction from rat liver. The function and organization of these structures is still not clear: they are probably devoted to the synthesis of specific (e.g., some intracellular) proteins [e.g. 67,68]; but free ribosomes may function as a freely exchangeable pool for the ribosomes bound to the endoplasmic reticulum [69] or may even be artefacts [67]. Free ribosomes have such a small sedimentation velocity that only about half of them are sedimented by the 105 000 *g* centrifugation in 0.25 M sucrose generally used to collect microsomes. Free ribosomes can be removed from the microsomal fraction either by sedimenting the microsomes onto a cushion of 2 M sucrose, in which case only the free ribosomes pass through the cushion to the bottom of the tube because of their high density [70]; or alternatively, by carrying out isokinetic centrifugation of the microsomes on a 5–20% gradient of sucrose, in which case the free ribosomes are left behind in the supernatant because of their low sedimentation velocity [19].

After washing with 0.15 M KCl or with 0.15 M Tris, pH 8, microsomes from rat liver are virtually uncontaminated by components of the soluble cytoplasm. Such preparations contain no detectable activity of lactate dehydrogenase or of triose phosphate dehydrogenase, two soluble enzymes.

In conclusion, it is interesting to consider the computations carried out by the Louvain group [24] concerning contamination of their microsomal fraction, even though – in part because of the considerations discussed above – there are difficulties with such computations. These investigators calculate that 7–8% of their total microsomal protein originates from the plasma membrane, 6% from intact mitochondria, 4–5% from the Golgi complex, 3% from outer mitochondrial membranes, 1% from lysosomes, 1% from peroxisomes, and the remainder – or no more than 77% – from the ER. For reasons already explained, we think that our total microsomal fraction is less contaminated than this.

## 4.3 Subfractionation of the total microsomal fraction into rough and smooth microsomes

### *4.3.1 Separation of rough and smooth microsomes*

As is well known, the ER of the rat hepatocyte is morphologically heterogeneous. In rats which have received no special treatment about 60% of the surface area of this membrane network is dotted with ribosomes, while the rest of the surface area is bare [4]. Initial attempts to subfractionate microsomes were designed to separate these two morphologically distinct regions, i.e., to obtain rough and smooth microsomes. More detailed subfractionation schemes have since revealed that the endoplasmic reticulum is also biochemically heterogeneous in certain respects (see below).

Unlike other organelles, the lamellar and cisternal structures of the ER cannot be isolated intact [18]. Homogenization procedures disrupt this organelle into small vesicles. The size distribution of rough and smooth microsomes overlap seriously, both populations having vesicle diameters of approximately 150–300 nm [71]. Thus, effective separation of these two subfractions based on their sizes using rate differential centrifugation is not possible. In addition, the difference in the density distributions of rough and smooth microsomal vesicles is small [71], so that separating them effectively with density gradient centrifugation requires careful control of the conditions employed.

Because, as mentioned above, sedimenting microsomes increases their tendency to aggregate, rough and smooth microsomes are isolated directly from the post-mitochondrial supernatant without first collecting the total microsomal fraction. Care should be taken, in accordance with the considerations described above, to minimize the aggregation of particles in this supernatant. In particular, the concentration of microsomal protein should be maintained at 4–5 mg ml$^{-1}$. The degree of microsomal aggregation in the post-mitochondrial supernatant can

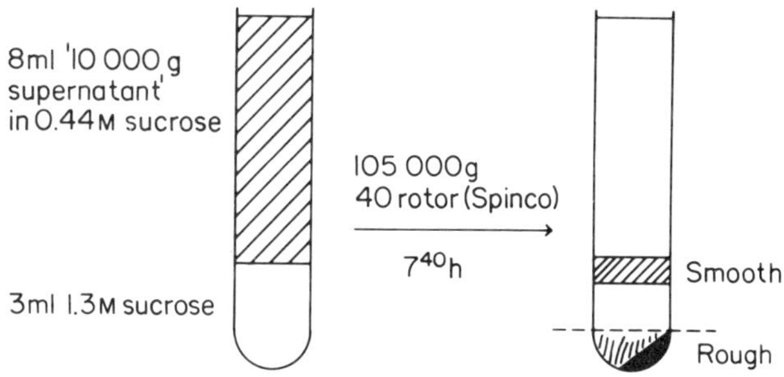


Fig. 4.2 Schematic representation of the modified Rothschild procedure for preparing rough and smooth microsomes.

be ascertained by the simple Millipore filter test already mentioned or by using light scattering techniques.

Initial attempts to isolate rough and smooth microsomes were carried out using rate differential centrifugation [72]. The difference in the average sedimentation rates of these two subfractions (which are influenced by a number of properties of which size is the most important) is sufficient to allow the preparation of fractions enriched in rough or smooth microsomes. However, these fractions are so highly cross-contaminated that the method is not a useful one.

Rothschild [73] then attempted to prepare rough and smooth microsomes using discontinuous gradient centrifugation. A modification of this procedure is illustrated in Fig. 4.2. Smooth microsomal vesicles have an equilibrium density in sucrose ranging from 1.06 to 1.18; while the equilibrium density of most of the rough microsomes in sucrose is greater than 1.20 [71,73]. The 10 000 *g* supernatant in 0.15 M sucrose was layered over 1.31 M sucrose and then centrifuged at 105 000 *g*. If this centrifugation was continued until all particles had reached their final position in the pellet or at the interface, i.e., for about 10 hours, too many smooth vesicles sedimented together with the rough; and the separation of rough and smooth microsomes was not acceptable. However, if centrifugation was terminated after 6–8 h, a lot of unsedimented particles remained in the 1.31 M sucrose layer; and the smooth microsomes at the interface between the sucrose layers and the rough microsomes in the pellet showed little cross-contamination. Free ribosomes were found partly in the smooth microsomal fraction, from which they could be removed by centrifugation.

The time of centrifugation at 105 000 *g* is critical in this approach. Using a Spinco 40 rotor we find 7 h 40 min to be optimal [74]. Rothschild reported that the concentration of sucrose in the upper layer of this discontinuous gradient is also critical; that if concentrations of 0.25 M or higher are used, separation of rough and smooth microsomes is less effective [73]. However, in our experience 0.44 M sucrose can be used in the upper layer and an effective separation of the two subfractions is still achieved [74]. Use of the higher sucrose concentration helps minimize microsomal aggregation which would interfere with the subfractionation.

It is important to control experimental details carefully if the Rothschild method is to be used successfully. For instance, the 1.3 M sucrose solution should be made up to volume at the temperature at which it will be used (usually 4°C, since this is the temperature at which centrifuges are generally run); making the solution at room temperature and then cooling it will cause significant changes in its density. At the same time centrifuges used in this procedure should have good temperature control. An increase in the temperature from 5 to 10°C during the run will decrease the density of 1.3 M sucrose by 0.0019 $gm^{-1}$ and decrease its viscosity by 20%. Since this large decrease in viscosity will increase the sedimentation coefficients of the microsomal particles by 20%, such relatively small temperature changes will affect the separation obtained. For the same reason it is best to use pre-cooled rotors for the centrifugation.

We have found it practical to begin isolation of rough and smooth microsomes according to the Rothschild procedure late in the evening, so that centrifugation of the discontinuous gradient is begun at about 1 am. Thus, when we arrive the next morning, the fractions can be collected and further subfractionated and important enzymes measured before the end of the day. In this way we minimize the time elapsed between homogenization of the liver and performance of the experiment and minimize time-dependent changes in the microsomes such as loss of enzyme activity.

Rothschild [73] also tried to separate rough and smooth microsomes by isopycnic gradient centrifugation on a continuous sucrose gradient, but there was considerable cross-contamination between the subfractions obtained. The discontinuous gradient procedure, although somewhat time-consuming, is the best available method for isolating rough and smooth microsomal fractions suitable for further subfractionation (see below). However, it should be remembered that this procedure is not quantitative. Particles present in the 1.3 M sucrose layer after centrifugation are discarded. These particles represent chiefly smooth microsomes of higher-than-average density and rough microsomes with few attached ribosomes and thus of lower density than the other rough vesicles. These fractions may originate from regions of the ER that are biochemically and functionally specialized. It is known, for instance, that rough microsomes with few attached ribosomes contain especially high levels of NADPH–cytochrome *c* reductase and of cytochrome $P_{450}$ [75].

Discontinuous sucrose gradients have also been used by Adelman and his co-workers [15,16] in a procedure designed to achieve a good recovery of rough microsomes. The RNA in the ribosomes bound to the endoplasmic reticulum represents about 60% of the total rat liver RNA [76]. In most procedures for isolating microsomes a large fraction of the total RNA sediments with the 'nuclear' and mitochondrial pellets, and much of this represents loss of rough microsomes [76–78]. The object of their method is to decrease this loss without worrying about possible increased contamination of the smooth microsomal fraction.

These investigators disrupt the liver with a tissue press and then homogenize it vigorously in 1.0 M sucrose. An equal volume of 2.5 M sucrose is added to the homogenate and this preparation is centrifuged at about 100 000 x *g* for 45 minutes. Only the nuclei, relatively uncontaminated by rough microsomes, are sedimented by this procedure. Other organelles either float, are isopycnic, or sediment more slowly.

The supernatant from this first step is diluted with water and centrifuged to obtain the mitochondrial pellet. This pellet is then washed twice with relatively ion-free sucrose solutions in order to remove trapped rough microsomes. Washing with a medium containing ions is much less effective, and washing more than twice does not remove any additional RNA from the mitochondria.

The combined post-mitochondrial supernatants are then placed over a discontinuous sucrose gradient consisting of two layers. The upper layer contains 1.6 M sucrose and low concentrations of Tris buffer, KCl, and $MgCl_2$; while the lower layer contains 2.0 M sucrose and the same ions in concentrations that are five times higher. This gradient is then centrifuged for 20 h at about 200 000 x *g* after which smooth microsomes are recovered at the interface between the supernatant and the layer of 1.6 M sucrose; rough microsomes are found at the interface between the layers of 1.6 M and 2.0 M sucrose; while free ribosomes are found in the pellet. Considerable amounts of free ribosomes are still present in the rough microsomal fraction as well, but these can be removed simply by diluting the rough microsomes with an ionic medium and sedimenting them at a centrifugal force that leaves free ribosomes behind in the supernatant. Also, there is an appreciable loss of rough microsomes to the smooth microsomal fraction, but this is the price that must be paid for decreased mitochondrial contamination of the rough microsomal fraction itself.

This procedure developed by Adelman and his co-workers is a modification of previous subfractionation schemes and yields a fraction containing 50% of the total rough microsomes. Actually, this yield does not seem to be much, if any, better than the yields of rough microsomes obtained by other methods (see above and below). This rough microsomal fraction has not been systematically examined for possible contamination by other organelles.

The great advantage of this subfractionation scheme is that it yields rough microsomes suitable for studies of *in vitro* protein synthesis [15,16]. However, this approach also has certain disadvantages. The vigorous homogenization (which may cause heat-denaturation of proteins) and the long centrifugation time probably result in a considerable loss of the activities of various microsomal enzymes. Furthermore, the vigorous homogenization and the washing of the mitochondrial pellet almost certainly increase contamination of the smooth microsomes with mitochrondrial membranes, pieces of the plasma membranes, and fragments of other organelles. Adelman *et al.* [15,16]

themselves point out that their smooth fraction is also contaminated with rough vesicles. Thus, the smooth microsomal fraction obtained by this procedure can not be used for most purposes. And finally, the presence of ions in the discontinuous gradient results in an aggregated rough microsomal fraction which is not suitable for further subfractionation.

The most widely used method for preparing rough and smooth microsomes is based on the finding that monovalent cations, and in particular $Cs^+$, can selectively aggregate rough microsomes [79,80]. Fig. 4.3 illustrates that rough microsomes have an appreciably higher affinity for $Cs^+$ ions than do smooth microsomes, free ribosomes, or rough microsomes stripped of their ribosomes [81]. Apparently, binding of limited amounts of $Cs^+$ by the rough microsomes decreases their negative surface charge, thereby diminishing the repelling forces between the vesicles and causing them to aggregate. This aggregation increases the mean radius of rough microsomal particles at least two- to three-fold and thus increases their sedimentation velocity four- to nine-fold [19].

We collect the 10 000 *g* supernatant in the usual manner and then add a small aliquot of a concentrated CsCl solution to give a final concentration of

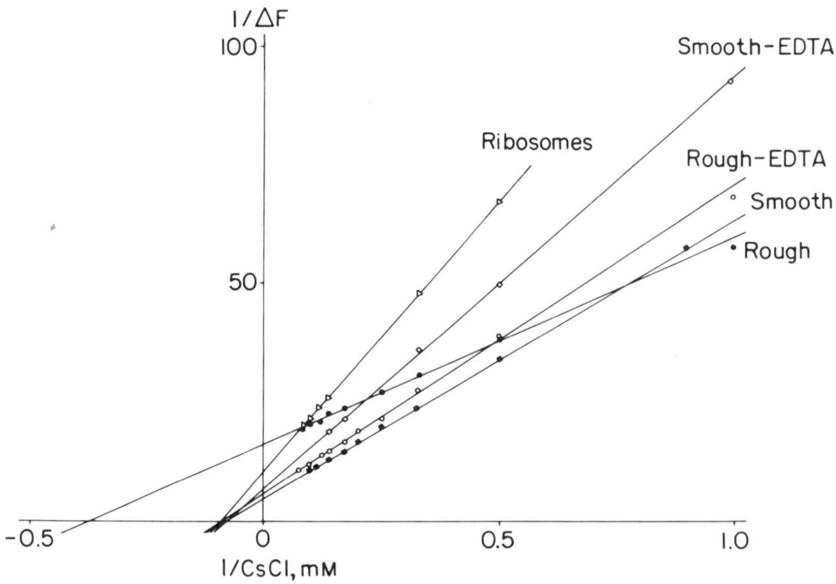


**Fig. 4.3** Affinity of microsomal membranes and free ribosomes for CsCl. Microsomal membranes or free ribosomes (1 mg $ml^{-1}$ in 0.25 M sucrose) were incubated with 1.5 μM ANS (1-anilinonaphthalene-8-sulphonic acid) and different concentrations of CsCl. The reciprocals of the resulting fluorescent increase (which is a function of the $Cs^+$ bound) were plotted against the reciprocals of CsCl concentration. For more details, see [81].

15 mM CsCl. This preparation is layered over a solution containing 1.3 M sucrose and 15 mM CsCl and centrifuged for 90 min at 102 000 *g* in a Spinco 40.2 rotor. The smooth microsomal fraction is recovered from the boundary between the two layers, while rough microsomes are found in the pellet. Free ribosomes and some soluble cytoplasmic protein are present in the smooth microsomal fraction obtained, but these contaminants can be removed by appropriate procedures. Centrifugation of the discontinuous gradient for longer than 2 h results in increased aggregation of smooth microsomes by $Cs^+$ and, consequently, increased amounts of smooth vesicles in the pellet.

Successful use of this procedure for isolating rough and smooth microsomes requires that the investigator exercise the various precautions discussed above for preventing microsomal aggregation in the post-mitochondrial supernatant. However, these precautions may not always be enough. For instance, even if there are not enough ions in this supernatant to cause aggregation by themselves, there may be enough ions present so that the addition of 15 mM CsCl aggregates both smooth and rough microsomes. Thus, even under the best of circumstances, this procedure occasionally fails to produce good rough and smooth microsomal fractions.

In order to improve this situation we have developed a modification of our original isolation procedure which is illustrated in Fig. 4.4. In this modification no CsCl is added to the 10 000 *g* supernatant. Instead, this supernatant is layered over a solution containing 0.6 M sucrose and 15 mM CsCl, which is in turn layered over the usual solution of 1.3 M sucrose – 15 mM CsCl. Centrifugation is then performed in the normal manner. Use of this middle layer in the discontinuous gradient results in removal of most of the non-particulate components of the supernatant from the microsomes before they are exposed to $Cs^+$ and apparently results in decreased aggregation of the smooth microsomes by $Cs^+$. To date, this modification has effectively separated rough and smooth microsomes every time we have employed it.

Other rotors, including swinging bucket rotors, may also be used in this

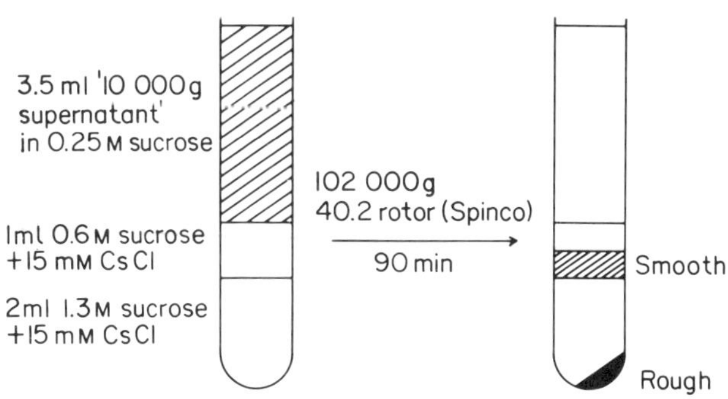


**Fig. 4.4** Schematic representation of the use of a $Cs^+$-containing discontinuous sucrose gradient for preparing rough and smooth microsomes.

procedure. However, the use of angle-head rotors with a relatively large (34° or more) tube angle of inclination to the axis of rotation, such as the Spinco 40.2, is more effective. Convections set up in such angle-head rotors decrease the time required for centrifugation without mixing the components in the interface layer and the pellet [78]. If a swinging bucket rotor is used, its tubes should be relatively narrow, since convection in wide tubes is too small to give an effective separation of rough and smooth microsomes even at large centrifugal forces.

This subfractionation of microsomes on a $Cs^+$-containing discontinuous sucrose gradient is widely used because of its simplicity, the relatively short time it requires, and the rather limited cross contamination of the rough and smooth microsomal fractions obtained. However, in certain types of investigations – for instance, where further subfractionation is to be attempted – aggregation of the rough microsomes is undesirable. In such cases it may be advisable to use the subfractionation scheme developed by Rothschild (see above). Alternatively, one might try to reverse the $Cs^+$-aggregation of rough microsomes prepared with our procedure by washing these microsomes. We have indications that the aggregation can be reversed in this manner, but we have not determined the best conditions for such washing. In addition, treatment of $Cs^+$-aggregated rough microsomes with EDTA removes their ribosomes and thus reverses the aggregation [81].

We would like to end this section of our discussion with a warning. If the procedures described here for preparing rough and smooth microsomes are applied to organs other than the liver or to species other than the rat, there is absolutely no justification for assuming that they will be effective. For example, the use of $Cs^+$-containing discontinuous gradients works with mouse liver [82] and with rat lung [83], but does not separate rough and smooth microsomes from rat kidney [84]. In addition, use of a different rotor in our procedure may change the nature of the fractions obtained.

In conclusion, then, when using a discontinuous density gradient consisting of two layers, distinct fractions will always be obtained at the interface and at the bottom of the tube. It is tempting to assume that these correspond to rough and smooth microsomes, respectively, but this assumption is not valid when a new tissue or modification of the reported procedures is used. In such cases subfractionation of the microsomes into rough and smooth vesicles must be verified by using at least some of the criteria discussed below.

### *4.3.2 Morphological characterization of rough and smooth microsomes*

It would seem at first thought that it should be relatively easy to distinguish between the rough and smooth microsomal fractions with an electron microscope. After all, procedures for obtaining these fractions were designed to separate ribosome-coated vesicles from other vesicles in the post-mitochondrial supernatant. And as a result, very many more ribosome-dotted profiles can be seen in electron micrographs of rough microsomes than of smooth.

However, it is difficult to quantitate cross-contamination between these two fractions morphologically. In the first place both rough and smooth microsomes are contaminated to a certain extent with free ribosomes. When sections are observed in the electron microsocope, these free ribosomes will always be between vesicles and may lie close enough to the membrane to be identified as bound. In this manner smooth vesicles may be mistaken for rough. This problem can be alleviated but not entirely eliminated by washing free ribosomes away from the microsomal vesicles before carrying out morphological studies. In the second place there are a substantial number of rough vesicles that bear only a few ribosomes on their surface. Most cross-sections through such vesicles will not reveal their ribosomes, so that rough microsomes can also be mistaken for smooth.

Finally, there are all of the well-known difficulties involved in using the electron microscope to evaluate quantitatively the purity of subcellular fractions. Perhaps the most serious of these is obtaining a representative sample. For instance, it is known that the morphological appearance at different levels in a pellet is not always the same. Rough microsomal pellets contain predominantly vesicles with many ribosomes at the bottom and vesicles with few or no ribosomes at the top. How does one choose a representative sample in such a case? Improved methods for obtaining a representative sample for electron microscopy are being developed using a Millipore filter technique [85].

With regards to other morphological properties such as vesicle size and fine structure of the membrane there are no useful differences between rough and smooth microsomes. Thus, the electron microscope can only be used semiquantitatively in determining the extent of cross-contamination between these fractions. Nonetheless, because the presence or absence of bound ribosomes is by definition the difference between rough and smooth microsomes, it would seem advisable and useful to characterize these fractions morphologically when they are prepared from a new tissue or using a new procedure.

### *4.3.3 Chemical characterization*

In order to get around the difficulties involved in using morphological criteria to characterize rough and smooth microsomes the content of RNA in these two fractions is often determined. Unfortunately, this approach, although very useful, also has its limitations.

In the first place measurement of RNA in rough and smooth microsomes is technically difficult. A method originally employed was to extract the RNA with hot trichloroacetic acid or perchloric acid and then to measure the optical density of this extract at 260 nm [86]. This method is convenient, but it is seriously interfered with by phospholipids, fatty acids, and other substances present in microsomes in appreciable quantities and is therefore not reliable.

The method now almost universally used for determining RNA in tissue fractions is based on reaction of the ribose it contains with orcinol [87]. The product of this reaction is coloured and thus can easily be quantitated. Reliable results can be obtained with this method only if certain pitfalls are recognized and avoided. The most serious problem is that orcinol also reacts with other carbohydrates such as sucrose and glycogen. Since rough and smooth microsomal fractions normally contain large amounts of sucrose, they must be thoroughly washed before application of the orcinol method. This wash is conveniently achieved by using a cold solution of trichloroacetic acid to precipitate (among other things) the RNA, centrifuging down this precipitate and washing the pellet several times with cold trichloroacetic acid and then water. If after this procedure significant amounts of sucrose still remain in the pellet, the reaction product with orcinol will be brown instead of green. In addition, liver microsomal fractions, even from starved animals, contain appreciable amounts of associated glycogen. This glycogen is also precipitated when trichloroacetic acid is added to the microsomes. Thus, the final stage in washing of the pellet should be two washes with water – one to remove the remaining trichloroacetic acid and the second to remove glycogen.

A second limitation involved in using RNA content to characterize rough and smooth microsomes is that the number and density of the ribosomes attached to the ER can be affected by the physiological and pharmacological state of the rats used. Previously, we have found much higher ratios of RNA to phospholipid in our rough microsomes than those reported by other investigators. However, we have recently moved to new facilities, including a new animal room with a more controlled environment; and we have begun using rats raised under more carefully controlled conditions. As a result, we now find RNA to phospholipid ratios in our rough microsomes that are only 50% of what we found before. All indications are that this decrease is not due to an increased contamination of our rough microsomes with smooth, but rather to a change in the nature of the rough microsomes themselves.

The last and most serious difficulty involved in using RNA content to characterize the rough and smooth microsomal fractions is that all of the RNA present is not contained in bound ribosomes. In particular, this is a problem with the smooth microsomal fraction. The RNA found in smooth microsomes may represent contaminating rough microsomes, free ribosomes, adsorbed *t*RNA, or 'membrane RNA' [88,89]. Consequently, the presence of a certain amount of RNA in the smooth microsomes is considered acceptable. We find an RNA: phospholipid ration [mg/mg] of 0.52 in our rough microsomes and of 0.11 in our smooth microsomes [21]. This five-fold difference is taken to indicate an effective separation of rough and smooth vesicles; but once again the exact extent of cross-contamination can not be determined.

It seems appropriate to comment here on the growing practice, which we heartily embrace, of relating the amount of a substance or the activity of an enzyme in microsomes to the phospholipid content of the preparation rather than to the protein content. We routinely express specific activities and specific contents on a phospholipid basis. The reason for this can be found in the foregoing discussion on adsorbed and luminal protein in microsomal fractions, where it was pointed out that as much as 40–50% of the protein of the original total microsomal pellet does not in fact belong to the membrane of the ER. Adsorbed and luminal proteins are removed to different extents by the different procedures used in preparing microsomal fractions, whereas the phospholipid content of the microsomal membranes remains essentially unchanged. Indeed, for similar reasons it might be advisable in studies with all sorts of biological membranes to express specific activities and specific contents on a phospholipid basis.

In addition to RNA other aspects of the chemical composition of rough and smooth microsomes have been investigated. Thus, after the removal of adsorbed proteins, the ratio of protein:phospholipid (mg/mg) is 3.5 in rough microsomes and 3.1 in smooth microsomes [21]. Even this small difference is partially due to the presence of substantial amounts of ribosomal protein in the rough fraction rather than to a quantitative difference in the protein content of rough and smooth membranes. Furthermore, studies using SDS disc gel electrophoresis have not revealed any differences in the kinds of proteins found in rough and smooth microsomes (e.g. [90]). Differences have been reported in the sugar content of the glycoproteins of these two fractions: the combined glycoproteins of rough microsomes contain more mannose, while those of smooth microsomes contain more galactose and sialic acid [91,92]. However, this is once again only a quantitative difference. And since it requires a week's work to prepare derivatives of these sugars and quantitate them by gas chromatography, it is not practical to carry out such assays routinely.

As can be appreciated simply by inverting the ratios given above, the ratios of phospholipid to protein in rough and smooth microsomes are also about the same. In addition, the phospholipids of both of these fractions consist of approximately 54% phosphatidylcholine, 25% phosphatidylethanolamine, 6% phosphatidylserine, 4% phosphatidylinositol, and 5% sphingomyelin [45]. The fatty acid compositions of rough and smooth microsomal phospholipids are very similar [93]. It has been reported that smooth microsomes contain twice as much cholesterol [45] and about 1.4 times as much vitamin K [94] on a phospholipid basis as rough microsomes.

### *4.3.4 Enzymatic characterization*

To date, no enzyme activity associated with the membrane of the ER has been shown to be localized exclusively in the rough or the smooth microsomes. For a while it was believed that GDP-mannosyl transferase was present only in rough microsomes [40,95,96], but this observation turned out to be an artefact. GDP-mannosyl transferase could not be measured in the smooth microsomes because of α-mannosidase activity associated with the Golgi fragments that contaminate this fraction [97]. As shown in Fig. 4.5, after removal of this Golgi

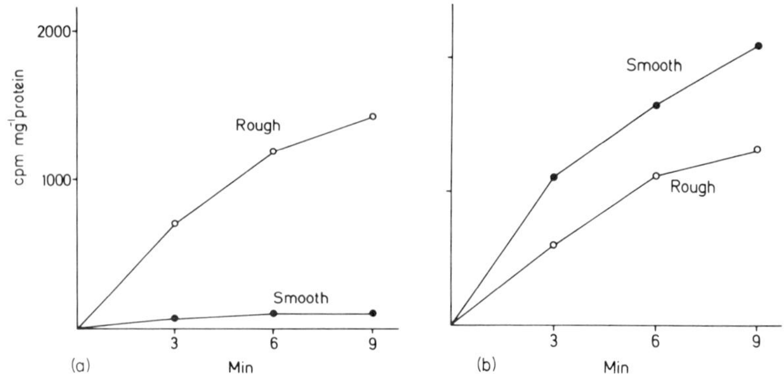


**Fig. 4.5** Distribution of GDP-mannosyl transferase (activity expressed as Cpm $mg^{-1}$ protein) in rough and smooth microsomes.
(a) Activities of rough and smooth microsomes prepared with a $Cs^+$-containing discontinuous sucrose gradient.
(b) Rough and smooth microsomes were prepared from ethanol-treated rats with a $Cs^+$-containing discontinuous sucrose gradient, after which smooth microsomes were flotated to remove contaminating Golgi fragments.

contamination by flotation, smooth microsomes demonstrate a specific activity of GDP-mannosyl transferase which is about the same as that found in rough microsomes. In general, then, claims for a qualitative difference in the enzyme activities localized in rough and smooth microsomes should be regarded with suspicion on the basis of what is already known.

On the other hand, numerous quantitative differences in the enzyme activities found in rough and smooth microsomes have been reported. The specific activities of NADPH-cytochrome *c* reductase, NADH-cytochrome *c* reductase, cytochrome $b_5$, aminopyrine demethylase and *p*-nitrophenyl phosphatase are approximately the same in rough and smooth microsomes; whereas the specific activities of glucose-6-phosphatase, IDPase, and cytochrome $P_{450}$ are somewhat higher in rough microsomes than in smooth [21]. The specific activity of glycerophosphate acyltransferase is about five times as high in smooth microsomes as in rough [98].

Even these quantitative differences must be questioned, especially in light of the fact that different investigators do not agree about some of them. Such discrepancies in the literature may be partly explained by different degrees of cross-contamination in the rough and smooth microsomal fractions used. Also, for the reasons already presented, expressing specific enzyme activity on a protein basis can give varying results for essentially the same microsomal preparation.

In addition, it seems likely that the problems involved in assaying microsomal enzymes will also contribute to these discrepancies. Among the most serious difficulties is one encountered with many membrane enzymes, namely, that maximal enzyme activity can be obtained only in the presence of a detergent. In some cases the detergent may stimulate enzyme activity simply by removing permeability barriers which prevent free access of substrate molecules to the active site, but in many cases the effect of the detergent is not yet understood.

For certain microsomal enzymes the concentration of detergent required for maximal activity is not critical. For instance, when assaying nucleoside diphosphatase one can add more detergent than is actually needed to disrupt the membrane so that the substrate can get in, since the extra detergent does not decrease the enzyme's activity [41]. However, the dependency of glucose-6-phosphatase activity on deoxycholate concentration shows a sharp maximum at about 0.03%; the exact concentration at which this maximum occurs depends on the protein concentration in the assay medium [41]. The dependency of glucosyl transferase activities on detergent concentration also shows a sharp maximum [99]. Consequently, whether or not to use a detergent when assaying a given microsomal enzyme and if so, how much, must be decided after careful study of the enzyme in question. In addition, other factors such as storage and dilution may affect the activity of certain microsomal enzymes (see Table 4.4) [100].

Another problem to be aware of when measuring microsomal enzymes is that microsomes contain a large number of hydrolytic and other activities which may interfere with the assay being used. This was seen above in the case of GDP-mannosyl transferase. Determination of microsomal ATPase activity by

TABLE 4.4
*Effects of storage, dilution, and DOC Treatment on Microsomal NADH-cytochrome c-reductase activity*

| Exp. no. | Procedure | NADH-cytochrome *c* red. ($\Delta E_{550}$/g liver) |
|---|---|---|
| 1 | Microsomes in sucrose: | |
| | Fresh | 292 |
| | After 6 h | 550 |
| | Microsomes in 0.026% DOC: | |
| | After 6 h | 711 |
| | Microsomes in 0.26% DOC: | |
| | After 0.5 h | 780 |
| | After 3.3 h | 588 |
| | After 3.3 h, dil. 50x: | |
| | Assayed after 7 min | 350 |
| | Assayed after 14 min | 198 |
| | Assayed after 21 min | 156 |
| 2 | Microsomes in sucrose: | |
| | Fresh | 558 |
| | After 8 h | 266 |
| | Microsomes in 0.26% DOC: | |
| | Dil. 5x, tested after 6 h | 148 |

Data taken from [100].

measuring the release of inorganic phosphate is another example of this problem, since the ADP produced may be further hydrolysed and the AMP resulting from this reaction may also be hydrolysed by microsomal 5′-nucleotidase.

Finally, certain microsomal enzymes may be in a different state in rough microsomes than they are in smooth microsomes. Thus, glucose-6-phosphatase has been shown to be 'activated' in rough microsomes and present in a less active or 'potential' form in the smooth microsomal fraction [101]. Furthermore, it has been observed that the glucose-6-phosphatase activity of rough microsomes is not denatured during long storage at low temperatures to the same degree as is the corresponding activity in smooth microsomes.

Because of these and other considerations, claims for quantitative differences in the enzymatic activities localized on rough and smooth microsomes must be examined carefully. There are fewer difficulties involved in the measurement of cytochromes $b_5$ and $P_{450}$, because one determines amounts rather than activities so that determinations of these cytochromes are probably more reliable than most measurements of microsomal enzymes. In general, however, because of the heterogeneous distribution of enzymes within the rough and smooth microsomal fractions themselves (see below), comparison of total rough microsomes with total smooth microsomes will yield information of only limited usefulness.

It should be pointed out that the enzymatic characterization of rough and smooth microsomes described here applies only when untreated adult rats are used. Microsomal enzyme activities in newborn rats appear first in the rough microsomal fraction and then increase in the smooth microsomes as well [102]. This is thought to be due to the fact that these enzymes are synthesized on the rough ER. A similar pattern is seen with the five-fold increases of NADPH-cytochrome *c* reductase and cytochrome $P_{450}$ caused by intraperitoneal treatment of adult rats with phenobarbitone [103]. Such findings represent yet another reason for carefully controlling the environment of animals to be used in the preparation of microsomes.

### *4.3.5 Other properties*

The rough and smooth microsomal fractions also differ in their physicochemical properties. This is indicated on a gross level by the observation that smooth microsomes aggregate much more readily than do rough microsomes [19]. This finding is probably related to the fact that, chiefly because of the negative charges on the phosphate groups in ribosomal RNA, the surface of rough microsomal vesicles is much more negatively charged than is that of smooth vesicles. Direct evidence for this difference in surface charge comes from electrophoretic studies [104]; whereas the binding of probes such as ANS and ethidium bromide has also provided indirect evidence [6]. Calculations of the zeta potentials (calculations which involve a number of assumptions) reveal a

TABLE 4.5
*Comparison of rough and smooth microsomes*

| | Rough | Smooth |
|---|---|---|
| Ribosomes | Present | Not present |
| Protein synthesis | Present | Not present |
| Transport of secretory proteins | Early time-point after label | Later time-point after label |
| Appearance of electron transport enzymes | First appearance | Later appearance |
| Ribosomal binding site | High | Low |
| RNA | High | Low |
| Cholesterol | Low | Higher |
| Dolichol-phosphate | High | Lower |
| Galactose, sialic acid | Low | High |
| Equilibrium density | High | Low |
| Affinity for monovalent cations | High | Low |
| Stability | Stable | Less stable |
| Net negative surface charge | High | Lower |

value of – 32 mV for rough microsomes and – 9 mV for smooth microsomes [6].

The question one would most like to answer by characterizing rough and smooth microsomes is whether or not there is any functional difference between these two fractions. It is, of course, well known that the rough ER is specialized for protein synthesis. At present there is little or no data suggesting other qualitative functional differences between rough and smooth microsomes. Table 4.5 summarizes some of the properties of rough and smooth microsomes.

### 4.4 Further subfractionation of microsomal fractions

The possibility of biochemical heterogeneity in the ER has also been investigated using centrifugal procedures designed to further subfractionate the total, rough, and smooth microsomal fractions. Of particular interest has been the possibility of enzymatic heterogeneity. There are two key considerations with respect to such studies.

In the first place histochemical investigations suggest that it will not be possible to find qualitative differences between microsomal vesicles, i.e., to isolate vesicles that contain certain microsomal enzymes and not others. Glucose-6-phosphatase can be localized by incubating microsomes with the substrate and a soluble lead salt; lead phosphate, which is electron-dense and can be seen in the electron microsocope, precipitates at places where inorganic phosphate is produced as a result of the enzymatic activity. Studies based on this method have revealed the presence of glucose-6-phosphatase throughout the endoplasmic reticulum and in virtually all microsomal vesicles originating from this organelle [105]. In addition, cytochrome $b_5$ can be localized by binding specific antibodies labelled with electron-dense ferritin to this protein. These studies show that cytochrome $b_5$ is also present in essentially all microsomal vesicles derived from the ER [106]. Such findings suggest that, at best, one should expect to observe quantitative heterogeneities in microsomes, i.e., vesicles enriched in one microsomal enzyme relative to another. As will be discussed below, partially successful attempts to get around this limitation have been carried out by sonicating the microsomes before further subfractionation.

The second thing to be considered in the further subfractionation of microsomes is the lack of any known relationship between the principles on which subfractionation procedures are normally based and the level of different enzymes present. There is no reason *a priori* to believe that separation of microsomal vesicles on the basis of size, density, surface charge, permeability, or other such physicochemical characteristics will necessarily reveal enzymatic heterogeneities, even if these exist. Vesicles enriched in glucose-6-phosphatase, for instance, might well have different sizes and densities. The main response to this consideration has simply been an empirical one – try it and see what happens. But recently, as discussed later, procedures designed to subfractionate microsomes based on the level of a given enzyme present are being developed.

### *4.4.1 Further subfractionation of the total microsomal fraction*

When performing subfractionation studies that involve subjecting subcellular particles to large centrifugal forces for relatively long periods of time, it is important to remember that such treatment may damage the particles. For instance, extensive biochemical and morphological alterations in rat liver mitochondria may occur during high-speed centrifugation in sucrose gradients [107]. Such procedures appear to disrupt lysosomes and peroxisomes as well [108]. The main cause of such damage seems to be the large hydrostatic pressures developed when large centrifugal forces are applied [109]. Fortunately for investigators working with microsomes, a study by Bronfman and Beaufay in 1973 [110] concluded that microsomal vesicles are relatively resistant to damage from hydrostatic pressure – more resistant than mitochondria, lysosomes, or peroxisomes. In addition it was found that the lower the temperature, the more resistant subcellular particles are to such damage.

One of the first attempts to subfractionate total microsomes involved zonal centrifugation of the post-mitochondrial supernatant [111]. The conditions of centrifugation and the sucrose gradient used were designed to achieve isopycnic banding. UDPase activity was found to distribute differently than ATPase or glucose-6-phosphatase.

In 1969 Murphy and his co-workers [112] layered the 10 000 *g* supernatant from rat liver above a continuous gradient of 10–40% sucrose and centrifuged for 150 min at 27 000 revs/min in a titanium B–XV rotor. Such rate-zonal centrifugation separates microsomal vesicles chiefly on the basis of size. After this procedure cytochrome $P_{450}$ was found to be concentrated in the small and medium-sized particles in the upper and middle regions of the gradient. Treatment of the rats with phenobarbitone resulted in an increased specific content of cytochrome $P_{450}$ in these smaller vesicles, which are thought to originate mainly from the smooth ER. On the other hand, treatment of the animals with methylcholanthrene induced an increase in cytochrome $P_{450}$ ($P_{448}$) that was more widely spread throughout the gradient, with some enrichment in the larger vesicles near the bottom.

Schulze and Staudinger [113] subfractionated total rat liver microsomes mainly on the basis of density. A linear sucrose gradient spanning the density range 1.05–1.18 g $ml^{-1}$ on top of a smaller gradient from 1.18–1.216 g $ml^{-1}$ was prepared. The microsomal fraction was layered over this gradient and centrifuged for 16 h at 121 750 x $g_{max}$ in the zonal rotor Ti–15 (Beckman). Glucose-6-phosphatase and the microsomal electron transport enzymes and cytochromes were found to be distributed in the same manner after this procedure. However, if the microsomes were sonicated for 100 s in 10 s intervals before subfractionation, differences in the distribution of these microsomal components began to appear; and these differences were even more pronounced if sonication was continued for 200 s. After this latter treatment the lighter particles in the upper region of the gradient were enriched in glucose-6-phosphatase, NADPH-cytochrome *c* reductase, NADH-cytochrome *c* reductase, and cytochrome $P_{450}$; whereas the specific content of cytochrome $b_5$ was found to be higher in heavier particles near the middle of the gradient. Most of the microsomal protein was recovered in the middle region.

Recently, Beaufay and his co-workers [24] in Louvain reported a detailed study of the subfractionation of total rat liver microsomes using a number of different gradients. The microsomes were centrifuged to near-equilibrium in density gradients prepared with sucrose and water (density range 1.1–1.25 g $ml^{-1}$), with sucrose and $D_2O$ (1.163–1.330 g $ml^{-1}$), with Ficoll containing 0.50 M sucrose (1.100–1.171 g $ml^{-1}$), and with Ficoll containing 0.25 M sucrose (1.070–1.160 g $ml^{-1}$). Centrifugations were performed in a special automatic rotor (designed by Beaufay) which is unfortunately not commercially available. This rotor has a short (1 cm) pathway located relatively far from the axis of rotation. As a result, the centrifugal force produced is large throughout the gradient and a relatively short time of centrifugation (3 h at 35 000 revs/min) was required to reach near equilibrium.

Beaufay and his co-workers were principally concerned with investigating contamination of the total microsomal fraction with particles derived from organelles other than the ER. However, their study also revealed biochemical heterogeneities in the ER itself. In all systems used group *b* enzymes (NADPH-cytochrome *c* reductase, NADH-cytochrome *c* reductase, aminopyrine demethylase, cytochrome $P_{450}$, and cytochrome $b_5$) were all distributed in a similar manner; group *c* enzymes (glucose-6-phosphatase, nucleoside diphosphatase, esterase, β-glucuronidase, and glucuronyl transferase) were all distributed in a similar manner; but the median density of vesicles enriched in group *c* enzymes was consistently higher than the median density of vesicles enriched in group *b* enzymes. In addition, treatment of the microsomes with conditions designed to detach ribosomes resulted in a more marked decrease in the median density of group *c* vesicles than of group *b* vesicles. This observation suggested that group *c* enzymes may be localized on ribosome-rich regions of the endoplasmic reticulum.

### *4.4.2 Further subfractionation of the rough microsomal fraction*

In 1968 Decloitre and Chauveau [114] layered the total microsomal fraction over a sucrose gradient (linear over the density range 1.21–1.26 g $ml^{-1}$ with a cushion of density 1.28 g $ml^{-1}$) and centrifuged for 40 h at 25 000 revs/min in a Spinco SW 25 rotor. After this procedure most of the smooth microsomes were found at the interface between the upper layer and the sucrose gradient, free ribosomes were recovered in the pellet, and rough microsomes were located in the linear portion of the gradient. The rough microsomes were collected as two fractions, and measurements of the ratio of RNA to phospholipids together with electron microscopy showed that these two fractions differed with respect to the number of ribosomes attached to each vesicle. Vesicles with small numbers of ribosomes had a specific activity of dimethylaminoazobenzene reductase that was 1.5 times that of the ribosome-rich vesicles; whereas the two fractions demonstrated almost identical specific activities of glucose-6-phosphatase.

Utilizing the fact that the tendency of microsomal vesicles to aggregate in the presence of $Cs^+$ is directly proportional to their number of bound ribosomes, investigations in our laboratory [75] have also resulted in subfractionation of rough microsomes on the basis of the number of ribosomes per vesicle. Rough and smooth microsomes were separated using a $Cs^+$-containing discontinuous sucrose gradient as described above. After removing the smooth microsomes, the 1.3 M sucrose layer down to the edge of the pellet was collected, but instead of discarding this layer in the usual manner, the particles present were sedimented by centrifugation and designated rough III microsomes. The remaining 1.3 M sucrose was collected separately instead of combining it with the pellet as was normally done, and the particles in this fraction were called rough II microsomes. The pellet contained rough I microsomes.

The RNA to phospholipid ratios in rough I, II, III, and smooth microsomes were found to be 0.62, 0.31, 0.20, and 0.11, respectively, demonstrating that subfractionation had occurred on the basis of the number of ribosomes per vesicle. The electron transport enzymes NADPH-cytochrome *c* reductase, NADH-cytochrome *c* reductase, cytochrome $P_{450}$, and cytochrome $b_5$ were found to be somewhat more concentrated in rough microsomes with few ribosomes. On the other hand, vesicles with many ribosomes were enriched in the phosphatases glucose-6-phosphatase, IDPase, ATPase, and *p*-nitrophenyl phosphatase.

A series of more detailed studies designed to subfractionate rough microsomes was begun in our laboratory in 1968 [74]. The initial experiments were performed with rough microsomes isolated using a modification of the Rothschild procedure. These microsomes were layered over a continuous gradient of 0.59–1.02 M sucrose and centrifuged for 60 min at 58 000 x *g* in an SW 25 rotor, a procedure which separates the microsomal vesicles mainly on the basis of size. NADPH-cytochrome *c* reductase, NADH-cytochrome *c* reductase, and glucose-6-phosphatase were found to be heterogeneously distributed, all being concentrated in the slowly sedimenting vesicles at the top of the gradient. On the other hand, cytochrome $b_5$ was found to be evenly distributed throughout the gradient.

Our next experiments involved sonicating the rough microsomes for ten 15 s intervals (with pauses in-between for cooling) before subfractionation [42]. This sonication reduced the rough microsomal vesicles to about 1/100 of their original mass, detached most of the ribosomes, and caused only limited denaturation of microsomal proteins. Subfractionation on the basis of size and density was accomplished by layering the sonicated microsomes over a continuous sucrose gradient (0.29–1.32 M) and centrifuging for 16 h at 58 000 x *g* in an SW 25 rotor. After this procedure NADPH-cytochrome *c* reductase and cytochrome $P_{450}$ were concentrated in vesicles at the top of the gradient; while vesicles in the bottom region were enriched in NADH-cytochrome *c* reductase and cytochrome $b_5$.

Since it was desired to sonicate the rough microsomes for a longer time before subfractionation and since even short periods of sonication are known to produce a great deal of heat at the tip of the probe, a special sonication chamber was designed (see Fig. 4.6) [43]. This chamber allowed the temperature very close to the tip of the probe to be measured and resulted in improved

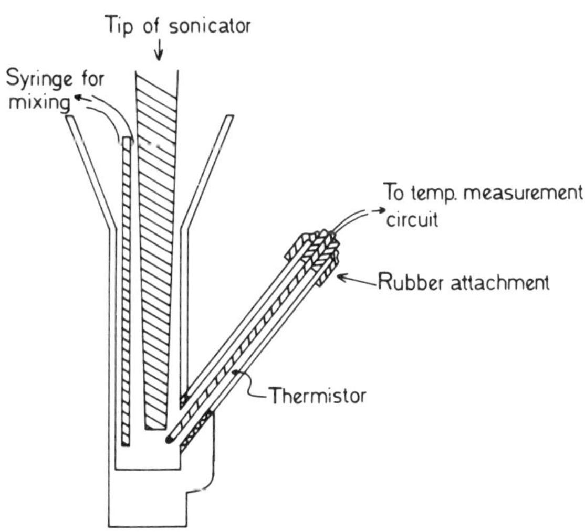


**Fig. 4.6** Chamber for continuous sonication. Taken from [43].

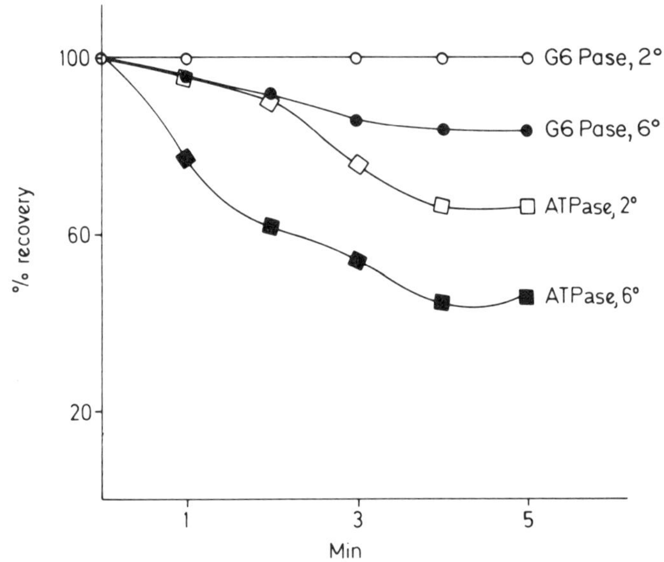


**Fig. 4.7** Effects of sonication on glucose-6-phosphatase and ATPase activities of rough microsomes. Batches of rough microsomes from 3.5 g liver suspended in 3.5 ml were sonicated in a specially designed chamber (see Fig. 4.6) and the temperature maintained at 2 or 6°C. Sonication was interrupted every minute for mixing and removal of a 0.1 ml aliquot for enzyme assay. For more details, see [43].

temperature control. Rough microsomes were sonicated continuously for 5 min at 2°C. This treatment caused about a 10% loss of the electron transport enzymes and, as illustrated in Fig. 4.7, a 10% loss of glucose-6-phosphatase and a 40% loss of ATPase. If the temperature was maintained 4 degrees higher, these losses were considerably greater. Controls demonstrated that this sonication procedure did not cause fusing of microsomal vesicles, a process which might have obscured the heterogeneities we were looking for.

The sonicated rough microsomes were separated mainly on the basis of size by layering them over a continuous sucrose gradient (0.29–0.58 M) and centrifuging for 60 min at 198 000 x *g* in an SW 40 rotor [43]. After this procedure, NADPH-cytochrome *c* reductase and cytochrome $P_{450}$ were concentrated in the vesicles at the top of the gradient; the vesicles in the bottom region were enriched in NADH-cytochrome *c* reductase, cytochrome $b_5$, and glucose-6-phosphatase; while *p*-nitrophenyl-phosphatase was evenly distributed throughout the gradient (see Fig. 4.8). Subfractionation mainly on the basis of density was also performed by layering the sonicated rough microsomes over a continuous sucrose gradient (0.29–1.32 M) and centrifuging for 16 h at 95 000 *g* in an SW 27 rotor. After this procedure NADPH-cytochrome *c* reductase and cytochrome $P_{450}$ again showed a distribution distinctly different

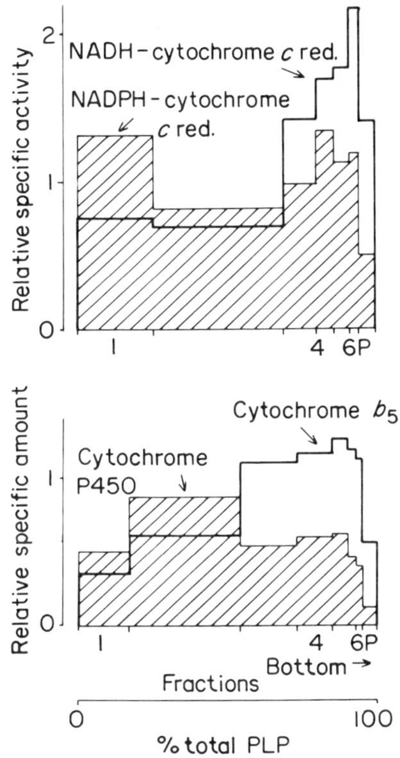


**Fig. 4.8** Distribution of electron transport enzymes in submicrosomal particles after rate differential centrifugation. Data taken from [43].

than that of NADH-cytochrome *c* reductase and cytochrome $b_5$. The first two components were concentrated in vesicles in the upper and middle portions of the gradient; while the latter two components were enriched for in the middle and low parts of the gradient.

The most striking differences between these two electron transport chains were obtained if vesicles separated on the basis of size were subfractionated again on the basis of density [13]. Thus, the fractions of the 0.29–0.58 M sucrose gradient containing slowly sedimenting particles and the corresponding pellet were collected, the pellet resonicated, and both preparations were recentrifuged to near-isopycnic equilibrium (see Fig. 4.9 and 4.10). The resulting low density vesicles were greatly enriched in NADPH-cytochrome *c* reductase activity, while NADH-cytochrome *c* reductase activity was much more concentrated in vesicles of higher density. In the case of the pellet (Fig. 4.10), the ratio between NADH- and NADPH-cytochrome *c* reductase activities varied from 0.1 at the top of the gradient to 2.3 at the bottom.

Finally, our original attempts to subfractionate whole rough microsomes by rate differential centrifugation on a linear stabilizing gradient were repeated using a zonal centrifugation system instead of a conventional swinging-bucket rotor [21]. A much more heterogeneous distribution of microsomal enzymes

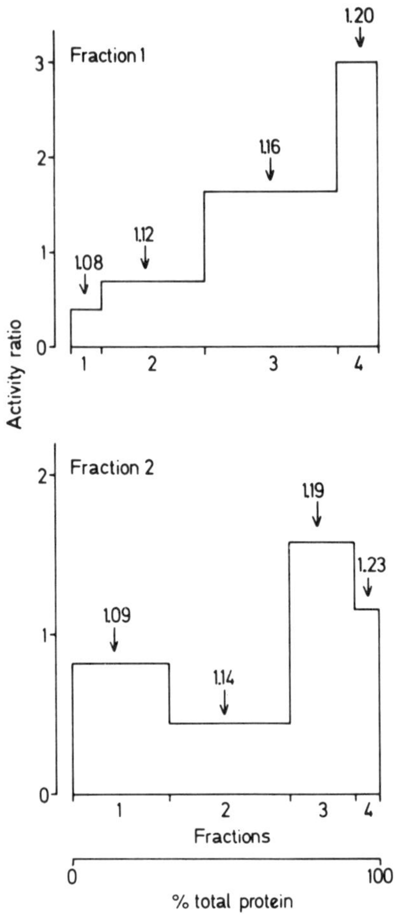


**Fig. 4.9** Ratio of NADH/NADPH-cytochrome *c* reductase activities of slowly sedimenting submicrosomal particles after recentrifugation on a density gradient. Data taken from [43].

were thereby obtained. As illustrated in Fig. 4.11, NADPH-cytochrome *c* reductase activity is concentrated in slowly sedimenting vesicles, glucose 6-phosphatase is concentrated in more rapidly sedimenting vesicles, whereas the most rapidly sedimenting rough microsomes are enriched in ATPase activity.

### *4.4.3 Further subfractionation of the smooth microsomal fraction*

Results similar to those last cited above for the further subfractionation of rough microsomes have been obtained with smooth microsomes [20]. Smooth microsomes prepared by the Rothschild procedure or using a $Cs^+$-containing discontinuous sucrose gradient were found to be aggregated to about the same degree; so the latter method was used to prepare smooth microsomes for further subfractionation. After separation mainly on the basis of size or mainly on the basis of density, NADPH-cytochrome *c* reductase, NADH-cytochrome *c*

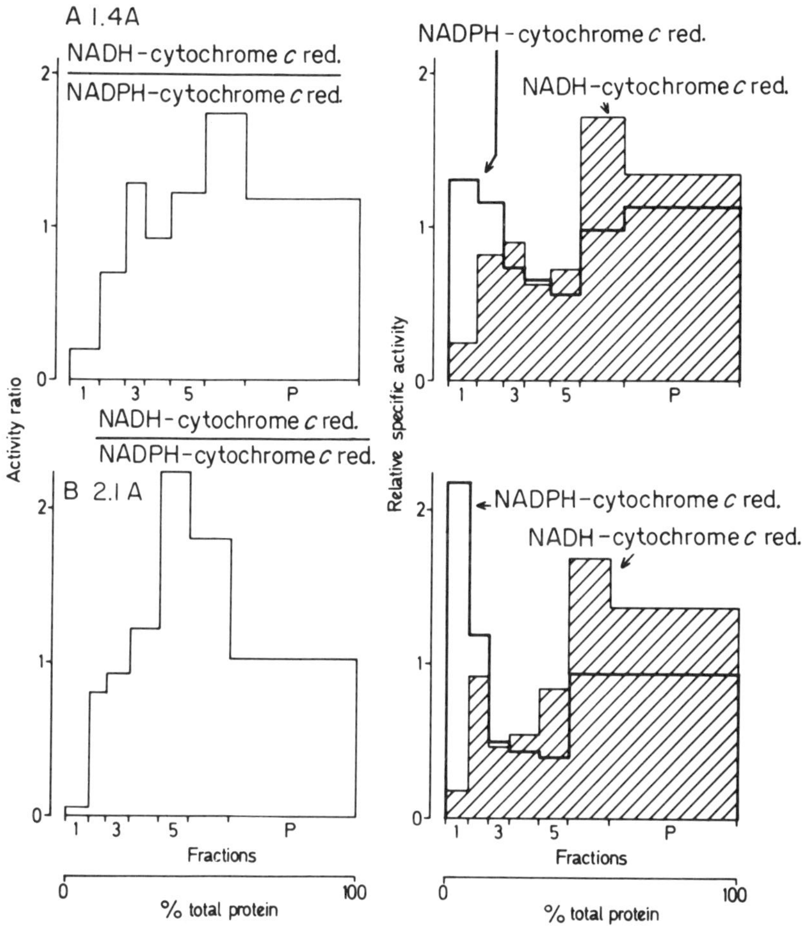


**Fig. 4.10** Cytochrome *c* reductases of the high sedimentation velocity submicrosomal particles after resonication and recentrifugation on a density gradient. Data taken from [43].

reductase, cytochrome $P_{450}$, and cytochrome $b_5$ were concentrated in the upper part of the gradient; glucose-6-phosphatase and IDPase were concentrated in the middle of the gradient; and ATPase, AMPase, and cholesterol were enriched for in the vesicles in the bottom region of the gradient (see Fig. 4.12).

Smooth microsomes can also be subfractionated by adding $Mg^{2+}$ and then centrifuging under conditions that sediment only those vesicles aggregated by this cation [79]. The pellet has been designated smooth I microsomes and unsedimented particles are called smooth II microsomes. The smooth II microsomes contain relatively little glucose-6-phosphatase; and although this

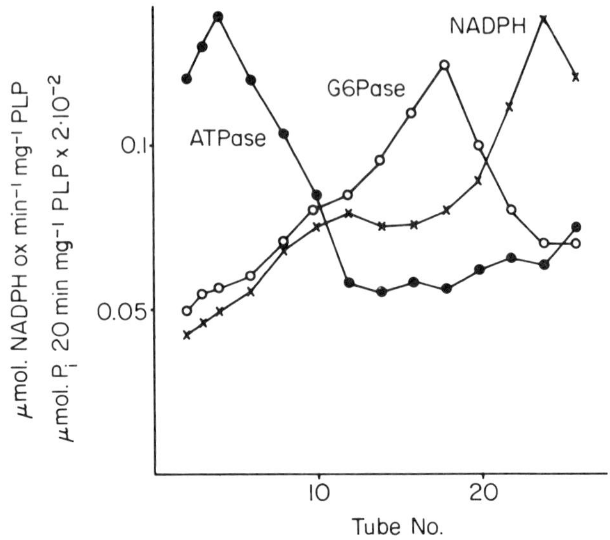


**Fig. 4.11** Enzyme activities of rough microsomal subfractions after zonal centrifugation on a stabilizing gradient. Data taken from [21].

fraction contains a normal amount of cytochrome $P_{450}$, its cytochrome $P_{450}$ is not induced by treatment of rats with phenobarbitone or methylcholanthrene nor is it altered to cytochrome $P_{448}$ by methylcholanthrene. It has been found that albumin is transported through those regions of the ER that give rise to smooth II vesicles in much the same manner as it is transported through regions from which smooth I microsomes originate [115]. Studies with ANS have revealed that smooth II microsomes have a relatively small negative surface charge, which probably explains why they are not aggregated by $Mg^{2+}$ [6].

### *4.4.4 New approaches to the further subfractionation of microsomes*

Lewis and Tata [116] have recently made use of a reaction employed in histochemical studies in their attempts to subfractionate rough and smooth microsomes. The microsomes were incubated with glucose-6-phosphate and $Pb(NO_3)_2$ in hopes that insoluble lead phosphate (formed when inorganic phosphate is produced by glucose-6-phosphatase activity) would 'attach' to the vesicles and increase their density to an extent proportional to their specific activity of glucose-6-phosphatase. The rough and smooth microsomes were then placed over discontinuous sucrose gradients consisting of seven layers and centrifuged for 3 hours at 94 000 x *g* in an SW 27 rotor. After this procedure the bands, found at all the interfaces, were assayed for glucose-6-phosphatase activity. With smooth microsomes there was a direct relationship between the density at which the particles were recovered and their specific activity of this

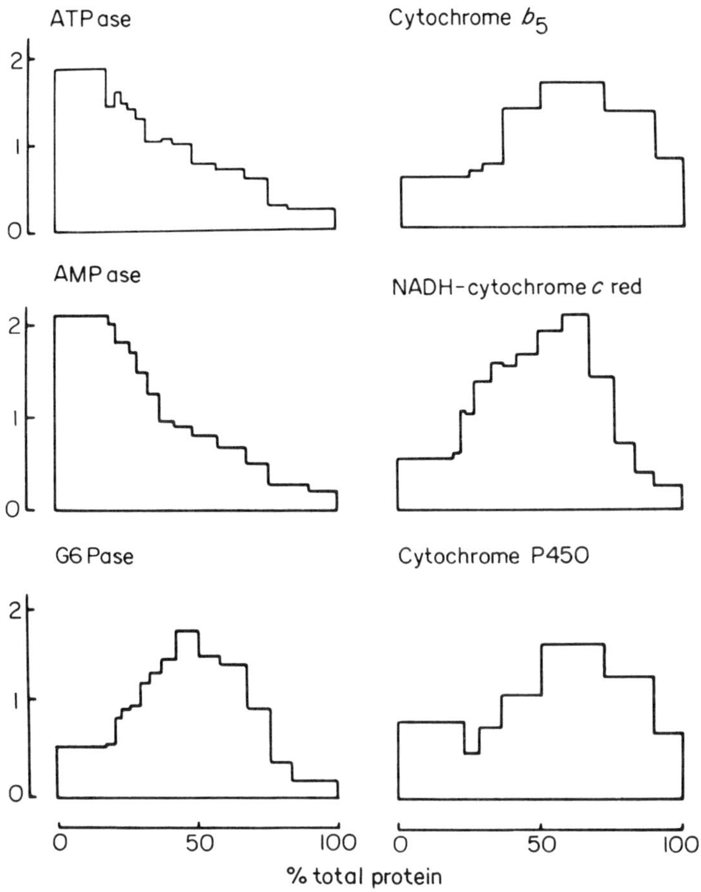


**Fig. 4.12** Distribution of some enzyme activities in subfractions of smooth microsomes after rate differential centrifugation. Data taken from [20].

enzyme; the particles found at the bottom of the gradient had a specific activity of glucose-6-phosphatase that was more than three times that of the particles at the top. A similar, but less clear relationship was seen with the rough microsomes; in this case detachment of ribosomes may have complicated the situation.

Upon examination in the electron microscope lead phosphate deposits were seen inside the vesicles. The frequency and amount of these deposits were directly related to the level of glucose-6-phosphatase measured biochemically. Leskes, Siekevitz, and Palade [105,117] had previously carried out an investigation very similar to that of Lewis and Tata. The latter investigators have suggested that this approach might also be applicable to other enzymes and other systems.

In 1972 Schulze, Pönnighaus, and Staudinger [118] subfractionated pig liver microsomes with rabbit antibodies to NADH-cytochrome $b_5$ reductase. The immunoglobins, which inhibited the membrane-bound enzyme, were covalently linked to Sepharose 4B. The bound antibodies were incubated with microsomes for 2 h at 37°C plus 13 h at 4°C and then removed by low-speed centrifugation. If whole microsomes were used, the fraction not sedimenting with the Sepharose beads contained NADH-cytochrome $b_5$ reductase, cytochrome $b_5$, NADPH-cytochrome c reductase, and glucose-6-phosphatase in approximately the same relative amounts as the microsomes before immunoadsorption. However, if sonicated microsomes were used, the fraction not bound was enriched in cytochrome $b_5$, NADPH-cytochrome *c* reductase, and glucose-6-phosphatase relative to NADH-cytochrome $b_5$ reductase. In this latter case the bound vesicles were presumably enriched in NADH-cytochrome $b_5$ reductase relative to the other three components.

### *4.4.5 General conclusions*

Three main conclusions may be drawn from these studies on the further subfractionation of rat liver microsomes. First, microsomal and submicrosomal vesicles are biochemically heterogenous, and the heterogeneities can be revealed both by application of the usual principles of subfractionation and by special approaches. The differences in the enzyme activities of different vesicles are quantitative and not qualitative, in agreement with histochemical studies.

The heterogeneity observed suggests that there are segments of the ER specialized for the performance of a given function. The NADPH-driven electron transport chain (composed of NADPH-cytochrome *c* reductase and cytochrome $P_{450}$) and the NADH-driven chain (NADH-cytochrome *c* reductase and cytochrome $b_5$) can be partially separated from one another, as well as from microsomal phosphatases. However, the components of a given chain have been seen to distribute in a similar manner in virtually all reported investigations. It would be of great interest to know how other functionally related enzymes – for instance, the enzymes of phospholipid synthesis – are distributed in microsomal and submicrosomal vesicles.

Thus, the ER may contain electron transport complexes similar to the complexes found on the inner mitochondrial membrane. These complexes may lie in the membrane side-by-side with other microsomal enzymes; large numbers of these complexes may be segregated together in a given region, for example, on an individual tubule; or the complexes may even be localized in certain hepatocytes and present in smaller amounts in others. It is not yet possible to distinguish between these three alternatives.

The second conclusion to be drawn is that the smaller the size of the microsomal vesicles, the more clearly heterogeneities can be revealed by subfractionation. Sonication, which may break the microsomal membrane randomly or at specific weak points, has been found to be an effective means of obtaining small submicrosomal particles.

We achieved best separation of the two electron transport chains after sonicating rough microsomes to give vesicles the smallest of which were 30 nm in diameter. It is of interest that Robertson considers this to be the minimum vesicle diameter compatible with the unit membrane theory [119], and it is the smallest size observed in tissue sections. We have calculated that the sum of the molecular weights of all protein molecules in such a vesicle would be about $7.5 \times 10^6$ daltons. Thus, making a number of assumptions (including the unlikely one that the only proteins present are the components of the electron transport chains), it can be estimated that the 30 nm vesicles most enriched in the NADPH-driven chain contain 15 complete NADPH-driven and 2 complete NADH-driven chains; whereas the vesicles most enriched in the NADH-driven chain contain 3 complete NADPH-driven and 25 complete NADH-driven chains. The object of these calculations is simply to demonstrate that best subfractionation is achieved with submicrosomal vesicles containing rather limited numbers of protein molecules.

The third conclusion to be drawn is that the larger the number of subfractionation steps employed, the more clearly microsomal heterogeneities can be revealed. The two electron transport chains are most effectively separated if the total microsomal fraction is first divided into rough and smooth microsomes; the sonicated rough microsomes are then subfractionated mainly on the basis of size; and fractions from this gradient are then further subfractionated on the basis of size or density. These chains are not at all separated if the total microsomal fraction is simply fractionated on the basis of density. Thus, the best approach here is analogous to the one used in protein purification, where one chromatographic separation is followed by another based on the same or a different principle.

Finally, it should be remembered that homogenization and sonication disrupt membranes only in the tranverse plane, and as a result, the subfractionation procedures discussed here reveal heterogeneities only in the lateral plane of the endoplasmic reticulum. In studies of transverse heterogeneities – such as the localization of an enzyme exclusively on the cytoplasmic or luminal surface – it would be very useful to have microsomal vesicles with an inside-out orientation like the inside-out vesicles that can be prepared from mitochondrial inner membranes and from red blood cells. However, it has not yet been possible to prepare inside-out microsomes.

It seems to us that the future of microsomal subfractionation does not lie in developing even better rotors, gradients, and centrifugation schemes. Rather, the most promising approaches are those allowing subfractionation on the basis of the level of a given protein present, such as the 'histochemical' approach of Lewis and Tata [116] or immunoadsorption along the lines of Schulze and his co-workers [118]. In a similar manner we have tried to use lectins (concanavalin

A, wheat germ agglutinin, and *Crotalaria*) covalently bound to Sepharose or Agarose to subfractionate microsomes, but the terminal sugar residues which these lectins bind specifically are not accessible in intact microsomes. It will be fascinating to see what new biochemical and functional heterogeneities in the endoplasmic reticulum are still waiting to be revealed.

## 4.5 Detailed experimental procedures

### *4.5.1 Homogenization of the liver*

The most reproducible preparations are obtained from rats weighing 180–200 g. The animals are starved for 20 h and then sacrificed by decapitation. The liver is removed, placed into cold 0.25 M sucrose, weighed, minced with a pair of scissors, and the pieces are rinsed several times with cold sucrose in order to remove as much blood as possible. (If it is important to minimize contamination of the microsomes by elements of the blood, the liver should be perfused through the portal vein with a solution of 0.25 M sucrose containing no cations before homogenization is performed.) The liver pieces are suspended in a volume of sucrose approximately equal to the volume of the liver itself and this suspension is transferred to a commercially available Potter-Elvehjem teflon homogenizer. The pestle is motor-driven at a speed of 400–500 revs/min and homogenization is carried out with four up-and-down strokes. This homogenate is then diluted with sucrose to give a concentration of 0.2–0.25 g original liver wet weight $ml^{-1}$.

### *4.5.2 Isolation and washing of the total microsomal fraction*

All centrifugations are performed at 4°C and preparations are otherwise stored in an ice bath. The homogenate is centrifuged at 10 000 x *g* for 20 min, the supernatant is decanted entirely, and the pellet obtained is discarded. The supernatant is centrifuged at 105 000 x g for 60–90 min and the resulting supernatant is discarded. The high-speed pellet is resuspended in a solution of 0.15 M Tris-Cl, pH 8.0, and recentrifuged at 105 000 x *g* for 60 min. The washed total microsomal pellet is resuspended in 0.25 M sucrose to give a concentration of approximately 1 g original liver weight wet $ml^{-1}$, which corresponds to 20–25 mg microsomal protein $ml^{-1}$.

### *4.5.3 Testing for aggregation* [19,21]

The 10 000 *g* supernatant prepared as described above can be used as it is; microsomal fractions are diluted with 0.25 M sucrose to give a concentration of about 5 mg microsomal protein $ml^{-1}$. In order to minimize aggregation ions, hypotonicity, and low concentrations of microsomal protein should be avoided in this dilution step. 1–3 ml of the preparation is filtered using a suction pump through a Millipore filter (Millipore Filter Co., Bedford, Massachusetts) of known uniform pore size. The amount of microsomal material passing through the filter can be estimated either by measuring the absorbancy of the filtrate at 280 nm (if only microsomal protein is present); or by assaying the percentage of a microsomal enzyme such as NADPH-cytochrome *c* reductase found in the filtrate (if non-microsomal proteins, e.g. soluble cytoplasmic proteins, are also present). In a preparation that is essentially unaggregated, 0% of the microsomes will pass through a filter with 0.22 $\mu$m pores; 63% will pass through 0.3 $\mu$m pores; and 91% will pass through 0.45 $\mu$m pores [19].

### *4.5.4 Removing adsorbed and luminal protein with cold-induced rupture* [45]

Unwashed microsomal pellets are resuspended in 0.15 M Tris-Cl, pH 8.0, and centrifuged at 105 000 x *g* for 1 h. The microsomes are then resuspended in distilled water to give a concentration of approximately 0.5–1 mg microsomal protein $ml^{-1}$ and incubated at 30°C for 15 min. The suspensions are then cooled in an ice water bath and centrifuged at 105 000 x *g* for 1 h. The resulting pellet is again washed with 0.15 M Tris-Cl, pH 8.0, to remove luminal protein which adsorbs to the surface of the microsomes after being released from the vesicles by the cold-induced rupture.

### *4.5.5 Removing luminal protein with low concentrations of detergent* [44,47]

A 0.54% stock solution of DOC is prepared by titrating a solution of sodium deoxycholate at 20°C to pH 7.5 using HCl. Washed microsomes are resuspended in a solution containing 50 mM Tris-Cl, pH 7.4, 50 mM KCl, and 5 mM $MgCl_2$. (500 mM KCl may also be used, in which case the stock solution of DOC should be 0.28%). This resuspension is accomplished by hand homogenization using a Potter-Elvehjem homogenizer and the protein concentration is adjusted to 3–4 mg $ml^{-1}$. (The protein concentration of rough microsomes can be estimated by dissolving them in 0.5% DOC and measuring the $OD_{260}$ of the ribosomal RNA which they contain; an $OD_{260}$ of 6.75 units $ml^{-1}$ corresponds to approximately 1 mg protein $ml^{-1}$. The protein concentration of suspensions of total and smooth microsomes should be determined chemically.) With vigorous shaking on a Vortex tube mixer a volume of the stock DOC solution equal to one-tenth of the volume of the microsomal suspension is added. The preparation is incubated at 0°C for 30 min. The microsomal membranes can then be collected by layering 8 ml of the suspension over 2 ml of a solution containing 20% sucrose, 50 mM Tris-Cl, pH 7.5, 50 mM KCl, and 5 mM $MgCl_2$ and centrifuging for 90 minutes at 50 000 revs/min in a Spinco 60 or an A321 IEC rotor. After this procedure membranes are found in the pellet and luminal protein remains in the upper layer.

### *4.5.6 Separating rough and smooth microsomes according to a modified Rothschild procedure* [74]

In this case homogenization of the liver is performed in 0.44 M sucrose but in all other respects preparation of the 10 000 *g* supernatant is carried out as described above. 8 ml of this supernatant is layered over 3 ml of 1.3 M sucrose and centrifuged at 105 000 x *g* for 7 h 40 min in a Spinco 40 rotor After this step the upper 0.44 M sucrose phase is sucked off and discarded. The milky band in the upper part of the 1.3 M sucrose layer contains the smooth microsomes and is collected. (If this fraction is not to be further subfractionated it may be diluted with 0.25 M sucrose and sedimented by centrifugation at 105 000 x *g* for 90 min. This procedure concentrates the smooth microsomes while removing cytoplasmic proteins, a part of the free ribosomes and large amounts of sucrose.) The 1.3 M sucrose layer is removed and discarded, taking care to leave the fluffy band just above the pellet behind. A few drops of water are added and this band and the pellet are resuspended by hand homogenization to give the rough microsomes. (The rough microsomes may also be concentrated and sucrose removed by dilution and recentrifugation.) The yield after recentrifugation is approximately 10 mg protein $g^{-1}$ liver in the rough microsomes and about 4 mg protein $g^{-1}$ liver in the smooth microsomes.

### *4.5.7 Separating rough and smooth microsomes in $Cs^+$-containing discontinuous sucrose gradients*

*(i) Spinco 40.2 rotor*

Using the Spinco 40.2 rotor [78] 1 ml of 0.6 M sucrose – 15 mM CsCl is layered over 2 ml of 1.3 M sucrose – 15 mM CsCl. Above this gradient is then layered 3.5 ml of the 10 000 *g* supernatant prepared as described above. Centrifugation is carried out at 102 000 x *g* for 90 min in a Spinco 40.2 rotor (tube angle 40°), after which the clear reddish upper phase can be removed with a Pasteur pipette either provided with a rubber bulb or connected to a water aspirator. This upper phase is discarded. The 0.6 M sucrose layer is also removed and discarded after which the fluffy double band (upper reddish, lower light brown) at and just under the interface between the 0.6 and 1.3 M sucrose layers is collected. This double band contains the total smooth microsomes. Finally, the 1.3 M sucrose layer is removed and the pellet is resuspended to give the total rough microsomes as described above for the modified Rothschild procedure. In addition, the rough and smooth microsomes may be washed as described above. The yield is approximately 12 mg protein $g^{-1}$ liver for the rough microsomes and 5–6 mg protein $g^{-1}$ liver for the smooth microsomes.

*(ii) SW 39 and Spinco 40 rotors* [82]

With the SW 39 rotor 2 ml of the 10 000 *g* supernatant prepared as described above is layered onto the same discontinuous gradient used with the Spinco 40.2

rotor. After 3 h of centrifugation at the highest speed allowable with this rotor (39 000 revs/min) the two microsomal subfractions are separated. In the case of the SW 40 rotor 9 ml of the 10 000 *g* supernatant is layered over a discontinuous gradient consisting of 1 ml 0.6 M sucrose – 15 mM CsCl and 3 ml 1.3 M sucrose – 15 mM CsCl. In spite of the high centrifugal force employed (198 000 x *g*), 3 h of centrifugation is required for separation of the rough and smooth microsomes.

With the Spinco 40 rotor 7 ml of the 10 000 *g* supernatant is layered over the same discontinuous gradient as described above for the Spinco 40.2 rotor [82]. Centrifugation for 140 minutes separates the rough and smooth microsomes but because the Spinco 40 rotor has a tube angle of only 26°, about 15% of the smooth microsomes are recovered in the pellet, probably due to side wall impaction.

*(iii) Christ Omega and International B-60 ultracentrifuges* [82]

The 60 rotor of the Christ Omega (with a tube angle of 34°) and the A-321 rotor of the International B-60 ultracentrifuge (with a tube angle of 35°) can both be spun at 60 000 revs/min in their respective centrifuges to provide a force of 250 000 x *g*. This situation is nearly ideal for the separation of rough and smooth microsomes on $Cs^+$-containing discontinuous sucrose gradients. In both cases gradients are formed with 1 ml of 0.6 M sucrose – 15 mM CsCl and 3.5 ml of 1.3 M sucrose – 15 mM CsCl and the tubes are filled with 10 000 *g* supernatant. 60 min of centrifugation at 60 000 revs/min yields an almost complete separation of rough and smooth microsomes with only limited amounts of particulate material remaining in the 1.3 M sucrose phase.

### *4.5.8 Preparation of rough microsomes by the method of Adelman* et al. [15,16]

The homogenization of the liver involved in this procedure is different from that described above. After cutting the liver into several large pieces, these pieces are blotted and forced through a tissue press consisting of a piston and a stainless steel plate with 1 mm perforations. The resulting pulp is weighed and 2 ml of 1.0 M sucrose is added for each gram. Homogenization is carried out with 8–10 passes in a motor driven homogenizer with the Teflon pestle turning at 1000–2000 revs/min. Somewhat more or less vigorous homogenization does not affect the fractionation significantly. About 17 g of liver pulp from 4 or 5 rats are conveniently processed in this manner at one time, and the following procedural details are for a preparation starting with this quantity of pulp.

The homogenate is filtered through a single layer of Nytex cloth (No. 130, Toblem, Ernest, and Traber, Inc., New York) to remove connective tissue and improve separation of the nuclei. An equal volume of 2.5 M sucrose is added to this filtrate with thorough mixing. 100 ml of this preparation are divided into

4 SB 110 tubes and overlaid with 1 ml of 1.0 M sucrose to prevent exposure of particles that float to the air-water interface. Centrifugation for 45 min at 24 000 revs/min (approximately 100 000 *g*) in IEC rotor SB 110 in the IEC B-60 centrifuge (International Equipment Co., Needham Heights, Massachusetts) results in a well packed, pinkish gray pellet, a tan supernatant, and a thick reddish tan pellicle. (In this and subsequent operations a Spinco centrifuge (Beckman Instruments) may be used instead of the IEC B-60 centrifuge, in which case rotors A 211 and SB 110 can be replaced with 40.2 or Ti 60 and SW 27 rotors, respectively.)

The pellicle and the supernatant are combined and homogenized with 2 or 3 passes to disperse all clumps. Approximately 100 ml of this preparation is thoroughly mixed with 50 ml of water, divided into 6 tubes and centrifuged for 15 min at 15 000 revs/min (approximately 22 000 x *g*) in IEC rotor A211. The pink, turbid supernatant is decanted and saved. The tan pellets are suspended and homogenized in 25 ml of a solution containing 9 parts 0.50 M sucrose and 1 part inhibitory supernatant. (Inhibitory supernatant is included to minimize nuclease attack on bound polysomes. It is prepared by homogenizing rat liver in 0.25 M sucrose (2 ml $g^{-1}$ tissue weight) and centrifuging the homogenate for 15 min at 20 000 revs/min in IEC rotor A211. The resulting supernatant is then centrifuged for 2 h at 44 000 revs/min (approximately 200 000 x *g* ) in IEC rotor A211. The clear red supernatant, designated inhibitory supernatant, is collected, being careful to avoid taking along the milky scum and the loose pellet, and is stored frozen.) This mixture is divided into two tubes and centrifuged for 15 min at 13 000 revs/min (approximately 17 000 x *g*) in the IEC rotor A211. The supernatant and fluffy, pinkish, loosely packed layer over the pellet are saved and this washing of the mitochondrial pellet is repeated once. The combined post-mitochondrial supernatants are then divided into 8 tubes and centrifuged in the same manner as was done for washing.

The resulting supernatant and fluffy layer over the pellet, representing a volume of approximately 180–190 ml, are mixed and divided evenly into 8 tubes. A syringe and a large steel cannula are then used to underlay the supernatant first with 4 ml of a solution consisting of 3 parts 2.0 M sucrose and 1 part inhibitory supernatant, and subsequently with 1 ml of a solution containing 2.0 M sucrose, 50 mM Tris-Cl, pH 7.5, 25 mM KCl, and 5 mM $MgCl_2$. (All pH's are measured at room temperature.) Centrifugation is carried out for 20 h at 44 000 revs/min (approximately 200 000 x *g*) in the IEC rotor A211, which is allowed to coast to a stop. Two reddish brown membrane bands are seen, one at each of the liquid-liquid interfaces. The crude smooth microsomes are found at the interface between the supernatant and the 1.56 M sucrose layer; while crude rough microsomes are found between the 1.5 and 2.0 M sucrose layers. Most of the free ribosomes are found in the small, pale orange pellet.

Everything down to and including the smooth microsomes are removed with a syringe and a large steel cannula. The rest of the fluid in the tubes is decanted into a graduated cylinder, and the tubes are rinsed with about 5 ml of a solution containing 0.25 M sucrose, 50 mM Tris-Cl, pH 7.5, 25 mM KCl, and 5 mM $MgCl_2$, taking care not to disturb the pellet. This suspension of rough microsomes is taken to a volume of 100 ml with the same solution, gently homogenized, divided into 6 tubes, and centrifuged for 15 min at 35 000 revs/min (approximately 125 000 x *g*) in IEC rotor A211. The rough microsomes are washed again with about 100 ml of the sucrose salt solution, the only difference being that this time centrifugation is performed at 30 000 revs/min. These washings remove free ribosomes and large amounts of sucrose from the rough microsomal fraction.

The yield of rough microsomes is about 15 mg protein $g^{-1}$ liver. This fraction can be stored as a pellet at – 20°C, but such storage results in aggregation of the vesicles. To minimize aggregation, resuspend the rough microsomes in 12 ml of the sucrose-salt solution described in the foregoing paragraph, add 4–6 ml glycerol, and store at – 20°C. When it is time for them to be used, the microsomes can be recovered by diluting them at least five times with the sucrose-salt solution and centrifuging.

### *4.5.9 Separation of rough I, II, and III microsomes* [21,75]

Rough and smooth microsomes are separated using $Cs^+$-containing discontinuous sucrose gradients in the Spinco 40.2 rotor as described above with two modifications: the liver is homogenized in 0.44 M sucrose and the volume of the lower sucrose phase is 3 ml instead of 2. The two clear upper phases and the smooth microsomes at the boundary between the 0.6 and 1.3 M sucrose layers are removed. The 1.3 M sucrose layer containing the rough III microsomes is sucked off down to the upper edge of the pellet. The remaining sucrose solution above the pellet contains the rough II microsomes and is taken up with a Pasteur pipette. The rough I microsomes in the pellet are resuspended in 0.25 M sucrose. Rough II and III microsomes may be pelleted after dilution with water or sucrose, and all three subfractions may be washed as described above.

### *4.5.10 Separation of smooth I and II microsomes* [79,82]

Rough and smooth microsomes are separated using $Cs^+$-containing discontinuous gradients in the Spinco 40.2 rotor as described above. The smooth microsomes from two tubes are combined, diluted dropwise with water to 4.5 ml and mixed with an aliquot of 1 M $MgCl_2$ to give a final concentration of 7 mM $MgCl_2$. This preparation is then layered over 2 ml of 1.15 M sucrose – 7 mM $MgCl_2$ and centrifuged at 102 000 x *g* for 45 minutes in the Spinco 40.2 rotor. The clear upper phase is discarded. The fluffy layer at the gradient boundary is collected, diluted with 0.25 M sucrose, and recentrifuged at 105 000 *g* for 90 min to give smooth II microsomes. The pellet contains the smooth I microsomes.

## References


1. Weibel, E. R., Stäubli, W., Gnägi, H. R. and Hess, F. A. (1969), *J. Cell Biol.*, **42**, 68.
2. Orrenius, S., Ericsson, J. L. E. and Ernster, L. (1965), *J. Cell Biol.*, **25**, 627.
3. Conney, A. H. (1967), *Pharmac. Rev.*, **19**, 317.
4. Greengard, O., Federman, M. and Knox, W. E. (1972), *J. Cell Biol.*, **52**, 261.
5. Dallner, G., Siekevitz, P. and Palade, G. E. (1966), *J. Cell Biol.*, **30**, 73.
6. Dallner, G. and Azzi, A. (1972), *Biochim. biophys. Acta*, **255**, 589.
7. Petermann, M. L. and Pavlovec, A. (1961), *J. biol. Chem.*, **236**, 3235.
8. Autuori, F., Svensson, H. and Dallner, G. (1974), *Biochem. biophys. Res. Commun.*, **56**, 1023.
9. Thomson, J. F. and Klipfel, F. J. (1957), *Archs. Biochem. Biophys.*, **70**, 224.
10. Coleman, R., Michell, R. H., Finean, F. B. and Hawthorne, J. N. (1967), *Biochim. biophys. Acta*, **135**, 573.
11. Fleischer, B., Fleischer, S. and Ozawa, H. (1969), *J. Cell Biol.*, **43**, 59.
12. Oliver, I. T., Blumer, W. F. C. and Witham, I. J. (1963), *Comp. Biochem. Physiol.*, **10**, 33.
13. Criss, W. E. (1970), *J. biol. Chem.*, **245**, 6352.
14. Deter, R. L. and de Duve, C. (1967), *J. Cell Biol.*, **33**, 437.
15. Adelman, M. R., Blobel, G. and Sabatini, D. D. (1973), *J. Cell Biol.*, **56**, 191.
16. Adelman, M. R., Blobel, G. and Sabatini, D. D. (1974), *Meth. Enzym.*, **31**, 201.
17. Amar-Costesec, A., Beaufay, H., Wibo, M., Thinès-Sempoux, D., Feytmans, E., Robbi, M. and Berthet, J. (1974), *J. Cell Biol.*, **61**, 201.
18. Palade, G. E. and Siekevitz, P. (1956), *J. biophys. biochem. Cytol.*, **2**, 671.
19. Dallner, G. and Nilsson, R. (1966), *J. Cell Biol.*, **31**, 181.
20. Glaumann, H. and Dallner, G. (1970), *J. Cell Biol.*, **47**, 34.
21. Eriksson, L C. (1973), *Acta pathol. Microbiol. Scand., A*, Suppl. 239.
22. Nilsson, R., Peterson, E. and Dallner, G. (1973), *J. Cell Biol.*, **56**, 762.
23. Leighton, F., Poole, B., Beaufay, H., Baudhuin, P., Coffey, J. W., Fowler, S. and de Duve, C. (1968), *J. Cell Biol.*, **37**, 482.
24. Beaufay, H., Amar-Costesec, A., Thinès-Sempoux, D., Wibo, M., Robbi, M. and Berthet, J. (1974), *J. Cell Biol.*, **61**, 213
25. Jellinck, P. H. and Perry, G. (1967), *Biochim. biophys. Acta*, **137**, 367.
26. Murdoch, J. A. and Heaton, F. W. (1968), *Comp. Biochem. Physiol.*, **26**, 121.
27. Porter, K. R. and Bruni, C. (1959), *Cancer Res.*, **19**, 997.
28. Luck, D. J. L. (1961), *J. biophys. biochem. Cytol.*, **10**, 195.
29. Arion, W. J. and Nordlie, R. C. (1965), *Biochem. biophys. Res. Commun.*, **20**, 606.
30. Kamath, S. A., Kummerow, F. A. and Narayan, K. A. (1971), *F.E.B.S. Letters*, **17**, 90.
31. Schenkman, J. B. and Cinti, D. L. (1972), *Life Sci.*, **11**, 247.
32. Bjørnstad, P. (1966), *Biochim. biophys. Acta*, **116**, 500.
33. Waite, M. and van Deenen, L. L. M. (1967), *Biochim. biophys. Acta*, **137**, 498.

34. Tangen, O., Jonsson, J. and Orrenius, S. (1973), *Analyt. Biochem.*, **54**, 597.
35. Sabatini, D. D., Tashiro, Y. and Palade, G. E. (1966), *J. molec. Biol.*, **19**, 503.
36. Sabatini, D. D. and Blobel, G. (1970), *J. Cell Biol.*, **45**, 146.
37. Peters, T. (1962), *J. biol. Chem.*, **237**, 1181.
38. Campbell, P. N. and Lawford, G. R. (1968), Structure and Function of the Endoplasmic Reticulum in Animal Cells, (Cambell, P. N. and Gran, F. N., eds), pp. 57, Oslo University Press, Oslo.
39. Schachter, H., Jabbal, I., Hudgin, R. L., Pinteric, L., McGuire, E. J. and Roseman, S. (1970), *J. biol. Chem.*, **245**, 1090.
40. Redman, C. M. and Cherian, M.G. (1972), *J. Cell Biol.*, **52**, 231
41. Ernster, L. and Jones, L. C. (1962), *J. Cell Biol.*, **15**, 563.
42. Dallman, P. R., Dallner, G., Bergstrand, A. and Ernster, L. (1969), *J. Cell Biol.*, **41**, 357.
43. Svensson, H., Dallner, G. and Ernster, L. (1972), *Biochim. biophys. Acta*, **274**, 447.
44. Kreibich, G., Pascale, D. and Sabatini, D. D. (1973), *J. Cell Biol.*, **58**, 436.
45. Glaumann, H. and Dallner, G. (1968), *J. Lipid Res.*, **9**, 720.
46. Schramm, M., Eisenkraft, B. and Barkai, E. (1967), *Biochim. biophys. Acta*, **135**, 44.
47. Kreibich, G. and Sabatini, D. D. (1974), *Meth. Enzym.*, **31**, 215.
48. Kreibich, G., Hubbard, A. L. and Sabatini, D. D. (1974), *J. Cell. Biol.*, **60**, 616.
49. Jamieson, J. D. and Palade, G. E. (1967), *J. Cell Biol.*, **34**, 597.
50. Kuff, E. L. and Dalton, A. J. (1959), *Subcellular Particles*, (Hayashi, T., ed.), pp. 114, Ronald Press, New York.
51. Rostgaard, J. and Møller, O. J. (1971), *Exp. Cell Res.*, **68**, 356.
52. Wallach, D. F. H. and Kamat, V. B. (1966), *Meth. Enzym.*, **8**, 164.
53. Gahmberg, C. G. and Simons, K. (1970), *Acta pathol. microbiol. scand. B*, **78**, 176.
54. Goodland, G. A. and Mills, G. T. (1957), *Biochem. J.*, **66**, 346.
55. Redman, C. M., Grab, D. J. and Irukulla, R. (1972), *Archs. Biochem. Biophys.*, **152**, 496.
56. de Duve, C. (1971), *J. Cell Biol.*, **50**, 20D.
57. DePierre, J. W. and Karnovsky, M. L. (1973), *J. Cell Biol.*, **56**, 275.
58. Evans, W. H. (1969), *F.E.B.S. Letters*, **3**, 237.
59. Evans, W. H. (1970), *Biochem. J.*, **116**, 833
60. Widnell, C. C. (1972), *J. Cell Biol.*, **52**, 542.
61. Coleman, R. and Finean, J. B. (1966), *Biochim. biophys. Acta*, **125**, 197.
62. Benedetti, E. L. and Emmelot, P. (1968), *The Membranes: Ultrastructure in Biological Systems*, (Dalton, A. J. ed.), Vol. 4, p. 33, Academic Press, New York.
63. van Hoeven, R. P. and Emmelot, P. (1972), *J. Memb. Biol.*, **9**, 105.
64. Davison, S. C. and Wills, E. D. (1974), *Biochem. J.*, **140**, 461.
65. Keenan, T. W., Morré, D. J. and Huang, C. M. (1972), *F.E.B.S. Letters*, **24**, 204.
66. Ehrenreich, J. H., Bergeron, J. J. M., Siekevitz, P. and Palade, G. E. (1973), *J. Cell Biol.*, **59**, 45.
67. Campbell, P. N. (1970), *F.E.B.S. Letters*, **7**, 1.
68. Tata, J. R. (1971), *Subcellular Biochem.*, **1**, 83.

69. Rolleston, F. S. and Mak, D. (1973), *Biochem. J.*, **131**, 851.
70. Blobel, G. and Potter, V. R. (1967), *J. molec. Biol.*, **26**, 293.
71. Dallner, G. and Ernster, L. (1968), *J. Histochem. Cytochem.*, **16**, 611.
72. Moulé, Y., Rouiller, C. and Chauveau, J. (1960), *J. biophys. biochem. Cytol.*, **7**, 547.
73. Rothschild, J. (1963), *Biochem. Soc. Symp.*, **22**, 4.
74. Dallner, G., Bergstrand, A. and Nilsson, R. (1968), *J. Cell Biol.*, **38**, 257.
75. Eriksson, L. C. and Dallner, G. (1971), *F.E.B.S. Letters*, **19**, 163.
76. Blobel, G. and Potter, V. R. (1967), *J. molec. Biol.*, **26**, 279
77. Howell, R. R., Loeb, J. N. and Tomkins, G. M. (1964), *Proc. natn. Acad. Sci. U.S.A.*, **52**, 1241.
78. Bergstrand, A. and Dallner, G. (1969), *Analyt. Biochem.*, **29**, 351.
79. Dallner, G. (1963), *Acta pathol. microbiol. scand.*, Suppl. 166.
80. Dallner, G., Ernster, L. and Azzi, A. (1971), *Chem.-Biol. Interactions*, **3**, 254.
81. Eriksson, L., Svensson, H., Bergstrand, A. and Dallner, G. (1972), *The Role of Membranes in Secretory Processes*, (Bolis, L., Keynes, R. D. and Wilbrandt, W. eds), pp. 3, North-Holland Publishing, Amsterdam.
82. Dallner, G. (1974), *Meth. Enzym.*, **31**, 191.
83. Johannesen, K., DePierre, J. W. and Bergstrand, A. (1975), Abstract, *Tenth Fed. Europ. Biochem. Soc. Meeting*, Paris.
84. Jakobsson, S. and Dallner, G., Unpublished observations.
85. Baudhuin, P., Evrard, P. and Berthet, J. (1967), *J. Cell Biol.*, **32**, 181.
86. Ogur, M. and Rosen, G. (1950), *Archs. Biochem. Biophys.*, **25**, 262.
87. Ceriotti, G. (1955), *J. biol. Chem.*, **214**, 59.
88. Pitot, H. C. (1964), *Perspect. Biol. Med.*, **8**, 50.
89. Tata, J. R. (1967), *Nature*, **213**, 566.
90. Hinman, N. D. and Phillips, A. H. (1970), *Science*, **170**, 1222.
91. Larsén, C., Dallner, G. and Ernster, L. (1972), *Biochem. biophys. Res. Commun.*, **49**, 1300.
92. Bergman, A. and Dallner, G. (1974), Abstract, *Ninth Fed. Europ. Biochem. Soc. Meeting*, Budapest, pp. 222.
93. Lee, T.-C. and Snyder, F. (1973), *Biochim. biophys. Acta*, **291**, 71.
94. Nyquist, S. E., Matschiner, J. T. and Morré, D. J. (1971), *Biochim. biophys. Acta*, **244**, 645.
95. Wagner, R. R. and Cynkin, M. A. (1971), *J. biol. Chem.*, **246**, 143.
96. Molnar, J., Tetas, M. and Chao, H. (1969), *Biochem. biophys. Res. Commun.*, **37**, 684.
97. Dallner, G. and Hemming, F. W. (1974), Abstract, *Ninth Fed. Europ. Biochem. Soc. Meeting*, Budapest, 222.
98. Yamashita, S., Hosaka, K., Taketo, M. and Numa, S. (1973), *F.E.B.S. Letters*, **29**,235.
99. Behrens, N. H., Parodi, A. J., Leloir, L. F. and Krisman, C. R. (1971), *Archs. Biochem. Biophys.*, **143**, 375.
100. Ernster, L., Siekevitz, P. and Palade, G. E. (1962), *J. Cell Biol.*, **15**, 541.
101. Stetten, M. R. and Ghosh, S. B. (1971), *Biochim. biophys. Acta*, **233**, 163.
102. Dallner, G., Siekevitz, P. and Palade, G. E. (1966), *J. Cell Biol.*, **30**, 97.
103. Ernster, L. and Orrenius, S. (1965), *Fedn. Proc. Fedn. Am. Socs. exp. Biol.*, **24**, 1190.
104. Blad, D. and Dallner, G. (1972), Abstract, *Eighth Fed. Europ. Biochem. Soc. Meeting*, Amsterdam, 193.
105. Leskes, A., Siekevitz, P. and Palade, G. E. (1971), *J. Cell Biol.*, **49**, 264.

106. Remacle, J., Fowler, S., Beaufay, H. and Berthet, J. (1974), *J. Cell Biol.,* **61**, 237.
107. Wattiaux, R. and Wattiaux-de Coninck, S. (1970), *Biochem. biophys. Res. Commun.,* **40**, 1185.
108. Wattiaux, R., Wattiaux-de Coninck, S. and Collot, M. (1971), *Archs. int. Physiol. Biochim.,* **79**, 1050.
109. Wattiaux, R., Wattiaux-de Coninck, S. and Rouveaux-Dupal, M. F. (1971), *Eur. J. Biochem.,* **22**, 31.
110. Bronfman, M. and Beaufay, H. (1973), *F.E.B.S. Letters,* **36**, 163.
111. El-Aaser, A. A., Reid, E., Klucis, E., Alexander, P., Lett, J. T. and Smith, J. (1966), *Nat. Cancer Inst. Monograph,* **21**, 323.
112. Murphy, P. J., Van Frank, R. M. and Williams, T. L (1969), *Biochem. biophys Res. Commun.,* **37**, 697.
113. Schulze, H.-U. and Staudinger, H. (1971), *Hoppe-Seyler's Z. physiol. Chem.,* **352**, 1659.
114. Decloitre, F. and Chauveau, J. (1969), *F.E.B.S. Letters,* **2**, 227.
115. Glaumann, H. (1970), *Biochim. biophys. Acta,* **224**, 206.
116. Lewis, J. A. and Tata, J. R. (1973), *Biochem. J.,* **134**, 69.
117. Leskes, A., Siekevitz, P. and Palade, G. E. (1971), *J. Cell Biol.,* **49**, 288.
118. Schulze, H.-U., Pönnighaus, J. M. and Staudinger, H. (1972), *Hoppe-Seyler's Z. physiol. Chem.,* **353**, 1195.
119. Robertson, J. D. (1964), *Cellular Membranes in Development*, (Locke, M., ed.), pp. 1, Academic Press, New York.

# 5 The isolation and characterization of the nuclear envelope

JAMES R. HARRIS and PAUL S. AGUTTER



## 5.1 Introduction

### *5.1.1 Terminology*

The *nuclear envelope* is morphologically defined as the double membrane which is situated between the nucleoplasm and cytoplasm of eukaryotic cells and which is studded with octagonal structures, known as the nuclear pore complexes. This specialized double membrane system is often simply referred to as the nuclear membrane, both this and the term nuclear envelope being accepted nomenclature. In this review the term *inner nuclear membrane* (INM) which is adjacent to the nucleoplasm, and *outer nuclear membrane* (ONM) which borders onto the cytoplasm (with the perinuclear space between the two membranes) will be used precisely when possible, and the term *nuclear envelope* reserved for the relatively intact isolated double membrane, which is synonymous with the expression nuclear ghost. The term *nuclear envelope* (NE) will also be used to refer to torn sheets of double nuclear membrane (which are thus to be distinguished from sheets of INM or ONM alone). Very often during an isolation procedure the NE undergoes considerable disruption and the INM and ONM separate from one another. In this instance, the term *nuclear envelope* cannot be applied and the expression '*fragmented nuclear membrane*' or '*inner and outer nuclear membrane fragments*' is more appropriate. The limit of disruption is when membrane vesicularization occurs probably with a concomitant destruction of the nuclear pore complexes.

### *5.1 2 Some reasons for interest in the nuclear envelope*

Interest in the NE as a morphological entity became widespread after the system was identified electron microscopically around 1950 [1–3]. In the last 7 or 8 years biochemical interest in the nuclear envelope has grown rapidly and during this period several widely differing isolation procedures have been published. It should be emphasized at the outset that work in this field is still in its infancy and that it is therefore improbable that any single isolation method can as yet be regarded as wholly satisfactory. Nevertheless, it is accepted that progress towards an understanding of the physiological function of the NE system, together with accurate information on its chemical composition and enzymatic properties is largely contingent on the production of satisfactory isolated preparations.

One major reason for the extensive interest in the NE is its probable involvement in the control of the movement of various metabolites into and out of the nucleus, that is, in nucleo-cytoplasmic interactions [4,5]. Moreover, during cell division, the NE breaks down during late prophase and reforms with a net increase in area and number of nuclear pore complex during late telophase. It is not known for certain whether pieces of NE are attached to the dividing chromosomes or whether the envelope dissociates into 'monomers' which in either case reassociate to give the NE's of the daughter cells.

Available evidence has led some workers to suggest that the ONM may be functionally related to the endoplasmic reticulum [6], as in many cases ribosomes have been detected on its cytoplasmic surface [7] and certain enzyme activites of NE and the endoplasmic reticulum appear very similar. The INM which under physiological conditions is in intimate association with the nucleoplasm, has a layer of heterochromatin along its nucleoplasmic surface termed the internal dense lamina. This layer, which appears to give structural rigidity to the INM, does not cover the inner surfaces of the nuclear pore complexes, which therefore apparently have direct access to the nuclear contents. A possible functional relationship between nucleoplasmic DNA and RNA and the inner nuclear membrane is thus indicated [6]. Much however, remains to be clarified experimentally. One of the main aims of current research on the NE must be to provide some explanation for the existence of the nuclear pore complex in terms of its involvement in the metabolism of the cell.

Permeability studies on intact nuclei [8,9] have shown that most small molecules and many macromolecules can enter and leave the nucleus freely, but such studies have not as yet provided information concerning the movement of RNA and ribosomal ribonuclear proteins from nucleoplasm to cytoplasm, or the associated feed back of information from the cytoplasm to control gene function. It is probable that knowledge of this and other functions of the NE will be gained from the study of the system in isolation.

### *5.1.3 Ultrastructural background*

The ultrastructure of the NE has recently been reviewed in articles by Feldherr [9], Kessel [10], Wischnitzer [11] and Franke [12]. Some discussion of the ultrastructure of the NE is appropriate at this point since its appearance in the electron microscope at present constitutes the major criterion of acceptability for an isolated preparation. Indeed a large proportion of the information currently available about the system is of an ultrastructural nature.

The term *nuclear envelope* was first used by Anderson in 1953 [3]. Prior to this the existence of the system had been suggested by several light microscopic studies; it was first shown to be a two layered membrane system by the electron microscopic studies of Callan and Tomlin [2]. These workers showed that the NE was associated with, and penetrated by, pore structures currently termed nuclear pore complexes. The discoveries of Callan and Tomlin which were obtained using air dried specimens of isolated oocyte NEs were subsequently confirmed by others using thin sectioning [13–17]. The intimate association of the nuclear pore complex with both INM & ONM has been deduced from many studies on intact tissues [18–22] and isolated NEs [5,23–25]. The double membrane state of the NE seems to be a universal phenomenon, as does the pore complex; but the number of pore complexes per unit area (pore density) varies greatly between species and tissues [14,26–33].

There is evidence that the two membranes of the NE differ considerably in their composition, ultrastructure and stability properties. The ONM is usually studded with ribosomes [7,34] and is removed in nonionic surfactants such as Triton X–100 [35] as well as in low pH media such as 1% citric acid [36], both treatments leaving the INM apparently intact. The closeness of contact between the internal dense lamina, which is also termed the granular perinuclear layer [37] or the peripheral heterochromatin layer [38], and the INM probably confers special properties of organization on the INM and generates a problem of intractable contamination of the NE with chromatin throughout any isolation procedure. The complementary contamination problem regarding the ONM with cytoplasmic material, notably endoplasmic reticulum [34], is also acknowledged. It should, nevertheless, be borne in mind that an apparent ultrastructural continuity between the ONM and the endoplasmic reticulum does not necessarily entail a similarity in composition and enzyme content. The heterogeneity of the latter membrane system within itself is, in any case well established. [39].

Some early accounts of the structure of the nuclear pore complex are remarkably detailed [16,19,40–42]. Recent reports have made use of high electron optical magnifications (derived from thin sectioning, negative staining and metal shadowing) in attempts to construct models of the nuclear pore complex [20–23,25,43–48]. Briefly, it appears that in all cases the lumen of the pore is surrounded by an annulus which overlaps the membrane margin. This annulus may further be resolved into an octagonal array of annular subunits approximately 25 nm in diameter. The lumen often, though by no means always, contains a central granule which may be attached by fibrils to the annular subunits or to the membrane margin itself. It has been proposed that both the central granule and the annular subunits have a *mini tubule* or hollow cylinder type of structure [20,25,40], and further, that the macromolecular organization underlying the hollow cylinder of the annular subunits may itself have a sixteen fold symmetry [25]. Despite these recent advances the details of the overall structure of the nuclear pore complex are far from clear, particularly when it is desired to express this structure in terms of a macromolecular arrangement. It is perhaps worth re-emphasizing the current view that the nuclear pore complex is likely to constitute more than a mere opening for passive diffusion of protein, RNA or polysomes [48].

The salient features of the NE described above are shown diagrammatically in Fig. 5.1. A selection of electron micrographs of intact nuclei are shown in Fig. 5.2 which illustrates the various aspects discussed.

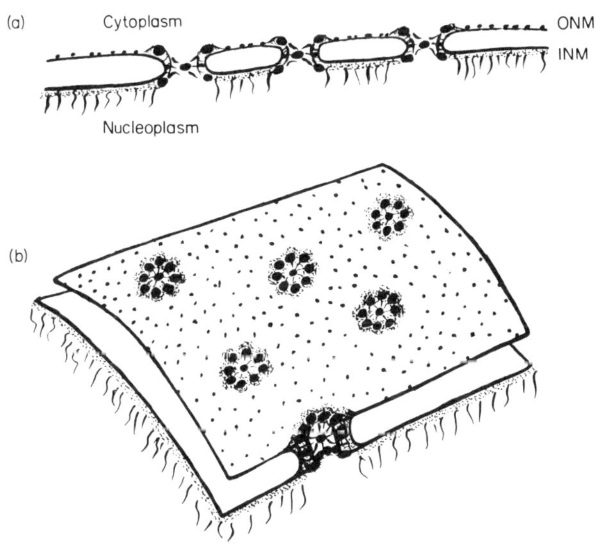


**Fig. 5.1** A diagrammatic representation of the nuclear envelope and the nuclear pore complexes. (a) Perpendicular to the two nuclear membranes. (b) Tangential to the two membranes. The cytoplasmic surface of the outer nuclear membrane (ONM) is studded with ribosomes. The annuli of the nuclear pore complexes project from the surface of the ONM and extend through to the inner nuclear membrane (INM). The internal dense lamina and chromatin fibrils are indicated on the nucleoplasmic surface of the INM.

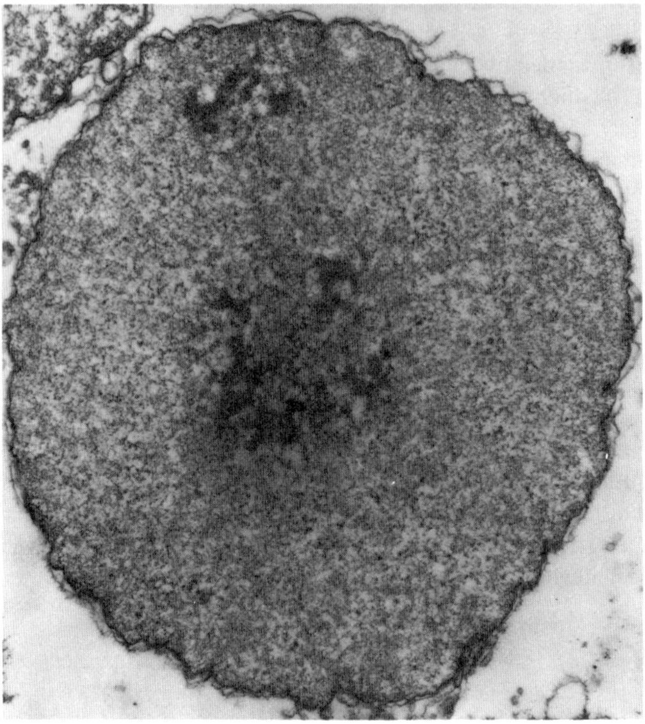

**Fig. 5.2a** A purified rat liver nucleus revealed by thin sectioning. (x20 000.)

### *5.1.4 Criteria for evaluating an isolated nuclear envelope preparation*

*(i) Enzymic markers*

To date no acceptable enzymic activity peculiar to the NE has been described; a positive enzymic marker is therefore not at present available. Markers for contamination can, however, usefully be employed. For example, nuclei lack succinate dehydrogenase activity (for assays see [49]), therefore mitochondrial contamination can be estimated. Acid phosphatase can similarly be used as a lysosomal marker [50].

A suitable marker for endoplasmic reticulum has hitherto proved elusive. The NE contains cytochrome $b_5$ and NADPH oxidase activity [51,52], and glucose-6-phosphatase activity seems to be present [52–54] although it has been reported to be absent in some preparations [55–57]. This absence has been attributed to the inherent instability of glucose 6-phosphatase (R. R. Kay personal communication) and this matter must therefore remain in some doubt. The authors have failed to demonstrate monoamine oxidase activity in their NE preparations, but this is contrary to the claims of Gorkin [58] who showed a

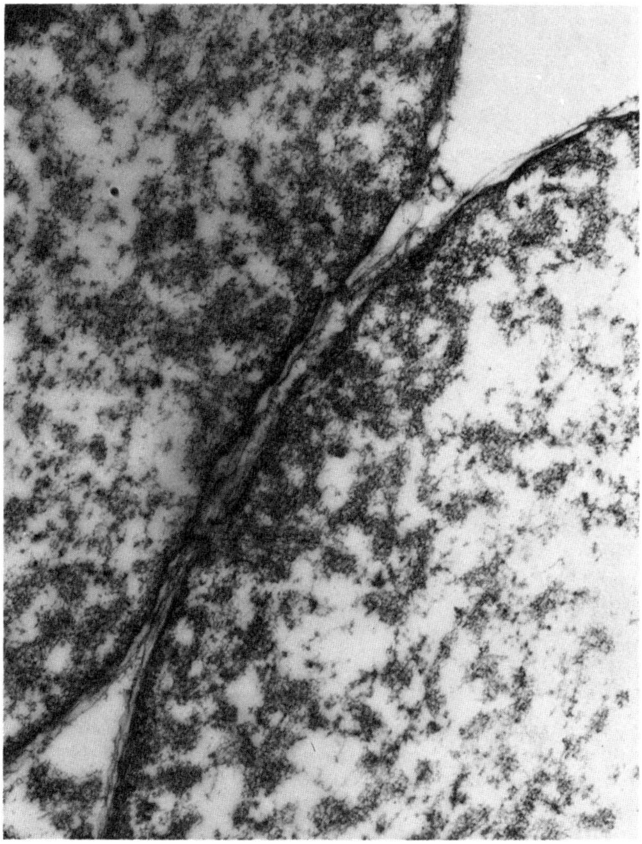

**Fig. 5.2b** Part of two thin section rat liver nuclei, showing clearly the ONM and the INM with its internal dense lamina of heterochromatin. (x46 000.)

significant increase in the specific activity of monoamine oxidase of rat liver NE over and above that of other cellular fractions. This major difference remains to be settled, but it does perhaps serve to underline the enzymic variability that may follow from the varying membrane isolation procedures. For example, the ultrasonication used by Agutter [57] could well inactivate monoamine oxidase. It must, nevertheless be pointed out that different substrates have been used for monoamine oxidase measurements, and that the enzymes assayed are not necessarily identical. The most comprehensive survey of the enzymes associated with the NE available to date is included in the review by Franke [6].

*(ii) Ultrastructural markers*

Since the pore complexes are peculiar to NE and the annulate lamellae (which are thought to be derived from the NE's), this morphological feature and the

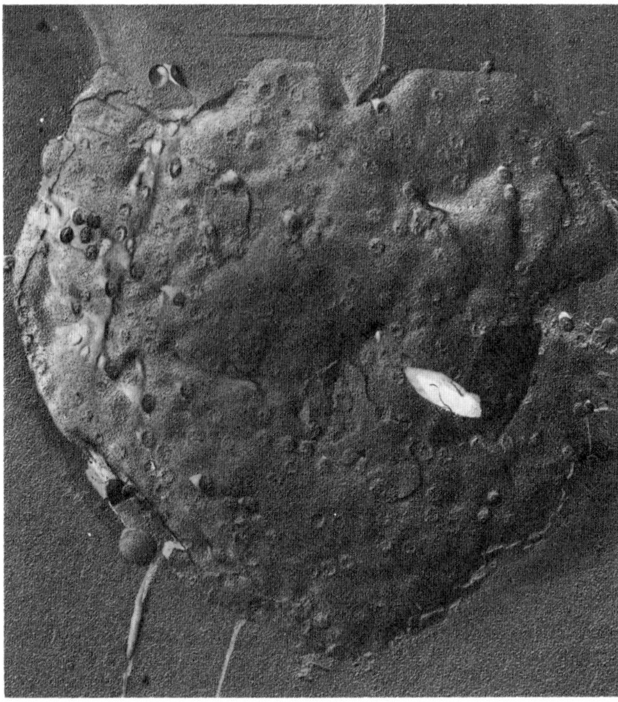

**Fig. 5.2c** A purified rat liver nucleus revealed by freeze-etching. The nuclear pore complexes are seen to project from the nuclear surface. (x23 000.)

double membrane status of the system have been taken in most laboratories to serve as structural markers for isolated NE in the absence of suitable enzymic markers. Indeed, it must be accepted that both the definition of the isolated NE, and the primary criteria for its evaluation, are based on ultrastructural data. For this reason it is at present essential for those working on the isolation of NE to have a detailed knowledge of those ultrastructural features which are, or should be, visible in the electron microscope under defined conditions of fixation and staining.

Negative staining has been employed with a considerable degree of success to reveal the ultrastructural detail of the NE [5,23–25,45,59]. It is claimed that this technique is able to reveal the macromolecular substructure of the nuclear pore complex [25], although the thickness of these structures presents a considerable obstacle and the results obtained are not as easily interpreted as those obtained with isolated macromolecules [60]. Of the negative stains commonly employed ammonium molybdate appears to provide the greatest detail within the nuclear pore complex [24]. Apart from the integrity of the nuclear pore complex, negative staining readily provides an assessment of the state of fragmentation of the isolated NE. Thus at low electron optical magnifications the technique can be used as an extension of light microscopy (phase contrast) when preparations of intact nuclear ghosts are obtained and it will also provide a qualitative assessment of the relative proportion of intact nuclear ghosts to torn sheets of NE. Of prime importance is knowledge of whether or not the INM and ONM are both present. This information is obtained from the negatively stained electron optical image of the folded edges of nuclear ghosts or torn sheets, two electron transparent lines being revealed [24,25]. Further to this, a classification can be assigned to a given membrane preparation as to whether it consists of intact nuclear ghosts, large torn sheets of double nuclear membrane (NE), or small fragments or vesicles of nuclear membrane (probably a mixture of both INM and ONM.)

Thin sectioning has been employed for the study of NE for a longer period than negative staining. The conventional fixation and staining methods reveal the double membrane nature of nuclear ghost preparations and the presence of nuclear pore complexes [24,52]. The pore complexes result in two electron optical images, one which shows a coming together of the two nuclear membranes is produced by transverse sections through the membrane, and the other which shows the octagonal densely staining annulus and central granule is obtained by sections parallel to the membranes. Small membrane fragments and vesicle preparations can likewise be appraised by thin sectioning [51,53,56,62–66] but in these cases it is more difficult to assess the state of integrity of the nuclear pore complexes and the presence of both nuclear membranes. Some qualitative indication of the presence of chromatin on the inner surface of the INM is obtained by thin sectioning [24,54,67].

Freeze cleavage and etching have been applied extensively for the study of nuclei within cells [27–33,68–74] and for isolated nuclei [24]. The technique has not yet been widely applied to isolated NE preparations [29,62]. It is doubtful whether, to date, freeze-cleavage has revealed as much structural detail of the NE as thin sectioning and negative staining. It most certainly is of value for the study of the nuclear pore density in different tissues at different stages of the cell cycle and growth, and following the application of stimulatory or inhibitory reagents to cells [28,69].

In general, NE preparations to be acceptable must consist of the two nuclear membranes linked by pore complexes, the annular subunits and central granules of the latter being visible under appropriate conditions of negative or positive staining in the electron microscope. The degree of morphological integrity required as the criterion of a satisfactory envelope preparation remains a matter for the individual worker to decide.

A strong case can be made for demanding that nuclear ghosts or *intact envelopes* are the most desirable state in which the isolated material should be obtained. By analogy more has been learnt about the functional properties of the erythrocyte membrane from studies using resealed ghosts than from studies with membrane fragments. Yet studies directed towards the biochemical composition of the erythrocyte membrane have made use ghosts and membrane fragments, the same situation apparently existing for studies with the NE.

The foregoing discussion has stressed the primacy of membrane integrity over purity. This would seem to be logically necessary in view of our present level of knowledge of the NE and appropriate in view of the comparative ease with which ultrastructural markers may be demonstrated in this system. Some discussion of purity, although an extremely difficult matter, is clearly relevant to the establishment of our criterion of a satisfactory NE preparation. Two possible sources of contamination will therefore be considered, the nucleoplasm and the cytoplasmic membrane systems.

### *(iii) Nuclear contamination*

On the basis of the assumptions that most or all of the lipid of the cell nucleus is in the NE, and that none or only a small fraction of the DNA is so intimately associated with the INM or pore complex as to be regarded as a structural part of the NE, most workers in this field have sought to produce preparations in which density and DNA content are minimized, and lipid content high. Similar considerations do not apply to RNA since ribosomes are known to be associated with the ONM [7,34]. Some evidence has been presented which suggests that the ONM [75] and the nuclear pore complexes [76] may be rich in RNA.

It must be noted in this context that, while available evidence is consistent with the association of most of the nuclear lipid with the NE [36], the extent of intimate DNA association with the system remains in some doubt [57,62]. The report by Kashnig and Kasper [53] has claimed a DNA-free preparation. In our hands, however, this method has yielded material containing 0.5–1.0% by weight of DNA. It should be pointed out in this context that the diphenylamine assay procedure for DNA developed by Burton [77] and used by Kashnig and Kasper is less sensitive than the modification of this procedure by Giles and Myers [78] used by ourselves; the amount of DNA in the preparation of Kashnig and Kasper is certainly below the detection limits of the Burton method. Further discussion of the problem of whether or not DNA is to be considered as a contaminant or a structural component of the system will be given below.

### *(iv) Cytoplasmic contamination*

The apparent continuity between the ONM and endoplasmic reticulum, by its very nature, constitutes an *a priori* problem of cytoplasmic contamination. Scanning electron microscopy (R. R. Kay personal communication) and freeze cleavage readily indicate that some membranous material often adheres to the surface of purified cell nuclei, which is less apparent by thin sectioning. It does not of course follow that any such contamination will remain attached to the NE throughout its subsequent purification. The problem of cytoplasmic contamination is made more acute by the lack of a suitable microsomal marker enzyme. Until the extent of microsomal contamination can be reliably estimated, detailed studies on NE composition must be regarded as premature. Apparent similarity between NE and microsomal preparations in terms of protein components may indicate no more than heavy endoplasmic reticulum contamination of the preparation [56,62]. In practice, when intact ghosts and large sheets of NE are isolated, the presence of any membrane material which lacks nuclear pore complexes is readily revealed by negative staining. It is perhaps pertinent that small amounts of collagen can also be detected by this method.

### *5.1 5 General comments*

To summarize the foregoing discussion the following steps are considered vital for the development of a satisfactory NE isolation procedure. (a) Cell nuclei should be obtained as free as possible from cytoplasmic contamination. (b) The nuclear contents (nucleoli, chromatin and dense lamina of heterochromatin) should be rendered soluble and thereby liberate the NE. (c) The required ultrastructural characteristics of the NE should be maintained (double membrane, pore complexes with intact annuli and central granules).

A parallel can be drawn between the release of haemoglobin from mammalian and avian erythrocyte ghosts by osmotic haemolysis [79–81] which leads to the production of haemoglobin-free erythrocyte ghosts, and procedures aimed at bringing about the release of chromatin from nuclei. This comparison suffers from the fundamental fallacy that chromatin, unlike haemoglobin, is not readily soluble under physiological or low ionic strength conditions. As will be seen from the experimental section, additional treatments have been devised to bring about the solubilization of chromatin and in doing so the situation with the lysed or disrupted nucleus may be considered at least partially analogous to the haemolysed erythrocyte.

Most studies on NE have been performed with mammalian liver, that of the laboratory rat being most commonly used, although pig, cattle and rabbit liver have been used in some cases. A small amount of work on the avian erythrocyte NE has been published [66,82]. These dormant nuclei have extremely condensed chromatin which in the experience of the authors is even more intractible than that of liver nuclei. For the sake of completeness it should be mentioned that a considerable amount of work has been done using the large amphibian oocyte NE which is isolated by micro-dissection [5,23,76,83].

## 5.2 A review of methodology

### *5.2.1 Isolation of nuclei*

All procedures described for the isolation of NE have begun with the isolation of purified cell nuclei (with one exception [61], later modified [84] to include initial purification of nuclei). Of all the methods available in the literature for nuclear isolation, only those which (a), make use of low shearing stresses during homogenization, ensuring a fairly high yield of unbroken nuclei and (b), utilize centrifugation through dense sucrose media (approx. 2.3 M specific gravity = 1.29) to separate nuclei from all other organelles are applicable. Thus a Dounce- or Potter-Elvehjen homogenizer is greatly preferable to a Waring blender for homogenizing tissue. The recently introduced blade homogenizers of the Polytron and Ultra Turrax type, which can be precisely controlled at low speeds have been found to be successful by the authors, and enable very large quantities of tissue to be rapidly processed. Ultrasonication has also been used to release avian erythrocyte nuclei from erythrocyte plasma membranes [81]. Detergent-containing and organic solvent systems which give highly purified nuclei clearly do so at the expense of the integrity of the NE and should therefore be avoided. Moreover, high ionic strength solution should not be used as these may remove the ONM [85] and undoubtedly disorganize and partially solubilize the chromatin [51,53,56,86]. As previously noted, low pH conditions also remove the ONM [36,75].

The pH of the medium used is most commonly in the range 7.0 to 7.5 [51–53,55,56,61,84] although pH values of about 6.0 have been used [54,57,59], since this minimizes the proteolytic activity associated with crude nuclear preparation [87]. It should, however, be noted that proteolytic activity is not detectable at any pH in nuclei once they have been purified by sedimentation through 2.3 M sucrose. To stabilize the chromatin a divalent cation ($Mg^{2+}$ or $Ca^{2+}$) should be included in the medium at 1 mM to 3 mM concentration. The original dense sucrose method for isolating nuclei [88] did not use buffer or a divalent cation; subsequent modifications in which pH is controlled and magnesium or calcium is added have undoubtedly improved the yield, integrity, and probably purity, of the preparations.

Homogenization of the tissue in 2.3 M sucrose can lead to undesirable local heating effects, as a result of applying shearing stress to a highly viscous medium. Since maintenance of the temperature in the range 0 to 5°C throughout the procedure is essential, it has become customary to homogenize in dilute sucrose solutions (0.25 to 0.44 M), suitably buffered and containing divalent cations and then to prepare a crude 'nuclear pellet' fraction by centrifugation at 600–1000 x *g*. This nuclear pellet may then be purified further by resuspension in a concentrated sucrose solution (approx. 3.0 M) to give finally 2.3 M sucrose (60% w/w). The nuclei are then separated from all other less dense organelles by centrifugation at approximately 60 000 x *g* for 1 h.

A large number of methods for the isolation of nuclei are available in the literature, all essentially fulfilling the requirements discussed above. The reader not familiar with the field who wishes to consider the relative merits of the most widely used isolation procedures may be recommended to study the purity of the preparations obtained by Blobel and Potter [89], Widnell and Tata [90], Sporn *et al.* [91], Harris and Agutter [59] and Harris [25].

### *5.2.2 Purity of the isolated nuclei*

During the course of isolation nuclei can be routinely examined by light microscopy using phase contrast or Nomarski interference optics and less routinely electron microscopy using thin sectioning or freeze cleavage, in order to obtain a qualitative assessment of purity. In general, mitochondria and other membranous contaminants are not found after pelleting through dense sucrose Visible contamination is normally restricted to small collagen fibres and occasional peroxisomes, which will usually be removed at a later stage if the NE is obtained by density gradient floatation. Glycogen granules can equally be eliminated by starvation of the animals for 16–24 h before sacrifice.

The absence of mitochondria can be confirmed by succinate dehydrogenase assay [49]. 5′-Nucleotidase [92] provides a suitable marker for plasma membrane contamination, and acid phosphatase [50] for lysosomes.

### *5.2.3 The disruption of nuclei*

Isolation of the nuclear envelope is necessarily preceded by disruption of the nuclei; the object of this process is to disperse the nucleoplasm while controlling conditions so as to retain the envelope is a state of maximum integrity. The principle methods used for disruption are outlined below.

Perhaps the simplest approach, which unfortunately is not widely applicable is that of manually bursting and washing the ghosts of nuclei following microdissection. This technique has in fact only been successfully applied to the large nuclei of amphibian oocytes [1,2,5,23,76,83]. Extensive ultrastructural studies have been performed with material isolated in this way, together with microanalysis [83]. An alternative approach using mechanical stress is ultrasonication [54–57,93]. Ultrasonication ruptures the envelopes and under controlled conditions of pH, ionic strength and temperature liberates the envelope fragments from appended heterochromatin without destroying the ultrastructural integrity of the pore complexes.

Treatment of nuclei in low ionic strength solutions in the absence of divalent cations brings about a pronounced swelling due to the change in condensation state of the chromatin [94,95]. The importance of calcium and magnesium in maintaining the condensed state of chromatin [96,97] implies that chelating agents may help to disperse chromatin while leaving the NE intact. This effect may in part account for the chromatin dispersal achieved by Kashnig and Kasper [53] using a 10% w/v potassium citrate solution as well as the heparin treatment of Bornens [64]. In the hands of the authors, low ionic strength treatments in the absence of chelating agents generally lead to the production of an extremely gelatinous suspension of nuclei (see Fig. 5.3), from which the chromatin is escaping, and which can nevertheless be seen to resemble nuclear ghosts. In the procedure devised by Price *et al.* [61], in which the nuclei were not purified, the crude 1000 x *g* nuclear pellet in 1 mM $NaHCO_3$ (pH 7.2) being used as the starting material for the rate zonal separation of nuclear ghosts the gelatinous state dispersed spontaneously during overnight storage at 4°C. This does not occur if purified nuclei that have become gelatinous and burst after washing in 1 mM $NaHCO_3$ are stored overnight. Dissolution of the gelatinous state can, however, be brought about by a brief treatment with DNAase I [84]. Recently, it has also been observed that when purified nuclei from rat liver are

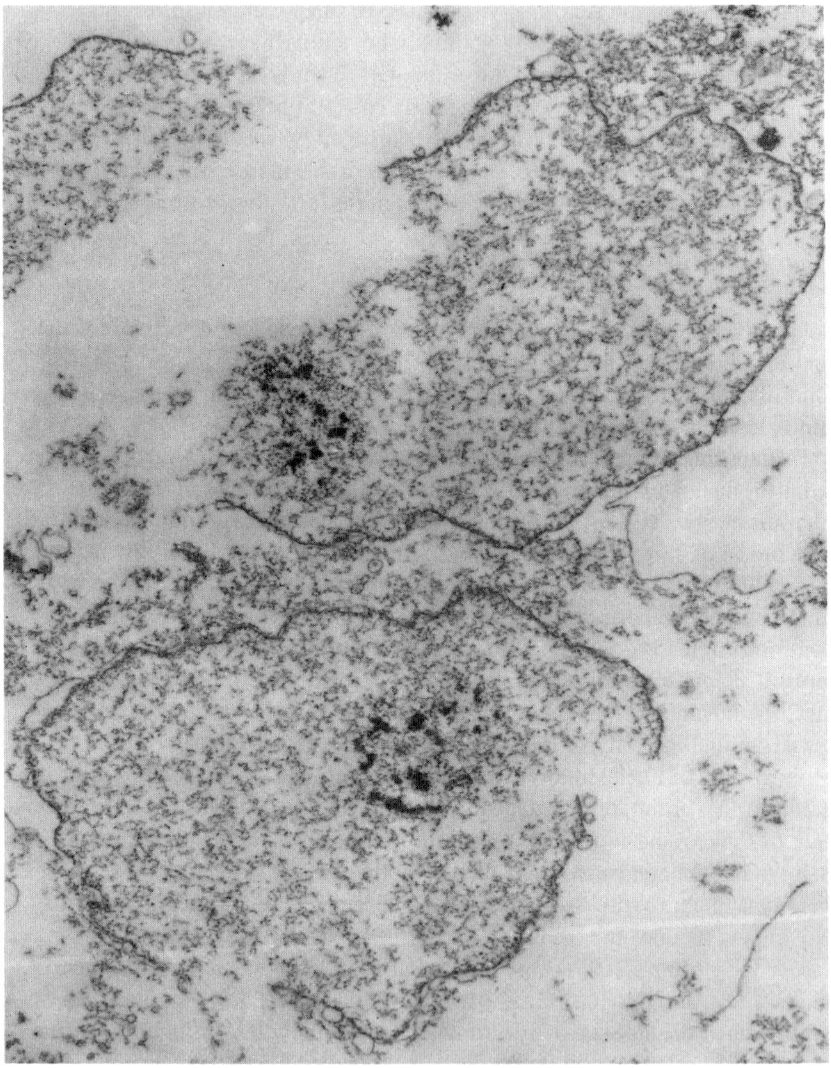

**Fig. 5.3** Rat liver nuclei after washing two times in 1 mM sodium bicarbonate (pH 7.2). The nuclei have burst, thus allowing the chromatin and nucleoli to escape. The nuclear envelope does not break up significantly. (x10 000.)

stored overnight at 4°C in 0.25 M sucrose, 2 mM $Mg\,Cl_2$, 10 mM Tris HCl (pH 7 4), they will then burst and loose their contents without going through a gelatinous phase when washed in 1 mM $NaHCO_3$ solution (J. R. Harris unpublished observations). Nuclear envelopes (ghosts and large torn sheets) have been isolated by differential centrifugation followed by density gradient floatation of a purified membrane band by both the above procedures (see Section 5.4.3 below). The suggestion may be advanced that in both the method of Price *et al.* [61] and the overnight treatment of purified nuclei in isotonic sucrose containing magnesium chloride, spontaneous breakdown of chromatin occurs due to endogenous DNAase but it is difficult to understand why this should not occur with the purified nuclei after bursting in 1 mM $NaHCO_3$. One possibility is that such endogenous DNAase is Mg-dependent and the sodium bicarbonate treatment has removed much of the chromatin bound magnesium.

An alternative approach stems from the assumption that salt bridges constitute the most significant stabilizing force in chromatin, which suggests that dispersal of the nuclear contents will occur in high ionic strength media Solutions containing 1 mM NaCl [98], 1.5 M KCl [56], 10% tripotassium citrate [53], and 0.5 M magnesium chloride [51] have been employed to disperse nuclei. The nuclear contents are usually made labile either by DNAase treatment or ultrasonication prior to the addition of the high ionic strength medium. Magnesium chloride alone at a concentration of 0.3 M to 0.5 M has been used by Monneron and her co-workers [62] Despite claims that the envelope fragments obtained after such treatments are ultrastructurally acceptable, it is clear the potassium ions in the medium destroy pore complexes [54,57,62] under conditions of neutral pH. Moreover high ionic strength media may remove both lipid and protein components from other membrane systems [99], and possibly also from the NE. After isolation of nuclei in phosphate buffer at pH 6.1 [54,57], high ionic strength treatment was found to destroy the nuclear pore complexes; this apparently is not the case if nuclei are isolated at neutral pH [51,62].

At pH 8.6 nuclei destabilized by mild ultrasonication or DNAase treatment swell irreversibly and the bulk of their contents are solubilized. The explanation for this effect probably lies once again in the disruption of salt bridges stabilizing the chromatin. Kay *et al.* [52] found that after treatment at pH 8.6 almost intact nuclear envelopes could be isolated, which contained ultrastructurally intact pore complexes. Such procedures might, however, be expected to result once again in the loss of membrane components [100], and the departure from physiological pH may produce inactivation of some enzymes, in particular the envelope Mg-ATPase (Agutter, unpublished observations).

Destabilization of chromatin by DNAase, which enters the isolated nucleus readily [101], but which may not have access to the DNA within condensed chromatin or be able to free the envelope of appended heterochromatin, has been employed in combination with the above mentioned approaches. Berezney *et al.* [51] used overnight incubation of nuclei with DNAase at 4°C prior to final dispersion of the nucleoplasm with 0.5 M magnesium chloride. Kay *et al.* [52] used DNAase at room temperature together with high pH disruption of the chromatin. Comparison of these two publications indicates that the nucleoplasm is very much more DNAase labile at room temperature (20°C) than at 4°C. Differences in enzyme activity alone do not wholly account for this observation; it is probable that the structure of chromatin differs sufficiently at the two temperatures and pH's to retard the digestion of DNA markedly under the conditions used by Berezney *et al.* [57]. The brief DNAase I treatment used by Harris and Milne [84], mentioned above in relation to the low ionic strength and magnesium-depleted conditions, may owe its success to the fact that the chromatin was considerably dispersed prior to the addition of DNAase due to washing in 1mM sodium bicarbonate solution, together with the fact that sufficient magnesium may remain bound to the chromatin to activate the exogenous enzyme.

Despite the success of DNAase treatment for the isolation of nuclear envelope, one report [57] suggests that this enzyme disrupts the nuclear pore complexes and that, concommitant with this disruption, collapse of the double-membrane structure into single-membrane vesicles occurs. Further investigation of this point suggests that the envelopes isolated from nuclei prepared in phosphate buffer at pH 6.0 have DNAase-labile pore complexes, while those from nuclei prepared in Tris buffer at pH 7.4 do not (c.f. ionic strength methods above). Again, alteration of the structure of the nuclear contents seems to be indicated; change of pH effects, presumably an irreversible structural change in the envelope and peripheral heterochromatin which could make the integrity of the nuclear pore complexes dependent on the presence of DNA.

From such observations it may be concluded that the environment of the nuclei during and after isolation has a marked effect on the structure of the nucleoplasm; some changes in this structure are apparently irreversible Such factors are of considerable importance in envelope isolation, since the structure of the nucleoplasm determines the ease with which it can be dispersed under any given conditions. Prior to any procedure for isolation of the nuclear envelope, particular attention must therefore be paid to the conditions under which the nuclei are prepared. Generally, it seems desirable to prepare nuclei in Tris buffer at pH 7.4 in the presence of 2 or 3 mM magnesium chloride, and to pass through the purification procedure as rapidly as possible to minimize degradation due to endogenous proteases in the homogenate and crude 1000 x *g* nuclear pellet

### *5.2.4 Isolation of the nuclear envelope from disrupted nuclei*

Following the disruption of cell nuclei by the above methods, centrifugation procedures are employed to separate the NE from the solubilized chromatin. Differential centrifugation has been employed by several workers [51,52,98], and also used in conjunction with density gradient centrifugation [25,84].

Repeated differential centrifugal washes can be given rapidly, which may be of advantage for enzymological investigations [52]. A major disadvantage is that trace contaminants (e.g. collagen and nucleoli) may be repeatedly packed along with the NE material, which is likely itself to become increasingly fragmented during the successive pelleting and resuspension treatments that are applied to liberate chromatin. Measurements of floatation density naturally require additional gradient centrifugation, usually in sucrose.

It is the opinion of the authors that the use of sucrose density gradient centrifugation following a series of differential centrifugal washes is likely to result in the production of NE of the greatest purity, since contaminants such as other membrane species, collagen and nucleoli as well as residual chromatin are unlikely to band at the same position as the NE on a gradient. This does of course depend on the absence of binding between NE and such contaminants. Nevertheless if one was satisfied that the nuclei used were of maximum possible purity and the disruptive treatment produced complete solubilization of the nucleoplasm, including the heterochromatin of the internal dense lamina then, in theory, repeated differential centrifugation might result in a final membraneous material not markedly more contaminated than that obtained by density gradient centrifugation; unfortunately, in practice these conditions cannot be achieved. In fact Zbarsky *et al.* [55] claimed that repeated isopycnic banding in density gradients was necessary to yield a satisfactory pure NE preparation.

The recent availability of zonal centrifugation rotors has enabled larger quantities of tissue to be processed for the isolation of cellular organelles than is possible with conventional fixed angle and swing out rotors. In addition a superior purification is generally achieved. Low speed rate zonal centrifugation can be considered roughly equivalent to differential centrifugation for purposes of the present discussion. Price *et al.* [61,102] employed this technique using the M.S.E. A–XII Zonal rotor to separate nuclear ghosts from the 1000 x g nuclear pellet of rat liver and hepatoma homogenized in 1 mM sodium bicarbonate (pH 7.2) and kept overnight at 4°C. Contamination of the nuclear ghosts with large sheets of plasma membrane (of equivalent sedimentation coefficient to the nuclear ghosts) necessitated a further isopycnic zonal centrifugation in a high speed B–XIV Zonal rotor, or alternatively in a swing out bucket rotor. In the former instance it was not necessary to pellet the nuclear ghosts taken from the initial rate zonal separation since a large sample volume can be applied on top of a sucrose gradient in the B Zonal rotor. The differing flotation densities of NE and plasma membrane enable a purification of intact nuclear ghosts to be achieved. One major drawback of the use of zonal rotors is the time required in setting up the gradients together with the dismantling and cleaning of the rotors and accessory equipment following each run. Nevertheless, if as suggested above, superior purification and increased quantities if intact nuclear ghosts can be obtained and repeated pelleting avoided, then the use of zonal rotors is amply justified.

### *5.2.5 Composition of the isolated nuclear envelope*

Marked differences in the ultrastructure, chemical composition and enzyme content have been found in the NE material obtained by the various published methods. A comparison of these variables is given in Tables 5.1, 5.2 and 5.3. As indicated previously, most of the procedures described have been established for mammalian liver. Studies have been performed on other tissues, e.g. Ehrlich ascites [55], chicken erythrocyte [67,82,103], calf thymus [66,104] and pea seedlings [54], but in general it is apparent that any one procedure is unlikely to be directly transferable to another tissue without some modification. The results tabulated have been taken only from studies on mammalian liver.

With reference to Table 5.1, protein has been estimated by the method of Lowry *et al.* [105] and DNA by the Burton procedure [77] usually in the modification by Giles and Myers [78]. Using mouse liver, NE prepared by the method of Agutter [57], it has been found that the ratio of satellite DNA to main band DNA in the envelope does not differ significantly from that found in whole nuclei (Agutter, unpublished observations). This suggests that the DNA associated with the NE is a random fraction of total nuclear DNA [57], although the studies of Franke *et al.* [106] have suggested a moderate enrichment of satellite DNA. The RNA present in most NE preparations is possibly, in view of the ribosomes universally found on the ONM, an rRNA fraction. The low RNA content of the NE reported by some authors [55,56,62] may suggest that some or most of these ribosomes have been removed during their isolation procedures.

Detailed studies of the protein [56,62,65] and lipid [107,108] contents of the NE have suggested that it differs from rough endoplasmic reticulum, but that the lipid content correlates well with that of whole nuclei. This latter observation is consistent with the assumption that all the nuclear lipid is associated with the NE, and therefore supports the use of phospholipid recovery from the nuclei as a measure of yield of envelope material.

The flotation densities of the various NE preparations range from 1.16 to 1.21 (Table I). Most values have been obtained in sucrose but determination in sorbitol have produced very much higher figures [54,57]. The range of values undoubtedly emphasizes the varied separation of INM and ONM and integrity of the nuclear pore complexes, together with nucleic acid content. At the moment it must be stated that no absolute figure for the density of NE can be given although the authors have reproducibly obtained the value of 1.21.

All attempts to produce NE have been accompanied by electron microscopic studies, by which means the integrity of the material, as well as the contamination with other organelles, can be qualitatively assessed. In Table 5.2 the ultrastructure of various NE preparations have been categorized in terms of the presence of double or single nuclear membranes, intact ghosts, large and small sheets, and vesicles. Also indicated is the integrity of the nuclear pore complex and the presence or absence of the central granule. Most investigators

TABLE 5.1
*Nuclear envelope composition*

| Tissue used | Percentage by weight | | | | | Density g ml$^{-1}$ | Remarks | References |
|---|---|---|---|---|---|---|---|---|
| | Protein | Lipid | | DNA | RNA | | | |
| | | Phospholipid | Neutral lipid | | | | | |
| Bovine liver | 75 | 14 | – | 0.9 | 8.9 | – | No density gradient used | [51, 102] |
| Rat liver | 62–70 | 25–33 | – | 3.5–4.0 | 3.5–4.0 | – | No density gradient used | [52, 113] |
| Rat liver | 59 | 29 | 6 | 0.0 | 3.1 | 1.16–1.18 | Light band | [53, 65] |
| Rat liver | 62 | 24 | 4 | 0.0 | 6.1 | 1.18–1.20 | Heavy band | [53, 65] |
| Rat liver | 24–27 | 21–35 | 7–15 | 0.2 | 2.1 | 1.16 | Light band | [55] |
| Rat liver | 46–48 | 21–35 | 7–15 | 1.3 | 3.3 | 1.19 | Heavy band | [55] |
| Pea seedlings | 78 | 4.6 | 4.0 | 2.9 | 7.2 | 1.24 | Gradient in sorbitol | [54] |
| Pig liver | 75 | 16–18 | 3 | 1–2 | 2.5–3.5 | 1.21 | – | [56, 108] |
| Rat liver | 75 | 16–18 | 3 | 1–2 | 2.5–3.5 | 1.21 | – | [56, 108] |
| Rat liver | 64 | 16 | 7 | 8 | 5 | 1.27 | Gradient in sorbitol | [57] |
| Rat liver | 73 | 23 | – | 0.6 | 3 | 1.17–1.18 | – | [62, 104] |
| Rat liver | 75–80 | 12–19 | – | 3–4 | 3–4 | 1.21 | without DNAase* | [25, 61, 84] |
| Rat liver | 75 | 15 | – | 3 | 7 | 1.21 | with DNAase* | [25, 61, 84] |

*Figures not previously published.

TABLE 5.2
*Yield and ultrastructure of nuclear envelope preparations*

| Yield (as % of phospholipid | Predominant state of envelope | Recovery of pore complexes: Annulus | Recovery of pore complexes: Central tubule | Chromatin contamination | References |
|---|---|---|---|---|---|
| – | Single-membrane vesicles | Trace | None | Very low | [51, 107] |
| 55 | Intact double-membrane sheets; nuclear ghosts | Complete | Present | Low | [52, 113] |
| – | Vesicles; some separation of INM and ONM | Trace | None | Negligible | [53, 65] |
| 13 | Envelope fragments | Apparently complete | Present | Extensive | [54] |
| Very low | Envelope fragments (sheets) | Complete | – | Some | [55] |
| | Vesicles; separation of INM and ONM | Some | Trace | Low | [56, 108] |
| 50 | Envelope fragments (sheets) | Complete | Present | Some | [57] |
| 55–60 | Large vesicles; a little separation of INM and ONM | Complete on larger vesicles | Some present | Very low | [62, 104] |
| 12–15 | Nuclear ghosts; some envelope fragments | Complete | Present | Low | [25, 61, 84] |

have employed both thin sectioning and negative staining to study their NE material [51–57,59–64]. It is considered desirable to detect the presence of both ONM and INM using thin sectioning, although with intact nuclear ghosts negative staining will provide this information [24,25]. Freeze cleavage, which has been applied extensively to intact nuclei has been used by Monneron and her colleagues to assess isolated NE, but it must be admitted that this technique does not provide additional knowledge on either contamination or ultrastructural integrity. It is the opinion of the authors that negative staining with ammonium molybdate reveals the greatest ultrastructural detail within the NE [24,25,59,61].

In Table 5.3, Mg-ATPase, glucose-6-phosphatase, acid and alkaline phosphatase are expressed as $\mu$mol Pi released $h^{-1}$ $mg^{-1}$ protein, and NADH, NADPH and glutamate dehydrogenase activities in $\mu$moles of substrate used or product formed $h^{-1}$ $mg^{-1}$ protein. The significance of the presence of most of these enzymes, and many others [6], in the NE is not clear. The basal Mg-ATPase, for example, is not stimulated by the presence of sodium and potassium [55,56,67], and its role in the NE is unknown at this time Discrepancies between certain preparation with regard to Mg-ATPase and glucose-6-phosphatase activity merit further investigation, as does the report by Gorkin [58] that mono-amine oxidase is present in the NE.

### *5.2.6 Subfractionation of the nuclear envelope*

If we accept the evidence that the ONM and INM are dissimilar in composition and possibly function, and that the nuclear pore complex is a distinct and universally found morphological structure which may in fact be considered as a separate organelle [20], there is obvious interest in the possibility of obtaining each of these component parts of the NE as separate entities. This may be achieved either from intact nuclei, or better to avoid nucleoplasmic contamination, from the isolated and purified NE. Zbarsky *et al.* [55] have reported that the ONM and INM can be separated, but details of the method are not yet available. Mizuno *et al.* [63] claimed to be able to isolate the INM, but their criterion of assessment was simply the absence of ribosomes. To the authors' knowledge at the time of writing, there have been no reports of isolation of the pore complex, though the fact that this structure becomes detached from the surrounding NE during certain negative staining procedures [25,109], suggests that isolation is in principle possible.

The major problem to separation of the two membrane leaflets of the NE lies in the difficulty of finding adequate markers (e.g ultrastructural, enzymic and varying flotation density) for these membranes and of assessing cross contamination. This is illustrated in the isolation of an ONM fraction by Busch and his co-workers [75]. The method, based on the finding [36] that citric acid removed the ONM, was applied to whole nuclei obtained by centrifugation of a tissue homogenate through dense sucrose. The objections to such a procedure, if

TABLE 5.3

*Enzymes of the nuclear envelope*

| Fraction | Mg-ATPase[a] | Glucose-6 phosphatase[a] | Alkaline phosphatase[a] | Acid phosphatase[a] | NADH-cytochrome *c* reductase[b] | NADPH-cytochrome *c* reductase[b] | Glutamate dehydrogenase[b] | Cytochrome $b_5$[c] | Cytochrome *P*-450[c] | References |
|---|---|---|---|---|---|---|---|---|---|---|
| – | 6.8 | 7.0 | – | – | 49 | 0.3 | – | 0.40 | 0.06 | [51, 107] |
| Heavy | 2.3 | 20.9 | – | – | 22.3 | 6.1 | – | 0.18 | 0.00 | [53] |
| Light | 1.7 | 15.0 | – | – | 22.4 | 6.1 | – | 0.18 | 0.22 | [53] |
| – | 0.67 | 0.52 | – | – | – | – | – | – | – | [54] |
| Heavy | 13.7 | 0.16 | – | – | 5.5 | 0.6 | 1.8 | – | – | [55] |
| Light | 23.1 | 0.0 | – | – | 1.0 | 0.2 | 3.0 | – | – | [55] |
| – | 5.5–6.8 | 0.1 | 1–3 | 3.1–9.8 | 5.8 | 1.2 | 1.8 | 0.03 | 0.03 | [56] |
| – | 0.1 | – | – | – | 10.0 | – | – | – | – | [67] |
| – | – | 8.5–11 | – | – | 12–18 | 3.0 | – | – | – | [52, 113] |

[a] $\mu$mol phosphate $h^{-1}$ $mg^{-1}$ protein

[b] $\mu$mol NAD(P) oxidized or reduced $h^{-1}$ $mg^{-1}$ protein

[c] nmol $mg^{-1}$ protein

the aim is to obtain an acceptable membrane preparation, are several. Firstly, low pH media are known to solubilize several membrane proteins [110,111]. Secondly, contaminants may be removed from the INM and nucleoplasm by the citric acid treatment and readsorbed onto the ONM. Thirdly, since only small membrane vesicles are recovered, it is not certain that the material isolated represent the whole of the ONM or only a specific fraction. Finally, the treatment may well inactivate several enzymes intrinsic to the membrane, for instance glucose-6-phosphatase is known to be inactivated at pH's of less than 5.5 [112].

It is unlikely that any procedure for subfractionation of the NE will meet with widespread acceptance until it uses as starting material an acceptable NE preparation, as previously defined. In addition satisfactory markers for the various fractions must be established and the conditions employed must be mild enough not to inactivate membrane bound enzymes irreversibly or remove components of the individual fractions.

### 5.3 General discussion

If we accept that the definition of an isolated cell membrane preparation must in some sense be operational, not being guided by such criteria as 'chemical purity', it is very difficult to defend the claim that any membrane isolation procedure is ever wholly satisfactory. Nevertheless, in the case of resealed erythrocyte ghosts and certain mitochondrial preparation, the impressive retention of functional integrity argues for acceptability of the methods involved, even though the detailed composition of the material may not be entirely reproducible. In the case of the isolated NE, claims of functional integrity cannot yet be made and the criterion of ultrastructural acceptability is bound to be subjective and may not satisfy the requirements for all investigations of the system.

The three methods described in detail below were chosen because they appear to be essentially satisfactory within the limitations of current understanding and present day research, and because they entail the use of more or less widely differing ionic conditions. Other methods to which reference has been made in earlier sections, are not necessarily to be dismissed. Nuclear membrane material essentially free from DNA [53], NE fragments still bearing appended DNA [54,57], and the other varieties of preparation which can be obtained from published methods, will certainly find a place in future research on the system. Low-DNA preparations, for example may be useful for studies on the protein components of the NE, since nucleic acids interfere with certain common protein fractionation techniques (e.g exclusion chromatography on Sephadex gel). High-DNA preparations may likewise be useful for investigations on the interrelations between NE and the chromatin.

Some attention must be paid to the difficulties inherent in procedures for isolating the NE. The affinity of DNA for membranes must be accepted as the most difficult problem in the present context. Any membrane that is isolated in the presence of DNA is liable to be contaminated by this substance, and the problem is particularly severe with the NE since organized heterochromatin is, in the interphase nucleus *in vivo*, contiguous with the INM in the form of the internal dense lamina. It is not clear, however, that the DNA content of isolated NE preparations is necessarily artifactual, or at least wholly artifactual. Dependence of the morphological integrity of the isolated NE on the presence of DNA [57] is found only under specific isolation conditions, but nonetheless DNA may be a genuine component of the NE *in vivo,* and the point remains as yet unsettled. An analogy may be drawn between the involvement of haemoglobin in the structure of the erythrocyte membrane, which likewise remained a matter of controversy for many years [25,116].

The question of the continuity of the ONM with the endoplasmic reticulum also remains as a major stumbling block. It is difficult to claim that any preparation of nuclei, and hence of NE, is wholly free of endoplasmic reticulum Such contamination would be undesirable even if the reticulum and the ONM were of the same composition, a position which is not easy to defend. Perhaps some improved microsomal enzymic marker or an increased use of freeze-cleavage electron microscopy to assess the extent of the endoplasmic reticulum contamination in a nuclear preparation, would assist in overcoming this problem.

The lability of the nuclear pore complex is also a fundamental problem. Excessive mechanical stress as produced by ultrasonication for example, extremes of pH and high ionic strength media, particularly if they contain potassium ions [57], are among the conditions which lead to loss of pore complex. The central granule and the annular subunits tend to disappear and, at best, only the remnant of the pore annulus is retained. Conditions of isolation must, therefore, be carefully defined with respect to pH, ion content and temperature of each medium used, as well as the time of exposure to each medium and the mechanical or other disruption conditions employed, for a completely satisfactory and reproducible NE preparation with stable pore complexes to be obtained. In the three techniques described in detail satisfactory pore complexes have been shown to persist in the final material. Complete loss of pore complexes is usually associated with collapse of the NE into double- or single-membrane vesicles. On the other hand if the pore complexes are present, intact ghosts, large torn sheets and large vesicles may be expected. The undesirable tendency of the membrane system to undergo vesicularization presents the possibility of contamination of the isolated material by foreign soluble proteins and nucleic acids trapped within the vesicles themselves.

In conclusion it must be re-emphasized that studies on the NE are still at an early stage and that considerable methodolgical developments may be expected which should provide a more satisfactory standard isolation and purification procedure. Once this is achieved the results obtained from the various approaches employed for the study of NE (e.g. electron microscopy, chemical and enzymic analysis, gel electrophoresis) will, hopefully, become more widely uniform and therefore scientifically more acceptable. Extending from the availability of an acceptable NE preparation considerable progress can be expected on the subfractionation of the NE, which in turn should provide some insight into the macromolecular organization of this extremely complex membrane system.

## 5.4 Experimental procedures for the isolation of the nuclear envelope (NE)

### *5.4.1 The DNAase/high pH treatment of Kay et al.* [52,113]

*(i) Procedure*

Nuclei are isolated from mouse or rat livers by the method of Widnell and Tata [90]. The nuclei from about 4 g liver are suspended in 1 ml 0.1 mM $MgCl_2$, and freshly dissolved DNAase I (Sigma, electrophoretically pure) is added to give a final concentration of 5 $\mu g\ ml^{-1}$. Four volumes of 10 mM Tris-HC1 (pH 8.5), 0.1 mM $MgCl_2$, 5 mM 2-mercaptoethanol, 10% w/v sucrose are added immediately and the suspension mixed and allowed to incubate at room temperature (22°C) for 15 min. This first digestion is terminated by the addition of an equal volume of ice-cold water and a crude nuclear envelope fraction obtained as a pellet by centrifuging at 38 000 x *g* for 15 min. The pellet is resuspended in 5 volumes of the previous digestion buffer, but at pH 7.4, and DNAase added to a concentration of 1 $\mu g\ ml^{-1}$. After incubation at 22°C for 20 min the digestion is terminated as before and purified nuclear envelope pelleted.

*(ii) Discussion*

The most striking advantages of this procedure are its rapidity (approximately 80 min starting with washed nuclei) and the well preserved ultrastructure of the final NE material which is in the form of ghosts and large sheets (see Figs 5.4 and 5.5). One disadvantage must be the very small yield of material from only 4 g liver. The method may well prove to be very useful for ultrastructural studies and some enzymic investigations. The primary interest of Kay *et al.* was in the DNA polymerase activity associated with the NE, and they claimed that about 1% of the total nuclear DNA polymerase was recovered in the envelope. While so low a yield of the enzyme may not appear significant, it should be noted that an unusually high proportion of the DNA associated with this NE preparation is late replicating. The unusually high glucose-6-phosphatase and NADH oxidase activities (cf. Table 5.3) in the NE may possibly be attributed to the rapidity with which it is obtained and the relative mildness of the procedure, these enzymes being somewhat unstable.

The significance of the inclusion of 2-mercaptoethanol in the digestion

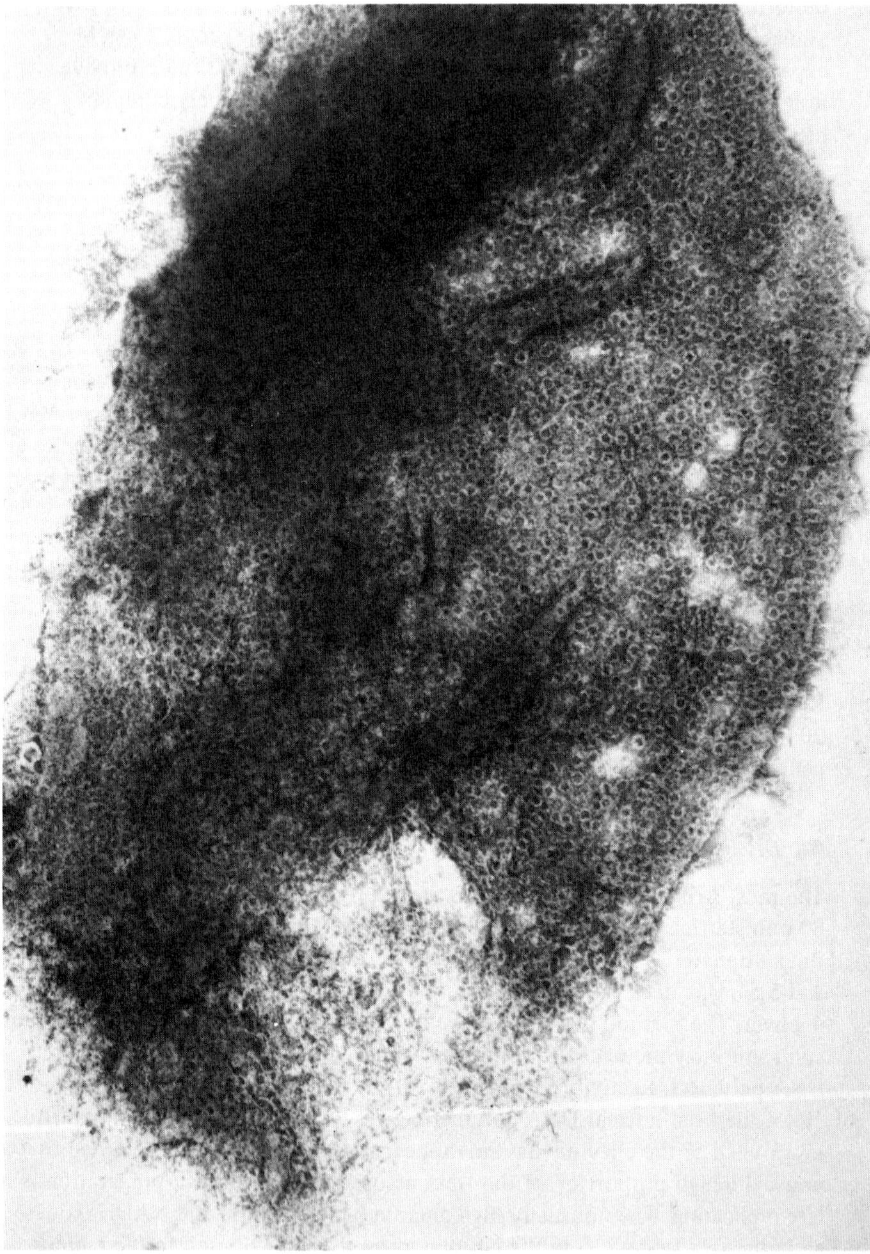

**Fig. 5.4** A rat liver nuclear envelope isolated by the method of Kay *et al.* Negatively stained with 2% ammonium molybdate (pH 7.0). (x15 000.)

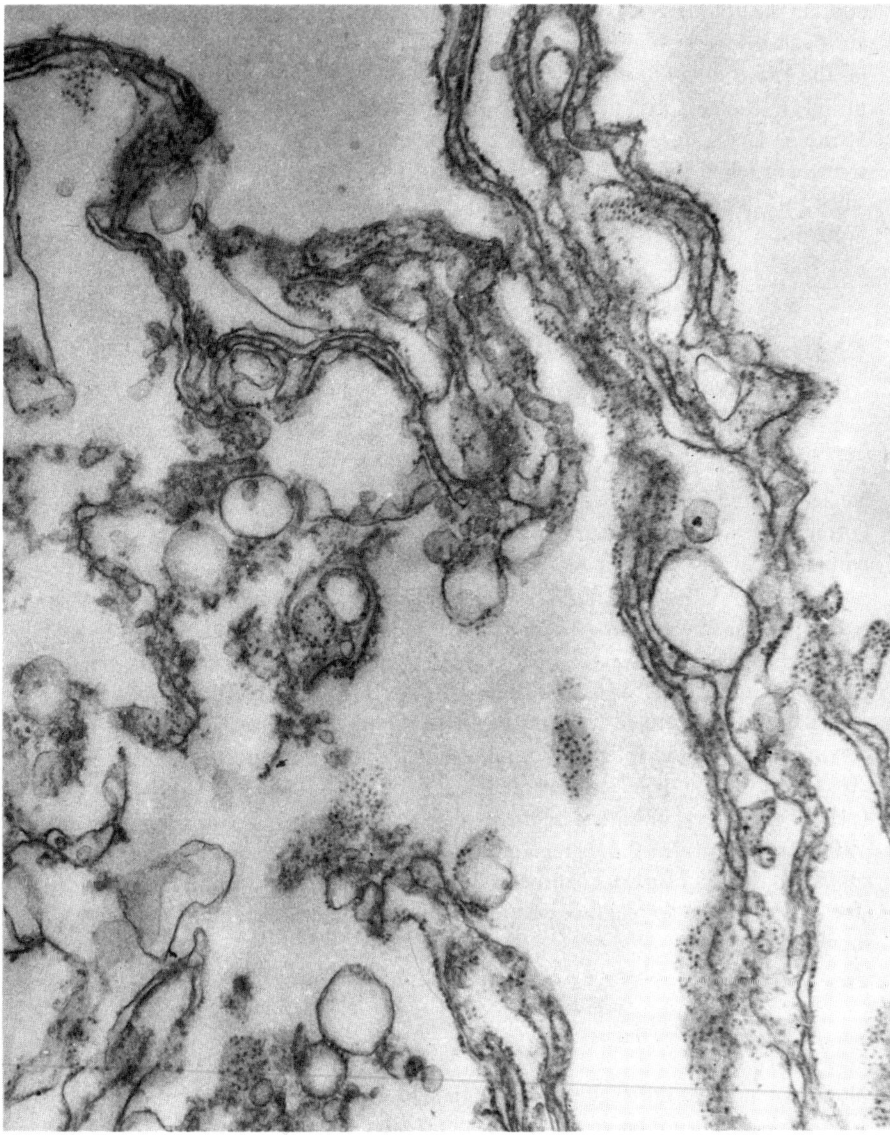

**Fig. 5.5** Rat liver nuclear envelope isolated by the method of Kay *et al.* This thin sectional field shows clearly the presence of ribosome particles, but there appears to be considerable separation of the INM and ONM. (x35 000.)

medium is puzzling. The morphology of the isolated nuclear envelope may be improved, but the authors have not found its inclusion to be necessary (see below). 2-mercaptoethanol may interfere with the activity of enzymes which contain disulphide bridges, the presence of which is essential for their activity although it will prevent oxidation of essential sulphydryl groups. Further data on the enzymology in this connection is obviously required. The high pH of the initial DNAase digestion (pH 8.5) could interfere with enzyme activities and, at the molecular level with the structural organisation of the NE. The DNAase treatment renders the quoted DNA content (Table 5.1) somewhat dubious since this value would presumably vary with the length of time of the incubation.

### 5.4.2 *The high concentration magnesium treatment of Monneron et al.* [62, 104]

#### *(i) Procedure*

Nuclei isolated from rat liver homogenates by the method of Blobel and Potter [89] are resuspended to a final concentration of about 30 $A_{260}$ units $ml^{-1}$ in ice-cold 0.05 M Tris-HCl, 0.025 M KCl, 5 mM $MgCl_2$, 0.25 M sucrose (pH 7.5, adjusted at 20°C). 5 ml aliquots of this suspension are centrifuged at 1000 x *g* for 5 min and the pellets resuspended in 1.8 ml ice-cold 0.05 M Tris-HCl, 0.5 M $MgCl_2$, 1.8 M sucrose (pH 7.5). To ensure homogeneous resuspension the pellet is first stirred with 2 or 3 drops of glycerol. The suspension is then mixed 'vigorously' three times, centrifuged at 1000 x *g* for five minutes, and the foamy layer on top of the liquid removed. A linear 20% (w/v) to 55% (w/v) sucrose gradient in 0.05 M Tris-HCl, 0.5 M $MgCl_2$ (pH 7.5) is generated on top of the suspension and the gradient centrifuged at 190 000 x *g* (ave) for 2 h. Nuclear envelope material floats up during this centrifugation and forms a band at a density of approximately 1.179 g $ml^{-1}$, (see Fig. 5.6). The appearance of the material in the electron microscope is shown in Fig. 5.7.

#### *(ii) Discussion*

Monneron and her colleagues indicate that solutions containing high concentrations of mono and divalent metal salts such as KCl, $CaCl_2$, $MgCl_2$ and $MnCl_2$ can be used for the complete dispersal and subfractionation of nuclei. Their NE isolation procedure is best viewed, therefore in the context of total nuclear fractionation The recovery of pore complexes is more convincing than in the closely comparable method of Berezney, Funk and Crane [51], the DNA content is very low (around 0.6%), and the recovery of NE is acceptably high

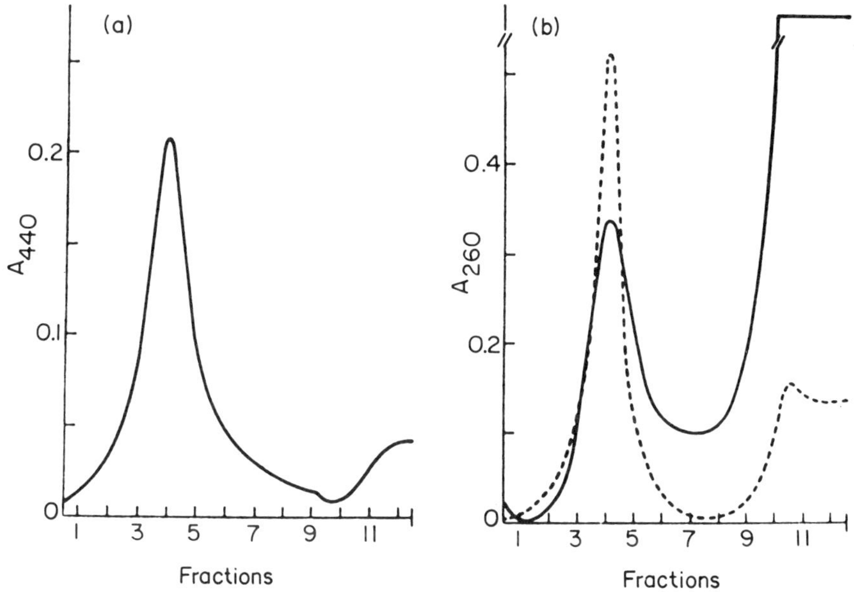


**Fig. 5.6** Sucrose gradient analysis of rat liver nuclei treated with 0.5 M $MgCl_2$. Recording was at 440 nm (a) or 260 nm (b). Dotted line represents absorbancy profile at 260 nm of a membrane band collected from gradient, pelleted, and refloated on a sucrose gradient.

(55–60%). As yet, nothing has been reported about the enzyme content of this material, though the preparation time of 3 hours from isolated nuclei and the high ionic strength $MgCl_2$ treatment suggest that any glucose-6-phosphatase and DNA polymerase activities are likely to be lost [52].

Certain features of the procedure indicate that it may be made more generally useful by further modification. Firstly, the NE density varies slightly as the magnesium chloride concentration is varied. (0.3 M $MgCl_2$ gives NE of density 1.169 and 0.7 M gives density 1.184 g $ml^{-1}$) suggesting that the composition of the envelope material is varying. Secondly, the NE vesicles and sheets produced are of varying size and there appears to be considerable separation of the INM and ONM. In addition, the nuclear pore complexes appear to be at least partially disrupted, as judged by their irregular shapes and lack of central granules in the published electron micrographs.

The destructive effect of high potassium concentration on NE material [54,57,62] is avoided here, and evidence is offered from SDS-polyacrylamide gel electrophoresis that the composition of rough endoplasmic reticulum is very different from that of the NE, in agreement with Bornens and Kasper [65]. Simple treatments such as the one developed by Monneron *et al.* which provide ultrastructurally acceptable envelope fragments with very low DNA content, will probably prove most useful for further

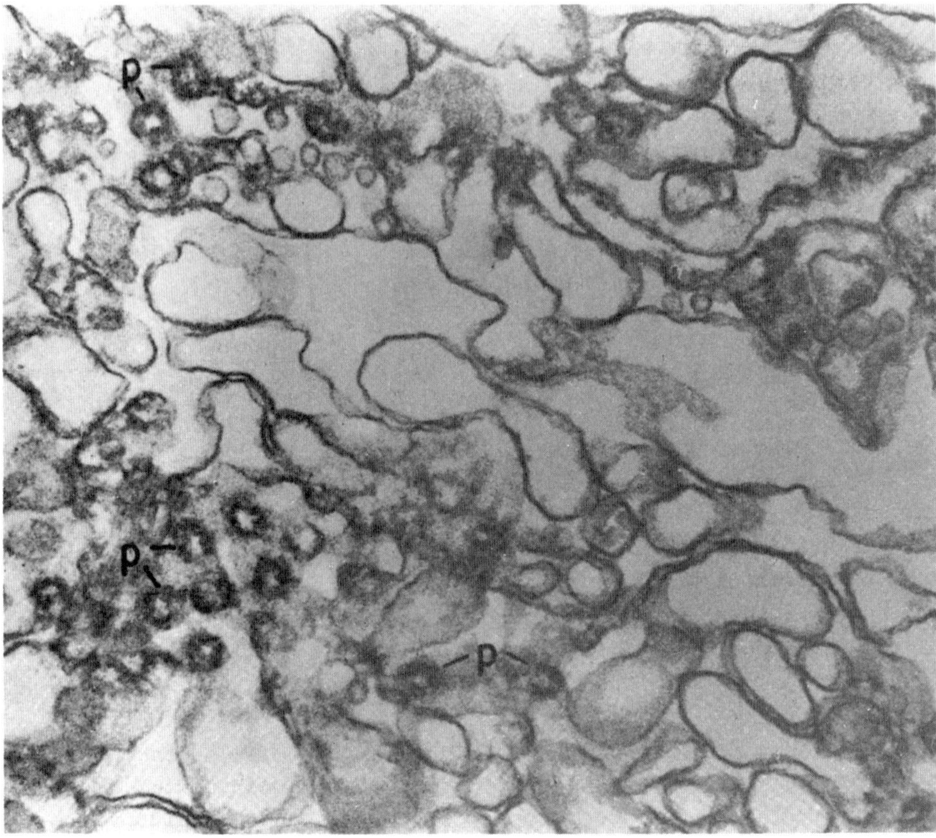


**Fig. 5.7** Nuclear envelope material isolated by the method of Monneron *et al.* Recognizable nuclear pore complexes (p) can be seen on many grazing sections of membrane vesicles. There appears to be extensive separation of the two nuclear membranes. (x72 000.)

investigation of the protein components of the NE, since the presence of a significant quantity of DNA renders many techniques of protein fractionation impracticable through its tendency to cause aggregation.

### *5.4.3 The low ionic strength treatment of Harris et al.* [25,61,84)

In its initial form, the low ionic strength treatment developed by Price Harris and Baldwin [61] for the isolation of nuclear ghosts from rat liver and hepatoma is at considerable variance to all other published procedures Firstly, it employs rate zonal centrifugation and isopycnic zonal centrifugation. Secondly, these centrifugation techniques are not applied to purified cell nuclei, but to the crude 1000 x *g* nuclear pellet obtained following homogenization in 1 mM sodium bicarbonate (pH 7.2) and overnight incubation at 4°C.

Because of the fact that zonal centrifugation facilities are still not widely available, the technique has recently been modified for use with purified rat liver nuclei and conventional centrifuge rotors [84]. It is this latter modification which will be presented below, but discussion will also be made of the original method.

*(i) Procedure*

The livers from six starved rats are homogenized in cold 0.25 M sucrose, 2 mM $MgCl_2$, 10 mM Tris-HCl (pH 7.4) using an Ultra Turrax type TP18/2 blade homogenizer. The 1000 x *g* nuclear pellet is prepared and resuspended in saturated sucrose to give a concentration of of 2.2 M (59% w/v) with respect to sucrose. The nuclei are then pelleted at 100 000 x *g* for 2 h and then washed two times at 1000 x *g* with the homogenizing buffer.

The purified nuclei are then resuspended in 40 ml 1 mM sodium bicarbonate solution (pH 7.2) by shaking and allowed to equilibrate for 5 min before centrifuging at 31 000 x *g* for 5 min. The pellet of slightly swollen nuclei is again resuspended in 1 mM sodium bicarbonate by gentle syringing through a large diameter needle and after a further period of 5 minutes incubation is recentrifuged as above. A very gelatinous pellet is obtained which cannot be dispersed by further centrifugal washing in 1 mM sodium bicarbonate. Incubation with DNAase 1 (Sigma type DN–100) for 15–20 min at room temperature (22°C) at a concentration of 10 $\mu g\ ml^{-1}$ will however readily disperse the gelatinous suspension of bursting nuclei. Thus during this brief incubation the DNA, which is dispersed and readily available for enzymic degradation, is solubilized, thereby allowing the NE to become freely suspended and therefore separable from the chromatin. It may be noted that the conditions under which the incubation is performed are less than optimal for maximum DNAase activity, which requires the presence of 5 mM $MgCl_2$.

The NE's are then repeatedly washed in 1 mM sodium bicarbonate with centrifugation at 31 000 x *g* for 5 min. The release of chromatin throughout this sequence of washes is shown in Fig. 5.8, which clearly shows the large percentage release of chromatin following the DNAase 1 treatment. The partially purified NE obtained from the final differential centrifugal wash is resuspended in 2 ml 10 mM Tris-HCl (pH 7 4) and is centrifuged for 5 min at 500 x *g* to remove any aggregates. The small discoloured pellet is discarded and the white membrane rich supernatant layered over a discontinuous sucrose gradient made up of 10 ml 2.0 M sucrose, 10 ml 1.8 M sucrose, 10 ml 1.5 M sucrose and 6 ml 0.25 M sucrose (all solutions made up in 10 M Tris-HCl, pH 7.4). The gradient is then centrifuged at 100 000 x *g* for 90 min. A major band of purified NE forms at the 1.5 M/1.8 M interface, as shown in Fig 5.9. Prolonged centrifugation does not alter the position of the membrane band and

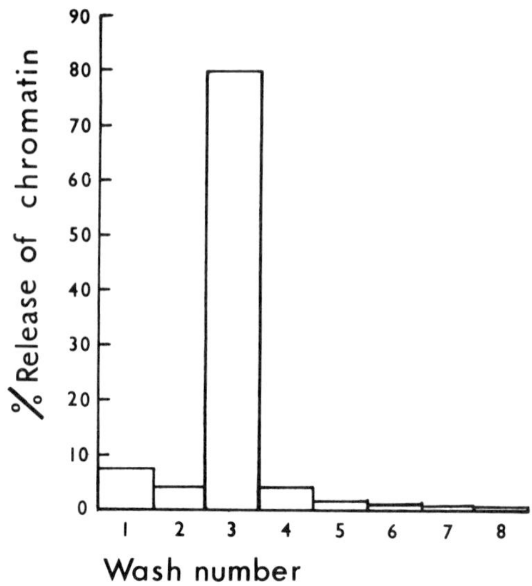


**Fig. 5.8** The release of solubilized chromatin (estimated by the absorbance of the 31 000 x g supernatants at 260 nm) throughout the 1 mM sodium bicarbonate (pH 7.2) washing and DNAase I incubation sequence employed for the preliminary separation of rat liver NE from chromatin. The histogram is the mean of three experiments. Note the large percentage release of chromatin following the DNAase I treatment (wash number 3).

isopycnic centrifugation on a continuous sucrose gradient for 20 h at 100 000 x *g* likewise gives a single membrane band of flotation density 1.21 ± 0.01, in agreement with the value obtained by zonal centrifugation method [61]. Fig. 5.10 summarizes the various steps in the above procedure.

Phase contrast examination of the final material reveals a free suspension of nuclear ghosts and large torn sheets of NE, as shown in Fig 5.11. This assessment is supported by electron microscopy using thin sectioning and negative staining after removal of the sucrose by differential centrifugal washing with 10 mM Tris-HCl (pH 7.4) (see Figs. 5.12–5.15).

*(ii) Discussion*

The underlying premise of the low ionic strength treatment presented above is that in the absence of divalent cations, nuclei will swell and burst thus

---

**Fig. 5.9** The purification of rat liver NE by centrifugation on a discontinuous sucrose gradient. (a) The appearance of a gradient after centrifugation at 100 000 x *g* for 90 min. (b) The absorbance profile of the gradient obtained by passing the tube contents through an LKB UV analyser. The main band of membrane material is located at the 1.5 M/ 1.8 M interface.

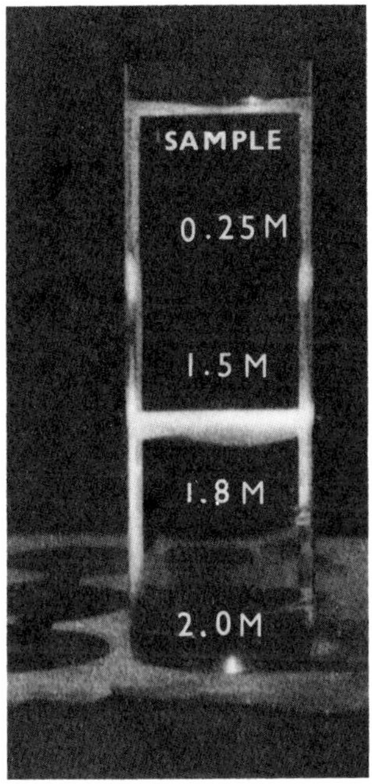
SAMPLE
0.25M
1.5M
1.8M
2.0M

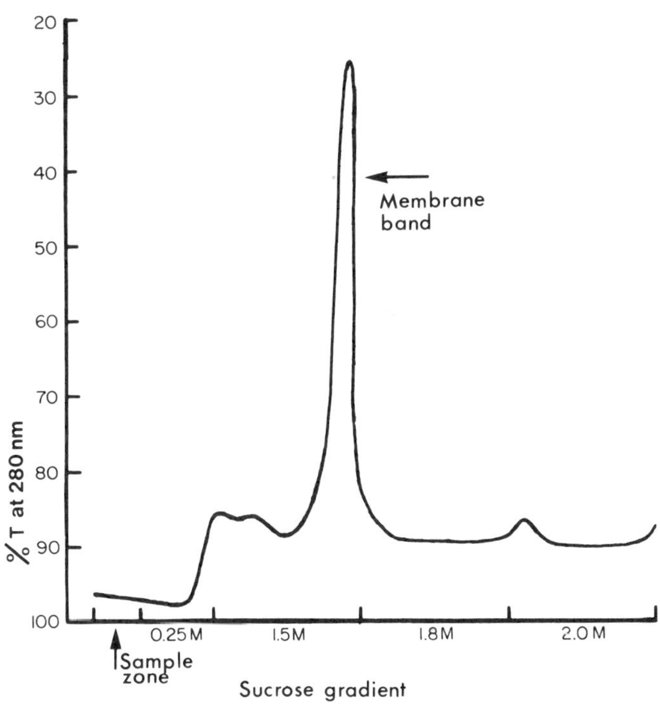
% T at 280 nm
20
30
40
50
60
70
80
90
100
Membrane band
0.25M
1.5M
1.8M
2.0M
Sample zone
Sucrose gradient

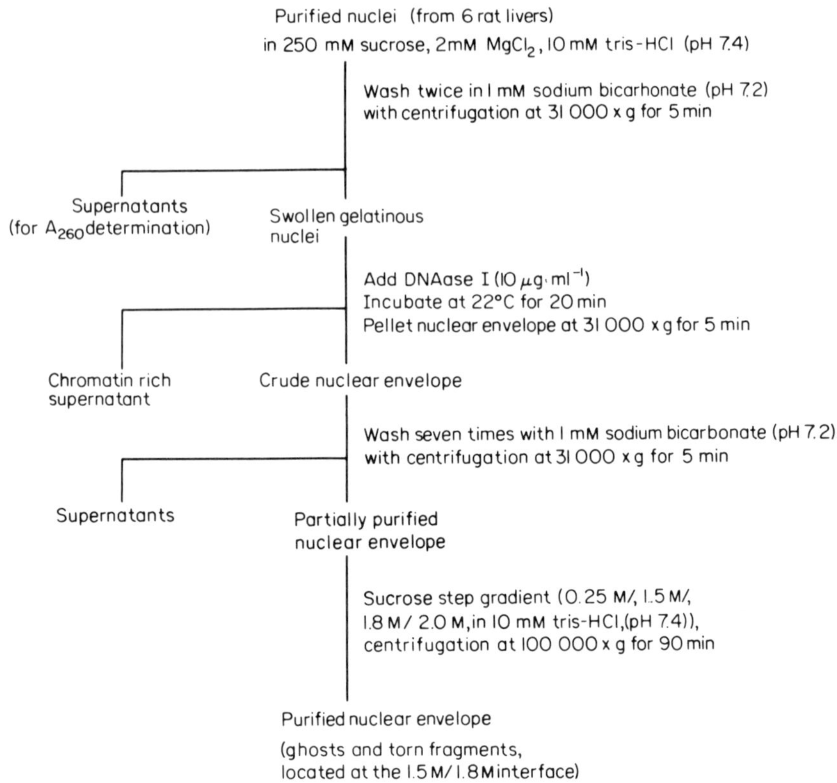


**Fig. 5.10** The isolation and purification of rat liver nuclear envelope.

presenting the possibility of isolating relatively intact nuclear ghosts. The studies of Olins and Olins [95] have underlined the changes in the condensation state of the chromatin that can be brought about in rat liver and chicken erythrocyte nuclei by varying the pH, total ionic strength and magnesium concentration and are in general agreement with the concepts presented here. The zonal centrifugation procedure [61], which itself extended from the 1 mM sodium bicarbonate procedure of Neville [114] for the isolation of rat liver plasma membrane, has the considerable advantage that repeated centrifugal packing of the NE material throughout the various steps of the procedure is reduced to a minimum. The disadvantage of the zonal procedure is that it is time-consuming and uses much larger quantities of gradient materials than are required in the variant presented above. Nevertheless, for those with zonal centrifugation facilities available, it is suggested that considerable advantages may possibly be gained as much larger quantities of material can be processed and purified. The authors therefore propose in the future to incorporate the use of zonal centrifugation into the method presented here, in the hope of being able to scale

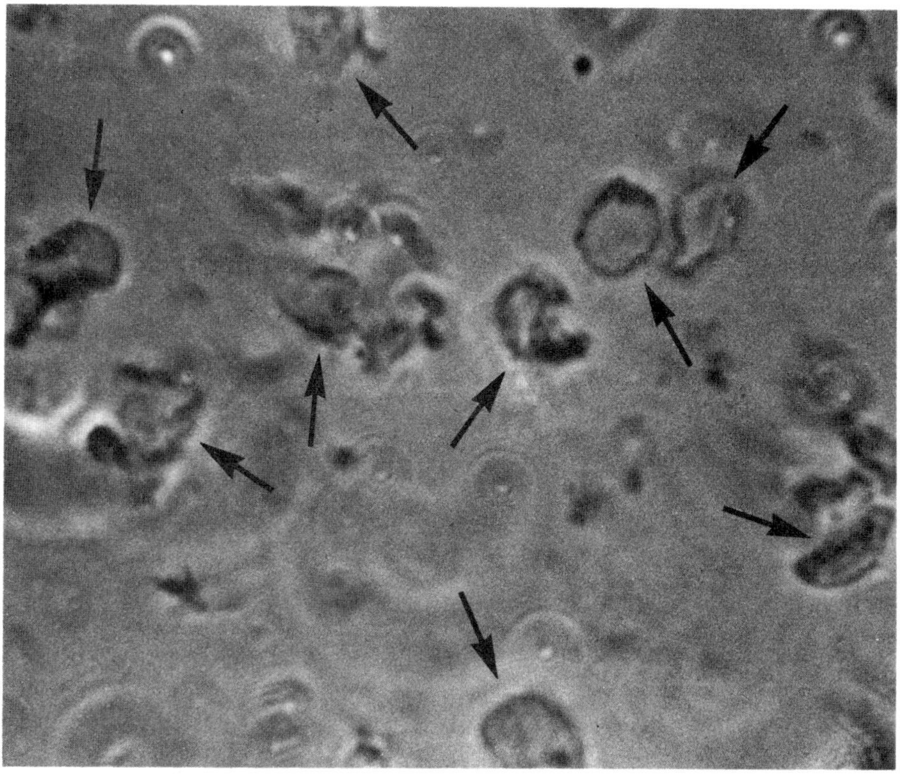

**Fig. 5.11** A phase contrast micrograph of purified rat liver nuclear envelope. Arrows indicate intact ghosts.

up the quantity of tissue that can be processed, while retaining satisfactory purification.

Recently, it has been found that the DNAase 1 treatment of the purified nuclei can be omitted if they are stored overnight at 4°C in the homogenization buffer (unpublished observations of the authors). On commencing the sequence of 1 mM sodium bicarbonate washes no gelatinous pellet of nuclei is produced, the material being readily resuspendable at each successive wash. The release of chromatin under these conditions is shown in Fig. 5.16. The behaviour of the material on the sucrose gradient is identical to that obtained following the DNAase I treatment. Further analytical and enzymic studies are in progress to assess whether or not this methodological variant is superior to that requiring the addition of DNAase I, but it may be noted that the overall conditions are brought closer to those employed in the original zonal procedure [61]. It can reasonably be proposed that autolytic digestion of the nuclear DNA will occur during an overnight incubation of 4°C, since it is known that there is an active

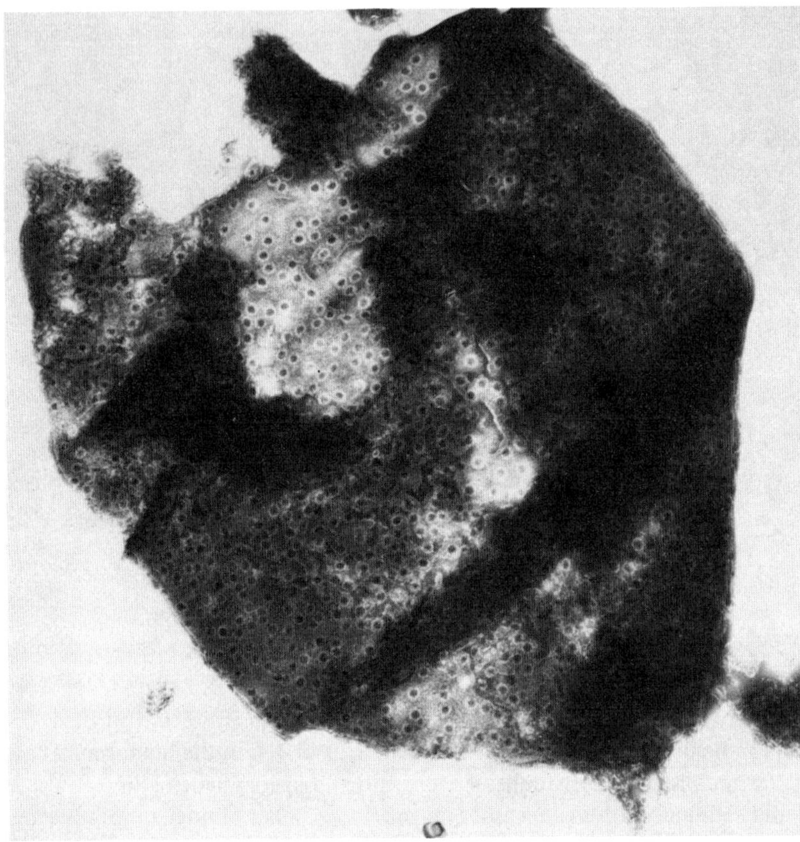

**Fig. 5.12** A rat liver nuclear ghost, negatively stained with 2% ammonium molybdate (pH 7.0). x15 000.

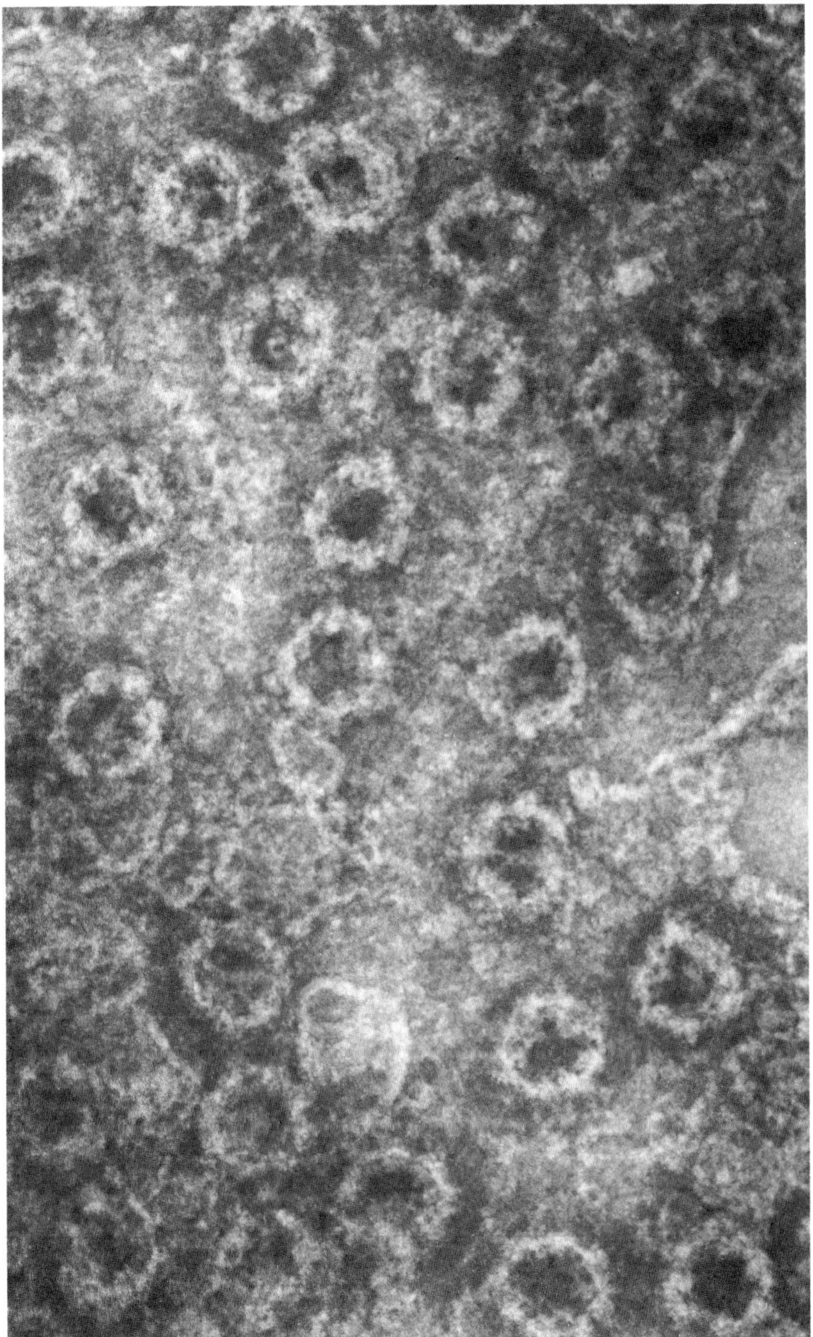

**Fig. 5.13** A torn sheet of rat liver nuclear envelope, negatively stained with 2% ammonium molybdate (pH 7.0). x72 000.

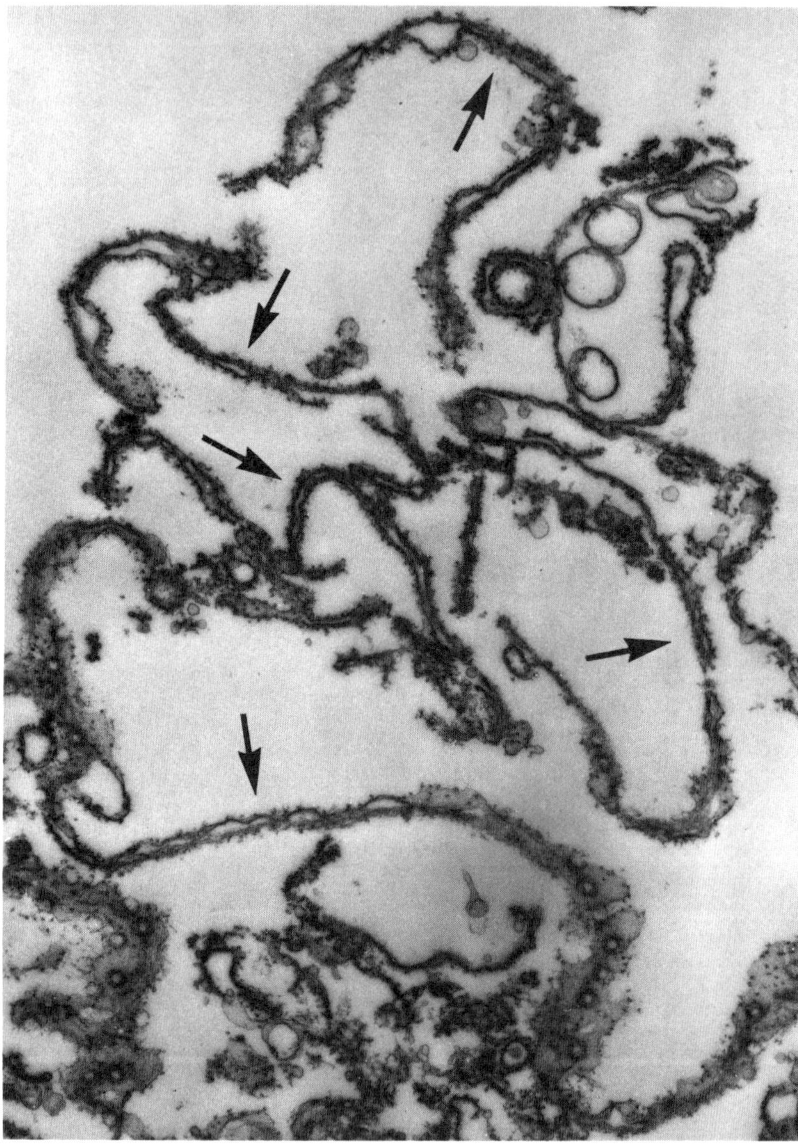

**Fig. 5.14** Thin sectioned rat liver nuclear envelope. Regions showing both the INM and ONM are arrowed. x15 000.

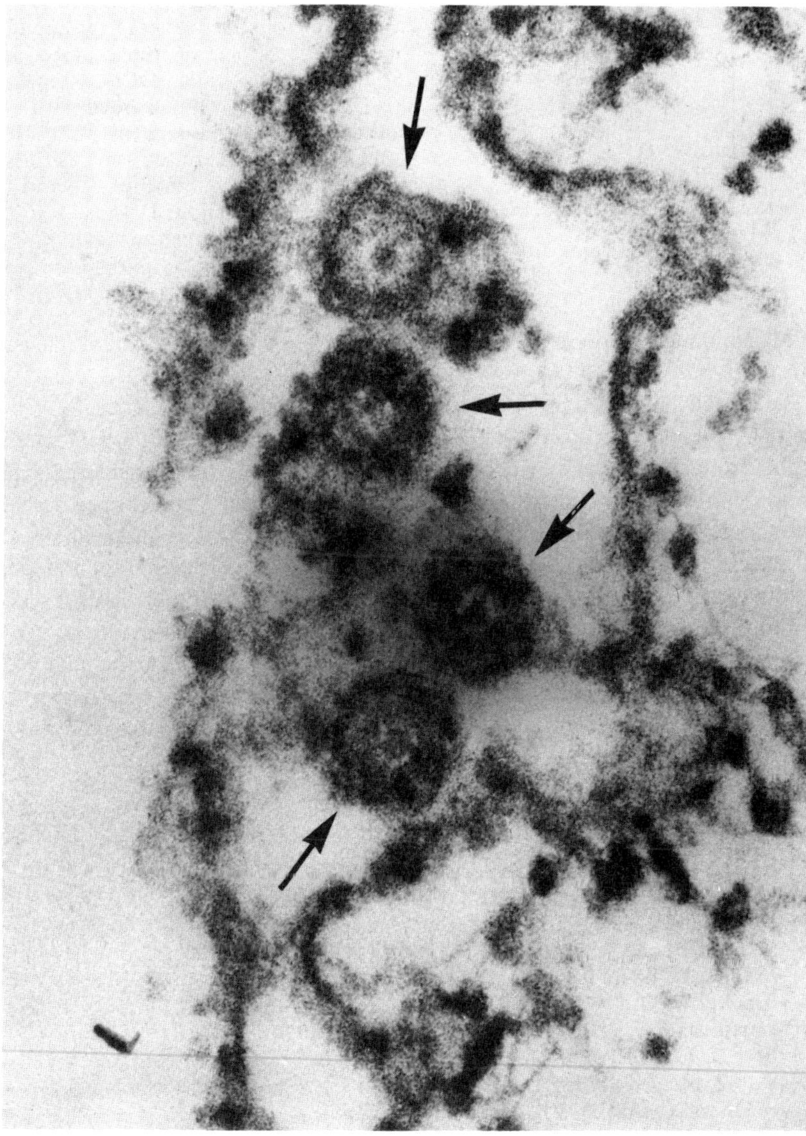

**Fig. 5.15** A region of tangentially sectioned rat liver nuclear envelope showing the nuclear pore complexes (arrowed). x72 000.

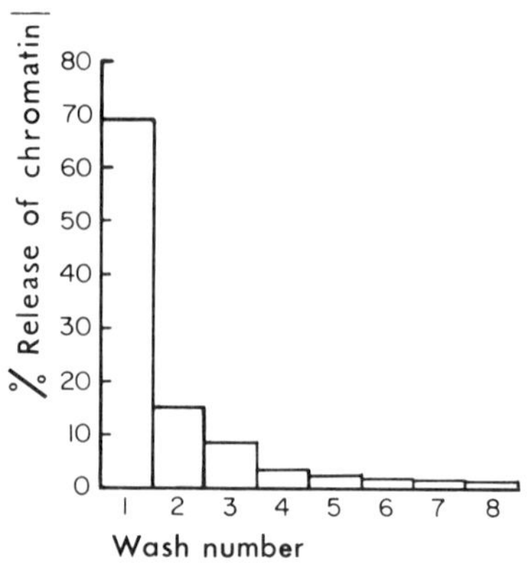


**Fig. 5.16** The release of solubilized chromatin from purified rat liver nuclei that had been stored overnight at 4°C in 250 mM sucrose, 2 mM $MgCl_2$, 10 mM Tris-HCl (pH 7.4), when washed with 1 mM sodium bicarbonate (pH 7.2). The washing conditions are the same as those described in the text (cf. Figs. 5.8 and 5.10), but no exogenous DNAase is added. The histogram is the mean of three experiments.

Ca–Mg endonuclease present [115], which will readily attack and digest nuclear DNA *in situ* at 37°C.

The overall aim of the procedure presented is to obtain morphologically intact nuclear ghosts or torn sheets of NE with a low DNA content as rapidly as possible from the time of sacrificing the animals, and this has been achieved. The procedure must be considered to be relatively mild in that it is performed throughout at a pH close to neutrality and that conditions are maintained at 4°C, except for the brief DNAase I treatment. The final NE material is obtained by flotation on a sucrose gradient, which is the most satisfactory procedure for separating NE from trace amounts of contaminants such as collagen, nuclei, peroxisomes as well as solubilized chromatin. One criticism that can be put forward is that the low ionic strength conditions used throughout may release components of the NE. This point can only be answered by extensive comparison of the available methods for the isolation of NE.

## References

1. Callan, H. G., Randall, J. T. and Tomlin, S. G. (1949), *Nature,* **163**, 280.
2. Callan, H. G. and Tomlin, S. G. (1950), *Proc. R. Soc. Ser.,* (*B*), **137**, 367.
3. Anderson, N. G. (1953), *Expl. Cell Res.,* **4**, 306.
4. Sharrer, B. and Wurzelmann, S. (1969), *Z. Zellforsch.,* **96**, 325.
5. Franke, W. W. and Scheer, U. (1970), *J. Ultrastruct. Res.,* **30**, 317.
6. Franke, W. W. (1974), *Phil. Trans. R. Soc. Ser. B,* **268**, 67.
7. Wartiovaara, J. (1971), *J. Ultrastruct. Res.,* **36**, 558.
8. Fry, D. J. (1970), *Membranes and Ion Transport,* (Bittar, E. E., ed.), Vol. 2, pp. 259, Wiley Interscience, New York.
9. Feldherr, C. M. (1972), *Adv. Cell and molec. Biol.,* **2**, 273.
10. Kessel, R. G. (1973), *Prog. Surface Memb. Sci.,* **6**, 243.
11. Wischnitzer, S. (1973), *Int. Rev. Cytol.,* **34**, 1.
12. Franke, W. W. and Scheer, W. (1974), *The Cell Nucleus,* (Busch, H., ed.), Vol. 21, Academic Press, New York and London.
13. Hartman, J. F. (1953), *J. comp. Neurol.,* **99**, 201.

14. Palay, S. L. and Palade, G. (1955), *J. biophys. biochem. Cytol.*, **1**, 69.
15. Afzelius, B. A. (1955), *Expl. Cell Res.*, **8**, 147.
16. Rebhun, L. I. (1956), *J. biophys. biochem. Cytol.*, **2**, 93.
17. Watson, M. W. (1959), *J. biophys. biochem. Cytol.*, **6**, 432.
18. Wischnitzer, S. (1958), *J. Ultrastruct. Res.*, **1**, 201.
19. Vivier, E. (1967), *J. Microscopie*, **6**, 371.
20. Abelson, H. T. and Smith, G. H. (1970), *J. Ultrastruct. Res.*, **30**, 558.
21. Roberts, K. and Northcote, D. H. (1971), *Microscopica Acta.*, **71**, 102.
22. LaCour, L. F. and Wells, B. (1974), *Phil. Trans. R. Soc. Ser. B*, **268**, 95.
23. Franke, W. W. and Scheer, U. (1970), *J. Ultrastruct. Res.*, **30**, 288.
24. Harris, J. R., Price, M. R. and Willison, M. (1974), *J. Ultrastruct. Res.*, **48**, 17.
25. Harris, J. R. (1974), *Phil. Trans. Roy. Soc. Ser. B*, **269**, 109.
26. Franke, W. W. (1967), *Z. Zellforsch.*, **80**, 585.
27. Branton, D. and Moor, H. (1964), *J. Ultrastruct. Res.*, **11**, 401.
28. Maul, G. G., Price, J. W. and Liebermann M. W. (1971), *J. Cell Biol.*, **51**, 405.
29. Kartenbeck, J., Zentgraf, H., Scheer, U. and Franke, W. W. (1971), *Adv. Anat. Embryol. Cell Biol.*, **45**, 1.
30. Teighler, D. J. and Baerwald, R. J. (1972), *Tiss. Cell*, **4**, 447.
31. Lott, J. N. A., Larsen, R. L. and Whittington, C. M. (1972), *Can. J. Bot.*, **50**, 1785.
32. La Fountain, J. R. and La Fountain, K. L. (1973), *Expl. Cell Res.*, **78**, 472.
33. Markovics, J., Glass, L. and Maul, G. G. (1974), *Expl. Cell Res.*, **85**, 443.
34. Parks, H. F. (1962), *J. Cell Biol.*, **14**, 221.
35. Sadowski, P. D. and Howden, J. A. (1968), *J. Cell Biol.*, **37**, 163.
36. Gurr, M. I., Finean, J. B. and Hawthorne, J. N. (1963), *Biochim. biophys. Acta.*, **70**, 406.
37. Bruni, C. and Porter, K. R. (1965), *Am. J. Path.*, **46**, 691.
38. Davies, H. G. (1968), *J. Cell Sci.*, **3**, 129.
39. Leskes, A., Siekevitz, P. and Palade, G. E. (1971), *J. Cell Biol.*, **49**, 264.
40. Wischnitzer, S. (1958), *J. Ultrastruct. Res.*, **1**, 201.
41. Gall, J. G. (1956), *J. biophys. biochem. Cytol.*, **2**, 393.
42. Gall, J. G. (1967), *J. Cell Biol.*, **32**, 391.
43. Roberts, K. and Northcote, D. H. (1970), *Nature*, **228**, 385.
44. Hoeijnakers, J. H. H., Schel, J. H. N. and Wanka, F. (1974), *Expl. Cell Res.*, **87**, 195.
45. Kessel, R. G. (1969), *Z. Zellforsch.*, **94**, 441.
46. La Cour, L. F. and Wells, B. (1971), *Z. Zellforsch.*, **123**, 178.
47. Engelhardt, P. and Pusa, K. (1972), *Nature*, **240**, 163.
48. Stephens, B. J. and Swift, H. (1966), *J. Cell Biol.*, **31**, 55.
49. King, T. E. (1967), *Meth. Enzym.*, **10**, 322.
50. Babson, A. L., Read, P. A. and Philips, G. E. (1959), *Am. J. clin. Path.*, **32**, 83.
51. Berezney, R., Funk, L. K. and Crane, F. L. (1970), *Biochim. Biophys. Acta*, **203**, 53
52. Kay, R. R., Fraser, D. and Johnston, I. R. (1972), *Eur. J. Biochem.*, **30**, 145.
53. Kashnig, D. M. and Kasper, C. B. (1969), *J. biol. Chem.*, **244**, 3786.
54. Stavy, R., Ben-Shaul, Y. and Galun, E. (1973), *Biochim. biophys. Acta.*, **323**, 167.

55. Zbarsky, I. B., Perevoshchikova, K. A., Delektorkaya, L. N. and Delektarsky, V. V. (1969), *Nature,* **221**, 257.
56. Franke, W. W., Deumling, B., Ermen, B., Jarasch, E. D. and Kleinig, H. (1970), *J. Cell Biol.,* **46**, 379.
57. Agutter, P. S. (1972), *Biochim. biophys. Acta.,* **255**, 397.
58. Gorkin, V. Z. (1971), *Experientia,* **27**, 30.
59. Harris, J. R. and Agutter, P. S. (1970), *J. Ultrastruct. Res.,* **33**, 219.
60. Harris, J. R. (1969), *J. Molec. Biol.,* **46**, 329.
61. Price, M. R., Harris, J. R. and Baldwin, R. W. (1972), *J. Ultrastruct. Res.,* **40**, 178.
62. Monneron, A., Blobel, G. and Palade, G. E. (1972), *J. Cell Biol.,* **55**, 104.
63. Mizuno, N. S., Stoops, C. E. and Sinha, A. A. (1971), *Nature,* **229**, 22.
64. Bornens, M. (1973), *Nature,* **244**, 28.
65. Bornens, M. and Kaspar, C. B. (1973), *J. biol. Chem.,* **248**, 571.
66. Jarasch, E. D., Reilly, C. E., Comes, P., Kartenbeck, J. and Franke, W. W. (1973), *Hoppe-Seyler's Z. physiol. Chem.,* **354**, 974.
67. Zentgraf, H., Deumling, B., Jarasch, E. and Franke, W. W. (1971), *J. biol. Chem.,* **246**, 2986.
68. Meyer, H. W., Roth, J. and Block, F. (1972), *Protoplasm,* **75**, 313.
69. Speth, V. and Wunderlich, F. (1970), *J. Cell Biol.,* **47**, 772.
70. Maul, G. G. (1971), *J. Cell Biol.,* **51**, 558.
71. Lui, T. P. (1972), *J. Parasit.,* **58**, 1151.
72. Cole, G. T. and Wynne, M. J. (1973), *Cytobios.,* **8**, 161.
73. Wunderlich, F., Batz, W., Speth, V. and Wallach, D. F. H. (1974), *J. Cell Biol.,* **61**, 633.
74. Moor, H. K., Mühlethaler, H., Waldner, H. and Frey-Wyssling, A. (1961), *Biophys. Biochem. Cytol.,* **10**, 1.
75. Smith, S. J., Adams, H. R., Smetana, K. and Busch, H. (1969), *Expl. Cell Res.,* **55**, 185.
76. Scheer, U. (1973), *Devl. Biol.,* **30**, 13.
77. Burton, K. (1956), *Biochem. J.,* **62**, 315.
78. Giles, K. W. and Myers, A. (1965), *Nature,* **206**, 93.
79. Dodge, J. T., Mitchell, C. D. and Hanahan, D. J. (1963), *Archs. biochem. Biophys.,* **100**, 199.
80. Harris, J. R. (1969), *Biochim. biophys. Acta.,* **188**, 31.
81. Harris, J. R. and Brown, J. N. (1971), *J. Ultrastruct. Res.,* **36**, 8.
82. Blanchet, J. P. (1974), *Expl. Cell Res.,* **84**, 159.
83. Scheer, U. (1972), *Z. Zellforsch.,* **127**, 127.
84. Harris, J. R. and Milne, J. F. (1974), *Trans. Biochem. Soc.* In press.
85. Callan, H. G. (1952), *Symp. Soc. Expl. Biol.,* **6**, 243.
86. Hunter, A. S. and Hunter, F. R. (1961), *Expl. Cell Res.,* **22**, 609.
87. Dounce, A. L. and Umana, R. (1962), *Biochemistry,* **1**, 811.
88. Chauveau, J., Moulé, Y. and Rouiller, Ch. (1956), *Expl. Cell Res.,* **11**, 312.
89. Blobel, G. and Potter, V. R. (1966), *Science,* **154**, 1662.
90. Widnell, C. C. and Tata, J. R. (1964), *Biochem. J.,* **92**, 313.
91. Sporn, M. B., Wanko, T. and Dingman, W. (1962), *J. Cell Biol.,* **15**, 109.
92. Emmelot, P. and Bos, C. J. (1966), *Biochim. biophys. Acta.,* **120**, 369.
93. Franke, W. W. (1966), *J. Cell Biol.,* **31**, 619.
94. Brasch, K., Seligy, V. L. and Setterfield, G. (1971), *Expl. Cell Res.,* **65**, 61.
95. Olins, D. E. and Olins, A. L. (1972), *J. Cell Biol.,* **53**, 715.
96. Mirsky, A. E. and Osawa, S. (1961), *The Cell,* (Brachet, J. and Mirsky, A. E., eds.), Vol. 2, pp. 677, Academic Press, New York and London.

97. Anderson, N. G. and Wilbur, K. M. (1951), *J. gen. Physiol.,* **34**, 647.
98. Ueda, K., Matsuura, T. and Date, N. (1969), *Biochem. biophys. Res. Commun.* **34**, 322.
99. Mitchell, C. D. and Hanahan, D. J. (1966), *Biochemistry,* **5**, 51.
100. Mazia, D. and Ruby, A. (1968), *Proc. natn. Acad. Sci. U.S.A.,* **61**, 1005.
101. Anderson, N. G. (1953), *Expl. Cell Res.,* **5**, 361.
102. Price, M. R', Harris, J. R. and Baldwin, R. W. (1973), *Methodological Developments in Biochemistry,* (Reid, E., ed.), Vol. 3, pp. 159, Longmans, Harlow.
103. Harris, J. R. In preparation.
104. Monneron, A. (1974), *Phil. Trans. R. Soc. Ser. B,* **268**, 101.
105. Lowry, O. H., Rosebrough, N. J., Farr, A. L. and Randall, R. J. (1951), *J. biol. Chem.,* **193**, 265.
106. Franke, W. W., Deumling, B., Falk, H. and Rae, P. M. M. (1973), *Expl. Cell Res.,* **81**, 365.
107. Keenan, T. W., Berezney, R., Funk, L. K. and Crane, F. L. (1970), *Biochim. biophys. Acta.,* **203**, 547.
108. Klenig, H. (1970), *J. Cell Biol.,* **46**, 396.
109. Yoo, B. Y. and Bayley, S. T. (1967), *J. Ultrastruct. Res.,* **18**, 651.
110. Schiechl, H. (1973), *Biochim. biophys. Acta.,* **307**, 65.
111. Maddy, A. H. and Kelly, P. G. (1970), *FEBS Letters,* **8**, 341.
112. Beaufay, H. and De Duve, C. (1954), *Bull. Soc. Chim. biol.,* **36**, 1525.
113. Kay, R. R., Haines, M. E. and Johnston, I. R. (1970), *FEBS Letters,* **10**, 113.
114. Neville, D. M. (1960), *Biochem. Biophys. Cytol.,* **8**, 413.
115. Hewish, D. R. and Burgoyne, L. A. (1973), *Biochem. biophys. Res. Commun.,* **52**, 504.
116. Weed, R. I., Reed, C. Fand Berg, G. (1963), *J. clin. Invest.,* **42**, 581.

# SECTION TWO

# 6 The solubilization of membranes

A. H. MADDY and M. J. DUNN



## 6.1 Introduction

The interest of biological membranes lies in their biological character, in the roles they have to play in the functioning and survival of living organisms. Therefore the study of their component molecules, although it can be fascinating in itself, should ultimately be directed towards a deeper understanding of these roles. This chapter and the following are primarily concerned with the methods available for the study of proteins in an isolated state, but this restriction should not draw attention from the value and feasibility of the complementary study of protein within the intact membrane.

In theory, membrane proteins are all those proteins required for the functioning of a membrane – in practice membrane proteins are those proteins present in a membrane preparation! Consequently, before starting the analysis of the proteins every effort should be made to ensure that during the process of membrane isolation there has been a minimal loss of true components, and minimal adsorption of non-membranous contaminants. Membrane proteins fall into three categories. (a) The first contains the most loosely bound proteins which might be only transiently located in the membrane. They may be washed away by physiological saline during preparation of the membrane and therefore not be readily recognized as membrane proteins even though they are essential for the complete functioning of the membrane. (b) A more easily recognized category are those proteins which are not released from the membrane during its isolation under physiological conditions but come away when the conditions are altered only slightly as by affecting the electrostatic environment of the membrane e.g. low ionic strength. Removal of these 'extrinsic' or 'peripheral' proteins [1] leaves a residue retaining a recognizable membranous morphology

and composed of virtually all the lipid and the more tightly bound proteins.
(c) Separation of these more tightly bound – 'intrinsic' or 'integral' – proteins from the lipid is only possible under conditions that disturb hydrophobic interactions and it leads to a complete disruption of membrane structure.

The first part of any programme for the study of the isolated molecules is, not surprisingly, their isolation. The extraction of lipids from membranes presents no particular difficulties (see Chapter 8) but many problems surround the extraction of proteins. This chapter considers the theoretical and practical factors that give rise to these problems and the various attempts to resolve them. Aspects relating specifically to glycoproteins and immunological analyses are discussed further in Chapters 9 and 11.

## 6.2 A statement of the problem

Solubilization of a membrane would, in the strict sense of the term, require the dispersion of its component molecules in a solvent as isolated molecules in a state where solute-solute interactions were negligible. If the process were to be of the fullest biological significance the isolated molecules would retain their native, i.e. biologically active, conformations.

In membrane studies the term 'solubilization' is rarely used in its strict sense, and while deploring the misuse of any term, it is preferable to tolerate a limited abuse so as to attain a practical goal rather than to pursue some unattainable ideal. In the context of membrane study, 'solubilization' can be defined as the conversion of the complex, intact, intractable membrane into relatively simpler states which are amenable to analysis by methods that cannot be applied to the intact membrane. (In many instances 'isolation', 'disaggregation' or 'extraction' would be preferable terms). Taking this very practical attitude towards the process of 'solubilization' does not imply a relaxation of standards but a recognition of the fact that it is prudent to use different criteria to judge the efficacy of 'solubilization' in different circumstances. These circumstances are determined by the purpose of the investigation. Thus, if the prime aim of the investigation is the constitutional analysis of a membrane then it is imperative to ensure that the membrane components in the 'solution' are fully dissociated from each other, for otherwise some supramolecular complex may be incorrectly identified as a distinct molecular species. The conformational states of the proteins are of secondary importance in this analysis. If, on the other hand, the prime aim is the investigation of some biological activity of a membrane, then the conformation of the isolated proteins is of crucial significance and the persistance of some intermolecular forces may be essential, e.g. if the biological activity depends on the integrity of a lipoprotein complex.

The difficulty of handling membrane proteins may be illustrated by initially considering lipids. When a lipid chemist undertakes the analysis of a mixture of lipids, the lipids are dissolved in a *non-polar solvent* where the individual lipid molecules are dissociated from each other and they may then be fractionated on the basis of their different chemical properties. With sufficient ultrasonic irradiation these same lipids could be dispersed in an *aqueous medium* and 'solubilized' to the extent that the dispersion does not scatter light and the lipids are not pelleted by low speed centrifugation. But the molecules would actually be present in multimolecular aggregates containing several different types of molecule and consequently the different molecular species of lipids could not be separated from each other. When a protein chemist attempts the fractionation of a protein mixture, as long as the proteins are water soluble he can be confronted with the converse situation. The mixture can be taken into solution in an *aqueous medium* of suitable pH and ionic strength and the constituent proteins fractionated from this solution. If the same proteins are exposed to a *non-polar solvent* the result would probably be an intractable protein precipitate of tightly associated molecules which would probably not be dissociated from each other by ultrasonic dispersion of the precipitate. Unfortunately, one cannot draw the simple conclusion that lipid chemists should work in non-polar solvents and protein chemists in aqueous solvents. All proteins are not water soluble and the notion that all proteins should be studied in aqueous media is not derived from the inherent properties of proteins, but rather from technical limitations which have largely confined the study of proteins to those which are water soluble.

Some membrane proteins may naturally exist in an aqueous domain and be water soluble. Others may exist in the non-aqueous regions of a membrane and be insoluble in water. It follows that the suitability of a solvent for any protein will depend on the location of the protein within its membrane. The different locations of proteins with respect to the lipid bilayer of a membrane are shown schematically in Fig. 6.1.

At one extreme (Type 1) there are those proteins which are bound to the membrane predominantly, or exclusively, by electrostatic interactions and which are probably located within the aqueous domain of the membrane: their native conformation should be retained in aqueous solutions, e.g. erythrocyte glyceraldehyde-3-phosphate dehydrogenase [2–4]. At the other extreme are proteins which may be located within the non-aqueous hydrocarbon core of the membrane (Types 4 and 5) and these could be expected to retain their native conformations in non-aqueous solvents, e.g. *Staphylococcus aureus* isoprenoid alcohol phosphokinase [5]. If these were the only two types of membrane proteins solubilization would be relatively simple, the former could first be removed by extraction with aqueous solvents and the latter subsequently dissolved by transferring the water insoluble lipid-protein residue into a non-polar solvent. Unfortunately, the problem is compounded by proteins which straddle both the aqueous and non-aqueous domains of the membrane, having part of their polypeptide sequences within the aqueous domain and other parts buried amongst the hydrocarbon chains of the lipids, e.g. erythrocyte glycophorin [6–8] and microsomal cytochrome $b_5$ [9,10]. The stability of the native conformation of such proteins depends on their being in this biphasic environment and it is difficult to see how this dual environment may be

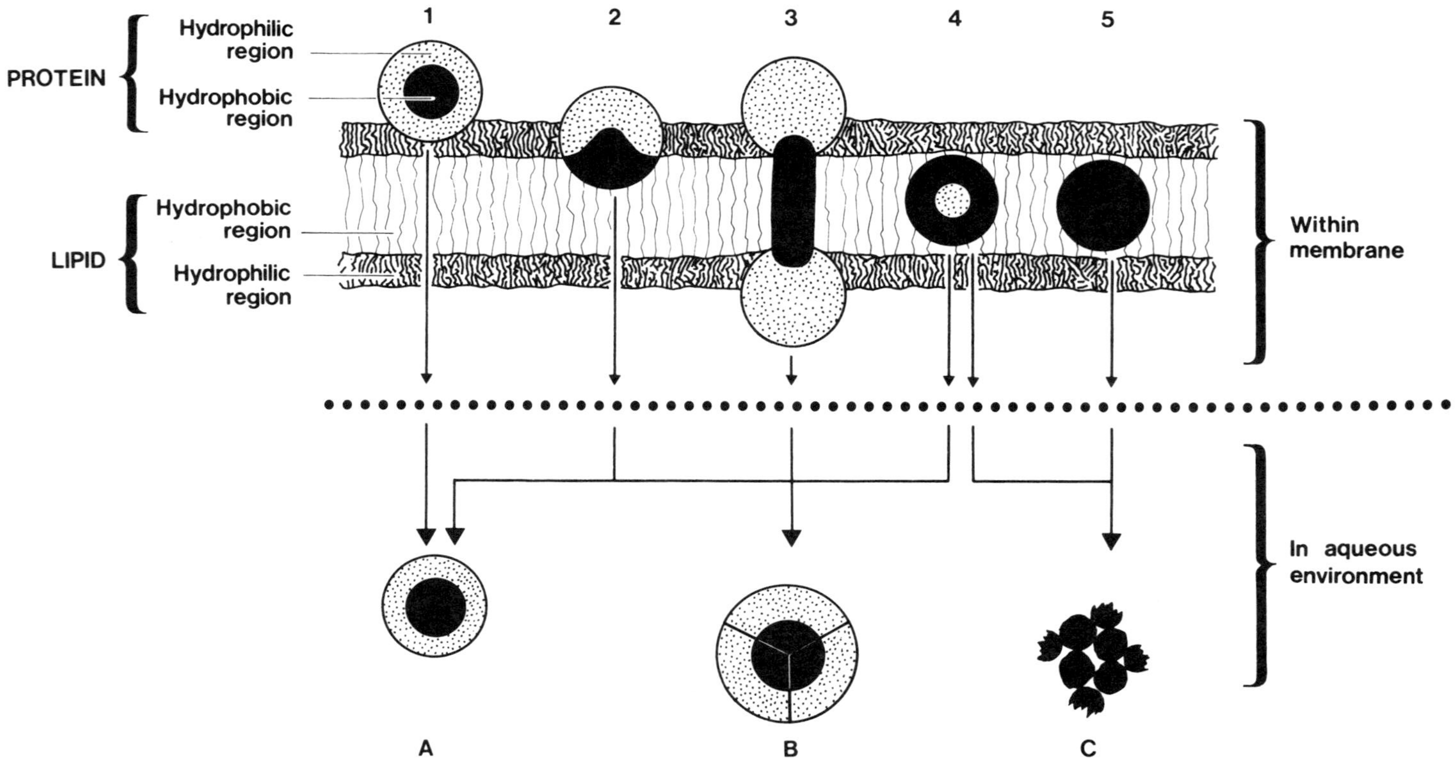
PROTEIN
Hydrophilic region
Hydrophobic region
LIPID
Hydrophobic region
Hydrophilic region
1
2
3
4
5
Within membrane
In aqueous environment
A
B
C

mimicked in a bulk solution. Possibly, a high level binding of detergent offers the opportunity of preserving this heterogeneous environment by encasing each protein molecule in a layer of detergent and thus providing it with a microenvironment that could be partly hydrophilic and partly lipophilic. Unfortunately although the biologically functional conformation may be preserved, its expression might be inhibited by the effect of the detergent on other components in the biological assay.

It was argued earlier that the assessment of any solubilization procedure should depend upon the aim of the investigation. What criteria are available in the various circumstances that can arise? No matter what the circumstance, the process of solubilization, if it is to be of any value, must convert the protein into a state in which it is accessible to techniques that could not be used to produce unequivocal data when it was integrated into the intact membrane. The crudest, but nevertheless essential, criterion of any successful solubilization is that the soluble particle has a relatively low sedimentation coefficient – in practical terms it should not be pelleted by 100 000 x *g* for 1 h. (This value applies to a medium with a density near that of water. Claims to successful solubilization have been made when the 'solubilized' material did not sediment in 8 M urea which itself has a density almost equal to that of the intact membrane). A low sedimentation coefficient is interpreted primarily as evidence of a decrease in size (gel exclusion chromatography is another possible way of

---

**Fig. 6.1** Possible arrangements of proteins relative to a lipid bilayer and the probable effects of extraction of the proteins into aqueous solutions.
1. Proteins and lipid interact exclusively through their hydrophilic regions and the protein can be transferred into an aqueous environment with the minimum perturbation (state A).
2. Proteins and lipids interact through both hydrophilic and hydrophobic regions and on separation and exposure of the hydrophobic facets of the proteins to water they rearrange, either within a molecule (state A) or as a multimolecular complex (state B).
3. Here the protein traverses the lipid but the consequences of aqueous solubilization are as in 2 (i.e. states A or B).
4. Interaction between the protein and lipid is purely hydrophobic as the hydrophilic amino acids have been internalized (a possible but improbable arrangement). In an aqueous environment the protein could invert to states A or B, or could form an insoluble conglomerate as in state C except that the subunits would have hydrophilic cores, or a mixed complex of B and C could arise.
5. The protein is composed exclusively of hydrophobic amino acids and would form an insoluble conglomerate in aqueous environments (state C). A protein of this type would be soluble in non-aqueous media.
It is assumed in this scheme that proteins are incorporated into membranes by direct interaction with lipids. Some membrane proteins may not, in nature, interact with lipid, but be bound to the membrane by interaction with those proteins that do. This additional possibility extends, but does not alter, the concepts illustrated in the diagram.

estimating the size of the disaggregated membrane particles – see Chapter 7), but decrease in size is, itself, of little value unless it is accompanied by a decrease in the complexity of the solubilized particles relative to the intact membrane. The situation can arise where a membrane has been disrupted into particles which are small enough to stay in suspension and which also retain biological activity, and yet they are merely small pieces of membrane, not representing any pre-existing subunit, and the molecular interactions within the particles are not significantly less complex than those within the whole membrane. Sonication is particularly prone to result in this type of product [11] and the disadvantages of this technique are compounded by the generation of lysophosphatides [12] which have a detergent action. Sonication is not to be dismissed out of hand, for under favourable circumstances carefully controlled sonication can be used to strip off specific membrane components as particles suitable for further investigation, e.g. the inner membrane of the mitochondrion [13]. However the technique must be applied with great care and the product well characterized before any conclusions can be drawn.

As small size is a necessary, but insufficient, criterion of solubilization it must be supplemented by other criteria. If it is intended to analyse the polypeptide composition of a membrane the abolition of intermolecular bonds is essential, and the demonstration of total dissociation obligatory. Complete dissociation can only be assumed in fractions shown by classical methods of protein chemistry to contain only one polypeptide chain. Some of these classical methods are discussed in Chapter 7.

Although such disruption of intermolecular forces is a prerequisite for establishing the polypeptide composition of a membrane, the complete destruction of the interactions of membrane proteins with each other and with lipids converts the proteins into denatured polypeptide chains. However, native proteins frequently contain more than one polypeptide chain, and this approach cannot therefore reveal the number of functional proteins in the membrane. Once the range of polypeptides has been established the investigation can proceed to the study of the functional proteins using milder conditions in order to isolate proteins with minimum perturbation of structure, and different criteria of success may then be adopted.

Consideration of the isolation of proteins in their native states necessarily leads on to the value of the persistance of biological activity as a criterion for the success of a solubilization method. If the central aim is the isolation of a biologically active component from a membrane then the retention of this activity is, of course, essential. As many activities associated with membranes are dependent on the integrity of multimolecular complexes, disruption of intermolecular bonds can now be counter-productive, yet must be pursued to the extent of releasing the active complex from the membrane matrix. This quandary is unavoidable when the biologically competent moiety may be a lipoprotein. The mere presence of lipid in an active preparation has little significance, e.g. plasma membrane 5′-nucleotidase isolated by Triton X-100 extraction contains lipid [14], but a lipid free active enzyme can be prepared by

*N*-dodecyl sarcosine extraction [15]. The biological necessity for lipid in a fraction may only be concluded when, as with sodium/potassium dependent ATPase, it has been demonstrated that the activity of an inert protein can be restored by the replacement of lipid [16].

Biologically significant associations need not necessarily be exhibited by overt biological activity. Some of the complications that arise are illustrated by the various attempts to isolate histocompatability antigens. Extraction of antigenically active products is relatively easy, but entirely different products can possess this activity, e.g. large complexes (200 000 mol. wt.) solubilized by NP 40 [17] and much smaller fragments extracted by 3.0 M KCl or papain digestion (circa 45 000) [18,19]. The former could contain many unrelated, inactive adventitious molecules and the latter type could be antigenically active degradation fragments; indeed the salt product is glycosylated and the papain fragment is not. Deoxycholate extraction helps to resolve at least some of the problems, for the activity of the deoxycholate extract resides in a sugar-bearing component (it is bound to an affinity column bearing a lectin which specifically binds carbohydrate) and the active particle is slightly bigger than the protein released by papain [20]. But the problem has taken a further recent twist with evidence now emerging that in the membrane the antigenically active polypeptide is intimately associated with a $\beta$-microglobulin which, of course, would be assiduously removed in the course of the isolation of the 'pure' antigen [21].

Loss of biological activity can only be used as an indicator of the effects of solubilization for those proteins with overt biological activity, and consequently is not applicable to the significant proportion of membrane proteins of unknown function. If some membrane proteins are structural in the sense that they contribute, with lipids, to the formation or stability of a membrane matrix then no activity will ever be apparent in the proteins once they have been separated from this matrix. In these proteins, and all others for that matter, some evaluation of the extent of any perturbation caused by extraction may be obtained by methods that analyse protein conformation, provided that these methods can be applied to both the intact membrane and the isolated protein. For example, infra-red spectroscopy of erythrocyte ghosts indicates the absence of any $\beta$ sheet. Protein extracted from these ghosts by a butanol treatment has an identical protein spectrum, with no indication of a $\beta$ structure [22], but protein extracted by dimethyl formamide is seen to possess a $\beta$ component which must have been induced by the solvent [23]. Electron spin resonance (E.S.R.) spectroscopy is another suitable method. Thus after labelling the sulphydryl groups of proteins within an intact membrane with an E.S.R. label (*N*-(1-oxyl-2,2,6,6-tetramethylpiperidinyl)-maleimide), the effects of several extraction procedures may then be assessed by observing the changes they produce on the E.S.R. spectrum [24].

In order to disperse a membrane, either into molecular or biological components, the forces that bind the components into the membrane matrix must be overcome. The whole gamut of molecular interactions is implicated, but for practical purposes they can be divided into electrostatic and hydrophobic forces. The proteins most easily detached are those bound predominantly, perhaps exclusively, by electrostatic forces, i.e. the 'extrinsic' or 'peripheral' proteins. However it is unwise to make facile deductions regarding the forces which bind proteins to membranes from what might be considered to be predictable effects of extraction procedures. Thus, it might be concluded that as proteins are extracted from membranes by dilute solutions of chelating agents [25–30] these proteins are bound to the membrane by divalent cation bridges, and furthermore as EGTA is equally effective [31], these are presumably calcium bridges. Although this explanation is consistent with the release of the same proteins by dilute acetic acid at pH 2.8 [27], when presumably the bridge is destroyed by protonation of its anionic component, it does not adequately explain a number of other observations. If dilute EDTA releases protein simply by the rupture of divalent cation bridges why should the amount of protein released by markedly diminished by the addition of as little as 10 mM sodium chloride, or why should the same proteins be released by the reducing reagents $\beta$-mercaptoethanol and dithiothreitol (Maddy and Dunn – unpublished observation)? Conversely, at higher concentration sodium chloride can release a different set of proteins, e.g. glyceraldehyde-3-phosphate dehydrogenase from the erythrocyte membrane [2–4]. This may be regarded as evidence of electrostatic binding, but what of the release of proteins by 3.0 M potassium chloride [18], e.g. $H_2$ antigens? Is extraction by salt of this concentration to be attributed to the diminishing of electrostatic forces by the high ionic strength or might the salt be having a chaotropic effect?

## 6.3 Attempts to solve the problem

Although it is conceptually valuable to classify proteins as 'extrinsic' or 'intrinsic', in practice the dividing line between the two categories is somewhat arbitary. As the extraction conditions become more vigorous there is a progressive release of protein. For example, treatment of the erythrocyte membrane with 0.1 N sodium hydroxide liberates those proteins soluble in chelating agents together with additional proteins [32]. Likewise additional proteins are released by altering the balance of ionogenic groups within the membrane instead of affecting the balance of charges by changes in the membrane's environment. Substitution of the protein sulphydryl groups with charged residues, e.g. *p*-chloromercuribenzoate, *p*-chloromercuribenzenesulphonate, or 5,5′-dithio*bis*-(2-nitrobenzoic acid), releases proteins while comparable uncharged reagents such as *N*-ethyl maleimide do not release proteins [2–4,35]. A similar result is observed when anionic groups are introduced by substitution with suitable acidic anhydrides, e.g. succinic, maleic, citraconic or dimethylmaleic anhydrides [32,36–38). Extraction by these procedures results in the substitution and modification of the proteins, and clearly if modification is unavoidable reagents which introduce labile substituents are greatly preferable to those which modify proteins irreversibly. The succinyl residue is difficult to remove but maleyl, citraconyl and dimethymaleyl residues are progressively more easily removed [39].

Although alteration of the charge distribution of membrane components may indirectly affect interactions other than those of an electrostatic nature, some proteins are too tightly bound to the lipid to be released by such techniques and are only released under conditions which directly affect hydrophobic interactions. Chaotropic salts [40] have a wide range of effectiveness in this respect, varying from potassium thiocyanate, which releases a small and selective portion of proteins from erythrocyte ghosts [41], through guanidine hydrochloride which releases all but one or two polypeptides from erythrocyte membranes [42–44], to lithium diiodosalicylate which brings about the complete dissolution of these same membranes [6–8]. Non-aqueous protein solvents, e.g. 2-chloroethanol [45], dimethyl formamide [23,46] hexafluoroacetone [47], butanol etc. [48–51], offer another approach to the problem of membrane dissolution. The variety of organic solvents potentially available suggests that solvents preferable to water may be found for membrane proteins, and such solvents for proteins in categories 4 and 5 (Fig. 6.1) should be relatively easy to discover, e.g. isoprenoid alchol phosphokinase from *Staphylococcus aureus* is readily soluble in *n*-butanol [5]. Alternatives to water for the other type of proteins may be more elusive. In spite of their enormous potential value these solvents all suffer from two major disadvantages, (a) virtually all techniques of protein chemistry are most effective in aqueous conditions, and (b) the demonstration of biological activity is almost exclusively confined to aqueous media. Thus conceivably an optimal solvent might be discovered for a protein, but only very limited studies could be pursued in this solvent and a retrogression into theoretically less desirable aqueous media be unavoidable for technical reasons. Occasionally organic solvents solubilize membrane proteins not by acting as simple solvents but by disrupting the membrane and converting the proteins into a water soluble state (cf. detergents). Treatment of an aqueous suspension of erythrocyte ghosts with *n*-butanol produces an aqueous phase containing virtually lipid free proteins with the lipids dissolved in the butanol phase [48–51]. Rather surprisingly the close homologue *n*-pentanol converts the whole membrane into water soluble lipoprotein complexes [52].

The final group of solubilizing agents that needs to be considered is that of detergents. Until recently the selection of a detergent for the isolation of a membrane component was purely empirical and consequently a vast and disjointed literature exists which is almost impossible to summarize. Although an understanding of some of the principles involved is beginning to emerge, a fully rational basis for the utilization of detergents remains to be established. The solubilizing effects of detergents stem from their amphiphilic character which enables them to interact with both hydrophilic and lipophilic regions of proteins in an essentially disruptive fashion. Two sets of interactions are of cardinal significance – interaction of the detergent molecules with membrane components and interactions between detergent molecules [53–55]. The binding to the membrane and its protein will depend on both the number of binding sites and the affinities of those sites for the detergent molecules. Sites of high affinity are few, and although they are saturated at low detergent concentrations this binding frequently has a minimal effect on protein conformation. As the concentration of detergent increases, sites of progressively lower affinities become saturated and as more and more molecules are bound the effect on protein conformation tends to increase. The interaction of detergent molecules with themselves is significant because the concentration of detergent effective in protein binding is not the total concentration of detergent but the concentration of free molecules, i.e. molecules not complexed into micelles. Hence the more avidly the detergent reacts with itself, the fewer will be the molecules available to interact with the proteins, and if the critical micellar concentration (c.m.c.) of the detergent is sufficiently low the concentration of free molecules will only be sufficient to saturate protein sites of high affinity for detergent with minimal effect on protein conformation. These facts possibly explain why Triton X-100 and deoxycholate, which have low c.m.c. values, have been repeatedly used for the isolation of biologically active molecules from membranes, while SDS, which has a high c.m.c., characteristically exhibits high level binding and causes the complete disruption of the native conformation of proteins (see Chapter 7). Nevertheless it must be borne in mind that the binding of detergents with high c.m.c. values can be restricted to protein sites of high avidity at sufficiently low concentrations of detergent. Hence active glyceraldehyde-3-phosphate dehydrogenase can be released from the erythrocyte membrane by low concentrations of SDS, but it is inactivated by very low concentrations of Triton X-100 which is generally regarded as a 'milder' detergent [4]. It should also be pointed out that it is sometimes more effective to use a detergent above its c.m.c., or at least at the c.m.c. as determined in the absence of protein. This is because the tendency of a detergent to associate with itself is counteracted by its tendency to associate with proteins, i.e. a greater overall concentration of detergent is necessary to form micelles in the presence of protein than in its absence.

When a detergent is selected for the isolation of a particular membrane protein the investigator must first select the detergent and then determine the optimal concentration of that detergent required to produce the desired end. Selection remains predominantly a matter of trial and error with perhaps the one guiding principle that detergents with a low c.m.c. will tend to be less disruptive than those with a high c.m.c. If, therefore, biological activity is to be retained, bile salts or non-ionic detergents will be preferable, but if disintegration to the molecular level is the prime object then a detergent with a more vigorous action may be required. In the past mixtures of detergents have been found to be more effective than a single detergent [56] but in recent years this approach seems to have passed out of favour. In selecting any detergent it should be remembered that the purity of commercial preparations is variable, but as any one detergent gets widely used so does the quality of the available preparations improve. Impurities can include heavy metal ions and extremes of pH, and a given sample may contain molecules with chain lengths varying by one or two carbons from its purported value. A detergent may be available in more than one form, e.g. lithium dodecyl sulphate is now available as a more soluble alternative to the more commonly used sodium salt. The insolubility of the potassium salt of this detergent precludes its use in the presence of potassium ions.

The optimal detergent concentration for any particular investigation can only be established by systematic study. Any given detergent may have quite different effects on the same membrane at different detergent concentrations. Two systematic studies illustrate the importance of detergent concentration; the dissolution of the Semliki virus membrane by Triton X-100 [57] and the disruption of mitochondria by digitonin and Lubrol [58]. The initial binding of Triton to the virus at low detergent concentration only ruptures the membrane and releases the nucleocapsid. As the detergent concentration is increased the membrane is first dissociated into lipo-protein-detergent complexes and then at high detergent concentrations the lipids and proteins are dissociated from each other. A similar progression is observed with the mitochondrion. At low concentrations digitonin ruptures the outer membrane; at higher concentrations this membrane is solubilized leaving an insoluble residue consisting of the inner membrane enclosing the proteins of the matrix. In this particular investigation the matrix proteins were released by rupture of the inner membrane with Lubrol.

If possible the detergent that has been used to solubilize a protein should be removed before undertaking any further study of the protein. Removal of detergent is desirable not only because of any untoward effects the bound detergent may have on the protein's conformation, but also because residual detergent can interfere with subsequent analysis of the protein, especially in any studies of biological activity. Techniques for the rapid removal of the more frequently used detergents are available and can result in restoration of biological activity. *Trans*-farnesyl pyrophosphate squalene synthetase is inactive after isolation from yeast membranes by deoxycholate, but activity can be regenerated by removal of the deoxycholate by passage down a column of Amberlite XAD-2 [59]. Even the extensive changes in conformation caused by SDS have been reversed in several enzymes by removal of the detergent by Dowex 1-X2 in buffered urea and renaturation of the enzymes from the urea [60]. It has also been reported that Triton X-100 is removed from proteins by the neutral styrene-divinyl benzene polymer Bio-beads SM-2 [61] (see Chapter 9).

## 6.4 Practical procedures for solubilization of membrane proteins

Solubilization frequently involves no more than suspending the membranes in the extracting medium for a specified period and then pelleting the insoluble residue. Many of the procedures listed in Table 6.1 are of this nature and little

TABLE 6.1

| | | |
|---|---|---|
| (1) Modification of the electrostatic conditions in the membrane's environment | | |
| (a) Low ionic strength = water | Harris [65] | A multiunit protein complex released from erythrocyte membranes |
| | Clarke [66], Hamaguchi and Cleve [67] | High molecular weight polypeptides released from erythrocyte membranes |
| (b) Low ionic strength + chelating agent, e.g. EDTA | Tillack *et al.* [26] | 40% erythrocyte membrane proteins released, mostly high molecular weight |
| | Neville [68] | Extracts a specific protein component from liver plasma membranes |
| | Thorley–Lawson and and Green [29] | Calsequestrin solubilized from sarcoplasmic reticulum membranes |
| | McCarty and Racker [30] | Extracts $F_1$ coupling factor from inner membrane of mitochondrion |
| | McCarty and Racker [30] | Chloroplast thylakoid protein, comparable with mitochondrial $F_1$, solubilized |
| (c) Low ionic strength + reducing agent, e.g. dithioerythritol | Maddy and Dunn (unpublished) | Releases those proteins also released by EDTA from erythrocyte ghost |
| (d) High ionic strength | Kant and Steck [3] Tanner and Gray [2] Shin and Carraway [4] | Specific extraction of glyceraldehyde-3-phosphate dehydrogenase from erythrocyte ghost |
| | Jacobs and Sanadi [69] | Cytochrome *c* extracted from mitochondria |
| | Reisfeld and Pellegrino [18] | Extraction of $H_2$ antigens by 3.0 M potassium chloride |
| (e) pH. | | |
| (i) Low pH Dilute acetic acid | Zahler, Saito and Fleischer [70] | Extraction of mitochondrial proteins |
| | Maddy and Kelly [27] | Solubilization of those erythrocyte ghost proteins also solubilized by EDTA |
| 0.2 N Sulphuric acid | Gagnon *et al.* [71] | Extraction of the basic protein of nerve myelin |
| (ii) High pH Potassium carbonate | Neville [72] | Removes certain proteins from liver plasma membrane |
| Dilute sodium hydroxide | Penevsky and Tzagaloff [56] | Extraction of protein from beef hear mitochondria, rat liver microsomes, *Bacillus subtilis* membrane |
| | Steck and Yu [32] | Solubilization of many erythrocyte ghost proteins including those also soluble in EDTA |
| (2) Modification of ionogenic groups within membranes | | |
| (a) Ionogenic sulphydryl reagents | Carter [34], Godin and Schrier [33], Cantrell [35] | Solubilization of many erythrocyte membrane proteins |

TABLE 6.1 (*Continued*)

| | | |
|---|---|---|
| (b) Acid anhydrides e.g. succinic | Maclennan, Tzagaloff and Rieske [36] | Solubilizes proteins of mitochondria |
| Maleic Citraconic Dimethylmaleic | Moldow *et al.* [37], Lundahl and Hjerten [38] | Extensive solubilization of erythrocyte membrane proteins |
| (3) Protein pertubants – chaotropic agents | | |
| Sodium iodide | Fischer, Lingsch and Urbina [73] | Extensive solubilization of squid retinal axon membrane proteins |
| Sodium *iso*thiocyanate | Fischer, Lingsch and Urbina [73] | Extensive solubilization of squid retinal axon membrane proteins |
| Potassium thiocyanate | Morrison and Neurath [41] | Releases active proteases from erythrocyte membranes |
| Sodium perchlorate | Davis and Hatefi [74] | Extraction of succinic dehydrogenase from mitochondria |
| Urea | Lauf and Poulik [75] | Proteins of erythrocyte membrane solubilized in urea and separated by starch gel electrophoresis |
| Guanidine hydrochloride | Gwynne and Tanford [42], Steck [43], Maddy and Kelly [44] | Almost total solubilization of erythrocyte membrane proteins, but some protein-lipid interactions resist its effect |
| Guanidine *iso*thiocyanate | Moldow, Robertson and Rothfield [76] | Solubilization of *Escherichia coli* envelope proteins. The *iso*thiocyanate should have a greater chaotropic effect than the hydrochloride |
| Lithium diiodosalicylate | Marchesi and Andrews [6], Segrest *et al.* [7], Segrest, Jackson and Marchesi [8], Marchesi [63] | Erythrocyte glycoprotein-glycophorin-solubilized and purified by a phenol treatment |
| | Choules *et al.* [77] | Solubilization of *Acholeplasma* membrane proteins |
| (4) Enzymes | | |
| Papain | Emmelot, Visser and Benedetti [78] | $\beta$-naphthylamidase containing particles liberated from liver plasma membranes |
| | Sanderson and Welsh [19] | Release of biologically active fragment of histocompatiability antigens |
| | Cogoli *et al.* [79] | Solubilization of hamster intestinal sucrase-maltase complex |
| $\beta$-glucosidase | Baldwin, Bowen and Price [80] | Tumour specific antigens solubilized from hepatoma plasma membrane |
| (5) Sonication | | |
| | Kahan and Reisfeld [81] | HL antigens of guinea pig released |

TABLE 6.1 (*Continued*)

| | | |
|---|---|---|
| | Horstaman and Racker [13] | Mitochondrial inner membrane coupling factor $F_1$ isolated |
| (6) Organic solvents | | |
| Aliphatic alcohols | Nachbar, Winkler and Salton [82] | Solvent effect of several alcohols compared on *Micrococcus lysodiekticus* membranes. About 50% of membrane proteins extracted |
| *n*-Butanol | Maddy [48] Zwaal and van Deenen [51], Rega *et al.* [50] | Membranes solubilized with proteins passing into aqueous phase and lipids into the butanol phase |
| | Salton and Schorr [83] | Extraction of ATPase from *Micrococcus lysodiekticus* membranes |
| | Higashi, Siewert and Strominger [5] | Hydrophobic proteins from *Staphylococcus aureus* membrane dissolved in butanol |
| *n*-Butanol + EDTA | Maddy, Dunn and Kelly [49] | The presence of EDTA facilitates the fractionation of the proteins solubilized by the action of butanol |
| *n*-Butanol + urea | Kundig and Roseman [84] | Solubilizes part of the Enzyme II complex of the phosphotransferase system of *Escherichia coli* |
| 2-Chloroethanol | Zahler and Weibel [45] | Complete solubilization of erythrocyte membrane proteins reported in this highly toxic solvent |
| Dimethyl formamide | Chapman, Kamat and Levine [23] | Solubilization of erythrocyte membrane proteins |
| | Schnaitman [46] | Mitochondrial and microsomal proteins solubilized in acidic dimethyl formamide |
| Hexafluoroacetone | Juliano [47] | Extensive solubilization of erythrocyte proteins in this toxic solvent |
| Chloroform/methanol mixture | Curtis [85] | Extracts % of mitochondrial membrane proteins |
| | Eng *et al.* [86] | Chloroform/methanol used for the extraction of proteins from myelin |
| 90% Acetic acid | Schubert, Poensgen Werner [87] | At this concentration acetic acid can dissolve those erythrocyte membrane proteins that are not soluble in dilute acetic acid |
| (7) Special methods for glycoproteins | | |
| Phenol extraction | Kathan, Winzler and Johnson [88], Howe, Lloyd and Lee [89] | Phenol treatment of erythrocyte membranes leads to an aqueous phase containing the glycoproteins |
| Lithium diiodosalicylate | Marchesi and Andrews [6], Marchesi [63] | |
| Aqueous pyridine | Blumenfeld and Zvilichovsky [90], | Treatment of membranes with aqueous pyridine results in the |

TABLE 6.1 (*Continued*)

| | | |
|---|---|---|
| | Zvilichovsky, Gallop and Blumenfeld [91] | preferential solubilization of the glycoproteins. Works well with human erythrocytes but less effective on human hepatic plasma membranes |
| Chloroform/methanol | Hamaguchi and Cleve [64] | Aqueous solution of glycoproteins prepard by extraction of erythrocyte ghosts with chloroform/methanol mixture |
| (8) Non-ionic detergents | | |
| Triton X-100 | Simons, Helenius and Garoff [57] | A systematic study of the solubilization of Semliki Forest virus membranes and proteins |
| | Makino, Reynolds and Tanford [54] | A study of the binding of Triton X-100 and deoxycholate to proteins |
| | Yu and Steck [92] | Extensive extraction of erythrocyte ghosts leaving an insoluble residual ghost composed of fibrillar proteinaceous structures |
| | Butters and Hughes [93] | Selective extraction of glycoproteins from plasma membranes of KB cells |
| | Meunier *et al.* [94] | Isolation of cholinergic receptor from *Electophorus electrius* electric organ |
| | Cuatrecasas [95] | Solubilization of insulin receptor from liver and fat cell plasma membranes |
| Tween 20 | Liljas, Lundahl and Hjerten [95] | Solubilization of the EDTA insoluble proteins of *Acholeplasma* membranes |
| Lubrol WX | Schnaitman and Greenwalt [58] | Solubilization of proteins of inner mitochondrial membrane |
| | Bjerrum and Lundahl [97] | Preparation of erythrocyte ghost proteins for immuno-electrophoresis |
| Berol EMU–043 | Bjerrum *et al.* [98] | Preparation of erythrocyte ghost proteins for immunoelectrophoresis |
| | Bjerrum and Lundahl [97] | Preparation of erythrocyte ghost proteins for immunoelectrophoresis |
| N.P. 40 | Schwartz *et al.* [17] | Solubilization of $H_2$ antigens |
| | Gavin, Buell and Rath [99] | Isolation of insulin receptor from lymphocytes |
| Digitonin | Salton and Netschey [100] | Solubilization of bacterial membrane proteins |
| | Erwin and Hellerman [101] | Extraction of monomine oxidase from mitochondrial outer membrane |
| | Schnaitman and Greenwalt [57] | The progressive disruption of mitochondria |
| A comparative study of many detergents | Hagopian and Eylar [102] | Solubilization of glycosyl transferases |

TABLE 6.1 (*Continued*)

| | | |
|---|---|---|
| (9) Ionic detergents | | |
| Deoxycholate and/or cholate | Penefsky and Tzagoloff [56] | A review of many applications, particularly to mitochondrial membranes |
| | Emmelot, Feltkamp and Vas Dias [103] Philippot [104] | Systematic studies of the action of the detergent on membranes |
| | Makino, Reynolds and Tanford [54] | A study of the binding of deoxycholate to proteins |
| | Snary *et al.* [20] | Combined with affinity chromatography for the isolation of HL–A antigens |
| | Dulaney and Touster [105] | A mild procedure for the isolation of several erythrocyte membrane enzymes |
| | Salton and Schmitt [106] | Application to bacterial membranes |
| Cetyl trimethyl ammonium bromide | Nozaki, Reynolds and Tanford [53] | A discussion of the mechanisms of binding to proteins |
| | Fairbanks and Avruch [107] | Employed as a cationic equivalent to SDS |
| | Davenport [108] | Solubilization of mitochondrial and microsomal proteins |
| | Heller [109] | Solubilizes bovine visual pigment without spectral damage. This protein is resistant to most solvents including 6 M guanidine hydrochloride |
| Sodium dodecyl sulphate | | Virtually every membrane isolated has has been investigated by SDS-gel electrophoresis. This application of the detergent has been discussed in some detail in Chapter 7 |
| Sodium dodecyl sarcosine | Evans and Gurd [110] | Solubilization of plasma membrane marker enzyme 5′-nucleotidase |

purpose would be served in listing them here again. Practical details are provided below for three methods where the procedure is rather more elaborate.

### 6.4.1 *Butanol method (as developed for bovine erythrocyte membrane proteins)* [49]

Ghosts prepared by the Dodge procedure [48,62] are washed three times with 0.5 mM EDTA in 2.5 mM phosphate buffer adjusted to pH 8.0 and the pellet finally suspended (8–10 mg ml$^{-1}$) in this medium. The suspension is cooled on ice, 0.75 vol. of *n*-butanol added, and the mixture shaken. After 20 min on ice the two phases are separated by centrifugation (22 000 x *g*, 5 min), the lower (aqueous) phase removed with an ice-cold hypodermic syringe and dialysed overnight against the dilute EDTA buffer, containing ice, to remove butanol. The resulting protein solution which contains about 90% of the membrane's protein is stored at 4°C in the presence of 0.01% sodium azide as a bacteriostatic.

It is essential for the ghosts to be suspended in a medium of low ionic strength before addition of the butanol, and once the butanol has been added the preapration must be kept at 0°C, i.e. in the presence of ice, until the butanol has been dialysed away.

### *6.4.2 Lithium diiodosalicylate extraction (as developed for human red cell membrane glycoproteins)* [6,63]

The ghosts are suspended (25 mg ml$^{-1}$ of ghost protein) in 0.3 M lithium diiodosalicylate in 0.5 M Tris buffer, pH 7.5, and stirred at room temperature for 15 min. Two volumes of water are added. The mixture is brought to 4°C by the addition of 2 vols of cold water and kept at 4°C for a further 10 min. All subsequent steps are performed at 4°C. The suspension is centrifuged at 45 000 x *g* for 90 min, the supernatant mixed with an equal volume of freshly prepared 50% (w/v) phenol in water, and stirred for 15 min before a further centrifugation for 1 h at 4 000 x *g*. The upper aqueous phase, which contains the glycoproteins, is dialysed against several changes of water over 36 h and then freeze dried. The dry product is washed three times in 100% ethanol, suspended in water, and dialysed against water overnight. The resulting turbid solution is centrifuged at 10 000 x *g* for 1 h to obtain the clear solution of glycoprotein.

### *6.4.3 Chloroform/methanol extraction (as developed for red cell membranes)* [64]

Ghosts are first washed in 10 mM Tris-HCl buffer, pH 7.4, containing 0.1 mM EDTA, then homogenized in a Dounce homogenizer and adjusted to a protein concentration of 2 mg ml$^{-1}$ in the Tris buffer. 9 volumes of a chloroform/methanol mixture (2:1, v/v) are added, the whole stirred vigorously for 30 min at 20°C and then centrifuged at 10 000 x *g* for 10 min. The aqueous layer is carefully removed and recentrifuged to remove any interfacial material before being concentrated on a rotary flash evaporator to ¼ – $^1/_{10}$ of the original ghost suspension volume. The solution will remain clear unless it is contaminated with interfacial material. The glycoproteins are precipitated by addition of 9 vol. of ethanol, centrifuged off and dried. They may be redissolved in water.

## References

1. Singer, S. J. (1974), *A. Rev. Biochem.*, **43**, 805.
2. Tanner, M. J. A. and Gray, W. R. (1971), *Biochem. J.*, **125**, 1109.
3. Kant, J. A. and Steck, T. L. (1973), *J. biol. Chem.*, **248**, 8457.
4. Shin, B. C., Carraway, K. L. (1973), *J. biol. Chem.*, **248**, 1436.
5. Higashi, Y., Siewert, G. and Strominger J. L., (1970), *J. biol. Chem.*, **245**, 3683.
6. Marchesi, V. T. and Andrews E. P. (1971), *Science*, **174**, 1247.
7. Segrest, J. P., Kahane, I., Jackson, R. L. and Marchesi, V. T. (1973), *Archs. Biochem. Biophys.*, **155**, 167.
8. Segrest, J. P., Jackson, R. L. and Marchesi, V. T. (1972), *Biochem. biophys. Res. Commun.*, **49**, 964.
9. Spatz, L. and Strittmatter, P. (1971), *Proc. natn. Acad. Sci. U.S.A.*, **68**, 1042.
10. Ozols, J., (1974), *Biochemistry*, **13**, 426.
11. Kamat, V. B., Wyatt, P. J. and Davis, M. A. F. (1972), *Chem. Phys. Lipids*, **8**, 341.
12. Hauser, H. O. (1971), *Biochem. biophys. Res. Commun.*, **45**, 1049.
13. Horstmann, L. L. and Racker, E. (1970), *J. biol. Chem.*, **245**, 1336.
14. Widnell, C. C. and Unkeless, J. C. (1968), *Proc. natn. Acad. Sci. U.S.A.*, **61**, 1050.
15. Evans, W. H. and Gurd, J. W. (1973), *Biochem. J.*, **133**, 189.
16. Dahl, J. L. and Hokin, L. E. (1974), *A. Rev. Biochem.*, **43**, 327.
17. Schwartz, B. D., Kato, K., Cullen, S. E. and Nathenson, S. G. (1973), *Biochemistry*, **12**, 2157.
18. Reisfeld, R. A. and Pellegrino, M. A. (1972), In: *Transplantation Antigens*, (Kahan, B. D. and Reisfeld, R. A. eds.), p. 259, Academic Press, New York.
19. Sanderson, A. R. and Welsh, K. I. (1972), In: *Transplantation Antigens* (Kahan, B. D. and Reisfeld, R. A. eds.), p. 273, Academic Press, New York.
20. Snary, D. Goodfellow, P., Hayman, M. J., Bodmer, W. F. and Crumpton, M. J. (1974), *Nature*, **247**, 457.
21. Ostberg, L., Lindblom, J. B. and Peterson, P. A. (1974), *Nature*, **249**, 463.
22. Maddy, A. H. and Malcolm, B. R. (1965), *Science*, **150**, 1616.
23. Chapman, D. Kamat V. B. and Levene R. J. (1968), *Science*, **160**, 314.
24. Schneider H. and Smith I. P. (1970), *Biochim. biophys. Acta*, **219**, 73.
25. Marchesi, S. L., Steers, E., Marchesi, V. T. and Tillack, T. W. (1970), *Biochemistry*, **9**, 50.
26. Tillack, T. W., Marchesi, S. L., Marchesi, V. T. and Steers, E. (1970), *Biochim. biophys. Acta*, **200**, 125.
27. Maddy, A. H. and Kelly, P. G., (1971), *Biochim. biophys. Acta*, **241**, 290.
28. Abrahams, A. (1965), *J. biol. Chem.*, **240**, 3675.
29. Thorley-Lawson, D. A. and Green, N. M. (1973), *Eur. J. Biochem.*, **40**, 403.
30. McCarty, R. E. and Racker, E. (1966), *Brookhaven Symp. Biol.*, **19**, 202.
31. Maddy, A. H. and Dunn, M. J. (unpublished result).
32. Steck, T. L. and Yu, J. (1973), *J. supramolec. Struct.*, **1**, 220.
33. Godin, D. V. and Schrier, S. L. (1972), *J. molec. Biol.*, **7**, 285.
34. Carter, J. R. (1973), *Biochemistry*, **12**, 171.
35. Cantrell, A. C. (1973), *Biochim. biophys. Acta*, **311**, 381.

36. MacLennan, D. H., Tzagoloff, A. and Rieske, J. S. (1965), *Archs. Biochem. Biophys.*, **109**, 383.
37. Moldow, C. F., Zucker-Franklin, D., Gordon, A., Hospelhorn, V. and Silber, R. (1972), *Biochim. biophys. Acta*, **255**, 133.
38. Lundahl, P. (1975), *Biochim. biophys. Acta*, **379**, 304.
39. Dixon, H. B. F. and Perham, R. N. (1968), *Biochem. J.*, **109**, 312.
40. Hatefi, Y. and Hanstein, W. G. (1969), *Proc. natn. Acad. Sci U.S.A.*, **62**, 1129.
41. Morrison, W. L. and Neurath, H. (1953), *J. biol. Chem.*, **200**, 39.
42. Gwynne, J. T. and Tanford, C. (1970), *J. biol. Chem.*, **245**, 3269.
43. Steck, T. L., (1972), *Biochim. biophys. Acta*, **255**, 553.
44. Maddy, A. H. and Kelly, P. G. (1971), *Biochim. biophys. Acta*, **241**, 114.
45. Zahler, P. and Weibel, E. R. (1970), *Biochim. biophys. Acta*, **219**, 326.
46. Schnaitman, C. A. (1969), *Proc. natn. Acad. Sci. U.S.A.*, **63**, 412.
47. Juliano, R. L. (1972), *Biochim. biophys. Acta*, **266**, 301.
48. Maddy, A. H. (1966), *Biochim. biophys. Acta*, **117**, 193.
49. Maddy, A. H., Dunn, M. J. and Kelly, P.G. (1972), *Biochim. biophys. Acts*, **288**, 263.
50. Rega, A. F. Weed, R. I., Reed, C. F., Berg, G. G. and Rothstein, A. (1967), *Biochim. biophys. Acta*, **147**, 297.
51. Zwaal, R. F. A. and van Deenen, L. L. M. (1968), *Biochim. biophys. Acta*, **163**, 44.
52. Zwaal, R. F. A. and van Deenen, L. L. M. (1968), *Biochim. biophys. Acta*, **150**, 323.
53. Nozaki, Y., Reynolds, J. A. and Tanford, C. (1974), *J. biol. Chem.*, **249**, 4452.
54. Makino, S. Reynolds, J. A. and Tanford, C. (1973), *J. biol. Chem.*, **248**, 4926.
55. Ito, A. and Sato, R. (1968), *J. biol. Chem.*, **243**, 4922.
56. Penefsky, H. S. and Tzagoloff A. (1970), *Meth. Enzym.*, **22**, 204.
57. Simons, K., Helenius A. and Garoff J. (1973), *J. molec Biol.*, **80**, 119.
58. Schnaitman, C. and Greenwalt, J. W. (1968), *J. Cell Biol.*, **38**, 158.
59. Shechter, I. and Block, K. (1971), *J. biol. Chem.*, **246**, 7690.
60. Weber, K. and Kuter, D. J. (1971), *J. biol. Chem.*, **246**, 4504.
61. Holloway, P. W. (1973), *Analyt. Biochem.*, **53**, 304.
62. Dodge, J. T., Mitchell, C. and Hanahan, D. J. (1963), *Archs. Biochem. Biophys.*, **100**, 119.
63. Marchesi, V. T. (1973), *Meth. Enzym.*, **28**, B, 252.
64. Hamaguchi H. and Cleve, H. (1972), *Biochim. biophys. Acta*, **278**, 271.
65. Harris, J. R. (1971), *Biochim. biophys. Acta*, **229**, 761.
66. Clarke, M. (1971), *Biochem. biophys. Res. Commun.*, **45**, 1063.
67. Hamaguchi H. and Cleve, H. (1971), *Biochim. biophys. Acta*, **233**, 320.
68. Neville, D. M. (1969), *Biochem. biophys. Res. Commun.*, **34**, 60.
69. Jacobs, E. E. and Sanadi, D. R. (1966), *J. biol. Chem.*, **235**, 531.
70. Zahler, W. L., Saito, A. and Fleischer, S. (1968), *Biochem. biophys. Res. Commun.*, **32**, 512.
71. Gagnon, J., Finch, P. R., Wood, D. D. and Mosarello, M. A. (1971), *Biochemistry*, **10**, 4756.
72. Neville, D. M. (1967), *Biochim. biophys. Acta*, **133**, 168.
73. Fischer, S., Lingsh, C. and Urbina, C. (1973), *J. Memb. Biol.*, **12**, 317.
74. Davis, K. A. and Hatefi, Y. (1971), *Biochemistry*, **10**, 2509.
75. Lauf, P. K. and Poulik, M. D. (1968), *Br. J. Haemat.*, **15**, 191.

76. Moldow C., Robertson, J. and Rothfield, L. (1972), *J. Memb. Biol.,* **10**, 137.
77. Choules, G. L., Sandberg, R. G., Steggall, M. and Eyring, E. M. (1973), *Biochemistry,* **12**, 4544.
78. Emmelot, P., Visser, A. and Benedetti, E. L. (1968), *Biochim. biophys. Acta,* **150**, 364.
79. Cogoli, A., Mosimann, H., Vock, C. V., Bathazar, A. K. and Semenza, G. (1972), *Eur. J. Biochem.,* **30**, 7.
80. Baldwin, R. W., Bowen, J. G. and Price, M. R. (1974), *Biochim. biophys. Acta,* **367**, 47.
81. Kahan, B. D. and Reisfeld, R. A. (1971), *Transplant Proc.,* **1**, 483.
82. Nachbar, M. S., Winkler, W. J. and Salton, M. R. J. (1972), *Biochim. biophys. Acta,* **274**, 83.
83. Salton, M. R. J. and Schor, M. T. (1972), *Biochim. biophys. Res. Commun.,* **49**, 350.
84. Kundig, W. and Roseman, S. (1971), *J. biol. Chem.,* **246**, 1407.
85. Curtis, P. J. (1970), *Biochim. biophys. Acta,* **211**, 575.
86. Eng, L. F., Chao, F. C., Gerstl, B., Pratt, D. and Tavaststjerna, M. G. (1968), *Biochemistry,* **7**, 4455.
87. Schubert, D., Poensgen, J. and Werner, G. (1972), *Hoppe-Seyler's Z. physiol. Chem.* , **353**, 1034.
88. Kathan, R., Winzler, R. J. and Johnson, C. A. (1961), *J. exp. Med.,* **113**, 37.
89. Howe, C., Lloyd, K. O. and Lee, L. T. (1973), *Meth. Enzym.,* **28** B, 236.
90. Blumenfeld, O. O. and Zvilichovsky, B. (1973), *Meth. Enzym.,* **28** B, 245.
91. Zvilichovsky, B., Gallop, P. M. and Blumenfeld, O. O. (1971), *Biochem. biophys. Res. Commun.,* **44**, 1234.
92. Yu, J. and Steck, T. L. (1973), *J. supramolec. Struct.,* **1**, 233.
93. Butters, T. D. and Hughes, R. C. (1974), *Biochem. J.,* **140**, 469.
94. Meunier, J. C., Olsen, R. W. and Changeux, J. P. (1972), *F.E.B.S. Letters,* **24**,63 .
95. Cuatrecasas, P. (1972), *J. biol. Chem.,* **247**, 1980
96. Liljas, L., Lundahl, P. and Hjerten, S. (1974), *Biochim. biophys. Acta,* **352**, 327.
97. Bjerrum, O. J. and Lundahl, P. (1974), *Biochim. biophys. Acta,* **342**, 69.
98. Bjerrum, O. J., Bhakdi, S., Knüfermann, H. and Bøg-Hansen, T. C. (1974), *Biochim. biophys. Acta,* **373**, 44.
99. Gavin, J. R., Mann, D. L., Buell, D. N. and Roth, J. (1972), *Biochem. biophys. Res. Commun.,* **49**, 870.
100. Salton, M. R. J. and Netschey, A. (1965), *Biochim. biophys. Acta,* **107**, 539.
101. Erwin, G. and Hellerman, L. (1967), *J. biol. Chem.,* **242**, 4230.
102. Hagopian, A. and Eylar, F. H. (1969), *Arch. Biochem. Biophys.,* **129**, 447.
103. Emmelot, P., Feltkamp, C. A. and Vas Dias, H. (1970), *Biochim. biophys. Acta,* **211**, 43.
104. Philippot, J. (1971), *Biochim. biophys. Acta,* **225**, 201.
105. Dulaney, J. T. and Touster, O. (1970), *Biochim. biophys. Acta,* **196**, 29.
106. Salton, M. R. J. and Schmitt, M. D. (1967), *Biochem. biophys. Res. Commun.,* **27**, 529.
107. Fairbanks, G. and Avruch, J. (1972), *J. supramolec. Struct.,* **1**, 66.
108. Davenport, J. (1962), *Biochem. J.,* **84**, 18p.
109. Heller, J. (1968), *Biochemistry,* **7**, 2906.
110. Evans, W. H. and Gurd, J. W. (1972), *Biochem. J.,* **128**, 691.

# 7 Techniques for the analysis of membrane proteins

M. J. DUNN and A. H. MADDY



## 7.1 Introduction

The study of the chemical nature of membrane proteins falls into two categories. First, the isolation and purification of the individual proteins and secondly, the characterization of these fractions, to detect any heterogeneity, or if pure, to elucidate the structure of the component.

Classical methods of protein fractionation exploit differences in size, shape, solubility and electrical charge between different molecular species. More recently, differences in specific binding affinities of proteins, e.g. immunological affinities, have been used as a basis for affinity chromatography. We shall confine our discussion to those techniques of particular relevance to the fractionation of membrane proteins and even in this limited selection it will be necessary to assume that the reader has some familiarity with the theoretical backgrounds to these techniques. Gel electrophoresis [1,2] is by far the most widely used technique and has the advantages of high resolving power and sensitivity, but is less effective for preparative fractionation of large amounts of protein. Gel exclusion chromatography [3,4,5], by contrast, is of relatively low resolution, but large amounts of protein can be separated in high yield. Recently, isoelectric focusing [6,7] and affinity chromatography [8,9] have been used, the latter with spectacular success in the isolation of specific trace components of membranes such as various receptor molecules.

Largely as a consequence of the imponderables of membrane protein fractionation, the application of classical methods for the structural analysis of the proteins is still in its infancy. In the second part of this chapter those techniques which have already shown themselves to be of value, and some of those that may soon prove their worth, are discussed.

## 7.2 Fractionation

### *7.2.1 Gel electrophoresis*

#### *(i) The gel medium*

The ideal gel medium would be electrically neutral, chemically inert, mechanically stable and with a homogeneous pore structure that could be varied over a wide size range. Electrical neutrality is usually desirable not only to avoid interaction between the gel and buffer components or proteins but also in order to minimize endosmotic flow of water. (Occasionally, such effects can improve the fractionation of a mixture). Mechanical stability, especially at low gel concentrations, facilitates the manipulation of the gel, but when protein is to be removed from the gel a matrix that can be liquefied is highly desirable. Pore characteristics are of vital significance as fractionation depends as much, and possibly more, on the differential sieving effect of the gel matrix on the proteins as on charge differences between the proteins.

Polyacrylamide has now largely supplanted other media as a matrix for electrophoresis of membrane proteins largely because of its superiority as a molecular sieve. Pore size may be varied over a wide range by variation of the acrylamide concentration and the extent of cross linking. Nevertheless the many advantages of polyacrylamide should not obscure its disadvantages, or its extensive application preclude consideration of other gel matrixes. For instance, starch has been used successfully to separate membrane proteins [10,11] and has an advantage over acrylamide in that its solubility facilitates the recovery of proteins after electrophoresis.

As polyacrylamide is composed of a covalently cross-linked matrix it is not easily dispersed and extraction of proteins may be troublesome. Liquefaction of the matrix is possible by piperidine treatment when the acrylamide is cross-linked with ethylene diacrylate but the solvent damages the protein [12]. In principle proteins could be electrophoresed out of a gel, but there is a pronounced tendency for membrane proteins to be lost on the dialysis membrane used in conjunction with electrophoretic extraction. For this reason Hjerten [13] has developed a method that avoids using a dialysis membrane and which concentrates the protein zones as they move out of the gel by increasing the viscosity and conductivity of the medium and so retarding the protein migration. Alternatively, proteins can be simply extracted after homogenization of the polyacrylamide, but unfortunately certain, incompletely characterized, contaminants diffuse out of the gel along with the proteins [14]. These contaminants are Lowry positive, absorb U.V. and are only partially dialysable. The difficulties of extraction have led to the use of an agarose slurry as the electrophoretic medium. Recovery of the proteins is then simple, but the molecular sieving effect of a solid matrix has to be forfeited [15].

Polyacrylamide gels have been used at both the analytical and preparative level. Analytical applications are based on either rods of gel contained in glass or plastic tubes, or on thin slabs. The latter are perhaps preferable for the comparison of many different samples but the former is, in our experience, preferable for the analysis of dilute solutions as relatively large samples may be applied with less trouble than with a slab. With the exception of a power pack the apparatus required for either the rod or slab techniques may be readily constructed at little expense in any workshop and commercially available versions have little if anything to offer in preference.

Two basic types of equipment are available for preparative work with polyacrylamide gel, either a large column of gel on which the proteins are separated electrophoretically and eluted sequentially from the end of the gel by a cross flow of buffer (several commercial variants are marketed), or the proteins can be separated on a large slab of acrylamide about 200 x 200 x 4 mm, which is then cut into slices at 90° to the electrical field and the separated proteins eluted from each slice. We, ourselves, have been relatively unsuccessful in our efforts to fractionate complex mixtures of erythrocyte membrane proteins using the gel column method. Recoveries are poor and such protein as is recovered is obtained in a very dilute solution. The record of the 280 nm absorbance of the eluate tends to be featureless and gives little or no indication of the course of the fractionation, which must therefore be ascertained by monitoring fractions of the eluate on an analytical gel system. However it has been used successfully to separate myelin proteins which are smaller than proteins from the erythrocyte membrane [15]. The method is also effective for the isolation of some highly characteristic components from a complex mixture, e.g. a tumour specific antigen from a dye-induced rat hepatomata [16].

Preparative fractionation using a large slab has, in our hands, been more convenient and more effective even though it suffers from the severe difficulties associated with extraction of protein from the gel slices [17]. No expensive apparatus need be purchased for slab work as the slabs may be cast in chambers constructed from two 200 x 200 mm glass plates separated by 4 mm plastic spacers and held together by steel clips [18]. (Commercial assemblies are available.) The slab has the additional advantage that it can be adapted for two dimensional separations. A sample is first fractionated on a 4 mm diameter gel rod, then this gel is squeezed onto the top of the slab and electrophoresed in the second dimension. Two-way electrophoresis was first developed for ribosomal proteins [19] and has been applied to erythrocyte membrane proteins [20,21]. Alternatively the initial separation can be based on isoelectric focusing [22]. The information obtained by two way separation can also be obtained by monitoring the fractions obtained from a preparative slab in a second (or more) buffer [17].

As size is one factor controlling the mobility of proteins through gels, many efforts have been made to derive information relating to molecular size from mobility measurements. This requires some means of compensating for differences in charge and shape between proteins (the other factors controlling mobility). If the mobility of a protein in gels of different pore sizes is measured, the change in pore size would be expected to have a greater effect on large proteins than small ones, and this is the basis of a method originally suggested by Ferguson [23] for molecular weight determination. The mobility of a protein is a linear function of the acrylamide concentration of the gel (provided that all other factors, including the amount of cross-linking, are kept constant) and the slope of the curve is a measure of the molecular weight of the protein. Although, theoretically, shape differences, as well as size differences, between molecules will affect their retardation, in practice with this actual method, shape has been found to be relatively unimportant. Thus the correct molecular weight is determined for fibrinogen by electrophoresis, although an erroneous value is obtained by gel filtration where size has a much more pronounced effect [24]. This type of electrophoretic analysis is a useful alternative to SDS gel electrophoresis (see below) and in the case of the EDTA soluble proteins of the erythrocyte membrane produces results consistent with those obtained by SDS electrophoresis [17].

Electrophoresis into a polyacrylamide gradient is a related technique [25]. Here a concentration gradient of acrylamide and cross-linking agent is polymerized to produce a gel with a gradually decreasing pore size. Proteins are electrophoresed into the gel and eventually they reach that part of the matrix where the pores are small enough to reduce their mobility to zero. Proteins of different sizes therefore stop in different regions of the gradient gel. Protein shape has a marked effect on this separation as the final position of a molecule will depend on its smallest dimension. Long thin molecules will therefore behave anomalously and as their mobility depends more on their orientation in the electrical field than on their size they tend to form smeared bands in the gel. No useful purpose is served by attempting to separate proteins on a gradient gel in the presence of SDS as this detergent is believed to convert all proteins into prolate ellipsoids with the same minor axis.

### *(ii) The selection of a buffer for electrophoresis*

Many of the factors that have been considered in Chapter 6 are also operative during the fractionation of membrane proteins. The difficulties of solubilization arise out of the propensity of membrane proteins to aggregate. In so far as aggregation determines the nature of the particles in solution it also determines the potential fractions which may be derived from the solution. Furthermore, as the actual fractionation procedure can change the state of aggregation, the fractions produced may not have been present in the original, unfractionated mixture.

If a highly disaggregating solvent has been necessary to isolate the proteins from the membrane there will probably be little value in attempting to fractionate under less disruptive conditions, but with this proviso, a general guideline in selecting a buffer system is to start with the mildest and only to apply more vigorous solvents when the milder systems have been shown to be inadequate. Thus we find that application of the Takayama phenol-acetic acid-water [26] system to erythrocyte membrane proteins may be an example of overkill. It is true that ghosts may be dissolved in phenol-acetic acid-water and electrophoresed in a Takayama system [27] to produce a series of protein bands, but a considerable amount of protein is left at the top of the gel. However, if an EDTA extract of the ghost is electrophoresed under the same conditions the same bands appear in the gel, but now there is no protein left at the origin. Consequently it may be concluded that the system in fact only separates EDTA soluble proteins and these can be analysed under much milder conditions, e.g. using a Tris/glycine buffer [28].

Apart from this general principle of ensuring that the mildest buffer possible is used there is little advice that can be offered in the selection of a buffer other than to search the literature for any previous successful separation of relevant membrane proteins. A vast range of buffers, including high [28] and low pH [29], buffers containing urea at high and low pH [30,31], buffers containing detergents, e.g. deoxycholate [32], Tween [21], SDS (see below), have been used. It would not be practical to list all of the these buffers as each system usually exists in a number of variations. This diversity does not arise out of the rugged independence of the workers in the field, still less any perversity on their part, but signifies the real need for careful experimentation to establish the precise conditions required for the optimal electrophoretic separation of each protein mixture.

Experimental details of three widely used buffers systems are presented, Tris/glycine, phenol-acetic acid-water, and electrophoresis in the presence of SDS. Tris has many undesirable features as a buffer but many laboratories find it preferable to phosphate buffers for gel electrophoresis, and Tris/glycine gives more reproducible results than Tris/hydrochloric acid. The phenol-acetic acid-water system was first developed for mitochondrial membrane proteins [26], and has been subsequently adapted to other proteins and modified by the addition of reducing agent [27], addition of detergent [33], and replacement of acetic by formic acid [34].

We have already considered some attempts to derive information relating to size from electrophoretic mobilities in gels. Unequivocal molecular weight determination is only possible if all proteins can be converted into particles of equal charge density and of equivalent shape, for then electrophoretic mobility is determined solely by size. It has been claimed that interaction of proteins with dodecyl sulphate (usually the sodium salt) converts all proteins into particles of this type. Therefore, electrophoresis of these complexes on polyacrylamide gels in the presence of the detergent results in a separation based exclusively on size differences between the polypeptide chains and consequently the mobilities of the complexes can be used for the determination of polypeptide molecular weight. Three conditions must be satisfied for this claim to be valid. (a) All polypeptide sequences bind an identical weight of SDS per unit length. (b) All the SDS-protein complexes are of equivalent shape – a prolate ellipsoid has been suggested [35]. The length of the ellipsoid is postulated to vary with the length of the polypeptide chains, but ellipsoids can all be treated as equivalent as long as a certain axial ratio is exceeded. Polypeptides of molecular weight smaller than 15 000 are not believed to attain this ratio and the method is not applicable to these chains [36]. (c) All polypeptide chains must be dissociated from each other, interchain bridges, e.g. disulphide bonds, must be cleaved by suitable treatment but it is assumed that the action of the detergent itself is sufficient to disrupt all non-covalent interactions.

Electrophoresis in the presence of SDS now holds a central position in membrane protein technology, and is of such exceptional importance that it requires a careful scrutiny. Although much valuable information has undoubtedly been obtained by this method, its theoretical basis barely justifies its dominant position. Therefore it is worth examining the evidence for the assumption that the three conditions listed above are always satisfied.

### *Condition (a) All polypeptides bind the same amount of detergent per unit amino acid chain length*

Even if all the proteins that have been studied had the same binding capacity the generalization from this relatively small number to all known proteins would be questionable, but, as the measured capacities vary significantly, the generalization retains little credibility. The first comparative survey from Tanford's laboratory was encouraging as a binding of 1.4 g detergent per gram protein was recorded for the 12 proteins analysed [37]. But one of these proteins was the mixture of proteins present in the human erythrocyte ghost, and since one component of this mixture, the glycoprotein, has been recently found to bind 8–10 g of detergent per gram protein [38] presumably some other components of the mixture must have unusually low binding capacities. These workers now concede the possibility of abberrant detergent binding by membrane proteins [39]. Other deviations were reported [40] and a second survey by Nelson [41] of 19 proteins, in contrast, found that the binding ranged from 0.8–2.2 g per gram of protein. It is difficult to interpret Nelson's further observation that in most cases a correct value for molecular weight was obtained regardless of the amount of detergent bound. It has been implied from a presumed equality of binding that binding (at a high level) is independent of amino acid composition. This is not so, for basic proteins bind a disproportionally large amount of detergent and acidic proteins a correspondingly small amount [41]. (Polylysine and polyarginine bind large quantities [42] and polyglutamic acid very little [43].) Binding by uncharged polypeptides is a matter of controversy [44,45]. Glycoproteins are a special case as relative mobility and 'apparent molecular weight' depend on the gel concentration [46]. Here the deviation may be due to the carbohydrate and does not impute the value of the method for unconjugated proteins.

*Condition (b) All SDS–protein complexes are of equivalent shape*

The shape of a small number of SDS–protein complexes has been examined by Tanford's group and they conclude that the complexes are prolate ellopsoids varying in length according to the length of the polypeptide chains. Although only a few complexes were analysed there is no direct contradictory evidence from other sources.

However the behaviour of SDS–protein complexes on hydroxylapatite (HAP) columns indicates that differences in shape and/or conformation occur between different SDS–protein complexes. The complexes adsorb onto HAP columns at low ionic strength and can be selectively eluted in the continued presence of the detergent by a phosphate gradient [47]. It has been pointed out that this technique complements separation by electrophoresis as the peptides are not eluted in order of their molecular weights. This may be a sound practical observation, but it undermines the theoretical basis of SDS–electrophoresis. As the elution of peptides from HAP is not a simple function of their size then they *must* differ in some other respect. Either the proteins bind varying amounts of detergent or it is bound in different ways to produce complexes of non-uniform shape.

*Condition (c) SDS dissociates all but covalent bonds between peptide chains*

If the above objections to SDS-electrophoresis are as grave as the arguments suggest then the reader might well wonder why it is ever used. It should therefore be made clear that for the large majority of soluble proteins which have been examined the mobility of their SDS–protein complexes is a linear function of the logarithm of their molecular weights [48–50]. A few anomalies have been reported [51–53], but it may be argued that the probability favours a reasonable estimate, and in the present state of knowledge of membrane proteins the risk of an erroneous value is well worth taking. As far as molecular weight determination of dissociated polypeptides is concerned one cannot disagree with this attitude. However, the assumption that, provided covalent bridges between polypeptides are broken, SDS disrupts all non-covalent interactions between membrane polypeptides, and therefore reveals the degree of heterogeneity of a mixture, is more controversial.

The empirical relationship between mobility and molecular weight which is the basis of the SDS method was established using soluble proteins. Soluble proteins were used because these were the proteins for which unequivocal molecular weight data was available. This data was available because these proteins do not undergo intractable interactions, i.e. they are 'soluble', but the difficulties of studying membrane proteins arises from the very fact that they *do* form intractable, insoluble complexes. To assume, therefore, that the data collected from soluble proteins can be applied directly to membrane proteins implicitly ignores this major and crucial difference. Some experimental support for the dissociation of membrane proteins is provided by the observation that SDS-dissociated membrane proteins are not dissociated further by additional treatment with other dissociating agents [54], and, where it has been applicable, proteolytic digestion of membrane proteins indicates that they are fully dissociated by the detergent. For example, a portion of the 100 000 molecular weight polypeptide of the human erythrocyte membrane can be digested away by pronase treatment of the intact cells leaving only *one* 70 000 segment buried in the membrane [55,56]. On the other hand, several examples of membrane protein aggregates that are resistant to dissociation by SDS have been observed. Two laboratories have reported that organic solvents can bring about the aggregation of membrane proteins into complexes that are not dissociated by SDS [54,57]. One of the earliest papers relating to the application of SDS gel electrophoresis to membrane proteins warned that prior treatment with EDTA was necessary for complete dissociation of the proteins [53], and more recently the formation of resistant aggregates in the presence of calcium ions has been reported [58].

The presence of experimental details for SDS-electrophoresis in the experimental section should dispel doubts of the authors' appraisal of the usefulness of this method, which is probably the premier technique for the comparison of protein fractions. At the same time it is hoped that the above discussion provides an adequate justification for their doubts over the simple interpretation of the information which the technique has yielded on membrane proteins.

### *7.2.2 Isoelectric focusing*

Isoelectric focusing separates proteins of different isoelectric points by focusing each protein at its isoelectric pH along a pH gradient. The gradient is generated by a series of ampholytes of overlapping isoelectric points produced commercially by Pharmacia under the trade name Ampholine.* The method has a very high resolving power but suffers from the drawback that proteins tend to

*Suitable ampholytes are now also marketed by Bio-Rad.

come out of solution at their isoelectric pH. The pH gradients were originally generated in free solution and stabilized by a sucrose gradient, and some membrane proteins, e.g. bovine rhodopsin [59] and *Torpedo* acetylcholine receptor protein [60], have been purified in this way. The tendency of proteins to precipitate at their isoelectric pH can be minimized by increasing the concentration of Ampholine, but the cost of increasing the Ampholine concentration in the relatively large sucrose gradients can be prohibitive. Partly for this reason pH gradients have been generated within polyacrylamide gels and used successfully for membrane proteins [61,22]. The small size of the gradient is not the only advantage, for the gels are much less trouble to set up, especially if several protein samples are to be compared, and the focusing is performed in rods of gel in glass tubes using the same apparatus used for gel electrophoresis. Alternatively the gel can take the form of a thin layer which can be efficiently cooled and consequently a higher electrical power can be used to complete focusing in a short time. Focusing in gels has its disadvantages. The molecular sieving effect of the gel which facilitates the electrophoretic separation of proteins is an impediment to focusing as it hinders the passage of the proteins towards their isoelectric pH. Gels of large pore size must therefore be used [62]. Increasing the time necessary to focus is more than just an inconvenience for, in contrast to pH gradients in free solution, and for reasons not fully understood, a pH gradient in a gel matrix has a limited lifetime and focusing must be completed before the gradient breaks down [63]. The life of the gradient can be extended by including sucrose in the gel. As an experiment cannot be allowed to continue indefinitely, completion of focusing must be determined empirically and assessed to be complete when the protein bands cease to move on prolongation of the run. Incorporation of protein markers of known isoelectric points is of value in this respect and these markers are also useful in monitoring a pH gradient [64]. The form of a gradient within a gel may be measured directly by cutting the gel into segments and measuring the pH of the Ampholine eluted from each segment. Focusing can be carried out in the presence of dissociating agents such as urea and non-ionic detergents [61], e.g. Emulphogene [59], Brij 35 [65], Triton X–100 (see Section 7.4.3). After completion of a run the proteins in a gel matrix are visualized by the same staining methods used in electrophoresis with the slight complication that as components of Ampholine can pick up Coomassie Blue and Amido Black, they must be allowed to diffuse out of the gel before staining.

As previously discussed in relation to gel electrophoresis, extraction of proteins from a polyacrylamide gel is difficult, and while this is no disadvantage in analytical applications it is a severe drawback in any preparative procedure. But, with this proviso, focusing in a gel is readily adapted to a preparative scale by generating the gradient in a 200 x 200 x 4 mm slab similar to that described for gel electrophoresis. Carbon dioxide must be excluded by degassing solutions and overlaying.the upper gel buffer chamber and the top of the gel slab with Stoddard's reagent. Should a wick be used to connect the upper buffer reservoir with the gel it must be encased so as to prevent its being a gateway for carbon dioxide. Surgical lint sewn into a ploythene bag with the ends of the lint protruding at either end serves admirably.

A slurry of polyacrylamide or dextran beads has been proposed [66] as a medium for the pH gradient and would appear to combine the advantages of a liquid gradient [ease of recovery] with the small size and versatility of a solid gel matrix. This adaptation could be of considerable value in the future.

### *7.2.3 Gel exclusion chromatography*

Gel exclusion chromatography or gel filtration is a liquid chromatographic method which separates molecules according to differences in molecular dimensions. The separation is achieved by passing the sample through a porous matrix of uncharged gel beads of fixed pore size. Molecules too large to diffuse into the gel migrate rapidly and are eluted quickly, whereas smaller molecules are able to penetrate the matrix and are retarded. The most widely used media for gel filtration are cross-linked dextrans (Sephadex), polyacrylamide gels (Bio-Gel P), and agarose gels (Sepharose, Bio-Gel A). Gel filtration can be used as a preparative technique for the separation and purification of the components of a mixture and also as an analytical tool to estimate molecular weight and molecular size (Stokes' radius). There are several reviews which provide details of the theory and practice of the technique [3–5].

Although there are few limitations on the use of gel filtration for the analysis of many proteins [4], the technique has proved singularly difficult to adapt for the study of membrane proteins. The choice of a suitable buffer system, although a relatively easy matter for many proteins, presents considerable difficulties for the study of the majority of membrane proteins. As we have explained previously these difficulties can be largely attributed to the characteristic tendency of membrane proteins to aggregate, and thus those factors which have been considered for the solubilization and the electrophoretic fractionation of membrane proteins are again relevant here. The application of aqueous, non-denaturing eluant systems to the analysis of membrane proteins by gel chromatography has been found to be unsatisfactory in the majority of instances. Therefore various attempts have been made to facilitate the gel filtration of membrane proteins by the use of eluants of increasing severity, from moderately disruptive conditions (e.g. nonionic detergents, deoxycholate), to systems which are claimed to totally abolish interactions between proteins.

A few successful applications of gel filtration to the study of membrane proteins without the need to resort to disaggregating conditions have been reported. For example a proteinase has been partially purified from erythrocyte ghosts [67]. This protein was not homogeneous for up to five bands were observed by gel electrophoresis, only one or two of which possessed proteolytic activity. But unfortunately membrane proteins are usually excluded from gel filtration columns using standard aqueous eluant systems. Some examples of this phenomenon are the almost total exclusion by gel filtration of proteins from squid retinal axon membranes [68], *Acholeplasma laidlawii* membrane proteins [69], and those proteins extracted from erythrocyte membranes by various procedures [70–75]. The exclusion of many membrane proteins by gel filtration media, even by Sepharose 4B with a reported exclusion limit of $20 \times 10^6$ daltons, does not necessarily imply that these molecules are enormous. It could be due to a high degree of molecular asymmetry but, in many instances it would appear to be a consequence of protein aggregation in aqueous media. There is also evidence suggesting that proteins may become aggregated during their passage through a gel filtration column, presumably due to some undefined interaction with the gel matrix. When an EDTA extract of bovine erythrocytes, which in the electron microscope consists of small particles of the order of 5 nm, is passed through a gel filtration column, the material in the void volume ($V_0$) contains large bundles of fibres together with some 5 nm particles [79]. Further evidence for possible protein aggregation occurring during gel filtration has been provided by the finding that components of an EDTA extract of the erythrocyte membrane, with apparent molecular weights of 45 000 [76,77] and 110 000 [78], appear to be converted to material with an apparent molecular weight of 200 000 simply by passage through a gel filtration column.

The inadequacies of the gel chromatographic behaviour of membrane proteins under mild conditions have stimulated attempts to improve the technique by resorting to more vigorous, disaggregating solvents. It is our experience that the properties of gels of the cross-linked dextran type can be adversely affected by such eluants, resulting in columns with very low flow rates and altered elution characteristics. Fortunately gel matrixes of the polyacrylamide type are generally not disrupted by organic reagents and tolerate solutions of urea, detergents and guanidine hydrochloride, although at high concentrations these reagents may alter exclusion limits. Some degree of success has been achieved in the purification of membrane glycoproteins using 33% aqueous pyridine as an eluant. The pyridine extractable proteins of erythrocytes are largely excluded from a Sephadex G-200 column in 33% aqueous pyridine, but the retarded peak, which represents only 8% of the protein applied to the column, contains the membrane sialoglycoproteins [80]. The membrane receptor for wheat germ agglutinin from mouse leukaemia line L1210 cells can also be purified in this way [81], but the method is not universally applicable as the glycoprotein-rich, pyridine extract of liver cell plasma membranes is totally excluded on G-200 in 33% aqueous pyridine [82].

Eluants containing non-ionic detergents have been successfully employed for the purification of several membrane proteins. These include an acetylcholinesterase [83] and a proteinase [84] from erythrocyte membranes, a polypeptide:*N*-acetylgalactosaminyl transferase from bovine sub-maxillary glands [85], and a glycoprotein:galactosaminyl transferase from Ehrlich ascites tumour cell membranes [86]. Sodium deoxycholate has also been used in gel filtration studies. It has been claimed to be as effective as SDS for the fractionation of *Acholeplasma laidlawii* plasma membrane proteins [69], but it does not appear to provide adequate resolution when applied to erythrocyte membrane proteins [87,88].

By far the most widely exploited of the anionic detergents employed for gel chromatography has proved to be SDS. Gel filtration in SDS has been recommended [89] for the determination of both the effective hydrodynamic radii and the molecular weights of polypeptide chains. However, we have already reviewed at some length in the discussion of SDS gel electrophoresis (Section 7.2.1) our reasons for believing that many membrane proteins do not fulfill the basic criteria necessary for the general validity of the method. Suffice it to say that extreme caution must be exercised when using this detergent to investigate the heterogeneity and molecular weights of a mixture of membrane proteins. Katzman's work exemplifies the sort of anomalies which might be encountered when using SDS as an eluant for gel chromatography [52]. When proteins extracted from bovine brain were chromatographed on Sepharose 4B in SDS, three protein peaks were resolved. Each peak was, however, found to be identical by gel electrophoresis and amino acid analysis. Moreover when the second peak was rechromatographed on the same column, it redistributed to give three identical peaks. In geneal membrane proteins are eluted from gel filtration columns containing SDS in order of their molecular weights as determined by SDS gel electrophoresis [57,68,69,71,82,90,91]. There is no reason to suppose that data derived in this way is any more reliable than that obtained by SDS gel electrophoresis alone, and in comparison with gel electrophoresis the resolving power of the technique is extremely poor. As a preparative technique the method suffers from the disadvantage that fractions are contaminated by detergent.

Many proteins undergo a marked transition upon the addition of guanidine · HCl, which is usually complete at between 6 M and 8 M. This transition reflects the dissociation of proteins, providing that disulphide bonds are cleaved, into their constituent polypeptide chains, in the form of linear random coils [92]. The molecular radius of a randomly coiled polypeptide chain is a function of the number of amino acid residues in that chain, and is therefore a function of molecular weight. On this basis guanidine·HCl has been advocated [93] as an eluant for gel chromatography. It is argued that because molecules of the same gross conformation are compared there is a valid correlation between molecular weight and elution volume, and in a study of nineteen polypeptide chains from standard proteins no anomalous behaviour was noted on agarose columns in guanidine·HCl [93]. Tanford and his coworkers [94,95], in a study of human erythrocyte membranes, found that a significant amount of protein with an apparent molecular weight of 210 000 was eluted from a 4% agarose column in the presence of 6 M guanidine · HCl. This material was considered to be equivalent to the 'spectrin' doublet which, by SDS gel electrophoresis, also has an apparent molecular weight of 200 000 [95]. However, incomplete disruption of protein interactions by guanidine·HCl has been demonstrated using proteins derived from both bovine [96] and human [97,98] erythrocyte membranes. It therefore appears that generalizations concerning the behaviour of globular proteins in guanidine·HCl cannot necessarily be extended to all membrane proteins.

The possibility of analysing membrane proteins in non-aqueous media has already been considered in Chapter 6. In principle, gel filtration using organic solvents as eluants is perfectly feasible, but in practice, application of this technique has suffered from technical difficulties. Sepharose LH-20 is suitable for use in organic solvents, but it possesses too low an exclusion limit to be useful for protein fractionation [79]. It can be profitably used to separate membrane proteins from lipids [97]. Sephadex gel filtration media can be rendered suitable for use with organic solvents by methylation [99] and they retain their ability to separate macromolecules according to their size [100]. A partial fractionation of the proteins of the rat liver mitochondrial membrane has been achieved using methylated Sephadex in chloroform–methanol solvent [100].

A recently developed method which may have considerable potential when applied to membrane proteins has been termed 'Hydrophobic chromatography.' Homologous series of hydrocarbon-coated agaroses that vary in the length of their alkyl (Sepharose–$NH(CH_2)_nH$) or ω-aminoalkyl (Sepharose–$NH(CH_2)_nNH_2$) side chains have been synthesized [101]. The hydrocarbon side chain presents proteins with the opportunity for interaction with the column and proteins may therefore be expected to be separated during passage through the column on the basis of their differing tendencies to interact hydrophobically. The hydrophobic character of many membrane proteins may well render them amenable to analysis using such columns, but to date no such studies have been reported.

### 7.2.4 Ion exchange chromatography

The separation of proteins on ion-exchange columns depends primarily on differences in their charge. The separation relies on the differential elution by a change either in pH or ionic strength of the components of a mixture which has been adsorbed onto a column. Elution can be achieved using either gradient or step elution methods. The gradient elution technique possesses a high resolving power, but the step elution method is often to be preferred for protein purification procedures. The matrix should ideally be stable to organic solvents and extremes of pH. Suitable matrixes for ion-exchange chromatography include cellulose, high molecular weight dextrans and polyacrylamide. The most commonly used anion exchangers are aminoethyl (AE)- and diethylaminoethyl (DEAE)-derivatives, while of the cation exchangers carboxymethyl (CM)-, phospho (P)-, and sulphoethyl (SE)-derivatives have proved the most useful. Although the behaviour of proteins on ion-exchange media is not as well understood as that of small molecules such as amino acids, ion-exchange chromatography has proved to be a very useful technique for the fractionation and purification of many proteins. The theory and practice of ion-exchange chromatography have been recently reviewed in detail by Himmelhoch [102].

Although ion-exchange chromatography is a powerful tool in the fractionation and purification of many proteins, it unfortunately has not proved so useful for the initial fractionation of membrane proteins. This deficiency is primarily because many membrane proteins tend to aggregate in the presence of high salt concentrations and to be precipitated at pH values approaching their isoelectric points [79]. These disadvantages are illustrated by the finding that virtually all of the protein extracted from bovine erythrocyte ghosts by butanol is adsorbed to DEAE–cellulose and cannot be eluted either by variations of pH [70] or by a salt gradient [103]. Similarly proteins extracted from human erythrocytes with 33% aqueous pyridine [104] and proteins solubilized from *Acholeplasma laidlawii* membranes [69] are irreversibly bound by DEAE-cellulose.

While only limited success has been achieved in the fractionation of complex mixtures of membrane proteins, ion-exchange chromatography has proved useful in the final purification of some isolated membrane components, particularly when non-ionic detergents have been incorporated in the eluants. A 46-fold purification of the enzyme, glycoprotein:galactosyl transferase, from Ehrlich ascites tumour cell membranes has been achieved using DEAE-cellulose in the presence of 0.1% Triton X–100 [86]. This enzyme which transfers galactose from UDP-galactose to specific glycoprotein receptors, was found to be intimately associated with the smooth membranes, probably the Golgi apparatus, of the cells. DEAE-Sephadex, again in the presence of Triton X–100, has been employed to purify acetylcholinesterase from human erythrocyte membranes [83]. Two peaks of enzymic activity were found, eluted at 0.15 M and 0.25 M NaCl respectively, which were suggested to be separate forms of the enzyme, yet both fractions had an apparent molecular weight of 142 000 by gel filtration, and the $K_m$ values for the two components were not significantly different. Some success has been achieved in the purification by ion-exchange chromatography of some of the proteins of the inner mitochondrial membrane [105,106]. Kagawa [106] describes suitable conditions for the purification of (a) the $F_1$ mitochondrial ATPase on DEAE-cellulose, (b) a basic protein of molecular weight 18 000, termed the 'oligomycin-sensitivity-conferring protein' (OSCP), on CM-cellulose, and (c) cytochrome *c* on Amberlite-IRC50 or -XE64 ($H^+$ form).

Ion-exchange chromatography can also be useful in some circumstances for the removal of unwanted contaminants from a mixture of membrane proteins. An example of this application is the removal by CM-cellulose (in 2.5 mM phosphate, 0.5 mM EDTA, pH 6.0) of haemoglobin from proteins extracted from bovine erythrocyte membranes with dilute solutions of EDTA [103]. This is possible as most erythrocyte membrane proteins appear to have isoelectric points substantially below that of haemoglobin.

An interesting ion-exchange chromatographic variant has recently been developed [107]. Lipophilic resins were prepared by cross-linking fatty acyl groups onto the amino groups of a polyacrylic acid resin (Bio-Rex 70). This column was found to be effective for the fractionation of the insoluble cytochromes of yeast mitochondria dissolved in deoxycholate.

### 7.2.5 *Affinity chromatography*

Most techniques for the purification of proteins depend for their effectiveness on differences in gross physicochemical properties between the various proteins present in a mixture. Affinity chromatography, however, utilizes selective adsorbents which have a biological specificity for the molecule under investigation. The technique therefore involves a 'functional' approach and is suitable for molecules such as enzymes, antigens, antibodies, vitamin- and drug-binding proteins, biological receptors, and transport proteins. Two recent reviews [8,9] provide detailed discussion of the technique.

Briefly the method involves the preparation of a selective adsorbent by the covalent immobilization of a specific interacting species, usually a small ligand, to a suitable insoluble matrix. When a mixture containing the protein which reacts specifically with the bound ligand is passed through a column of the adsorbant, this protein is selectively bound by the column. All the other proteins remain unbound and can therefore be washed out of the column. The selectively-bound protein can then be eluted under conditions which are known to disrupt the specific ligand-protein interaction.

It is an essential prerequisite for affinity chromatography that the proteins of interest should be present in a disaggregated state. Unfortunately, as has already been repeatedly discussed, membrane proteins exhibit a pronounced tendency to aggregate. Most workers have therefore found it necessary to prepare membrane proteins for affinity chromatography under disaggregating conditions [81], the use of detergents often being efficacious in this respect [108,110,242]

Cuatrecasas [8] lists the properties which must be possessed by a carrier suitable for use in affinity chromatography: (a) it must interact very weakly with proteins in general to minimize non-specific adsorption, (b) it should exhibit good flow properties that are retained after coupling, (c) it must possess chemical groups that can be activated or modified to allow chemical linkage of ligands, (d) it must be mechanically and chemically stable, and (e) it should form a very loose, porous network that allows uniform and unimpaired entry and exit of large macromolecules throughout the entire matrix.

The use of polystyrene or cellulose derivatives as carriers is not usually recommended [8]. Polyacrylamide gels appear to have many desirable features as supports for affinity chromatography, and highly substituted derivatives can be produced [11]. Agarose would appear to have nearly all the properties of an ideal adsorbant [111] and is available commercially (e.g. Sepharose). It readily undergoes substitution reactions when activated with cyanogen halides [112] and CNBr-activated Sepharose 4B is available commercially (Pharmacia). Procedures have been described [111] for the preparation of numerous agarose derivatives, facilitating the attachment of a wide variety of ligands. Agarose derivatives with varying spacer 'arm' lengths and terminal reactive groups are available from Bio-Rad Laboratories [Affi-Gels] and Pharmacia [AN-and CH–Sepharose 4B]. Concancavalin A-agarose which is potentially useful in the study of membrane glycoproteins, providing they contain α-D-glucosyl or sterically similar residues, is available from Pharmacia and Miles-Seravac.

One of the principal problems in developing an affinity chromatography system is the selection of a suitable ligand or 'acceptor' for the particular membrane component or 'receptor' under investigation. For the investigation of membrane components acting as hormone receptors, the hormones themselves obviously represent suitable ligand species. Liver cell membranes are known to possess specific binding structures for glucagon. However glucagon attached directly to an agarose backbone does not interact with liver cell membrane fragments [113]. This illustrates that although a ligand can be directly linked to a column matrix, it must be coupled in such a way that its ability to be 'recognized' is not impaired. It is often necessary to position the ligand at a distance from the matrix backbone to minimize steric restrictions on the interaction between the ligand and the macromolecule to be purified. In these cases a hydrocarbon extension 'arm' is interposed between the ligand and the support [111]. In this way it was found that when a hydrocarbon extension 'arm' was interposed between the agarose backbone and glucagon, liver cell membrane fragments could then be specifically retained [113]. Similarly insulin binding material extracted from liver and fat cell membranes using Triton X–100 or Lubrol-PX, has been subjected to affinity chromatograpy on an insulin-agarose column possessing spacer 'arm', and an 8000-fold purification of the insulin receptor achieved [114]. These receptors were eluted with buffers containing urea and having a relatively low pH, conditions known to dissociate the insulin-receptor complex. This purification has allowed further analysis of the insulin receptors, which have been suggested to be highly asymmetric molecules with a molecular weight of 300 000 and to be localized exclusively at the cell surface [115].

The purification of membrane glycoproteins by affinity chromatography has been facilitated by the occurrence of plant lectins or agglutinins which have specificities directed towards different carbohydrate moieties. It may be possible, using the lectins as ligands, to subfractionate membrane glycoproteins. This approach has been used to fractionate the glycoproteins of the human erythrocyte membrane extracted with Triton X–100. The major erythrocyte sialoglycoprotein was isolated using a wheat germ agglutinin-Sepharose column, while the non-sialic acid containing glycoproteins were isolated using a *Ricinus communis* lectin-Sepharose column [108].

The effectiveness of affinity chromatography for the purification of proteins depends primarily on the specificity of the selected ligand or 'acceptor', which determines the nature of the membrane protein or 'receptor' which is selectively adsorbed. However the affinity of the ligand can also be an important factor in determining the recovery or yield of proteins from the column. This phenomenon is illustrated by the work of Crumpton and his colleagues on the glycoproteins of pig lymphocyte plasma membranes. Their initial approach [242] was to use a concanavalin A- Sepharose 4B column. This column selectively adsorbs the membrane glycoproteins which can be displaced with α-methyl-D-glucopyranoside, but unfortunately the total recovery of protein (80%) and particularly the yield of glycoprotein (5%) is poor. As phytohaemagglutinin (PHA) from *Lens culinaris* beans binds the same residues as concanavalin A but with a 50-fold lower affinity, the decrease in the strength of carbohydrate binding by this lectin resolves the recovery problems. Consequently when pig lymphocyte plasma membranes solubilzed with deoxycholate are applied to a *Lens culinaris* PHA-Sepharose 4B column, 95% of the added protein is recovered; 83% of this material is present in the unadsorbed fraction, while 12% of the protein is retained by the column [110]. The adsorbed fraction represents the major portion of the membrane glycoproteins and can be eluted from the column with methyl-α-D-mannopyranoside. A similar method has been used to partially purify the HL-A2 antigen activity of the plasma membranes from a diploid lymphoblastoid cell line (BRI8) [116].

Unfortunately the biological functions of the majority of membrane proteins are unknown and it is therefore difficult to select suitable 'receptor'-ligand pairs to facilitate the analysis of these proteins by affinity chromatography. A possible approach which may overcome this difficulty is to incorporate a receptor, for which a ligand is available, into the proteins under investigation. For example, the proteins exposed at the erythrocyte cell surface can be selectively labelled with 2,4,6-trinitrobenzene sulphonic acid. These proteins, solubilized with deoxycholate, can then be selectively isolated using rabbit, antidinitrophenyl antibodies linked to a Sepharose column [117]. The use of antidinitrophenyl antibodies which cross react with trinitrophenyl analogues is preferred, because the high affinity of rabbit antibodies for their specific nitrophenylated derivatives prevents elution of the trinitrophenylated proteins except under vigorous conditions where denaturation or release of antibodies might occur.

## 7.3 Chemical methods

### *7.3.1 Protein concentration*

The availability of an effective method of protein concentration is of great assistance in any work on proteins. As might be predicted from the tendency of membrane proteins to aggregate, great care must be exercised during their concentration. The most suitable method for any application will depend on the use that is made of the resulting preparation which can range from amino acid analysis, where freeze-drying may suffice, to the preparation of a sample with the minimum perturbation of protein conformation, e.g. for enzyme assay. In our experience the most reliable, gentle method is vacuum dialysis in ordinary Visking tubing. The method is unfortunately slow, but the more rapid ultrafiltration that can be achieved with modern, commercially available membranes can lead to irrecoverable loss of protein on the surface of the membrane. Some proteins can be effectively concentrated by adsorption and subsequent elution from a cellulose ion exchanger. Others can be concentrated by precipitation, either with ammonium sulphate or by pH, but precipitation can result in irreversible aggregation. A simple, rapid, and frequently satisfactory method for concentration prior to gel electrophoresis, is to enclose the dilute solution in a dialysis bag and bury it for a short time in crystalline sucrose. Some commercial products based on this principle can contaminate the protein and therefore must be used cautiously.

### *7.3.2 Protein estimation*

The Lowry method [118] is, in our experience, the most suitable method for the estimation of membrane proteins. Direct measurement of 280 nm absorbance can be impaired by traces of insoluble material scattering light, and the Buiret method is rather insensitive. Estimation by a ninhydrin method of the amino acids produced from a protein sample by acid hydrolysis is an acceptable but laborious alternative. The major disadvantage of the Lowry method is that colour development varies between proteins, and a calibration curve based on a standard protein such as bovine serum albumin is not altogether satisfactory. As significant differences occur between different fractions, especially if the fractions contain glycoproteins [28], each protein fraction should be calibrated against itself (Fig. 7.1). Carbohydrate reduces the extinction coefficient by a larger factor that can be attributed to the dry weight of the carbohydrate component itself. It should be noted that, as the original publication [118] reported, the calibration curve is not linear. The method described in the experimental section is adapted from the original procedure [119] by the addition of deoxycholate to facilitate the dissolution of the proteins. Lipids and EDTA do not interfere with colour development.

### *7.3.3 Terminal amino acid determination*

Terminal amino acid analysis, although a reasonably straightforward procedure for many proteins, is complicated in the case of membrane proteins by three factors: (a) Blocked terminal amino acids occur in some membrane proteins. For example, bovine retinal rhodopsin lacks both reactive N- and C-terminal residues [120], and no N-terminus has been found for either of the two subunits of *Streptococcus faecalis* ATPase [121]. (b) Terminal amino acid determination will be complicated by the occurrence of either branched or cyclic polypeptides. There is evidence for the occurrence of covalent interchain cross-linkages between some membrane proteins [122,123], and although there is at present no evidence suggesting the presence in membranes of cyclic polypeptides this possibility should not be discounted. (c) As discussed in some length in Section 7.2, it is often difficult to fractionate and purify membrane proteins in sufficient quantities to permit chemical analysis. It is therefore important to

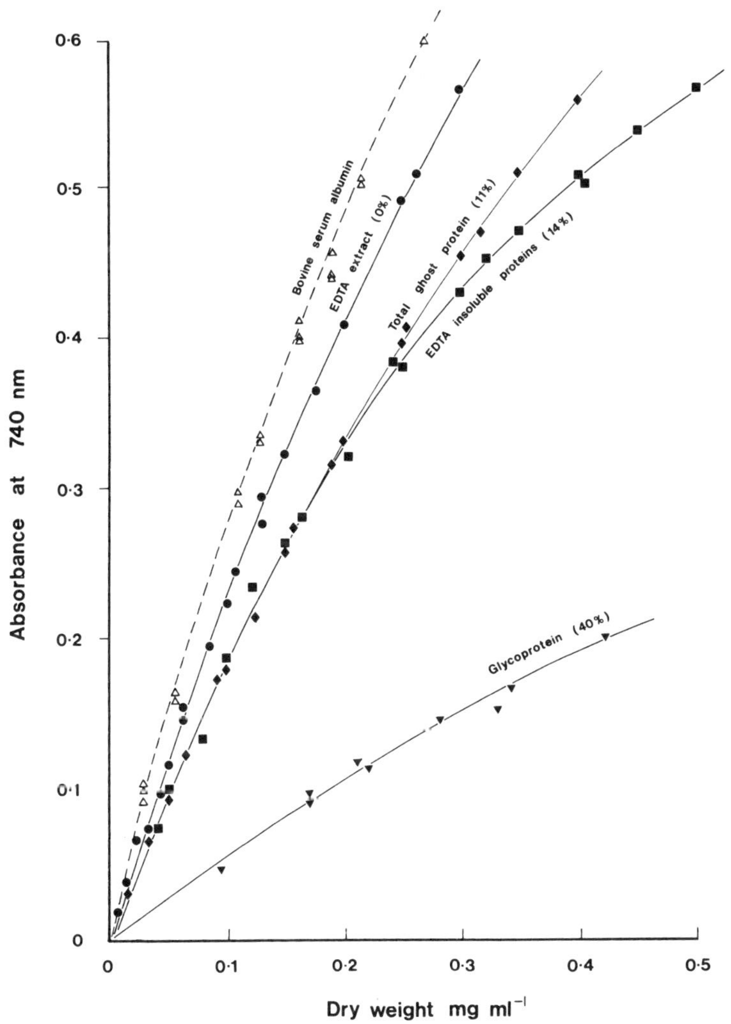


**Fig. 7.1** Calibration curves of the Lowry method for membrane proteins to illustrate the variation in colour intensity between different protein fractions. Reproduced from [28] with permission of the editors.

employ the most sensitive technique available for the determination of terminal amino acid residues. It should be noted that these factors should also be considered when protein sequencing studies are contemplated.

*(i) N-terminus*

Although the Edman degradation technique using phenyl *iso*thiocynate [124,125] can be used for N-terminal amino acid determination, it is primarily used in sequencing studies and is discussed in that section (Section 3.6).

Until recently the most widely used method for the determination of N-terminal residues has been the technique employing 1-fluoro-2,4-dinitrobenzene (DNFB) [126,127], based on the method developed by Sanger. Unfortunately, the DNFB method suffers from one major drawback in that for its successful application it requires substantial amounts of protein. It has been stated that about 100 mg of DNP-protein are usually required for the estimation of one end group per mole unit of 100 000. The method has been applied successfully to those membrane proteins which can be obtained in a high state of purity and at high yield. The method may be useful for demonstrating the heterogeneity of membrane protein fractions, and in this way it has been shown that the protein fraction isolated from bovine erythrocyte membranes by dilute solutions of EDTA contains at least eight N-terminalamino acids [17,77]. Although a method has been described for the analysis by DNFB of proteins fractionated by polyacrylamide gel electrophoresis [128], the technique requires material from at least 20–30 identical gels to be pooled for a successful assay to be performed.

Similarly, the cyanate method developed by Stark [129] is of limited applicability in the field of membrane protein biochemistry for it was originally designed for use with 50 mg of protein. Although it can be scaled down to a certain extent, it still requires considerable quantities of material. The method does have the advantage that it can be performed in denaturing solvents, e.g. 8 M urea, 6 M guanidine•HCl, which enhance the possibility that the protein will remain soluble throughout the procedure. The method has been employed in a study of erythrocyte membrane glycoproteins [57].

Dimethylamino-naphthalene-5-sulphonyl chloride (DNS–Cl) was developed as a fluorescent end-group reagent for proteins and peptides by Gray and Hartley, and its use has been recently reviewed [130,131]. The advantages of DNS–Cl are that DNS-amino acids are more stable to acid hydrolysis than the corresponding DNP-amino acids and above all it provides a hundred-fold increase in sensitivity. The dansylation of proteins is complicated by the decreased accessibility of terminal groups caused by steric factors, necessitating the use of relatively large amounts of DNS–Cl. Optimal labelling conditions have been found to be in the range pH 9.5 to 10.5, for above this range the reagent is hydrolysed too rapidly and below pH 9.5 the unreactive protonated form of the amino acid predominates. DNS–Cl is unfortunately sparingly soluble in water, consequently the reaction must be performed in acetone-water mixtures. The use of performic-acid-oxidized protein [132] is recommended as DNS-cysteic acid and DNS-methionine sulphone are both stable in acid hydrolysis [133]. Glutamine and asparagine are converted to the respective acids during acid hydrolysis, and leucine aminopeptidase has been recommended for the identification of amides in the N-terminal position [130]. It should be noted that peptides in the form DNS-val-X and DNS-ile-X are only partially split during acid hydrolysis, so that both DNS-amino acid and DNS-dipeptide are produced.

Two general methods are recommended for the separation of DNS-amino acids. The first method involves the use of high voltage paper electrophoresis. A flat plate apparatus must be used as equipment employing white spirit as a coolant is unsuitable due to the solubility of most DNS-amino acids in the cooling solvent. Although good separations can be achieved by flat plate paper electrophoresis [133,134], a substantial capital investment is required in the equipment. This drawback has encouraged the development of chromatographic techniques, both on paper [130] and on thin layers [130,135,136]. At present the method of choice would appear to be the two-dimensional separation of DNS-amino acids on 'polyamide' thin layers [137].

The high sensitivity of DNS–Cl makes it the preferable reagent for use with membrane proteins. Using this reagent, methionine has been determined to be the only N-terminal amino acid found in the major chromatophore protein component from *Rhodopseudomonas spheroides* [138]. A protein possessing D-glyceraldehyde 3-phosphate dehydrogenase activity has been isolated from human erythrocyte membranes [139,140] and this protein has been found to possess glycine as its only N-terminal amino acid [139,140].

The total membrane protein from human erythrocyte plasma membranes has been found to contain several N-terminal amino acids [141,142] demonstrating the polypeptide heterogeneity of this membrane. The sensitivity of the DNS–Cl technique has made it possible to perform N-terminal analysis on membrane proteins separated by polyacrylamide gel electrophoresis. Langdon [141] labelled the total erythrocyte membrane proteins with DNS–Cl prior to electrophoresis, excised the labelled bands from the gels, and performed N-terminal assays on each band. Each protein band was found to display N-terminal heterogeneity. On the basis of the distribution and nature of these N-terminal amino acids, it was argued that the apoproteins of the soluble plasma lipoproteins also constitute the major proteins of the human erythrocyte membrane. Langdon has supported this claim by immunological analysis, but the validity of this immunological data has recently been questioned [143]. He also claimed [123] that these apoproteins are apparently covalently cross-linked both to each other and probably to other polypeptides. These cross-links were claimed to be of the type characteristic of collagen and elastin formed from lysine and its oxidation product $\alpha$-amino adipic semialdehyde.

Recent studies on the N-terminal analysis of those proteins which can be extracted from erythrocyte ghosts by dilute solutions of EDTA, termed 'spectrin' [75], have advanced our understanding of the molecular nature of this fraction. It was originally claimed that this fraction represented a single protein species with a molecular weight of 140 000 [75], but it has since been found to be heterogeneous both by electrophoretic [17,78,144] and immunological [145,146] techniques. The polypeptide heterogeneity of this fraction has been conclusively demonstrated by the presence of several N-terminal amino acids in the proteins extracted by EDTA from both bovine [17,77] and human [78,144] erythrocytes. The protein bands which are produced by this fraction when it is subjected to polyacrylamide gel electrophoresis have also been shown to display N-terminal heterogeneity both in the absence [17] and in the presence [17,144] of detergent. It seems certain, then, that this protein fraction is composed of several polypeptide chains, but precisely how these are arranged is at present a matter of speculation. It is interesting in this respect that no $\epsilon$($\gamma$-glutamyl)lysyl interchain isopeptide cross-linkages have been detected in the 'spectrin' fraction by Fuller [78], although these have been shown to be present in an aqueous soluble membrane protein from L cells [122]. Unfortunately, Fuller [78] did not search for the other types of cross-linkages which have been found in the total erythrocyte membrane by Langdon [123].

*(ii) C-terminus*

Carboxypeptidases which sequentially remove amino acids one residue at a time from the C-terminal end of a polypeptide chain can be used to identify C-terminal residues [127,147]. As amino acids are sequentially released until a resistant residue is exposed, the original C-terminal amino acid is established by investigating the rate of amino acid release. An interesting modification of the method has been described [148] which relies on carboxypeptidase digestion in $H_2{}^{18}O$. The only amino acids which do not contain excess $^{18}O$ are those which were initially in a C-terminal position. There are two major problems in the use of carboxypeptidases. The first is that it is possible that the C-terminal amino acid may be resistant to attack by a carboxypeptidase, which would erroneously suggest the presence of a blocked C-terminal residue. This situation can be alleviated by the use of carboxypeptidases with differing specificities. Secondly, the method depends on measurement of the sequential release of amino acids at differential rates. This is possible as long as only one polypeptide chain is present, but if a sample contains a heterogeneous mixture of proteins it becomes very difficult, if not impossible, to identify the C-terminal amino acids with any certainty. Some successful applications of this technique to the study of membrane proteins have been reported [57,138].

A method using tritium labelling suitable for use with about 2 mg of protein has been described [149]. A selective $^3$H-labelling reaction at the C-terminal amino acid is achieved through a C-terminal racemization mechanism. This selective tritiation is based on C-terminal oxazolone formation by the action of acetic anydride or dicyclocarbodiimide, followed by base-catalysed hydrophobic ring-opening in tritium oxide. Using this technique, glycine has been unequivocally confirmed as the only C-terminal amino acid of the major protein component of *Rhodopseudomonas spheroides* chromatophore [150].

Hydrazinolysis is the most frequently used chemical method for the identification of C-terminal amino acids. Heating proteins in hydrazine causes transamidation of all peptide bonds to hydrazides and only C-terminal acids are liberated as free α-amino-α-carboxylic acids. Despite the superficial simplicity of the technique its quantitative use is complicated by the large and uncertain correction factors which are involved [151]. Full practical details are provided by Schroeder [151]. The method is applicable to most amino acids, but arginine is partially destroyed and it is unsuitable for asparagine and glutamine. The C-terminal amino acids of several purified membrane proteins have been investigated by this technique [57,138,150]. The method has also been used to demonstrate C-terminal heterogeneity among the total membrane proteins of the human erythrocyte [141].

### *7.3.4 Fingerprint techniques*

This is a technique by which a protein may be characterized by the nature of the peptide mixture derived from the protein either by enzymic or by selective chemical [152] cleavage. The peptides are usually characterized by a two dimensional combination of paper electrophoresis and paper chromatography, the resultant characteristic pattern of peptides being known as the peptide 'map' or 'fingerprint'. The method was developed initially for the study of abnormal human haemoglobins.

Peptide mapping can be used as an analytical tool to investigate the homogeneity of an isolated protein. One of the most useful techniques for this application is the selective cleavage of peptide bonds of methionine residues with cyanogen bromide [152,153]. The reaction is very specific for at acidic pH the ϵ-amino groups and other basic groups are protonated and therefore protected from reaction with cyanogen bromide. The use of an acid medium is also beneficial in that the protein under investigation is unfolded, ensuring exposure of the methionine residues. Cysteine, however, is slowly attacked and therefore should be converted to an unreactive species by carboxymethylation or benzylation. As long as the methionine content of a protein has been established by amino acid analysis, it is possible to test for homogeneity by comparing the number of peptides actually found with the number expected from the methionine content. Marchesi and co-workers [75] have used this approach to characterize human erythrocyte 'spectrin'. The peptide analysis indicated a single protein of molecular weight 146 000, which was consistent with data obtained by equilibrium ultracentrifugation and SDS gel electrophoresis.

Unfortunately this data is now suspect as 'spectrin' has been found to be heterogeneous by gel electrophoresis with several bands in the 220 000 molecular weight region [17,78,144]. The heterogeneity has been confirmed by N-terminal analysis [17,77,78,144], and by immunological methods [145,146].

If a pure protein can be obtained it is often useful to combine peptide mapping with purification and subsequent sequence analysis of the separated peptides. A protein of molecular weight 33 000 purified from human erythrocyte ghosts has been investigated in this way [139]. Trypsin digests of this protein produced maps with twenty major peptide spots. On the basis of amino acid analysis, cation exchange chromatography and N-terminal analysis of these peptides, the protein was identified as D-glyceraldehyde 3-phosphate dehydrogenase.

An interesting use of peptide mapping procedures in the field of membrane protein research is in those studies aimed at determining the spatial distribution, or 'sidedness', of proteins within membranes. From the specific radioactive labelling of intact erythrocytes, pronase-treated cells and ghosts, and the subsequent fingerprinting of the radioactive peptides derived from components purified by gel electrophoresis, it has proved possible to establish [46,56] the spatial organization in the membrane of the major glycoprotein and the protein of molecular weight 105 000 from human erythrocyte membranes. These studies are discussed at length elsewhere in this volume (Chapter 9).

Due to the difficulties encountered in obtaining homogeneous membrane proteins, the most useful application of peptide mapping in the field of membrane protein biochemistry would appear to be in the study of differences and similarities among groups of proteins. This is possible as each peptide occupies a position on the final map which is characteristic of its amino acid composition and, to a much smaller degree, of its sequence. Trypsin, which catalyses the hydrolysis of only the peptide bonds between the carboxyl group of lysine or arginine and the amino group of the adjacent amino acid, is the enzyme most widely used for the preparation of peptide digests suitable chymotrypsin [154], it is usual to pretreat either with diphenylcarbamylchloride [155] or *N*-tosylphenylalanine chloromethylketone [156] to eliminate chymotryptic activity. Chymotrypsin and mixtures of chymotrypsin and trypsin are also commonly used. Enzymes of more liberal action, such as subtilisin and thermolysin, are not generally as useful since they generate large numbers of peptides which complicate the analysis and comparison of peptide maps.

As most proteins in their native states are attacked slowly and only to a limited extent by proteolytic enzymes, it is often necessary to denature the protein under investigation prior to enzymic digestion [157]. This denaturation is usually accomplished by heating at 100°C, but it is also possible to treat the protein with 8 M urea, then diluting to 2 M urea in which trypsin and chymotrypsin are both fully active. Any disulphide bonds present in the protein should be cleaved and blocked by carboxymethylation [158] prior to enzymic hydrolysis.

Most methods which have been described for fingerprinting involve the use of two-dimensional paper chromatography and electrophoresis. Solvents frequently employed for chromatography are *n*-butanol–glacial acetic acid–water (4:1:5) [159] or *n*-butanol–pyridine–glacial acetic acid–water (90:60:18:72) [160]. Electrophoresis is usually carried out in acetic acid–pyridine–water at pH 3.6 [159] or pH 6.4 [161]. The resulting map is usually stained with ninhydrin, although the use of a cadmium–ninhydrin reagent [162] may increase sensitivity.

The major drawback with paper methods for fingerprinting membrane proteins is that they require rather high sample loadings, about 1.0 to 1.5 mg [160]. This situation can be improved by using more sensitive staining techniques or radioactively-labelled peptides, but in most cases where the amount of material available is small it is better to employ methods developed for the two-dimensional separation of peptides on thin layer plates of silica gel [163] or cellulose [164]. Unless radioactively-labelled peptides can be obtained [165], these methods still rely on ninhydrin staining. As the sensitivity of this reagent can vary considerably for different peptides, the minimum amount of material which can be used for thin layer techniques is of the order of 0.01 $\mu$mol. Highly sensitive fluorometric techniques are currently being developed to complement both ninhydrin e.g. fluorescamine [239], *o*-phthalaldehyde [240] and autoradiography i.e. fluorography [241]. The new techniques are extremely promising, but the optimal conditions for their use in the generality of cases have not yet been established. They have not, therefore, been discussed in detail here, but the reader should take note of any rapid developments in the near future.

Using thin layer techniques, trypsin digests of membrane proteins from *Rhodopseudomonas spheroides* have been fingerprinted [166]. Although a few neutral peptides were found to be common between the proteins, the majority of peptides from each digest appeared to be unique to that digest. On the basis of the distinct fingerprints, and supported by amino acid analysis, it was concluded that there is a heterogeneous mixture of proteins in the membrane. Rather different conclusions have been reached concerning the proteins of *Micrococcus lysodeikticus* membranes. Tryptic digests of ATPase, NADH dehydrogenase and an electrophoretically 'fast-moving' component of unknown function, from these membranes revealed a number of major peptides common to all three fractions [167]. The principal differences occurred in the 'background' of minor peptides and in the relative proportions of the major peptides. It has been estimated [168] that 92% of the 'fast-moving' component and 53% of the ATPase consist of common peptides. These results reinforce the claim, using immunological techniques, that these three components possess a common antigenic activity in addition to a specific unique activity. These workers suggest that membrane proteins may possess a very similar preferred sequence of peptides, perhaps of a hydrophobic character, facilitating their incorporation into the membrane. The possible existence of a common contaminant protein bound to each purified protein was not totally excluded but considered improbable. In a recent comparative study of the membrane proteins which can be radio-iodinated by lactoperoxidase in human platelets and lymphocytes [169], it has been shown that at least one protein from both cell types produced very similar peptide maps after thermolysin digestion. It was concluded that despite their different specialized functions, some of the membrane proteins from these two cell types are the same or are very closely related.

The constraint imposed on the fingerprinting of membrane proteins by the relatively large quantities of protein digest which are required can be overcome by the use of 1-dimethylaminoaphthalene-5-sulphonyl chloride (DNS–Cl)[170,171]. In this method a normal trypsin digest is dansylated and the dansyl-peptides are subsequently separated by two-dimensional chromatography. The use of DNS–Cl not only allows the detection of millimicromole quantities of peptides but also renders them more hydrophobic, thereby facilitating their separation by two-dimensional chromatography on silica gel plates with organic solvents.

There are several limitations inherent in the fingerprinting method for the characterization of proteins. The number of tryptic peptides observed on a map should exceed the number of lysine and arginine residues by one. However, there is usually some deviation from this ideal state due to the occurrence of adjacent lysine and arginine residues, and also due to the slow cleavage of some lysine and arginine bonds, both daughter peptides and unsplit parent peptides being formed. Spurious peptides can also be produced by trace enzymic activities, such as the chymotrypsin activity present in most trypsin preparations. Peptide mapping will not detect very slight differences between proteins unless more arginine and/or lysine is present in one than in the other. Enzymic hydrolysis can result in a 'resistant' core but this can often be analysed by subjecting it to alternative enzymic procedures. As paper or thin layer techniques are usually not absolutely reproducible it is desirable, if possible, to prepare comparative maps simultaneously.

### *7.3.5 Amino acid analysis of proteins*

The analysis of the amino acid composition of proteins has been well reviewed by several authors [172,173] and as the analysis of membrane proteins presents no unique technical difficulties, the reader is referred to these reviews for full details of the technique. The principal aspects of the methodology will be briefly described here.

*(i) Preparation of a sample suitable for analysis*

Light and Smith [173] state that, 'A protein suitable for analysis must satisfy the standard tests for homogeneity'. This is precisely the major problem in the amino acid analysis of membrane proteins. As will be readily apparent from Section 7.2, mixtures of membrane proteins are notoriously difficult to fractionate and, with a few exceptions such as *Streptococcus faecalis* ATPase [121], bovine rhodopsin [120], liver microsomal cytochrome $b_5$ [174], and mitochondrial cytochrome *c* [175] which have been successfully isolated and analysed, it has proved difficult to prepare membrane proteins which adequately satisfy criteria for homogeneity. This fundamental difficulty has not, however, deterred workers from subjecting various membrane preparations containing complex mixtures of proteins to amino acid analysis. Although the information derived by this approach is of limited value, it has, as will be discussed later, thrown some light on the nature of some membrane proteins.

*(ii) Hydrolysis*

The use of alkaline hydrolysis is restricted to the estimation of acid labile amino acids (e.g. tryptophan, tyrosine) as it results in the complete degradation of arginine, serine, threonine, cystine and cysteine. Hydrolysis with 6 M HCl at 110°C is usually preferred due to the high rate of peptide bond cleavage, and the volatility of the reagent facilitates its removal at the completion of hydrolysis. Various hydrolysis times up to 70 h should be conducted so that corrections can be made for the partial degradation of amino acids by extrapolation of the content of each amino acid to zero time of hydrolysis. Tryptophan, cystine and cysteine are particularly sensitive to acid hydrolysis and must be estimated by other methods. It should be noted that as glutamine and asparagine are converted to the respective acids by acid hydrolysis they cannot be distinguished from glutamic and aspartic acids.

*(iii) Separation and estimation of amino acids*

The most widely used methods for the separation of amino acids are based on the ion-exchange method developed by Moore and Stein, using derivatives of polystyrene or polymethacrylic acid [176]. Fully-automated systems have been developed using either one [177] or two [172] columns. Commercial equipment of both types is available, and using the more sophisticated versions of this apparatus complete analyses can be performed very rapidly. Recently gas liquid chromatography (GLC) has been increasingly employed for amino acid analysis [178–180]. This has been possible due to the development of methods for the production of amino acid derivatives possessing the requisite volatilities. GLC may eventually replace ion-exchange chromatography for it is a very rapid technique and its high sensitivity facilitates the assay of much smaller quantities of amino acids.

### *(iv) Determination of acid labile amino acids*

Cysteine and cystine are extensively destroyed by acid hydrolysis and are best estimated after oxidation with performic acid and subsequent chromatography of cysteic acid on Dowex 1-X8 or on an amino acid analyser [181]. A quantitative conversion of methionine to methionine sulphone also occurs, permitting the simultaneous estimation of this amino acid. Tryptophan is very unstable in a strongly acid environment and must be assayed either after alkaline hydrolysis or by a colourimetric method. Recently, *p*-toluenesulphonic acid has been used for hydrolysis [182] and it is claimed that tryptophan can then be analysed together with the other amino acids.

### *(v) Amide nitrogen*

Acid hydrolysis liberates ammonia from the amide linkages of glutaminyl and asparaginyl residues. Liberated ammonia is assayed by the Conway microdiffusion technique or by its reaction with ninhydrin or Nessler's reagent. It is essential to correct for the free ammonia content of proteins prior to amide determination. Amide nitrogen can be estimated from the ammonia value found after automated separation of acid hydrolysates, but this must be measured as a function of time to correct for ammonia formed by deamination of certain amino acids, especially serine and threonine. Enzymic procedures are available for estimating the content of glutamine, asparagine, glutamic acid and aspartic acid in a single protein sample.

### *(vi) Expression of results*

In order to avoid confusion, care should be exercised when comparing data from various laboratories as results can be expressed in many ways, for example as moles per 100 moles (mol %), moles per 1000 moles, moles per 100 000 g, and as residues per molecule of protein. It is useful to express the results in terms of residues per molecule of protein, but frequently this is not possible as the molecular weights of many membrane proteins have not been unequivocally established.

### *(vii) Application to membrane proteins*

As has been stated earlier, few membrane proteins can be obtained in a worthwhile state of purity to justify investigation by amino acid analysis. The virtual futility of investigating the amino acid composition of a complex mixture of proteins is exemplified by the analyses of whole plasma membranes presented in Table 7.1. Similar data for other membrane proteins can be found in the reviews by Steck and Fox [183] and Guidotti [184]. The classification of amino acids used here, and in the following tables, is based on the hydrophobicity scale established by Nozaki and Tanford [185]. Although the hydrophobic nature of membrane proteins has often been stressed, it is apparent that the amino acid compositions of the total mixture of membrane proteins do not show an increase in the proportion of hydrophobic residues. Indeed the analyses do not differ markedly from that of the mixture of proteins present in a post-ribosomal supernatant of *Escherichia coli* [186]! Ketterer and co-workers [187] have produced a histogram illustrating the distribution of

TABLE 7.1
*Comparison of the amino acid analyses (mol %) of various plasma membrane preparations.*

| A.A. type | Amino acid | Human myelin | Human erythro-cyte | Liver | Ehrlich ascites | Rat bladder lumen | Amoeba | *E. Coli* cell supernatant |
|---|---|---|---|---|---|---|---|---|
| Basic | LYS | 6.2 | 5.5 | 7.5 | 6.3 | 5.4 | 4.8 | 6.4 |
| | HIS | 3.6 | 2.7 | 2.3 | 2.6 | 1.7 | 2.1 | 2.0 |
| | ARG | 5.6 | 5.0 | 5.0 | 4.7 | 4.7 | 3.6 | 4.1 |
| | Sum | 15.4 | 13.2 | 14.8 | 13.6 | 11.8 | 10.5 | 12.5 |
| Acidic | ASP | 5.8 | 8.6 | 8.4 | 8.8 | 8.9 | 12.0 | 10.4 |
| | GLU | 7.5 | 13.6 | 11.5 | 10.1 | 10.4 | 8.1 | 11.2 |
| | Sum | 13.3 | 22.2 | 19.9 | 18.9 | 19.3 | 20.1 | 21.6 |
| Neutral | THR | 6.4 | 5.7 | 4.5 | 5.5 | 5.5 | 10.1 | 5.7 |
| | SER | 6.8 | 8.1 | 6.1 | 6.6 | 6.5 | 7.5 | 4.7 |
| | PRO | 4.9 | 5.2 | 4.6 | 5.2 | 8.4 | 5.8 | 4.2 |
| | GLY | 11.4 | 7.1 | 8.5 | 8.5 | 8.7 | 10.8 | 8.5 |
| | ALA | 9.4 | 8.3 | 9.0 | 7.8 | 7.6 | 7.0 | 10.2 |
| | ½CYS | 1.9 | 0.9 | n.d. | trace | 1.9 | 4.5 | 1.1 |
| | Sum | 40.8 | 35.1 | 32.7 | 33.6 | 38.6 | 45.7 | 34.4 |
| Hydro-phobic | MET | 1.2 | 2.0 | 1.6 | 2.7 | 2.2 | 1.2 | 2.6 |
| | VAL | 5.4 | 5.9 | 6.2 | 6.6 | 6.5 | 6.2 | 7.9 |
| | ILE | 4.0 | 4.0 | 5.6 | 6.1 | 4.5 | 4.0 | 5.9 |
| | LEU | 9.0 | 11.4 | 10.5 | 10.1 | 8.9 | 6.1 | 8.1 |
| | TYR | 3.7 | 2.4 | 3.1 | 3.1 | 4.1 | 2.6 | 2.8 |
| | PHE | 6.7 | 4.2 | 4.6 | 4.8 | 4.1 | 3.1 | 3.5 |
| | Sum | 30.0 | 29.9 | 31.6 | 33.4 | 30.3 | 23.2 | 30.8 |
| | Ref | a | b | c | d | e | f | g |

[a] [236]
[b] [71]
[c] [237]
[d] [238]
[e] [187]
[f] [109]
[g] [186]

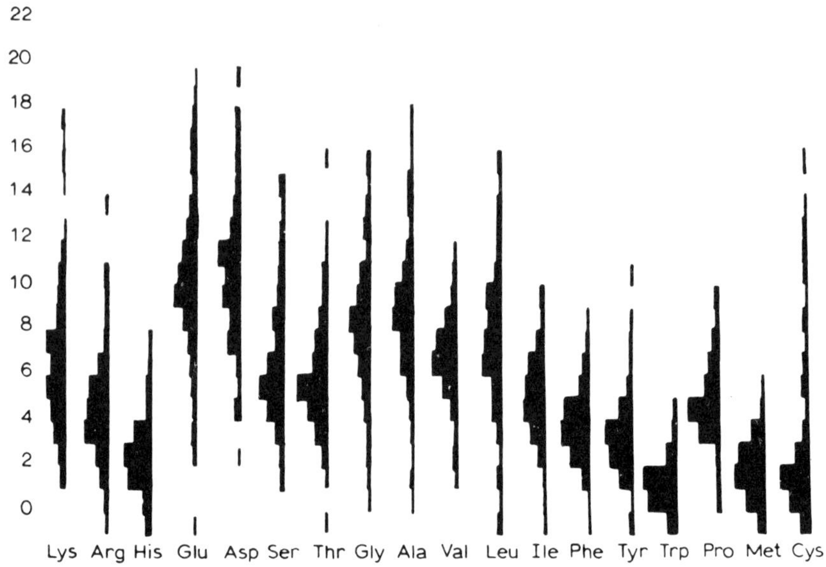


**Fig. 7.2** A comparison of the amino acid analysis of 200 pure proteins. The frequency of each amino acid in the proteins is represented as a separate histogram. In each histogram the ordinate represents the content of each amino acid in mol%, simplified so that values falling between two whole numbers are considered together. Thus the first category is 0 mol%, the second 0–1 mol%, the third 1–2 mol% and so on. The abscissa represents the number of proteins from 0–100 in which each content is found. Reproduced from [187] with kind permission of the authors and editors.

amino acid residues in 200 proteins [Fig. 7.2] and have recommended the use of this histogram to identify distinctive features of amino acid compositions of membrane proteins. If the data in Table 7.1 are compared with this histogram (Fig. 7.2) it will be readily observed that amino acid compositions of total membrane proteins exhibit no significant deviations from the norm when compared with many other proteins.

A survey of work performed on the proteins of myelin and erythrocyte membranes amply serves to illustrate the value and drawbacks of amino acid analysis when applied to membrane proteins. A discussion of the principal features of the compositions of certain other membrane proteins can be found elsewhere [183,184].

Myelin is one of the most well-characterized membranes with respect to its constituent proteins. This fortunate state of affairs has arisen because myelin contains only three major types of protein: (a) basic protein, (b) Folch-Lees proteolipid, and (c) Wolfgram protein. This simplicity of protein composition has facilitated the isolation of the constituents in a sufficiently pure state for amino acid analysis. The amino acid compositions of the proteins isolated from human [188,189] and bovine [188,190–192] myelin are strikingly similar.

TABLE 7.2
*Comparison of the amino acid analyses (mol %) of the protein fractions of human CNS myelin*

| A.A. type | Amino acid | Total protein | Basic protein | Folch–Lees proteolipid | Wolfgram protein |
|---|---|---|---|---|---|
| Basic | LYS | 6.2 | 8.1 | 4.4 | 7.8 |
| | HIS | 3.6 | 5.8 | 2.3 | 2.3 |
| | ARG | 5.6 | 9.8 | 1.8 | 5.0 |
| | Sum | 15.4 | 23.7 | 8.5 | 15.1 |
| Acidic | ASP | 5.8 | 7.0 | 4.8 | 6.5 |
| | GLU | 7.5 | 6.6 | 7.2 | 10.8 |
| | Sum | 13.3 | 13.6 | 12.0 | 17.3 |
| Neutral | THR | 6.4 | 4.4 | 8.4 | 6.1 |
| | SER | 6.8 | 9.9 | 6.3 | 5.9 |
| | PRO | 4.9 | 7.3 | 2.8 | 4.3 |
| | GLY | 11.4 | 15.0 | 11.2 | 9.0 |
| | ALA | 9.4 | 8.3 | 12.0 | 8.5 |
| | ½CYS | 1.9 | 0.0 | 4.2 | 2.5 |
| | Sum | 40.8 | 44.9 | 44.9 | 36.3 |
| Hydrophobic | MET | 1.2 | 1.1 | 1.6 | 1.4 |
| | VAL | 5.4 | 1.8 | 6.6 | 6.1 |
| | ILE | 4.0 | 1.9 | 4.4 | 4.7 |
| | LEU | 9.0 | 6.0 | 11.7 | 9.7 |
| | TYR | 3.7 | 2.3 | 4.9 | 4.0 |
| | PHE | 6.7 | 4.9 | 8.0 | 5.4 |
| | Sum | 30.0 | 18.0 | 37.2 | 31.3 |
| | Ref | a | b | c | b |

[a] [236]
[b] [188]
[c] [189]

Representative analyses for these proteins are presented in Table 7.2. The basic protein is enriched in basic and neutral amino acids but is deficient in hydrophobic residues. This composition is consistent with the suggestion that, as it can be extracted with acid or neutral salts [188,190], this protein is bound in the membrane primarily by ionic bonds with acidic lipids and/or proteins. The total absence of cysteine eliminates the possibility of disulphide bridges occurring in this protein, suggesting a rather flexible structure. The proteolipid, which contains carbohydrate [189] and fatty acids covalently bound by ester bonds [189,193], has a characteristically high content of hydrophobic residues and is rich in cysteine and methionine. On the basis of this composition, it has been suggested that the protein may be held in the membrane by hydrophobic interactions [184,188], an idea consistent with the release of the protein from the membrane by chloroform–methanol (2:1) [192]. In contrast, the Wolfgram protein contains a high proportion of acidic amino acids, although its content of hydrophobic residues is nearly as high as that of the proteolipid. This composition would appear to be consistent with the suggestion that this protein is bound in the membrane by a combination of both ionic and hydrophobic interactions [188].

The erythrocyte represents another instance where amino acid data is available for various subfractions of the total membrane protein. Although analyses of total proteins from human [71,104,139] camel [194], and bovine [103] erythrocytes are remarkable similar, a high variability in serine content is apparent if the results from different laboratories are compared. This raises a point of general relevance to the study of membrane proteins, for many preparations of membrane proteins may contain lipid and the presence of phosphatidylserine will artificially elevate the serine content. It should be noted that the presence of carbohydrate can also interfere with the analysis. The data presented in Table 7.3 is a comparison of some of the subfractions of the human erythrocyte membrane. It can be seen that the 'intrinsic' 90 000 molecular weight protein [183] is enriched in hydrophobic residues compared with the 'extrinsic' proteins 'spectrin' [75] and D-glyceraldehyde 3-phosphate dehydrogenase [139,140].

That it is sometimes possible to observe characteristic features of membrane proteins, even if these cannot be obtained in a homogeneous state, is illustrated by the amino acid analysis of those proteins extracted from the erythrocyte membrane by dilute solutions of EDTA. The composition of these proteins, termed 'spectrin', is remarkably similar when they are extracted from human [75,78], guinea pig [195], sheep [195], horse [195], and bovine103] erythrocytes. If the composition of human 'spectrin' is compared with the histogram (Fig. 7.2) it will be readily observed that 'spectrin' is characteristically rich in leucine and contains more glutamic acid than any of the 200 proteins used in the construction of the histogram. This would, at first sight, indicate that the 'spectrin' proteins are highly acidic in character, but due to the degradation of glutamine and asparagine during acid hydrolysis, no reliable figures are available for the degree of amidation of the acidic amino acids.

If a protein possesses characteristic features in its amino acid composition it is sometimes possible, by a comparison with well characterized proteins, to identify the protein by its amino acid composition. On this basis it has been suggested [184] that 'spectrin' represents an actomyosin-like protein. This idea appears to have gained a certain degree of general acceptance. However, 'spectrin' does not contain methylhistidine [103,196] which is a residue characteristic of both striated [197,198] and smooth [197] muscle actin and which is also usually present in myosin [197,199]. Neither does it contain methyllysine [103] which is often found in myosin [200,201]. It should also be noted that bovine 'spectrin', which is similar to human 'spectrin' by gel electrophoresis and amino acid analysis, does not react with antibodies directed against actomyosin and myosin from either smooth or striated muscle [103].

Amino acid analysis alone, in the absence of sequence data, is often

TABLE 7.3
*Comparison of the amino acid analysis (mol %) of various human erythrocyte membrane proteins*

| A.A. type | Amino acid | Whole membrane | 'Spectrin' | Glyceraldehyde-3-phosphate dehydrogenase | Major glycoprotein ('glycophorin') | mol. wt. x 90 000 |
|---|---|---|---|---|---|---|
| Basic | LYS | 5.5 | 6.7 | 8.4 | 4.4 | 4.7 |
| | HIS | 2.7 | 2.6 | 3.4 | 3.9 | 1.9 |
| | ARG | 5.0 | 5.8 | 3.3 | 4.4 | 4.8 |
| | Sum | 13.2 | 15.1 | 15.1 | 12.7 | 11.4 |
| Acidic | ASP | 8.6 | 10.9 | 11.0 | 5.9 | 7.5 |
| | GLU | 13.6 | 20.5 | 7.7 | 9.9 | 10.9 |
| | Sum | 22.2 | 31.4 | 18.7 | 15.8 | 18.4 |
| Neutral | THR | 5.7 | 3.6 | 5.9 | 10.8 | 6.2 |
| | SER | 8.1 | 4.1 | 6.3 | 13.8 | 5.8 |
| | PRO | 5.2 | 2.4 | 4.0 | 7.4 | 5.5 |
| | GLY | 7.1 | 4.9 | 10.4 | 6.4 | 7.5 |
| | ALA | 8.3 | 9.2 | 10.3 | 5.4 | 8.3 |
| | ½CYS | 0.7 | 1.1 | 1.0 | 0.0 | 1.0 |
| | Sum | 35.1 | 25.3 | 37.9 | 43.8 | 34.3 |
| Hydro-phobic | MET | 2.0 | 1.7 | 2.1 | 2.0 | 1.9 |
| | VAL | 5.9 | 4.7 | 8.3 | 8.4 | 7.3 |
| | ILE | 4.0 | 4.0 | 5.5 | 6.9 | 4.7 |
| | LEU | 11.4 | 12.4 | 7.1 | 6.4 | 14.1 |
| | TYR | 2.4 | 2.0 | 1.9 | 3.0 | 2.5 |
| | PHE | 4.2 | 3.0 | 3.8 | 1.5 | 5.6 |
| | Sum | 29.9 | 27.8 | 28.7 | 28.2 | 36.1 |
| | Ref | a | b | c | d | e |

[a] [71]
[b] [75]
[c] [140]
[d] [202]
[e] [183]

inadequate to identify the characteristic features of proteins, especially if these depend on the particular arrangement of the sequence of amino acid residues rather than on the overall amino acid composition. This is illustrated by the major human erythrocyte membrane glycoprotein, termed 'glycophorin' [202]. This protein is of interest in that it is claimed to extend through the membrane. It has been suggested that the carbohydrate-containing N-terminal segment and the proline-rich C-terminal protion of this molecule are exposed at the exterior and interior membrane faces respectively. A 23-residue long uncharged region of the molecule is said to reside in the hydrophobic interior of the membrane [203], and this has been found to be easily distinguishable from 774 segments of other proteins on the basis of its hydrophobicity [204]. Unfortunately, amino acid analysis of the whole glycoprotein (Table 7.3) gives no indication of this interesting molecular microarchitecture. (See p. 234).

### *7.3.6 Protein sequencing*

The sequencing of proteins can be accomplished either by enzymic methods or by chemical procedures. Enzymic methods have certain advantages over chemical methods in that, (a) asparagine and glutamine can be detected as such, (b) there is no tendency to lose a residue as the result of side reactions, and (c) the yield of sequentially released residues should not decrease. The main disadvantages of this approach, however, is that due to the varying rates of release of different amino acids, the technique may provide useful information for only a relatively short length of the protein or peptide under investigation.

Sequencing of proteins can be initiated either from the C-terminal or N-terminal extremities of the peptide or protein.

#### *(i) Sequencing from the C-terminus*

*Chemical methods*

Although most effort has been directed towards establishing chemical methods for sequencing proteins from their N-terminus, a method has been developed for sequential degradation from the C-terminus [205]. The peptide or protein is reacted with ammonium thiocyanate and acetic anydride to form peptidylthiohydantoins at the C-terminus. The peptidylthiohydantoin can be cleaved to a thiohydantoin characteristic of the C-terminus and to an acetylated peptide or protein in which the previously penultimate residue is terminal. Degradations can be performed sequentially and may be followed subtractively by amino acid analysis of the peptide product after each stage.

*Enzymic methods*

Pancreatic carboxypeptidases, which sequentially release amino acids from the carboxyl terminus, can be used to establish the partial sequence of a peptide or protein. Of the two enzymes which are available, carboxypeptidase A has the most liberal action, removing C terminal aromatic and long side-chain aliphatic residues most readily. However, glycine and acidic amino acids are only slowly attacked. Carboxypeptidase B has a much narrower specificity being restricted to terminal residues of arginine and lysine, although neutral amino acids may be removed in some instances [206]. Both enzymes must obviously be used to sequentially release amino acids from tryptic peptides. The C-terminal sequence is established by studying the time-dependent release of amino acids from the peptide or protein. This approach has been utilized to establish the C-terminal sequence of the major chromatophore protein isolated from the photosynthetic bacterium, *Rhodopseudomonas spheroides*. Its probable C-terminal sequence was found to correlate well with the overall amino acid composition of the isolated C-terminal tryptic peptide [150].

*(ii) Sequencing from the N-terminus*

*Enzymic methods*

A kinetic analysis of the amino acids released from the N-terminus of a peptide or protein by the action of leucine aminopeptidase can be used in sequencing studies. Difficulties arise in that the rates of hydrolysis for some amino acids can differ by as much as several orders of magnitude. In general the reliability of sequence information decreases as the number of amino acids released becomes large, limiting useful information to perhaps only the first two or three residues [207].

Dipeptidyl aminopeptidase I (DAP I) is another enzyme suitable for determining the N-terminal sequence or a peptide or protein. This enzyme catalyses the removal of dipeptide moieties from unsubstituted N-termini of polypeptide chains. It possesses a broad specificity but arginyl, lysyl or prolyl residues are not susceptible. The lack of susceptibility of arginyl and lysyl residues is not a problem if tryptic peptides are used, as these residues will now be C-terminal, but the appearance of a terminal prolyl residue will stop the reaction. Time course chromatography can be used to follow the reaction for at least several dipeptide products. Although this is often not sufficient to establish a sequence, it is possible to modify the substrate in such a way that the enzyme will generate the alternate, overlapping dipeptides [208].

*Chemical methods*

Edman degradation. The use of phenyl *iso*cyanate for N-terminal investigation was first developed by Bergmann, but the technique was largely neglected after the development of the dinitrophenylation method. It is due to the work of Edman, who substituted phenyl *iso*thiocyanate for phenyl *iso*cyanate, in developing a simple reliable technique suitable for use with many types of peptides, that the method has become one of the most powerful and versatile tools in protein sequencing. The N-terminal amino acid residue is removed in a two step process. In the first step the peptide is treated with phenyl *iso*thiocyanate to form a phenylthiocarbamyl peptide (PTC-peptide). The second step involves the acidic cleavage, using anhydrous trifluoroacetic acid, of the N-terminal residue in the form of a 2-anilino-5-thiazolinone derivative, simultaneously releasing intact the remainder of the peptide with a new N-terminal residue. It is usual, but not always essential, to convert the thiazolinone derivative to the 3-phenylthiohydantoin derivative (PTH-amino acid) by heating or by treatment with aqueous acid. The amino acid sequence is established by repeating the coupling and cyclization steps until all the residues, if possible, are removed, and identifying the residue at each stage.

Several variations of the Edman procedure have been developed. In the 'direct-Edman' technique, the PTH-amino acids released after each degradation cycle are directly identified. Although many methods are available for the identification of PTH-amino acids, a combination of gas liquid chromatography [209,210] and thin layer chromatography [126] is commonly used. Mass spectrometry is becoming increasingly popular as is amino acid analysis after the hydrolysis of the PTH-amino acids to the free amino acids using HI or sodium dithionite [211]. A modification of the method has been described in which the protein under investigation is absorbed on a filter paper strip which acts as a carrier for the protein throughout the reaction [127,212].

The alternative approach, termed the 'subtractive-Edman' procedure, is not to identify the released PTH-amino acid directly, but to subject a portion of the peptide remaining after each cyclization step to amino acid analysis [213]. In the related 'dansyl-Edman' technique a portion of the residual peptide can be dansylated, hydrolysed and the DNS-amino acids identified by any of the standard procedures [130,131]. PTH-amino acids can also be converted to DNS-amino acids [214] and so identified directly. The major disadvantage of the 'subtractive' techniques is that a portion of the peptide under study is removed for analysis at each step, so that the amount remaining may soon become a limiting factor.

Solid-phase degradations. The method developed by Laursen [215] involves the attachment of peptides by their carboxyl groups to a suitable resin. Peptide amino groups are blocked using *tert*-butyloxycarbonyl azide and the carboxyl groups are activated with carbonyldiimidazole. The resulting peptide imidazole is coupled to an aminopolystyrene resin and the amino groups are deblocked with trifluoroacetic acid. The peptide is then sequentially degraded and the thiazolinones liberated at each step are collected, converted to PTH-amino acids and identified in the normal manner. A problem with this technique is that the side chain carboxyl groups of aspartic and glutamic acid also attach to the resin so that no thiazolinone will be released when the degradation reaches that residue.

In the alternative method developed by Stark [216], it is the Edman reagent, not the peptide, which is prepared in an insoluble form. The insoluble reagent, analogous to phenyl *iso*thiocyanate, is synthesized from polystyrene beads (cross-linked with 0.25% divinylbenzene) by nitration, reduction, and reaction with carbon disulphide and ethyl chloroformate. To enable the NCS-polystyrene to swell in polar solvents, it is essential to reduce its hydrophobicity by coupling 60% of the NCS groups to glucosaminol. During sequencing the peptide initially attaches to the resin and is eluted during cleavage. In this method the thizaolinone derivatives are lost at each stage, so the reaction must be followed either by the 'subtractive-Edman' or 'dansyl-Edman' procedures. Although the solid-phase methods are simple, rapid and relatively inexpensive they are only suitable for sequencing relatively short peptides.

'Edman–Begg sequenator'. A sophisticated, fully-automated protein sequenator was developed by Edman and Begg [217] and subsequently several commercial instruments based on their technique have become available. Although this technique gives excellent results, the requisite instrumentation is prohibitively expensive unless routine sequencing studies are contemplated. In this method the protein is spread as a thin film on the wall of a spinning, cylindrical cup. Reagents and solvents enter at the bottom of the reaction vessel and extracting solvents are scraped off at the top. The reaction vessel is thermostated and can be evacuated. The sequence of reactions involving thirty steps is controlled by a programming unit. The extracted thiazolinone derivatives are collected, converted to PTH-amino acids and analysed by any of the standard techniques. A full discussion of the technique can be found in the article by Niall [218]. The sequenator is unsuitable for short peptides due to their potential solubility in the extracting solvents, but it is the ideal method for sequencing large polypeptides and proteins, degradation of up to 150 residues being possible.

Sequencing material from polyacrylamide gels. Of particular interest in the field of membrane protein research are those procedures which have been described for the N-terminal sequence analysis of proteins separated on polyacrylamide gels. The reaction can be carried out in the gel itself [219] or the proteins can be eluted and analysed by a modified 'dansyl-Edman' procedure [221].

Studies on membrane proteins. The Edman procedure in its various forms has been used to good effect in the sequencing studies which have been performed on the basic proteins purified from myelin membranes. This protein induces the autoimmune disease, experimental allergic encephalomyelitis. The 27 tryptic and the 16 peptic peptides derived from bovine basic protein have been utilized to establish the complete sequence of 170 residues [221]. This was accomplished by the use of both 'direct' and 'subtractive' Edman procedures. The overall sequence reveals no obvious periodicity, but several peptide segments of nine residues occur in which basic residues are absent. It has been suggested that the somewhat nonpolar character of these regions may facilitate participation in nonpolar interactions. An unusual feature is the occurrence at residue 107 of methylated arginine, present either as the di- or mono-methyl derivatives. Located near this methylated arginine residue is the characteristic sequence -pro-arg-thr-pro-pro-pro-. This sequence may induce a sharp U-shaped bend in the molecule and thereby induce the molecule to fold back on itself. An open chain structure for the molecule has been proposed rather than a random coil. The analogous basic protein of human myelin has been investigated [222,223]. Peptides produced by trypsin, pepsin, thermolysin, chymotrypsin and cyanogen bromide were analysed by the 'dansyl-Edman' procedure to reveal the sequence of 170 residues. The sequence of this protein is very similar to that of the bovine basic protein, containing mono- and di-methyl arginine and the characteristic sequence -pro-arg-thr-pro-pro-pro- [222,224]. In rat brain myelin two basic proteins are present. The larger protein (L protein) is similar to the basic protein of other CNS myelin [225]. The other protein (S protein) is smaller by 4000 daltons [225], which by a combination of the 'dansyl-Edman' and 'subtractive-Edman' procedures has been demonstrated to be due to the presence in the S protein of a major deletion of 40 residues equivalent to residues 120–160 in the L protein [226].

The sequences of the two rat myelin basic proteins have been compared with the sequence data available for myelin basic proteins from other species: human, bovine, rabbit, guinea pig, monkey and pig. It was concluded [220] that a large proportion of the basic myelin sequence is invariant and that the following characteristic features are always found: (a) a threonine residue (residue 100) that can be glycosylated *in vitro*, (b) an unusual triproline sequence (residues 98–103) (-pro-arg-thr-pro-pro-pro-), (c) a serine residue (residue 112) that can be phosphorylated *in vitro*, (d) a methylated arginine (residue 109), and (e) a sole tryptophan residue (residue 118) that is essential for the induction of auto-immune encephalomyelitis.

The 'dansyl-Edman' technique has been used [227] to sequence a low molecular weight glycopeptide containing triglucosylcysteine which was claimed to have been extracted from erythrocyte membranes. It is similar in sequence to a glycopeptide found in normal human urine and its possible involvement in glucose transport across cell membranes was also suggested [227].

An Edman-Begg sequenator has been used to establish the N-terminal sequence of a protein of molecular weight 33 000 which was extracted from human erythrocyte ghosts [139]. The N-terminal sequence of this protein was found to be strikingly similar to that of pig muscle D-glyceraldehyde-3-phosphate dehydrogenase. The presence of this enzyme in the membrane was confirmed by enzyme assays.

That sequencing studies can provide useful information on the organization of proteins in membranes is illustrated by the sequencing studies performed on the major human erythrocyte membrane glycoprotein, termed 'glycophorin' [203]. The glycoprotein yields five peptides after cyanogen bromide cleavage. On the basis of amino acid and carbohydrate analysis of these fragments, as previously discussed in the section on amino acid analysis, it was concluded that [203] this molecule has three distinct regions: (a) the N-terminal segment containing all the carbohydrate and receptor sites, (b) a hydrophobic region, (c) and a hydrophilic, proline-rich C-terminal portion. The hydrophobic region, consisting of 51 residues, has been sequenced using the relevant cyanogen bromide peptide and an isoluble tryptic peptide to provide overlap data. The sequence was found to be extremely hydrophobic, being rich in leucine and isoleucine and possessing no charged residues between residues 11 and 35. These studies support the suggestion [46] that the major glycoprotein spans the erythrocyte membrane. The N- and C-terminal segments being exposed at the exterior and interior membrane faces respectively, while the central hydrophobic region is situated in the hydrophobic interior of the membrane.

## 7.4 Practical procedures relating to fractionation techniques*

### *7.4.1 Polyacrylamide gel electrophoresis – general comments*

Acrylamide can be recrystallized from chloroform and *bis*acrylamide from acetone [228]. Such purification reduces the UV absorption of the gels but does not necessarily improve the separation of proteins; sharper bands are occasionally observed in gels made from acrylamide that has not been recrystallized. As acrylamide is a cummulative poison personal contact with it must be avoided.

*Removal of gels from tubes*

Gels may be removed from tubes by inserting a fine hypodermic needle with a smoothly ground tip between the gel and the tube wall and continuously injecting water into the tube while using the needle to rim the gel out of the tube. The removal of very dense gels (>10% acrylamide) is facilitated by siliconizing the tubes before use. Gels of low acrylamide content (around 3.0%) can usually be simply blown out of their tubes with a rubber bulb.

*Fixation*

Many methods of fixation have been published and it is advisable to test a number of them before embarking on any given routine. Although high molecular weight proteins may readily be fixed by a mixutre of methanol-acetic acid-water [49], smaller proteins may not be adequately fixed in this way, especially if detergent is present in the gels, and overnight fixation in trichloroacetic acid (12%) or sulphosalicyclic acid (20%) is more reliable. Some authors recommend the addition of isopropanol (25% v/v) to the trichloroacetic acid [54], while others recommend a mixture of trichloroacetic and sulphosalicylic acids [229]. In our experience the most effective and convenient routine is to immerse the gels overnight in either 12% trichloroacetic acid or 20% sulphosalicylic acid, changing the solution once if detergent is present. (If Triton X-100 is present trichloroacetic acid must be avoided as it precipitates the detergent in the gel).

*Staining*

Gels may be stained individually in test tubes, or each gel may be placed in a test tube which has a slot cut in its bottom and a batch of gels stained simultaneously by transferring the tubes in a rack between baths of staining solutions.

*All solutions described as % refer to w/v unless otherwise indicated.

In addition to the general stains described below numerous colourimetric methods for specific enzymes are available.

Protein. Coomassie blue (Coomassie Brilliant Blue R) is now almost universally used as a general protein stain. In the absence of detergent gels may be simply and effectively stained by immersing for about 24 h in a 0.05% solution of Coomassie blue in 12% trichloroacetic acid. Gels which contain detergent may be similarly stained, after fixation and removal of detergent, by immersion in 0.05% Coomassie blue in methanol–acetic acid–water (10:7:83, v/v). Staining in these dilute dye solutions has the advantage that destaining is unnecessary. The alternative is to overstain by immersion for 2 h at 37°C in 2.5% Coomassie blue in methanol–acetic acid–water (5:1:5, v/v) and to destain by frequent changes of methanol–acetic acid–water (10:7:83). Electrolytic destaining in methanol–acetic acid–water is also feasible. Gels are best stored in 3.5% acetic acid.

Periodic acid – Schiff's stain for glycoprotein [54]. The gels are washed with 10% acetic acid to remove any detergent or fixative. The gels are treated for 2 h with 0.5% periodic acid, and then given two 30 min incubations in 0.5% sodium arsenite in 5.5% acetic acid. The gels are then washed three times (20 min each wash) in 5.0% acetic acid before being transferred to Schiff's reagent where they are left overnight. (Schiff's reagent: 1.5 g fuchsin is dissolved in 500 ml boiling water. The solution is filtered at 55°C, cooled to 40°C, 25 ml of 2 M HCl is added followed by 3.75 g sodium metabisulphite. After 16 h at 4°C 1.25 g of activated charcoal is added, the mixture shaken and filtered. The reagent is stored at 4°C).

During the course of the next day the gels are washed repeatedly with 0.1% sodium metabisulphite in 0.01 M hydrochloric acid until the washings no longer turn pink on the addition of formaldehyde. The pink P.A.S. positive bands appear rapidly in the Schiff's reagent and the subsequent treatment removes unreacted reagent and retards fading. Nevertheless the staining is not permanent and if a permanent record is required a densitometric scan or photograph should be taken as soon as possible.

Quantitation. Some quantitative estimate of the relative amounts of each component of a mixture may be obtained by scanning a stained gel at the absorption maximum of the dye, but such data should be interpreted with caution. In some instances parallel radioactive counting of suitably labelled proteins has validated the simpler colourimetric method, but it should always be remembered that: (a) dye binding to protein may not be stoichiometric and (b) Beer–Lambert's relationship only applies over a limited concentration range which can be exceeded in a sharp, densely stained protein band. The more reliable method of radioactive labelling is more laborious, especially if it involves slicing a gel and counting each slice.

### 7.4.2 *Polyacrylamide gel electrophoresis – buffer systems*

*(i) Tris-glycine buffer system* [49]

The running buffer consists of 5.0 mM Tris(Tris(hydroxymethyl) aminomethane) + 5.0 mM glycine + 0.5 mM EDTA adjusted to pH 8.0. Gels are prepared from 12.0 ml of a solution of 8% acrylamide + 0.2% *bis*acrylamide which is degassed under vacuum. To polymerize the gels 0.12 ml of 10% ammonium persulphate and 0.12 ml of 10% (v/v) TEMED are added, and the mixture rapidly pipetted into 8 tubes which are filled to within 20 mm of the top. Before polymerization can occur the acrylamide is carefully overlaid with a small volume of the running buffer. (The tubes (i.d. 40 mm x 100 mm) are previously sealed at one end with a rubber bung and arranged vertically in a suitable rack). The gels are ready for use after 2 h.

The protein sample is first dialysed against the running buffer and 5% sucrose or 10% glycerol added before electrophoresis to increase the density of the solution and so facilitate its application to the top of the gels. The tops of the gels are washed with a little of the running buffer, the bottom seal removed, and the tubes placed in the electrophoresis apparatus. The two reservoirs are filled with the running buffer and the gels prerun for 2 h at 20V $cm^{-1}$. The buffer in the two reservoirs is mixed, replaced, and the samples applied to the tops of the gels. The samples (not more than 50 $\mu$l each) are applied with a micropipette, or, if accuracy is not essential, using a 1.0 ml hypodermic syringe with a needle which has had its end ground flat. The proteins are then electrophoresed for 45 min at 40V $cm^{-1}$.

*(ii) An acidic buffer* [29]

Dilute acetic acid (1.5%) makes a most effective mild buffer of low pH (pH 2.8). It is used as the running buffer and the gels are prepared in it. 12 ml of 7% acrylamide + 0.1% *bis*acrylamide in the acetic acid solution are polymerized by the addition of 0.12 ml of 10% ammonium persulphate and 0.12 ml of 10% (v/v) TEMED and by incubation at 50°C for 1 h – elevated temperature is necessary for polymerization to occur at low pH. The sample is prepared in 1.5% acetic acid. The gels are prerun for 2 h at 40V $cm^{-1}$, the sample applied, and electrophoresed for 20 min at 40V $cm^{-1}$.

*(iii) Phenol–acetic acid–water* – as modified by Ray and Marinetti [27]

The membranes are dissolved in a mixture consisting of 20% acetic acid, 24% urea, 40% phenol and 5% 2-mercaptoethanol. Gels are prepared by mixing 3 vol. of stock solution A, which consists of 6% acrylamide, 0.23% *bis*acrylamide, 28% urea, 37.3% acetic acid dissolved in water, to 1 vol. of stock solution B, which consists of 1.6% ammonium persulphate in 60% aqueous urea, and then adding

0.02 vol. TEMED to the mixture. Polymerization is completed in 60 min at room temperature. The running buffer is 10% acetic acid. The sample is applied and electrophoresed for 1 h at 1.5 mA per tube followed by 4.5 h at 4.0 mA per tube.

*(iv) Electrophoresis with sodium dodecyl sulphate (SDS)*

The protein solution at a concentration of about 0.5–5.0 mg ml$^{-1}$ is dialysed against the running buffer containing 0.5 mM EDTA. (The optimal protein concentration should be determined separately for each mixture and depends to a great extent on the complexity of the mixture). Solid dithioerythritol is added to a concentration of 20 mM and after 30 min at room temperature solid SDS is added to 5%. (Any potassium ions must be removed as the detergent forms an insoluble potassium salt. The recently available lithium salt (Searle) is more soluble than the sodium salt.) When the protein solution has cleared it is heated for 3 min at 100°C, cooled, and solid urea added to 8 M (760 mg to each ml of urea-free solution).

Two convenient continuous buffer systems are described, but discontinuous systems are also available [230].

*A phosphate buffer* [48,49,53]

Running buffer – 0.1 M phosphate buffer, pH 7.2 containing 0.1% SDS. Gel preparation – a 5% acrylamide, 0.135% *bis*acrylamide solution is prepared in the running buffer. 12 ml of this solution is degassed, 0.12 ml of 10% ammonium persulphate followed by 0.12 ml of 10% (v/v) TEMED added, the mixture stirred and pipetted into glass tubes as described previously. The acrylamide is carefully overlayered with a little of the running buffer before polymerization occurs. After at least 2 h at room temperature the tops of the gels are washed with running buffer, the bottom seal removed, and the tubes placed in the electrophoresis chamber. Running conditions – The gels are prerun at 10 mA per gel for 1.5 h, the sample applied to the tops of the gels and the proteins electrophoresed for 2 h at 10 mA per tube. (for 8 gels of the size described a potential of about 70V is generated.)

The 5% gel is suitable for the separation of proteins down to about 40 000 daltons; for smaller proteins a 10% gel is preferable.

*A Tris buffer* [54]

Running buffer – a solution containing 40 mM Tris (free base), 20 mM sodium acetate, 2.0 mM EDTA is adjusted to pH 7.4 with acetic acid and SDS added to 1.0%. Gel preparation – a 5.6% acrylamide and 0.15% *bis*acrylamide solution is prepared in the running buffer. The gels are polymerized by adding 0.1 ml of 15% ammonium persulphate and 0.1 ml of 10% (v/v) TEMED to 10 ml of the degassed acrylamide solution. Subsequently the same procedure is adopted as is described above for the phosphate buffer.

### *7.4.3 Isoelectric focusing in polyacrylamide gels*

#### *(i) General comments*

As discussed in Section 7.2.2 gels of large pore size must be used to minimize molecular sieving of proteins. Unfortunately, these gels have a poor adhesion to glass and are therefore prone to slip out of any vertical assembly. Adhesion is improved by cleaning the glass carefully or alternatively, and perhaps more reliably, the focusing gel can be supported by a denser gel (5% acrylamide – 0.1% *bis*-acrylamide) about 10 mm long cast in the bottom of the gel tube. The acrylamide for this gel is made up in the bottom reservoir buffer.

Unless solutions are carefully deaerated, carbon dioxide bubbles form in the gel during the course of a run. This is avoided by carefully deaerating, under vacuum, all solutions used in the experiment, including the protein sample, and overlaying the upper buffer reservoir (alkaline reservoir) with a thin layer of Stoddard's reagent (an inert liquid paraffin fraction).

#### *Sample application*

The protein may be incorporated into the gel by mixing the protein solution into the acrylamide solution before polymerization. Alternatively, and more conveniently, the sample can be applied to the prepared gel either before or after the pH gradient has been established. The density of the sample is increased by addition of urea up to the concentration of urea in the gel and then the sample applied to the top of the gel as described for gel electrophoresis. To protect the sample from the high pH in the reservoir a thin (1 mm) layer of urea containing Ampholine is first placed on the top of the gel. This solution has sufficient urea to give it a density between that of the sample and the upper reservoir solution.

#### *(ii) Method of Merz, Good and Uitman* [61]

#### *Preparation of gels*

A solution containing 10% acrylamide, 0.4% *bis*acrylamide, 25% sucrose is prepared in 8 M urea (either Analar grade, purified by stirring for 2 h with charcoal and Amberlite MB3 mixed bed resin to remove cyanate and other ions, or Aristar grade). To 2.5 ml of this solution are added 0.25 ml Ampholine (pH 3 – 10 range, 40%) 1.25 ml 50% sucrose in 8 M urea, 4.75 ml 8 M urea, 0.1 ml of 10% TEMED. This mixture is carefully degassed before addition of 1.25 ml 0.5 ml 0.5% ammonium persulphate in 8 M urea. The mixture is quickly, but with minimum agitation, pipetted into glass tubes (i.d.

4.0 mm x 120 mm) previously sealed at one end with a 10 mm long gel plug made from 5% acrylamide, 0.1% *bis*acrylamide dissolved in the acidic lower reservoir solution (polymerized as described for the acetic acid gels of Section 7.4.2). The focusing gel is overlaid with water before polymerization occurs. 70 mm is a suitable length for this gel.

*Preparation of the membranes*

The membranes in a low ionic strength medium are incubated at 37°C for 12 h in 8 M urea, 0.2% β-mercaptoethanol and 20 mM EDTA.

*Running conditions*

The top (cathode) buffer reservoir contains 0.4% ethanolamine and the bottom reservoir (anode) contains 0.2% sulphuric acid. The ethanolamine is carefully degassed before use and during the run is overlaid with a layer of Stoddard's reagent. The gels are prerun for 30 min at 0.5 mA per tube, 20 μl of 2% Ampholine in 4 M urea placed on top of each gel and a 10–100 μl sample of protein applied under this layer. Initially the proteins are focused at a constant current of 0.5 mA per gel until a potential of 200 V is reached (for a 70 mm gel), and the process is completed at this constant voltage. The whole procedure takes about 24 h. Beyond 36 h there is a danger of the gradient breaking down.

*(iii) Method of Chignell as developed for mitochondrial membranes**

*Preparation of gel*

To 1.0 ml of a solution of acrylamide (40%) and *bis*acrylamide (3.0%) in water are added (a) 650 μl of 40% Amphloline, pH 3.5–10.0 range, (b) 75 μl 40% Ampholine, pH 6.0–8.0 range, (c) 75 μl 40% Ampholine, pH 7.0–10.0 range, (d) 8 ml of 8 M urea (purified as in the previous method) containing 2.5% Trition X-100, and (e) 10 μl TEMED. This solution is carefully degassed, cooled on ice, 200 μl of 4.0% ammonium persulphate added and the mixture then rapidly pipetted into tubes and overlaid with water before polymerization has occured. The tubes are kept for 30 min at room temperature and 2 hr at 4°C before use.

*Preparation of sample*

The salt free proteins or whole membranes are dissolved in 8 M urea containing 2.5% Triton X-100 and made 10 mM with dithioerythritol. Ampholine,

*We are indebted to Dr D. Chignell of the University of Dundee for details of his method prior to their publication.

pH 3.0–10.0 range, is added to a final concentration of 2% with the precaution that the resulting pH of the solution is well above the isoleletric point of the proteins.

*Running conditions*

These are as in preceding method with the exceptions that the ethanolamine is replaced by 0.02 M sodium hydroxide and the sulphuric acid by 5.6 M urea in 0.5 M phosphoric acid.

### *7.4.4 Column chromatography*

*(i) Cyanogen bromide activation of agarose* [231]

This method is preferable to those previously reported as it does not require automatic or manual titration to maintain the requisite alkaline pH for maximum activation.

One volume of a slurry of washed agarose (usually a 4% gel) consisting of equal volumes of gel and deionized water is added to 1 volume of 2 M sodium carbonate and mixed by stirring slowly. The rate of stirring is increased and 0.5 volumes of an acetonitrile solution of CNBr(2 g per 1 ml acetonitrile) is rapidly added. The slurry is stirred vigorously for 1–2 min at room temperature, after which the slurry is poured onto a coarse sintered-glass funnel, washed with 5–10 volumes each of the following: (a) 0.1 M sodium bicarbonate (pH 9.5), (b) deionized water, and (c) the buffer which is to be used in the subsequent coupling reaction.

A stock solution of CNBr can be prepared by adding 50 ml of dry, redistilled acetonitrile to a 100 g bottle of cyanogen bromide and stored at –20°C when not in use. The amount of CNBr solution added to the reaction mixture can be varied to achieve the desired degree of activation.

*(ii) Preparation of aminoalkyl-agarose columns for 'hydrophobic chromatography'* [101,232]

Sepharose 4B is activated with CNBr (as above). The appropriate α-ω-diaminoalkane ($NH_2(CH_2)_nNH_2$) is dissolved (4 moles per mole of CNBr used for the activation of the Sepharose) in equal volumes of N,N-dimethylformamide and 0.1 M sodium bicarbonate (pH 9.5). The pH of the solution is checked and readjusted, if necessary, to pH 9.5. If an aminoalkane of more than an 8 carbon chain length is used a precipitate may form on standing and should this occur the mixture must be heated to 60°C and cooled rapidly with ice water before mixing with activated Sepharose.

Coupling is performed by mixing 4 volumes of the diamine solution with 3 volumes of the suspension of activated agarose in 0.1 M sodium bicarbonate (twice the settled volume of the gel) (pH 9.5). The pH is checked and if necessary readjusted to pH 9.5. The reaction is allowed to proceed for 10 h at 20–25°C. The agarose is then filtered and washed with 5 times its volume of each of the following; (a) deionized water, (b) 0.2 M acetic acid, (c) deionized water, (d) 0.05 M NaOH, (e) deionized water, (f) dioxane–water (1:1, v/v), (g) 0.2 M acetic acid, and finally with (h) 20 volumes deionized water. The column is then washed and equilibrated with the buffer selected for chromatography.

## 7.5 Practical procedures relating to chemical analysis

### *7.5.1 Protein assay by the Lowry* [118] *method modified by the addition of deoxycholate* [119]

#### *(i) Reagents*

Copper–alkali solution – 100 ml of 0.1 N sodium hydroxide and 100 ml of 4% (w/v) sodium carbonate (anhydrous) are mixed and 2.0 ml of a 2% (w/v) sodium potassium tartrate solution are added *followed by* 2.0 ml of a 1% (w/v) copper sulphate ($5H_2O$) solution. The mixture is prepared at least 30 min before use and made freshly each day.

Folin–Ciocalteau reagent – the commercial reagent is diluted with water until it is 1.0 M with respect to acid.

The protein sample – the protein solution or membrane suspension is made 0.25 M with respect to sodium hydroxide and 2% (w/v) with respect to sodium deoxycholate.

#### *(ii) Procedure*

2.5 ml of the copper alkali reagent are added to 0.5 ml of the protein sample containing about 0.1 mg ml$^{-1}$ (see Fig. 7.3), mixed, and stood at room temperature for 10 min. 0.25 ml of the diluted Folin–Ciocalteau reagent are added as rapidly as possible and mixed immediately. Colour is allowed to develop in the dark for 90 min and read against a reagent blank at 740 nm.

### *7.5.2 N-terminus methods*

#### *(i) Dinitrophenylation of proteins in acrylamide gels* [128]

The gels are electrophoresed normally and then placed in a 2.5% (w/v) solution of DNFB in ethanol. After reaction in the dark for 16 h the gels become opaque, shrink and no bands are visible. The gels are washed several times with ether and transferred containing 5% (v/v) 2 M HCl in ether. The flasks are shaken until no further colour is eluted from the gels. At this stage the gels attain their original length and the protein bands appear as dark yellow regions against a pale yellow background. Individual bands can be excised from the gels, but it should be noted that at least 20–30 bands from replicate gels had to be combined in the original report of the method to permit a satisfactory N-terminal analysis to be performed. The combined gel slices are hydrolysed, the hydrolysate filtered and the DNP-amino acids identified by any of the normal thin layer chromatographic procedures [126,127].

*(ii) Dansylation of proteins* [131]

Although the original method recommends the use of performic acid oxidized protein, we find that this can lead to irreversible aggregation of membrane proteins. It is better to block sulphydryl groups by reduction and carboxymethylation [158]. Then the protein (50–250 μg) is dissolved in 50 μl 1% (w/v) sodium dodecyl sulphate at 100°C for 2–5 min. After cooling, 50 μl N-ethylmorpholine are added and mixed. To this mixture is added 75 μl DNS–Cl (25 mg $ml^{-1}$) in anhydrous dimethylformamide. The reaction is allowed to proceed for at least 1 h in the dark at room temperature. At the completion of the reaction the protein is precipitated with 0.5 ml acetone, washed with 0.5 ml 80% (v/v) acetone, dried and hydrolysed with 0.5 ml 6 M HCl for 18 h at 105°C. After hydrolysis the HCl is removed *in vacuo* over NaOH pellets. The hydrolysate can be redissolved in 50% (v/v) aqueous pyridine, spotted on to filter paper and dried. The DNS-amino acids are extracted with water-saturated ethyl acetate leaving behind residual DNS–OH and salt. The extract is dried and redissolved in 50% (v/v) aqueous pyridine for identification of the DNS-amino acids. This extraction procedure can, however, result in the loss of some DNS-his, DNS-arg, DNS-$CySO_3H$, *o*-DNS-tyr, and α-or ε-DNS-lys.

*(iii) The separation of DNS-amino acids on 'polyamide' layers*

'Polyamide' layers are durable ε-poly-caprolactam layers bonded to solvent resistant polyester sheets. Commercially available 'polyamide' layers are expensive, but the plates can be used repeatedly if they are suitably washed soon after running the chromatograms [233]. A suitable solvent pair for the separation of DNS-amino acids on 'polyamide' layers is water–formic acid (200:3, v/v) in the first dimension, followed by benzene–acetic acid (9:1, v/v) or n-heptane–n-butanol–acetic acid (3:3:1, v/v) in the second dimension [137]. As some DNS-amino acids are not well resolved, additional runs in the second dimension are sometimes necessary. Re-running the chromatogram in the second dimension using ethylacetate–methanol–acetic acid (20:1:1, v/v) resolves DNS-thr, DNS-ser, DNS-glu, and DNS-asp [135], while a further run in the same dimension with 1 M ammonium hydroxide–ethanol (1:1, v/v) or 0.05 M trisodium phosphate–ethanol (3:1, v/v) [234] resolves α-DNS-his, DNS-arg and ϵ-DNS-lys.

### *7.5.3 C-terminus analysis by tritium labelling* [149]

To 2 mg of protein in 0.1 ml tritiated water are added 0.2 ml pyridine and the solution shaken vigorously. Acetic anydride, 0.05 ml, is added and the mixture incubated for 2–3 h at room temperature. Residual tritiated water is removed by distillation, the residue subjected to performic acid oxidation [132] and then hydrolysed in 5.7 M HCl for 24 h. The C-terminal tritiated amino acids are separated and identified by two dimensional chromatography and electrophoresis.

### *7.5.4 Fingerprinting*

#### *(i) Cadmium-ninhydrin reagent* [162]

This reagent is claimed to be more sensitive, to give a more stable colour, and to provide a lower background staining than the standard ninhydrin procedure.

Stock solution – 2.0 g cadmium acetate, 40 ml acetic acid, 200 ml distilled water.

24 ml of stock solution are added to 200 ml acetone, followed by 2.0 g ninhydrin. The chromatograms are dipped in this reagent and placed in a sealed enclosure containing a small vessel of concentrated sulphuric acid for 18 h at 25°C.

#### *(ii) Peptide mapping using DNS–Cl*

A normal trypsin digest is dried *in vacuo*, redissolved in 0.1 M $NaHCO_3$, and dansylated with a ten-fold (w/w) excess of DNS–Cl in acetone [235]. The large excess of DNS–OH can be removed by application of the dansylated digest to a Dowex 50-X8 column (2.0 mm x 20 mm) equilibrated with 0.01 M acetic acid [171]. DNS–OH passes through the column in this medium while the peptides can be displaced with 1 M $NH_3$ in 25% (v/v) aqueous acetone. Samples of the digest are applied to silica gel thin layer plates. Recommended solvents are methyl acetate–*iso*propanol–$NH_3$ (9:7:4, v/v) [170] in the first dimension followed by butan-2-ol–formic acid–water (15:3:2, v/v) [170], benzene–pyridine–acetic acid (16:4:1, v/v) [171], or chloroform–95% ethanol– acetic acid (38:4:3, v/v) [235] in the second dimension. An increase in sensitivity can be obtained by exposing the plates to dioxan vapour prior to examination [235]. Spraying the chromatograms with triethanolamine–*iso*propanol (1:4, v/v) is also claimed to enhance and stabilize the fluorescence [130].

## References


1. Gordon, A. H. (1969), *Laboratory Techniques in Biochemistry and Molecular Biology*, (Work, T. S. and E., eds.), Vol. I, pp. 1, North-Holland, Amsterdam.
2. Maurer, H. R. (1971), *Disc Electrophoresis*, de Gruyter, New York and Berlin.
3. Andrews, P. (1973), *Meth. biochem. Analysis,* **18**, 1.
4. Reiland, J. (1971), *Meth. Enzym.*, **22**, 287.
5. Fischer, L. (1969), *Laboratory Techniques in Biochemistry and Molecular Biology*, (Work, T. S. and E., eds.), Vol. I, pp. 151, North-Holland, Amsterdam.
6. *Ann. N.Y. Acad. Sci.* (1973), **209**.
7. Righetti, P. G. and Drysdale, J. W. (1974), *J. Chromat.*, **98**, 271.
8. Cuatrecasas, P. (1972), *Adv. Enzymol.*, **36**, 29.
9. Cuatrecasas, P. and Anfinsen, C. B. (1971), *Meth. Enzym.*, **22**, 345.
10. Azen, E. A., Orr, S. and Smithies, O. (1965), *J. Lab. clin. Med.*, **65**, 440.
11. Azen, E. A., Nazhat, R. A. and Smithies, O. (1966), *J. Lab. clin. Med.*, **67**, 650.
12. Choules, G. L. and Zimm, B. H. (1965), *Analyt. Biochem.*, **13**, 336.
13. Hjerten, S. (1971), *Biochim. biophys. Acta.*, **237**, 395.
14. Johansson, K. E. (1974), in *Protides of the Biological Fluids* (Peeters, H., ed.), Vol. 21, pp. 151, Pergamon, Oxford.
15. Waehneldt, T. V. (1971), *Analyt. Biochem.*, **43**, 306.
16. Baldwin, R. W., Harris, J. R. and Price, M. R. (1973), *Int. J. Cancer,* **11**, 385.
17. Dunn, M. J., McBay, W. and Maddy, A. H. (1975), *Biochim. biophys. Acta.*, **386**, 107.
18. De Wachter, R. and Fiers, W. (1971), *Meth. Enzym.*, **21**, 167.
19. Kaltschmidt, E. and Wittmann, H. G. (1969), *Analyt. Biochem.*, **30**, 132.
20. Bhakdi, S., Knufermann, H., Fischer, H. and Wallach, D. F. H. (1974), *Biochim. biophys. Acta.*, **373**, 295.
21. Liljas, L., Lundahl, P. and Hjerten, S. (1974), *Biochem. biophys. Acta.*, **352**, 327.
22. Bhakdi, S., Knufermann, H. and Wallach, D. F. H. (1974), *Biochim. biophys. Acta.*, **345**, 448.
23. Ferguson, K. A. (1964), *Metabolism*, **13**, 985.
24. Rodbard, D. and Chrambach, A. (1970), *Proc. natn. Acad. Sci. U.S.A.*, **65**, 970.
25. Margolis, J. and Kendrik, K. G. (1968), *Analyt. Biochem.*, **25**, 347.
26. Takayama, K., MacLennan, D. H., Tzagaloff, A. and Stoner, C. D. (1966), *Archs. Biochem. Biophys.*, **114**, 223.
27. Ray, T. K. and Marinetti, G. V. (1971), *Biochim. biophys. Acta.*, **233**, 787.
28. Maddy, A. H., Dunn, M. J. and Kelly, P. G. (1972), *Biochim. biophys. Acta.*, **288**, 263.
29. Maddy, A. H. and Kelly, P. G. (1971), *Biochim. biophys. Acta.*, **241**, 290.
30. Zwaal, R. F. A. and van Deenen, L. L. M. (1968), *Biochim. biophys. Acta.*, **163**, 44.
31. Neville, D. M. (1967), *Biochim. biophys. Acta.*, **133**, 168.
32. Salton, M. R. J. and Netschey, A. (1965), *Biochim. biophys. Acta.*, **107**, 539.
33. Panet, R. and Selinger, Z. (1970), *Eur. J. Biochem.*, **14**, 440.
34. Demus, H. and Mehl, E. (1970), *Biochim. Biophys. Acta.*, **203**, 291.

35. Reynolds, J. A. and Tanford, C. (1970), *J. biol. Chem.*, **245**, 5161.
36. Williams, J. G. and Gratzer, W. B. (1971), *J. Chromatog.*, **57**, 121.
37. Reynolds, J. A. and Tanford, C. (1970), *Proc. natn. Acad. Sci. U.S.A.*, **66**, 1002.
38. Grefarth, S. P. and Reynolds, J. A. (1974), *Fedn. Proc. Fedn. Am. Socs. exp. Biol.*, **33**, 1531.
39. Grefarth, S. P. and Reynolds, J. A. (1974), *Proc. natn. Acad. Sci. U.S.A.*, **71**, 3913.
40. Simons, K. and Kaariainen, L. (1970), *Biochem. biophys. Res. Commun.*, **38**, 981.
41. Nelson, C. A. (1971), *J. biol. Chem.*, **246**, 3895.
42. Satake, I. and Yang, T. T. (1973), *Biochem. biophys. Res. Commun.*, **54**, 930.
43. Fasman, G. D., Lindblow, L. and Bodenheimer, E. (1964), *Biochem.*, **3**, 155.
44. Igou, D. K., Lo, J. T. and Clark, D. S. (1974), *Biochem. biophys. Res. Commun.*, **60**, 140.
45. Murai, N., Makino, S. and Sugai, S. (1972), *J. Colloid Interface Sci.*, **41**, 399.
46. Bretscher, M. S. (1971), *Nature New Biology*, **231**, 229.
47. Moss, B. and Rosenblum, E. N. (1972), *J. biol. Chem.*, **247**, 5194.
48. Shapiro, A. L., Vinuela, E. and Maizel, J. V. (1967), *Biochem. biophys. Res. Commun.*, **28**, 815.
49. Weber, K. and Osborn, M. (1969), *J. biol. Chem.*, **244**, 4406.
50. Weber, K., Pringle, J. R. and Osborn, M. (1973), *Meth. Enzym.*, **26**, 3.
51. Tung, J. S. and Knight, C. A. (1971), *Biochem. biophys. Res. Commun.*, **42**, 1117.
52. Katzman, R. L. (1972), *Biochim. biophys. Acta.*, **266**, 269.
53. Lenard, J. (1970), *Biochemistry*, **9**, 1129.
54. Fairbanks, G., Steck, T. L. and Wallach, D. F. H. (1971), *Biochemistry*, **10**, 2606.
55. Bender, W. W., Garan, H. and Berg, H. C. (1971), *J. molec. Biol.*, **58**, 783.
56. Bretscher, M. S. (1971), *J. molec. Biol.*, **59**, 351.
57. Tanner, M. J. A. and Boxer, D. H. (1972), *Biochem. J.*, **129**, 333.
58. Triplett, R. B., Wingate, J. M. and Carraway, K. L. (1972), *Biochem. biophys. Res. Commun.*, **49**, 1014.
59. Huang, H. V., Molday, R. S. and Dreyer, W. J. (1973), *FEBS Letters*, **37**, 285.
60. Edelframi, M. E. and Edelframi, A. R. (1972), *Proc. natn. Acad. Sci. U.S.A.*, **69**, 1176.
61. Merz, D. C., Good, R. A. and Litman, G. W. (1973), *Biochem. biophys. Res. Commun.*, **49**, 84.
62. Finlayson, G. R. and Chrambach, A. (1971), *Analyt. Biochem.*, **40**, 292.
63. Chrambach, A., Doerr, P., Finlayson, G. R., Miles, L. E. M., Sherins, R. and Rodbard, D. (1973), *Ann. N.Y. Acad. Sci.*, **209**, 44.
64. Bours, J. (1973), *Sci. Tools*, **20**, 29.
65. Friesen, A. D., Jamieson, J. C. and Ashton, F. E. (1971), *Analyt. Biochem.*, **41**, 149.
66. Radola, B. J. (1973), *Biochim. biophys. Acta.*, **200**, 404.
67. Moore, G. L., Kocholaty, W. F., Cooper, D. A., Gray, J. L. and Robinson, S. L. (1970), *Biochim. biophys. Acta.*, **212**, 126.
68. Fischer, S., Lingsch, C. and Urbina, C. (1973), *J. Memb. Biol.*, **12**, 317.

69. Ne'eman, Z., Kahane, I., Kovartovsky, J. and Razin, S. (1973), *Biochim. biophys. Acta.*, **266**, 255.
70. Maddy, A. H. (1966), *Biochim. biophys. Acta.*, **117**, 193.
71. Moldow, C. F., Zucker-Franklin, D., Gordon, A., Hospelhorn, V. and Silber, R. (1972), *Biochim. biophys. Acta.*, **255**, 133.
72. Harris, J. R. (1971), *Biochim. biophys. Acta.*, **229**, 761.
73. Hamaguchi, H. and Cleve, H. (1971), *Biochim. biophys. Acta.*, **233**, 320.
74. Clarke, M. (1971), *Biochem. biophys. Res. Commun.*, **45**, 1063.
75. Marchesi, S. L., Steers, E., Marchesi, V. T. and Tillack, T. W. (1970), *Biochemistry,* **9**, 50.
76. Dunn, M. J. and Maddy, A. H. (1973), *FEBS Letters,* **36**, 79.
77. Maddy, A. H. and Dunn, M. J. (1973), in *Protides of the Biological Fluids,* (Peeters, H., ed.), Vol. 21, pp. 21, Pergamon, Oxford.
78. Fuller, G. M., Boughter, J. M. and Morazzani, M. (1974), *Biochem.*, **13**, 3036.
79. Maddy, A. H. (1972), *Sub-cell. Biochem.*, **1**, 293.
80. Blumenfeld, O. O., Gallop, P. M., Howe, C. and Lee, L. T. (1970), *Biochim. biophys. Acta.*, **211**, 109.
81. Janson, V. K. and Burger, M. M. (1973), *Biochim. biophys. Acta.*, **291**, 127.
82. Gurd, J. W., Evans, W. H. and Perkins, H. R. (1972), *Biochem. J.*, **126**, 459.
83. Shafai, T. and Cortner, J. A. (1971), *Biochim. biophys. Acta.*, **236**, 612.
84. Bernacki, R. J. and Bosmann, H. B. (1972), *J. Memb. Biol.*, **7**, 1.
85. Hagopian, A. and Eylar, E. H. (1969), *Archs. Biochem. Biophys.*, **129**, 515.
86. Caccam, J. F. and Eylar, E. H. (1970), *Archs. Biochem. Biophys.*, **137**, 315.
87. Shrager, P., Tosteson, D. C. and Lauf, P. K. (1972), *Biochim. biophys. Acta.*, **290**, 186.
88. Philippot, J. (1971), *Biochim. biophys. Acta.*, **225**, 201.
89. Fish, W. W., Reynolds, J. A. and Tanford, C. (1970), *J. biol. Chem.*, **245**, 5166.
90. Bragg, P. D. and Hou, C. (1971), *FEBS Letters,* **15**, 142.
91. Estrugo, S. F., Larraga, V., Corrales, M. A., Duch, C. and Munoz, E. (1972), *Biochim. biophys. Acta.*, **255**, 960.
92. Tanford, C. (1968), *Adv. Protein Chem.*, **23**, 121.
93. Fish, W. W., Mann, K. G. and Tanford, C. (1969), *J. biol. Chem.*, **244**, 4989.
94. Gwynne, J. T. and Tanford, C. (1970), *J. biol. Chem.*, **245**, 3269.
95. Trayer, H. R., Nozaki, Y., Reynolds, J. A. and Tanford, C. (1971), *J. biol. Chem.*, **246**, 4485.
96. Maddy, A. H. and Kelly, P. G. (1971), *Biochim. biophys. Acta.*, **241**, 114.
97. Triplett, R. B., Summers, J., Ellis, D. E. and Carraway, K. L. (1972), *Biochim. biophys. Acta.*, **266**, 484.
98. Steck, T. L. (1972), *Biochim. biophys. Acta.*, **255**, 553.
99. Nystrom, E. and Sjovall, J. (1965), *Analyt. Biochem.*, **12**, 235.
100. Curtis, P. J. (1972), *Biochim. biophys. Acta.*, **255**, 833.
101. Shaltiel, S., Ames, G. F. and Noell, K. D. (1973), *Archs. Biochem. Biophys.*, **159**, 174.
102. Himmelhoch, S. R. (1971), *Meth. Enzym.*, **22**, 273.
103. Maddy, A. H. Dunn, M. J. and Kemp, R. B. (unpublished work).

104. Blumenfeld, O. O. (1968), *Biochem. biophys. Res. Commun.*, **30**, 200.
105. Tsai, J. H. J., Tsai, H. and Ehrenstein, G. von (1970), *FEBS Letters*, **10**, 276.
106. Kagawa, Y. (1974), in *Methods in Membrane Biology*, (Korn, E. D., ed.) Vol. I, pp. 201, Plenum Press, New York and London.
107. Weiss, H. and Bucher, T. (1970), *Eur. J. Biochem.*, **17**, 561.
108. Adair, W. L. and Kornfeld, S. (1974), *J. biol. Chem.*, **249**, 4696.
109. Allen, H. J., Ault, C., Winzler, R. J. and Danielli, J. F. (1974), *J. Cell Biol.*, **60**, 26.
110. Hayman, M. J. and Crumpton, M. J. (1972), *Biochem. biophys. Res. Commun.*, **47**, 923.
111. Cuatrecasas, P. (1970), *J. biol. Chem.*, **245**, 3059.
112. Porath, J., Axen, R. and Ernback, S. (1967), *Nature*, **215**, 1491.
113. Krug, F., Desbuquois, B. and Cuatrecasas, P. (1971), *Nature New Biology*, **234**, 268.
114. Cuatrecasas, P. (1972), *Proc. natn. Acad. Sci. U.S.A.*, **69**, 1277.
115. Cuatrecasas, P. (1972), *J. biol. Chem.*, **247**, 1980.
116. Snary, D., Goodfellow, P., Hayman, M. J., Bodmer, W. F. and Crumpton, M. J. (1974), *Nature*, **247**, 457.
117. Tarone, G., Prat, M. and Comoglio, P. M. (1973), *Biochim. biophys. Acta.*, **311**, 214.
118. Lowry, O. H., Rosebrough, N. J., Farr, A. L. and Randall, R. J. (1951), *J. biol. Chem.*, **193**, 265.
119. Maddy, A. H. and Spooner, R. L. (1970), *Vox Sang.*, **18**, 34.
120. Heller, J. (1968), *Biochem.*, **7**, 2906.
121. Schnebli, H. P., Vatter, A. E. and Abrams, A. (1970), *J. biol. Chem.*, **245**, 1122.
122. Birckbichler, P. J., Dowben, R. M., Matacic, S. and Loewy, A. G. (1973), *Biochim. biophys. Acta.*, **291**, 149.
123. Langdon, R. G. (1974), *Biochim. biophys. Acta.*, **342**, 229.
124. Edman, P. (1950), *Acta. chem. scand.*, **4**, 283.
125. Edman, P. (1953), *Acta. chem. scand.*, **7**, 700.
126. Rosmus, J. and Deyl, Z. (1972), *J. Chromat.*, **70**, 221.
127. Fraenkel-Conrat, J., Harris, J. and Levy, A. L. (1955), *Meth. biochem. Analysis*, **2**, 359.
128. Catsimpoolas, N. (1967), *Analyt. Biochem.*, **19**, 592.
129. Stark, G. R. (1972), *Meth. biochem. Analysis*, **25**, 103.
130. Seiler, N. (1970), *Meth. biochem. Analysis*, **18**, 259.
131. Gray, W. R. (1972), *Meth. Enzym.*, **25**, 121.
132. Hirs, C. H. W. (1967), *Meth. Enzym.*, **11**, 197.
133. Gray, W. R. and Hartley, B. S. (1963), *Biochem. J.*, **89**, 59P.
134. Boyer, J. and Talalay, P. (1966), *J. biol. Chem.*, **241**, 180.
135. Crowshaw, K., Jessup, S. J. and Ramwell, P. W. (1967), *Biochem. J.*, **103**, 79.
136. Gros, C. and Labouesse, B. (1969), *Eur. J. Biochem.*, **7**, 463.
137. Woods, K. R. and Wang, K. T. (1967), *Biochim. biophys. Acta.*, **133**, 369.
138. Fraker, P. J. and Kaplan, S. (1972), *J. biol. Chem.*, **247**, 2732.
139. Tanner, M. J. A. and Gray, W. R. (1971), *Biochem. J.*, **125**, 1109.
140. Carraway, K. L. and Shin, B. C. (1972), *J. biol. Chem.*, **247**, 2102.
141. Langdon, R. G. (1974), *Biochim. biophys. Acta.*, **342**, 213.
142. Moore, G. L., Antanoff, R. S. and Rau, D. C. (1972), *Vox Sang.*, **22**, 481.
143. Green, J. R., Dunn, M. J. and Maddy, A. H. (1975), *Biochim. biophys. Acta.*, **382**, 457.

144. Knufermann, H., Bhakdi, S., Schmidt-Ullrich, R. and Wallach, D. F. H. (1973), *Biochim. biophys. Acta.*, **330**, 356.
145. Bjerrum, O. J., Bhakdi, S., Knufermann, H. and Bøg-Hansen, T. C. (1974), *Biochim. biophys. Acta.*, **373**, 44.
146. Green, J. R., Dunn, M. J., Spooner, R. L. and Maddy, A. H. (1974), *Biochim. biophys. Acta.*, **373**, 51.
147. Ambler, R. P. (1972), *Meth. Enzym.*, **25**, 143.
148. Kowalsky, A. and Boyer, P. D. (1960), *J. biol. Chem.*, **235**, 604.
149. Matsuo, H., Fujimoto, Y. and Tatsuno, T. (1966), *Biochem. biophys. Res. Commun.*, **22**, 69.
150. Huang, J. W. and Kaplan, S. (1973), *Biochim. biophys. Acta.*, **307**, 332.
151. Schroeder, W. A. (1972), *Meth. Enzym.*, **25**, 138.
152. Spande, T. F., Witkop, B., Degani, Y. and Patchornik, A. (1970), *Adv. Protein Chem.*, **24**, 97.
153. Gross, E. (1967), *Meth. Enzym.*, **11**, 238.
154. Smyth, D. G. (1967), *Meth. Enzym.*, **11**, 214.
155. Erlanger, B. F. and Cohen, W. (1963), *Biochem.*, **85**, 348.
156. Schoellman, G. and Shaw, E. (1963), *Biochem.*, **2**, 252.
157. Bennett, J. C. (1967), *Meth. Enzym.*, **11**, 211.
158. Hirs, C. H. W. (1967), *Meth. Enzym.*, **11**, 199.
159. Katz, A. M., Dreyer, W. J. and Anfinsen, C. B. (1959), *J. biol. Chem.*, **234**, 2897.
160. Bennett, J. C. (1967), *Meth. Enzym.*, **11**, 330.
161. Michl, H. (1951), *Monatsch. Chem.*, **82**, 489.
162. Dreyer, W. J. and Bynum, E. (1967), *Meth. Enzym.*, **11**, 32.
163. Wieland, T. and Georgopoulos, D. (1964), *Biochem. Z.*, **340**, 476.
164. Burns, D. J. W. and Turner, N. A. (1967), *J. Chromat.*, **30**, 469.
165. Bray, D. and Brownlee, S. M. (1973), *Analyt. Biochem.*, **55**, 213.
166. Huang, J. W. and Kaplan, S. (1973), *Biochim. biophys. Acta.*, **307**, 301.
167. Fukui, Y., Nachbar, M. S. and Salton, M. R. J. (1971), *Biochim. biophys. Acta.*, **241**, 30.
168. Fukui, Y. and Salton, M. R. J. (1972), *Biochim. biophys. Acta.*, **288**, 65.
169. Tanner, M. J. A., Boxer, D. H., Cumming, J. and Verrier-Jones, J. (1974), *Biochem. J.*, **141**, 909.
170. Langner, J. (1966), *Hoppe-Seyler's Z. physiol. Chem.*, **347**, 275.
171. Schmer, G. and Kreil, G. (1967), *J. Chromat.*, **28**, 458.
172. Moore, S. and Stein, W. H. (1963), *Meth. Enzym.*, **6**, 819.
173. Light, A. and Smith, E. L. (1963), in *The Proteins*, (Neurath, H., ed.), Vol. I, pp. 1, Academic Press, New York and London.
174. Spatz, L. and Strittmatter, P. (1971), *Proc. natn. Acad. Sci. U.S.A.*, **68**, 1042.
175. Dickerson, R. E., Takano, T., Eisenberg, D., Kallai, O. B., Samson, L., Cooper, A. and Margoliash, E. (1971), *J. biol. Chem.*, **246**, 1511.
176. Jacobs, S. (1966), *Meth. biochem. Analysis.*, **14**, 177.
177. Hamilton, P. B. (1967), *Meth. Enzym.*, **11**, 15.
178. Gehrke, C. W., Kuo, K. and Zumwalt, R. W. (1971), *J. Chromat.*, **57**, 209.
179. Gehrke, C. W. and Leimer, K. (1971), *J. Chromat.*, **57**, 219.
180. Moss, C. W., Lambert, M. A. and Diaz, F. J. (1971), *J. Chromat.*, **60**, 134.
181. Hirs, C. H. W. (1967), *Meth. Enzym.*, **11**, 59.
182. Liu, T.-Y. and Chang, Y. H. (1971), *J. biol. Chem.*, **246**, 2842.
183. Steck, T. L. and Fox, C. F. (1972), in *Membrane Molecular Biology* (Fox, C. F. and Keith, A. D., eds.), p. 27, Sinauer Assoc. Inc. Stamford.
184. Guidotti, G. (1972), *A. Rev. Biochem.*, **41**, 731.

185. Nozaki, Y. and Tanford, C. (1971), *J. biol. Chem.*, **246**, 2211.
186. Spahr, P. F. (1962), *J. molec. Biol.*, **4**, 395.
187. Ketterer, B., Hicks, R. M., Christodoulides, L. and Beale, D. (1973), *Biochim. biophys. Acta.*, **311**, 180.
188. Eng, L. F., Chao, F.-C., Gerst, B., Pratt, D. and Tavaststjerna, M. G. (1968), *Biochem.*, **7**, 4455.
189. Gagnon, J. Finch, P. R., Wood, D. D. and Moscarello, M. A. (1971), *Biochem.*, **10**, 4756.
190. Nakao, A., Davis, W. J. and Einstein, E. R. (1966), *Biochim. biophys. Acta.*, **130**, 163.
191. Carnegie, P. R., Bencina, B. and Lamoureux, G. (1967), *Biochem. J.*, **105**, 559.
192. Tenenbaum, D. and Folch-Pi, J. (1966), *Biochim. biophys. Acta.*, **115**, 141.
193. Stoffyn, P. and Folch-Pi, J. (1971), *Biochem. biophys. Res. Commun.*, **44**, 157.
194. Livne, A. and Kuiper, P. J. C. (1973), *Biochim. biophys. Acta.*, **318**, 41.
195. Tillack, T. W., Marchesi, S. L., Marchesi, V. T. and Steers, E. (1970), *Biochim. biophys. Acta.*, **200**, 125.
196. Steers, E. and Marchesi, V. T. (1969), *J. gen. Physiol.*, **54**, 65s.
197. Johnson, P. and Perry, S. V. (1970), *Biochem. J.*, **119**, 293.
198. Elzinga, M. (1971), *Biochem.*, **10**, 224.
199. Huszar, G. and Elzinga, M. (1971), *Biochem.*, **10**, 229.
200. Hardy, M. F. and Perry, S. V. (1969), *Nature*, **223**, 300.
201. Huszar, G. and Elzinga, M. (1969), *Nature*, **223**, 834.
202. Marchesi, V. T., Tillack, T. W., Jackson, R. L., Segrest, J. P. and Scott, R. E. (1972), *Proc. natn. Acad. Sci. U.S.A.*, **69**, 1445.
203. Segrest, J. P., Kahane, I., Jackson, R. L. and Marchesi, V. T. (1973), *Archs. Biochem. Biophys.*, **155**, 167.
204. Segrest, J. P. and Feldmann, R. J. (1974), *J. molec. Biol.*, **87**, 853.
205. Stark, G. R. (1972), *Meth. Enzym.*, **25**, 369.
206. Ambler, R. P. (1972), *Meth. Enzym.*, **25**, 262.
207. Light, A. (1972), *Meth. Enzym.*, **25**, 253.
208. Callahan, P. X., McDonald, J. F. and Ellis, S. (1972), *Meth. Enzym.*, **25**, 282.
209. Pisano, J. J. and Bronzert, T. J. (1969), *J. biol. Chem.*, **244**, 5597.
210. Lequin, R. M. and Niall, H. D. (1972), *Biochem. biophys. Acta.*, **257**, 76.
211. Smithies, O., Gibson, D., Fanning, E. M., Goldfleisch, R. M., Gilman, J. G. and Ballantyne, D. L. (1971), *Biochem.*, **10**, 4912.
212. Schroeder, W. A. (1972), *Meth. Enzym.*, **25**, 298.
213. Konigsberg, W. (1972), *Meth. Enzym.*, **25**, 326.
214. Fambrough, D. M. and Bonner, J. (1966), *Biochemistry*, **5**, 2563.
215. Laursen, R. A. (1971), *Eur. J. Biochem.*, **20**, 89.
216. Stark, G. R. (1970), *Adv. Protein Chem.*, **24**, 261.
217. Edman, P. and Begg, G. (1967), *Eur. J. Biochem.*, **1**, 80.
218. Niall, H. D. (1973), *Meth. Enzym.*, **27**, 942.
219. Schyns, R. (1969), *J. Chromat.*, **44**, 207.
220. Weiner, A. M., Platt, T. and Weber, K. (1972), *J. biol. Chem.*, **247**, 3242.
221. Eylar, E. H., Brostoff, S., Hashim, G., Caccam, J. and Burnett, P. (1971), *J. biol. Chem.*, **246**, 5770.
222. Carnegie, P. R. (1971), *Biochem. J.*, **123**, 57.
223. Carnegie, P. R. (1971), *Nature*, **229**, 25.
224. Baldwin, G. S. and Carnegie, P. R. (1971), *Biochem. J.*, **123**, 69.

225. Martenson, R. E., Deibler, G. E. and Kies, M. W. (1970), *Biochem. biophys. Acta.*, **200**, 353.
226. Dunkley, P. R. and Carnegie, P. R. (1974), *Biochem. J.*, **141**, 243.
227. Weiss, J. B., Lote, C. J. and Bobinski, H. (1971), *Nature New Biology*, **234**, 25.
228. Loening, U. E. (1969), *Biochem. J.*, **113**, 131.
229. Vesterberg, O. (1971), *Biochim. biophys. Acta.*, **243**, 345.
230. Neville, D. M. (1971), *J. biol. Chem.*, **246**, 6328.
231. March, S. C., Parikh, I. and Cuatrecasas, P. (1974), *Analyt. Biochem.*, **60**, 149.
232. Shaltiel, S. and Er-El, Z. (1973), *Proc. natn. Acad. Sci. U.S.A.*, **70**, 778.
233. Wang, K.-T. and Wu, P.-H. (1968), *J. Chromat.*, **37**, 353.
234. Hartley, B. S. (1970), *Biochem. J.*, **119**, 805.
235. Atherton, R. S. and Thomson, A. R. (1969), *Biochem. J.*, **111**, 797.
236. Gerstl, B., Eng, L. F., Hayman, R. B., Tavaststjerna, M. G. and Bond, P. R. (1967), *J. Neurochem.*, **14**, 661.
237. Evans, W. H. and Gurd, J. W. (1972), *Biochem. J.*, **128**, 691.
238. Wallach, D. F. H. and Zahler, P. H. (1968), *Biochim. biophys. Acta.*, **150**, 186.
239. Undenfriend, S., Stein, S., Bohlen, P., Dairman, W., Leimgruber, W. and Weigele, M. (1972), *Science*, **178**, 871.
240. Benson, J. R. and Hare, P. E. (1975), *Proc. natn. Acad. Sci. U.S.A.*, **72**, 619.
241. Bonner, W. M. and Laskey, R. A. (1974), *Eur. J. Biochem.*, **46**, 83.
242. Allan, D., Auger, J. and Crumpton, M. J. (1972), *Nature New Biology*, **236**, 23.

# 8 Techniques for the analysis of membrane lipids

J. H. VEERKAMP and R. M. BROEKHUYSE



## 8.1 Introduction

Membrane lipids vary extensively both in type and their proportional distribution. The organism, the cell type and even the particular (intra) cellular membrane determine the specific differences in lipid components [1,2]. The lipid composition of an isolated membrane can be used as an indication of the purity or cross-contamination with other membranes. The analysis of lipids has expanded greatly in the last two decades, especially by the development of chromatographic, spectroscopic and isotope methods. Excellent books and reviews in this field have been published [3–10]. In consequence, this chapter is not intended to be a review of all the possible analytical methods, but is primarily to discuss the procedures, which are relatively new or appropriate for analysis of micro amounts of lipids, such as are present in small samples of purified membranes, and to add some critical notes. Detailed procedures are given at the end of this chapter.

Preparation of membranes may cause selective loss of lipids or may be accompanied by degradation due to the procedures and time necessary to achieve purity [11]. The presence of certain chemicals used for the isolation of membranes may also impede lipid analysis [12]. Therefore careful control of lipid composition during membrane preparation is necessary. The localization of lipids in intact membranes has been carried out with group-specific labels [13] and specific phospholipases [14,15]. This subject will not be treated in this chapter.

## 8.2 Extraction of lipids

### *8.2.1 Treatment of biomembranes for lipid extraction*

Membrane preparations isolated for lipid analyses should be protected against autoxidation and breakdown by proteolytic and lipolytic enzymes [11]. The usual precautions during isolation are working at low temperatures, appropriate ionic strength and pH, and the prevention of foaming and intensive contact with oxygen. The influences of oxygen and possible enzymic actions have to be investigated for each type of preparation in which damage could be expected. Although the components of membranes are much less vulnerable *in situ* than in the isolated state, their natural protection could be decreased by membrane distortion. Data obtained from the storage of tissues also urge great caution. In intact tissues freezing may damage the cell membranes in such a way that the lipids come into contact with enzymes from which they are normally protected. Even at very low temperatures, or on thawing and contact with organic solvents, lipolytic enzymes may hydrolyse lipids. In fact, all conditions should be considered which are usually of importance for the successful preparation of a particular type of membrane. For a limited number of membrane types, these conditions are described in literature [16,17].

Ideally, membranes should be extracted as soon as possible after isolation so that there is little opportunity for changes to occur in the lipid components. However, if storage overnight is necessary, the membranes should either be frozen as quickly and deeply as possible under protection of nitrogen and/or slightly reducing conditions or they should be stored at a low temperature (–20°C) and homogenized in a polar lipid solvent containing an antioxidant, preferably in a nitrogen atmosphere. This solvent can be used next day as a (first) extraction fluid.

### *8.2.2 Lipid extraction*

It is now generally accepted, that a mixture of a relatively apolar lipid solvent and a polar alcohol is suitable as a general purpose lipid extractant. Chloroform has been selected as a good lipid solvent, ethanol and especially methanol are used as the alcohol component. Chloroform–methanol mixtures are probably the best solvents for complete extraction of lipids at present, although they do not remove all phosphoinositides, gangliosides nor probably some other lipids. Chloroform has serious disadvantages. However, a good substitute has not been found, although possibly toluene may be a substitute in combination with ethanol [18]. Chloroform–methanol as lipid extractant has been investigated for many years and excellent reviews about its application are given [6,7]. The problem of contamination and artefact formation has been described by various authors [19]. Small amounts of alkali catalyse methanolysis of lipid esters, with or without the presence of water. It is therefore important to avoid extraction and storage of lipids in the presence of alcohols together with alkali (or acid), even in traces. Analysis of lipids should be done as soon as possible. Contact of solvents with plastics should be avoided and they should be of high purity (reagent grade, preferably freshly distilled).

The procedure described below (A) is based on the frequently used method of Folch *et al.* [20] and may be applied to small amounts of tissues or membrane preparations. Procedure B is suitable if large amounts of water are present and avoids the use of large volumes of solvents. It is based on the procedure of Bligh and Dyer [21]. To prevent oxidation, addition of 50–100 mg BHT (butylated hydroxy-toluene) to the extraction fluids and working under a nitrogen atmosphere are recommended. Absence of oxygen does not appear always necessary if BHT is present [22]. This substance disturbs cholesterol determinations, yields a peak at the position of myristic acid during gas chromatography of fatty acid methyl esters and makes reliable dry weight determinations impossible. For certain purposes it can be removed by thin-layer chromatography [23].

### 8.2.3 Specific extraction procedures

#### (i) Polyphosphoinositides

Those present in nerve membranes can be extracted with chloroform–methanol after addition of $CaCl_2$ at an optimal level [24], followed by extraction with acidified chloroform–methanol [25,26].

#### (ii) Lysophospholipids

These are particularly prone to loss in the Bligh and Dyer procedure (B) and to a lesser extent in the chloroform–methanol procedure (A). If special attention has to be paid to these lipids, extraction with butanol may be appropriate [27].

#### (iii) Gangliosides

Gangliosides can be quantitively recovered according to the recommendations of Suzuki [28], who used a chloroform–methanol procedure (see procedure A), but continued with an additional extraction with chloroform–methanol 1:2 (v/v) containing 1/20 vol of water. The pooled extracts were adjusted to the chloroform–methanol ratio of 2:1 (v/v) and washed with KCl-solution (see procedure A). Most of the gangliosides were found in the upper phase.

*(iv) Glycosphingolipids*

Glycosphingolipids can be recovered in a pure state by a combination of the extraction procedure A and the further purification according to Saito and Hakomori [29]. Total lipids are acetylated, and the glycolipids are obtained by chromatography and deacylation.

*(v) Membrane lipids and proteins*

Those from erythrocyte ghosts have been separated by extraction with butanol–water mixtures [30]. This method gives a good (but not quantitive) recovery of proteins in the water layer and of lipids in the butanol layer, which may be suitable for recombination studies. 2-chloroethanol has also been studied for this purpose [31]. Selective extraction methods for loosely and strongly bound lipids have been applied to erythrocyte membranes of various species [32].

*(vi) Amino acyl phospholipids*

Those present in bacterial membranes must be extracted with a acidic medium at the pH of harvest [33].

### *8.2.4 Purification of lipid extracts*

Removal of non-lipid contaminants by solvent partitioning is most frequently used. This method has been described in procedures A and B. In the cited literature the effects of addition of water and various salt solutions to the chloroform–methanol extract are described in detail. If the partitioning is performed according to protocol, optimal purification of the lipid extract is attained. Wells and Dittmer [34] have described a column chromatographic method for the removal of impurities. This technique has been applied successfully by many investigators but seems less suitable for very small quantities of lipid extracts. Modifications more appropriate for large numbers of samples [35] or for batch-wise application [36] have also been described.

## 8.3 Hydrolytic procedures

Selective hydrolysis procedures to produce water-soluble phosphate esters of the diacyl phospholipids and the alkenyl and alkyl ether phospholipids were described by Dawson *et al.* [37,38], and improved by Wells and Dittmer [40]. The partial hydrolysis method can be used with great success for identification of lipids, because lipids give products, which can be separated on basis of their polarity and/or remaining apolar groups. The deacylation products of acidic phospholipids can sometimes better be resolved than the intact molecules. The plasmalogen and alkyl–ether analogues of the phospholipids may be separated from the diacyl compounds by the same procedure, the former will also separate after mild acid hydrolysis.

For quantitative determinations the method has serious disadvantages. The rate of hydrolysis varies with the phospholipid concerned and also with the position of the fatty acid. Methylation of the phosphate esters, degradation of the ester and alkenyl–ether bonds during subsequent procedures and cyclization of the glycerophosphate ester are other difficulties. Quantitative analysis of complex mixtures of phospholipids is better attempted by thin-layer chromatography.

Various improvements of the alkaline hydrolysis method are suggested. Addition of a small amount of water prevents the formation of methyl and ethyl esters of glycerophosphate [39]. Cyclization and further breakdown of the phosphate esters can be prevented by deacylation at moderate base concentration in a highly polar solvent [41]. Under these conditions the sphingolipids and the deacylated 1-alkenyl- and 1-alkyl-derivatives remain stable and can be extracted with chloroform.

Analysis of the phosphate esters can be performed by (two-dimensional) paper chromatography [37], electrophoresis [39], a combination of both [37,39] or by anion exchange chromatography [39,40,42]. Recently, a high pressure chromatographic separation of the different phosphate-containing degradation products of phosphatidylcholine was reported [43]. The glycerophosphate esters of the bases can also be converted to their trimethylsilyl derivatives and investigated by gas-liquid chromatography and mass spectrometry [44,45]. A two-step procedure also appears possible. A first very mild hydrolysis can be followed by a second more vigorous one on the organic layer [40]. We prefer this procedure to that recommended by Dawson [39] for carrying out both a mild and a strong alkaline alcoholysis on equal portions of the same sample.

Alkaline hydrolysis of glycosyldiglycerides may be performed in a mixture of chloroform–methanol (2:3, by vol.) containing 0.1 M sodium methoxide at room temperature for 10 min [46]. We found that quantitative deacylation of galactosyldiglycerides was best carried out by incubation in 0.1 M sodium methoxide in methanol for 4 h at 37°C [47]. Phosphoglycolipids also need a longer time for deacylation. The alkaline hydrolysis of a mixture of these compounds with phospholipids can best be repeated two or three times on the organic phase obtained after the Dowex column.

Different acid hydrolysis procedures on phospholipids are used for various purposes such as total or partial degradation (see analytical procedures).

Acetolysis is performed for the determination of molecular species [48]. Hydrofluoric acid appeared to be a good dephosphorylating reagent for phospholipids and phosphoglycolipids with no changes in the acyl and glycosidic linkages [49]. The glycosidic linkage of phosphogalactofuranosyllipids was not stable against this reagent [50].

Enzymatic hydrolyses may be used as well for analysis and identification of lipids [6,7]. These methods are mostly very specific, since they do not alter other portions of the molecules. Mixtures of glycerophospholipid analogues containing ether, ester, ketone and hydroxyl groups can be separated by thin-layer chromatography on the basis of these functional groups, while phosphate or phosphoryl base are removed with phosphatase and phospholipase C, respectively [51]. Fatty acids at a specific positon of the phospholipids may be assayed with phospholipase $A_1$, phospholipase $A_2$ and lipase isolated from pancreatic tissue or *Rhizopus arrhizus* [6,52]. Phospholipase C may be used to convert phospholipids to their diglycerides in order to investigate the molecular species [53]. The configuration of the glycosidic linkages of glycolipids may be investigated with specific glycosidases [54]. Use of these enzymes sometimes requires a preceding conversion of the glycolipids to water-soluble derivatives.

## 8.4 Chromatographic methods

In this section procedures for thin-layer and gas–liquid chromatography, in particular, will be described. Column chromatography is widely used for the separation of large samples. It has some advantages because samples can better be protected against light or oxygen, but the other techniques are superior in resolution, rapidity and sensitivity. Major uses of column chromatography are therefore for preparative purposes: removal of water-soluble contaminants like inorganic ions [55], proteins and sugars [34,35], separation of lipid classes on silicic acid [56–58], florisil [59,60], DEAE cellulose [61] and lipophilic Sephadex [62]. The three major lipid classes may easily be separated on silicic acid [63,64]. All neutral lipids are first eluted with chloroform, next the glycolipids with acetone and eventually the phospholipids with methanol and methanol–water. A quantitative isolation method of total glycolipids in microgram quantities from crude lipid extracts was described by Saito and Hakomori [29]. After acetylation glycolipids can be separated from non-glycolipids on a florisil or Sephadex G25 column. The recovered glycolipids are deacylated in chloroform–methanol–sodium methoxide and further analysed by thin-layer chromatography.

Chromatography on silicic acid impregenated paper [65–67] is applied in various separation procedures, because it gives good resolution, especially after development in two dimensions, and the chromatograms can be stored and used for autoradiography more simply. Use of paper chromatography as well as thin-layer chromatography is advantageous for identification purposes.

## 8.5 Thin-layer chromatography

### *8.5.1 Dimensions and composition of the thin-layer medium*

For micro-analytical studies of membranes, thin-layer chromatography is an excellent tool and superior to column chromatographic techniques with respect to rapidity and resolution. Practically all lipid classes can be separated by choosing the appropriate composition of thin-layer solvent sytem and special detection reagents help localization and identification after development [68]. For analytical purposes 0.25 mm-thick layers are widely used, while for preparative purposes 0.5 mm layers are a good compromise between the required resolution and the amount of lipid to be isolated. 20 x 20 cm plates serve most purposes, allowing one or two-dimensional separation and subsequent detection of microgram amounts or more on 0.25 mm layers and of milligram amounts on 0.5 mm layers. These 'thick' layers allow the one-dimensional separation of 20–40 mg of neutral lipids and of 5–10 mg of phospholipids, if the solution is carefully applied along the whole width of the plate.

For lipid chromatography the layer usually consists of silica gel with a very fine grain, which was acid-washed during manufacture. High quality silica gel, moreover, is treated with various other reagents in order to attain the high purity necessary for the analysis of microgram or submicrogram quantities. Plain silica gel ('silica gel H') is the standard material for thin-layer chromatography of lipids. Without addition of binder or other substance it allows the separation of most lipid classes. For one-dimensional resolution of neutral lipids and glycosphingolipids 'silica gel G' is used. This consists of plain silica gel with added gypsum as binder. This mixture is commercially available and gives very firm layers. Silica gel H is superior to the G type in phospholipid separations. Neutral magnesium silicate (10% w/w) has been used as binder [57]. Improved resolution of acidic lipids can be obtained if alkaline substances are added, e.g. sodium carbonate or sodium bicarbonate [68]. Both qualities can be combined by the addition of 3% (w/w) alkaline magnesium carbonate, which yields maximal and reproducible resolution of all phospholipid classes, even of phosphatidylserine and phosphatidylinositol, on two-dimensional development [69,70].

Details of the techniques are given in Section 8.9.3. Improved resolutions of special lipid classes are obtained by the addition of sodium tetraborate to silica gel G or H. Isomeric mono- and diglycerides, ceramides containing normal and hydroxy fatty acids, di- and trihydroxy ceramides [71] or ceramide glucoside and galactoside [72] have been resolved. For the study of molecular species the addition of 5–15% silver nitrate is applied. This layer enables not only the resolution of phosphatidyl cholines [73], phosphatidyl ethanolamines [74] or phosphatidyl inositols [75], but also that of fatty acid methyl esters [76], diglycerides [77] and triglycerides [78] according to the number of double bonds present in the molecules. This argentation chromatography is based on the property exhibited by silver compounds to form reversible complexes with the double bonds. The method is usually followed by the isolation of the molecular species from the plate after separation and subsequent gas-chromatographic analysis of their fatty acids as methyl esters. If not, the silver ions bound to phospholipids can be removed by washing in hexane–ether (1:1, v/v) solution with sodium chloride solution and then with 0.05 M Tris buffer (pH 9) [79]. *Cis*- and *trans*-olefinic compounds may also be separated. A special advantage in argentation chromatography is the protection of double bounds against autoxidation [80].

If thin-layer chromatography is followed by micro-analytical techniques, the purity of silica gel is not always sufficient. Gas chromatographic and densitometric detection, for instance, are of such high sensitivity, that preceding purification with lipid solvents may be necessary. For densitometric analyses, moreover, the layers have to be of absolutely uniform thickness.

### *8.5.2 Chromatographic analysis – general remarks*

For optimal chromatographic and analytical results some precautions have to be taken. It is advisable to line the tanks with filter paper and to moisten it with the solvent. This speeds up the development and often improves the resolution. The solvent systems have been selected during the past decades according to their successful application. Their eluting strength is based on the polarity of their components, which have been ordered according to the following eluotropic series: petroleum ether, cyclohexane, carbon tetrachloride, toluene, benzene, chloroform, diethyl ether, ethyl acetate, acetone, *n*-propanol, ethanol, methanol, water.

Apolar lipids are usually separated with mixtures of petroleum ether and ether containing 1–2% acetic acid in order to form free fatty acids from their salts and thus to obtain a good $R_F$ value. Polar lipids like phospholipids and glycosphingolipids are usually separated in solvent systems based on chloroform–methanol with addition of ammonia or dilute acetic acid. The choice of the latter two substances enhances the resolution of the acidic lipids. For instance chloroform–methanol–ammonia in the first direction and chloroform–methanol–acetic acid in the second direction separate phosphatidylserine and phosphatidylinositol. It has been shown that multiple development in one direction improves resolution [81]. In lipid analysis two-fold development appears to be adequate for resolution of gangliosides [82] and neutral lipids [83]. In the two-dimensional systems the use of an acid system in the first direction must be avoided, because decompositon of certain lipids during drying may occur. Use of systems containing ammonia in amounts of about 5 vol in the first direction is also preferred, because this causes the acidic phospholipids to be chromatographed as their ammonium salts, regardless of the ionic states of the applied lipid and the adsorbents. In acidic or neutral solvents a mixture of the monovalent and divalent salts of an acidic phospholipid isolated from membranes will result in two spots [84]. Some systems with high resolving capacity are described in Section 8.9.3. More examples are summarized in the literature [68,85].

In order to prevent autoxidation, for instance if the separated lipid has to be isolated for further study, air should be replaced by nitrogen as much as is possible. However, the presence of BHT (butylated hydroxy-toluene) in the extract and in the solvent systems seems to be sufficient for some purposes [22]. After development the plate is dried in air before detection. If the lipids have to be isolated, short drying may be accomplished in a stream of nitrogen, followed by a non-reactive detection, for instance with a water or rhodamine B spray (see Section 8.9.3) and viewing under ultraviolet light. The spots can be scraped from the plate and eluted in a micro-column or centrifuge tube with chloroform–methanol (2:1 or 1:1, v/v). For very polar phospholipids it may be necessary to add a few percent water to the eluting solvent for quantitative recovery. If silver nitrate is present, chloroform has to be replaced by diethyl ether. Specific, reactive detection sprays for identification purposes are given in the literature [6,68] and in Section 8.9.3. Autoradiography and scanning are applied on radioactive compounds [6].

Thin-layer chromatography preceding gas chromatographic analysis of glycolipids and cholesterol is described in the gas chromatography section. Thin-layer chromatography of tissue lipids without extraction was applied on frozen sections [86].

## 8.6 Quantification of thin-layer chromatograms

### *8.6.1 Spectrophotometric analysis*

The analysis of phospholipids (lipid–phosphorus) is described in Section 8.9.3. Resolution and subsequent determination of gangliosides, cerebrosides and sulphatides is the only possible method for analysis of microgram quantities of the individual glycolipids [87]. Similar techniques have also been used for the determination of glycolipid galactose with orcinol [88] and anthrone [89]. The analysis of neutral lipids is given in Section 8.9.3. A new method using *o*-phtalaldehyde was described for cholesterol. This procedure is also applicable after thin-layer chromatography [90].

#### *(i) Densitometry and fluorimetry*

These techniques require layers of high purity and very even thickness, while the measurement demands a great deal of skill. Some of the techniques have been reviewed [91]. A densitometric method for gangliosides after resorcinol detection requires various standard curves [92]. A method for densitometry of all neutral and phospholipids after charring was elaborated by Privett and co-workers [93]. The measured peak area was proportional to the amount of carbon between 5–25 μg carbon for all lipids if they had been hydrogenated before chromatography. By first making a photograph and scanning it manually several problems are overcome, especially the development of background color or fading of reagent color [94]. The range of measurement was 2–20 μg lipid. A described photoreflectometric method for neutral lipids requires simpler and inexpensive equipment but needs a minimum of 50 μg lipid [95]. Various problems of densitometry have been discussed in recent literature [96].

A relatively simple and very sensitive method for fluorimetric microdetermination of phospholipids (0.01–50 μg) allows the use of microplates [97]. This technique was simultaneously elaborated for neutral lipids for the ranges 0.0–3.0 μg and 3.0–130 μg [98]. After two-dimensional chromatography of phospholipids X-ray emission spectrometry may be applied [99].

## 8.7 Gas chromatographic analysis of lipids

Gas liquid chromatography allows the separation and detection of microgram to submicrogram quantities of lipids and their constituents. About the numerous applications and procedures several reviews have appeared [3,7,100]. Only the main techniques, relevant for fatty acid analyses of membrane lipids will be summarized here.

### *8.7.1 Fatty acids*

Analysis of fatty acid methyl esters is performed on total lipid extracts, containing very complex lipid mixtures or on pre-separated fractions. For esterification, the boron–trifluoride–methanol method reviewed by Morrisson [101] is usually applied. Although various objections have been made [102], the esterification by sulphuric acid–methanol [103], methanolic hydrogen chloride [104] or diazomethane [104] also causes certain problems, the introduction of artefacts being the most important. After esterification, the methyl esters are extracted with a volatile solvent, which is washed with water or bicarbonate solution and dried with anhydrous sodium sulphate. The solution is concentrated *in vacuo* or with a stream of nitrogen and the esters are analysed as soon as possible. Each of these steps may introduce unwanted changes in the sample by contamination from impure reagents and incomplete or selective recovery. If the original lipid mixture was pre-separated, e.g. by thin-layer chromatography, the sensitivity of the gas chromatographic detection usually has to be increased. For this reason, the silica gel and the solvents have to be of very high purity. The analysis of 'blanks' may help to detect impurities, especially when working with samples containing only a few micrograms or less of the total mixture of fatty acids. In this case the thin-layer plate is developed without a lipid sample and silica gel of the relevant $R_F$ value and appropriate area is analysed.

Gas chromatographic analysis of fatty acid methyl esters is performed on polar columns, e.g. containing poly- or di-ethylene glycol succinate (EGS or DEGS) or another material as liquid phase. However, not all components can be resolved and with complex mixtures several other methods have to be used in order to detect and identify all components. The main methods are pre-separation into molecular species by argentation chromatography, hydrogenation of the unsaturated components and the separation on apolar phases like the silicones SE–30, QF–1 or OV–1. These liquid phases separate the esters according to their molecular weights [100]. Hydrogenation was described by various investigators [105,106]. A very useful micromethod is that of Appelqvist [107], which allows the hydrogenation of lipids on the micro- and nanomole scale in a small tube. A Reacti-vial is appropriate. However, for quantitative hydrogenation full dispersion of the lipid is necessary and the addition of a good lipid solvent to the methanol is effective (R. M. Broekhuyse, unpublished result). An important aid to the identification of fatty acid methyl esters is the plotting the logarithms of the retention times of a homologous series with straight chains against the number of carbon atoms of the chain. Under identical operation conditions, the unknown components can be identified by measuring their retention times and finding the corresponding number of carbon atoms via the graph. For some types of column a modification of the method has been proposed by Nelson [108]. For quantification the areas under the peaks are measured. They are linearly proportional to the amounts of the components concerned. Working isothermally it is also possible to calculate the height multiplied by retention time [109]. For control purposes calibration mixtures are commercially available. Injection technique, losses on columns and detection errors may be investigated before analysis. Sometimes it may be necessary to remove impurities or antioxidants by thin-layer chromatography. A method for separation of BHT is described in Section 8.9.3. Pre-separation may also be helpful if hydroxy fatty acids (from glycolipids) [110] or aldehydes (from plasmalogens) are present [111]. For the analysis of short-chain fatty acids esterification to ethyl, propyl or butyl derivatives of higher boiling point may be suitable. Dimethyl formamide dialkyl acetals, which are commercially available (Pierce Chem. Co.), have been proposed [112,113] as reagents. Direct injection is possible. In a comparison of gas-liquid chromatographic methods Wayne-Moss *et al.* [114] demonstrated that saponification was required for analysis of total fatty acids from bacterial cells. Transesterification with boron-trihalide reagents may cause extensive degradation of cyclopropane acids [115,116] and artifacts [117]. A short methylation time and the use of boron trichloride–methanol reagent appeared to give reliable results on fatty acids released by acid or alkaline hydrolasis [114]. Methylation of extracted bacterial lipids with boron trifluoride–methanol could well be carried out under defined conditions [118]. Cyclopropane fatty acids can be identified because they are removed by bromination like unsaturated acids, but remain resistant to mild hydrogenation [119].

### *8.7.2 Other lipids and lipid components*

Because of their relatively low boiling points, a number of intact apolar lipids can be separated by direct gas chromatography, while molecular species are resolved to a limited extent. Preceding isolation of the particular lipid class is usually necessary, because molecular species of various classes would overlap.

#### *(i) Monoglycerides*

Monoglycerides have been separated as their trimethylsilyl (TMS) ether derivatives. These compounds are rapidly and quantitatively formed without isomerization [120]. Polar (diethyleneglycol succinate, DEGS) as well as apolar (Apiezon L) columns are suitable. 1- and 2-monoglycerides, and 1,2- and 1,3-diglycerides can be preseparated on silica gel G, containing boric acid [121].

#### *(ii) Diglycerides*

These are preferably isolated as described above and analysed as their TMS-ethers or acetates in order to avoid transesterification [122]. Apolar silicone columes may be used, programmed from 250–300°C or isothermally at about 300°C [123,124].

#### *(iii) Triglycerides*

Pre-separated on silica gel, plain or containing silver nitrate, triglycerides can be resolved on short columns loaded with an apolar silicone with high temperature stability. Temperature programmes from about 200°C to 350°C are used and separations according to the total number of carbon atoms are obtained [125,126].

### *8.7.3 Phospholipids*

Intact phospholipids have not been separated by gas–liquid chromatography. Horning and co-workers [127] separated TMS–diglycerides obtained from phospholipids. Polyphospho–inositides have been deacylated and separated as their TMS–ethers [128]. The same authors also measured monophospho inositide in microgram quantities as TMS–glycerylphosphorylinositol [129].

### *8.7.4 Long-chain bases, steroids and sterols*

The conversion of sterols, steroids, ceramides and long-chain bases into derivatives for gas chromatography has been studied by several investigators and reviews of this specialized type of analysis have been published [3,7]. Many details about ceramide and long-chain base analysis have been described by Karlsson *et al.* [130]. Recently, the groups of Sweeley [131] and Stoffel [132] have analysed sphingolipid components by combined thin-layer and gas chromatography and have given excellent instructions. Cholesterol determination by gas chromatography was reviewed by Thorpe *et al.* [133] and by Fumagalli [134]. The same procedures are also useful for cholesterolesters after hydrolysis. Recently the determination of cholesterol in plasma was thoroughly studied [135]. The results can serve as instruction for similar determinations in membrane lipid extracts. Intact cholesterol esters have been separated by Swell [136] on QF–1 at 290°C. After thin-layer chromatography, hydrolysis and conversion to trimethylsilyl ether, cholesterolesters were determined on OV–1 at 250°C [137].

### *8.7.5 Glycolipid hexose*

The gas chromatographic methods most commonly used to estimate sugars are those of the trimethylsilyl [138,139], trifluoroacetyl [140,141], methyl [142] and alditolacetate [143–145]. The latter derivatives have the advantage, that they form single peaks, resolve better and do not decompose easily. In contrast to the trimethylsilyl and trifluoroacetyl derivatives, the alditol acetate method involves the hydrolysis of the glycolipid with aqueous acid. The use of mannitol as an internal standard for analysis of carbohydrates is limited, because it should be added after the methanolysis step [146].

Mass spectrometry of methylated and methylated-plus-reduced derivatives of glycosphingolipids can give much information on the carbohydrate composition and sequence [147].

## 8.8 Analytical procedures

### *8.8.1 Phosphorus*

The determination of phosphorus in lipids has been described in Section 8.9.3. A new colorimetric method without acid digestion was recently described by Raheja *et al.* [148].

### *8.8.2 Inositol*

Most accurate determination of myo-inositol is by gas chromatography. The procedure usually consists of hydrolysis of the lipids with 6 M HCl and removal of HCl by drying over NaOH pellets *in vacuo* [149,150]. The deionization step with Amberlite MB–3 may introduce loss of myo-inositol and omission is sometimes possible [151]. Gas chromatography is performed on polar columns [152], after conversion to the TMS–ether. Analysis by periodate oxidation has also been described [153].

### *8.8.3 Cholesterol*

An appropriate method for the analysis of cholesterol and its esters has been described in Section 8.9.3. This method is also applicable to total cholesterol in lipid extracts, if no BHT is present. In the presence of this antioxidant and $FeCl_3$ very high blanks are obtained. BHT may be removed by thin-layer chromatography, or cholesterol may be determined by either gas-liquid chromatography, or according to Rodnight as described by Dittmer and Wells [154]. This method also allows the separate determination of free cholesterol. A relatively low sensitivity may be a limitation in membrane research.

### *8.8.4 Ester groups*

These groups can still best be assayed with hydroxylamine and ferric perchlorate according to the procedure of Snyder and Stephens [155] as modified by Renkonen [156].

### *8.8.5 Glycerol*

The assay of glycerol is impeded in particular by the lack of quantitative recovery after hydrolysis. Losses by volatilization and destruction have been reported [152]. The hydrolysis of glycerophospholipids except glycerylether bonds can best be performed according to Renkonen [157] in 2 M HCl at 125°C for 120 h. Glycosyldiglycerides are better saponified first in methanolic sodium hydroxide. The resulting glycosides are hydrolysed for 2 h in 2 M HCl at 100°C. The liberated glycerol can be assayed through the sodium periodate–chromotropic acid procedure [6,158], enzymatically with glycerokinase in combination with other enzymes [159–161] or by gas chromatography [152] in free form, as trimethylsilylether or triacetin. A separate method for determination of glycerol in phospholipids involves acetolysis, saponification and preparation of triacetin for gas chromatographic assay [162]. A thermal cleavage with diphenylether was used for the liberation and determination of glycerol as a measurement of the amount of glycerophospholipids [163].

### *8.8.6 Hexoses*

Total hexose in lipid extracts may be determined with the anthrone method according to Radin *et al.* [164] or with phenol [165]. One needs to use the appropriate standard sugar or sugar mixture, because the extinction coefficient of the various hexoses is different. Glycosyldiglycerides can be assayed separately after thin-layer chromatography by the procedure of Roughan and Batt [89]. The silica gel has to be carefully purified to obtain acceptable blanks [166]. For analysis of free hexoses, glycosyldiglycerides have to be hydrolysed in 3 M HCl at 100°C for 3 h. After saponification with methanolic sodium hydroxide a hydrolysis time of 2 h is sufficient. Ceramidehexosides, especially those containing glucose need a hydrolysis time of 12 h in 1 M aqueous HCl at 100°C [145]. Glucose and galactose may be assayed enzymatically with glucose oxidase or with hexokinase–glucose-6-phosphate dehydrogenase [167] and galactose dehydrogenase [168], respectively or together with other components using gas-liquid chromatography. Recently, the effectiveness of tritium-labelled borohydride as a reagent for the quantitative determination of sugars was reported [169].

### *8.8.7 Vicinal hydroxyl (amino) and terminal glycol (amino–glycol) groups*

Determination of total vicinal hydroxyl groups can be performed by analysis of the results obtained from the use of periodate for oxidation of intact lipids or their hydrolysis products. The amounts of periodate can be assayed spectrophotometrically at 225 nm [170] or by iodometric titration [6]. Terminal glycol groups are determined by periodate assay at oxidation under very mild conditions during short time intervals [171] or by determination of the formaldehyde liberated by action of periodate [172]. After destruction of the residual periodate with arsenite, chromotropic acid is reacted with the formaldehyde. Periodate oxidation of lipids is best carried out in mixtures of chloroform–ethanol–water (1:3:1, by vol) at a periodate concentration of 12.5 mM [47].

### *8.8.8 Nitrogen and nitrogen-containing components*

Total nitrogen determination in phospholipids will be relatively inaccurate, due to the presence of nitrogen-free lipids. Nitrogen content is always lower than 3%. A micro-Kjehldahl destruction, for which various digestion fluids (sulphuric acid, perchloric acid) may be used [6,173] is followed by an assay of the formed ammonia. This can be performed by titration, nesslerization [154] or reaction with phenol–hypochlorite [6,174].

#### *(i) Amino-nitrogen*

Amino-nitrogen can be assayed by the ninhydrin procedure [175], by reaction with dinitrofluorobenzene [6,176] or trinitrobenzenesulfonic acid [177].

#### *(ii) Choline*

In the intact lipids choline can be determined according to Böttcher *et al.* [178] with *cis*-aconitic anhydride and acetic anhydride. The preferred methods applied to the water phase of hydrolysates are the reineckate [6,154] and the periodide procedures [6] and a radiochemical assay with choline kinase [179].

*(iii) Ethanolamine and serine*

Both can be assayed after hydrolysis and paper or ion exchange chromatography [180] by ninhydrin [175,181] or by determination of the formaldehyde produced by periodate oxidation [154]. The hydrolysis of the *o*-phosphate ester of serine and ethanolamine presents special difficulties [174,180]. Hydrolysis with 6 M HCl for 3 h at 105°C appears a good method, if appropriate standards are included.

*(iv) Amino acids*

Amino acids such as *O*-amino acid esters linked to phospholipids, like alanine, lysine and ornithine can easily be released by mild alkaline hydrolysis. Incubation in a borate buffer (pH 9.0) for 30 min at room temperature causes a practically complete hydrolysis of lysylphosphatidylglycerol into lysine and phosphatidylglycerol without any detectable cleavage of the acyl ester bonds [33]. Amino acids can be estimated by amino acid analyser or by specific methods, such as for ornithine [182]. *N*-acyl amino acids have to be hydrolysed like ceramides and sphingolipids.

*(v) Hexosamines*

Hexosamines from glycosphingolipids and also from glucosaminylphosphatidylglycerol are only liberated after strong acid hydrolysis (6 M HCl at 105°C for 18 h). Glucosaminides are only slowly hydrolysed by strong acid. Assays can be performed by spectrophotometric methods based on the Elson-Morgan reaction [183]. Gas chromatography of the methyl glycosides is possible after heating the lipid sample with 2 M methanolic HCl for 5 h. The hexosamines and their derivatives, such as the *N*-acetyl- and alditol derivatives and the methylglycosides can quantitatively be analysed as TMS-derivatives [138,184]. When non-polar stationary phases are used, no *N*-acetylation is necessary to obtain separation of TMS-hexosamines [185]. The TMS-derivatives are not stable, cannot be preserved and give multiple peaks. Varma *et al.* [186] recently described a rather simple procedure, in which hexosamines are deaminated with nitrous acid. The 2,5–anhydro–mannose and 2,5–anhydrotalose formed from glucosamine and galactosamine, respectively are derivatized to the aldonitrile acetates [187]. These derivatives are very stable and give single well-separated peaks.

*(vi) Neuraminic acid*

Neuraminic acid can be determined with the thiobarbituric acid assay [188] or by spectrophotometric estimation of the complex formed with resorcinol [189]. Gas chromatographic determination is also possible [190].

*(vii) Sphingosine*

Sphingosine is determined by the procedure of Siakotos and Rouser [191], using trinitrobenzenesulfonic acid as reagent and Hyamine as solubilizer of the aqueous reagent in the organic solvents. The modified procedure also determines phosphatidylethanolamine and phosphatidylserine, and water-soluble substances with free aminogroups [9,178]. The same reagent has been used for fluorimetric determination of sphingosine [192]. A rapid and sensitive assay of sphingosine bases and sphingolipids with fluorescamine was recently described by Naoi *et al.* [193]. Interference from other primary amines such as hexosamines can be excluded by extraction with ether.

## 8.9 Procedures

### *8.9.1 'Folch' extraction*

The procedure of Folch *et al.* [20] is usually applied with modifications. The described extraction with 20 vol chloroform–methanol (2:1, v/v) is made more generally suitable if homogenization is performed in methanol or chloroform–methanol (1:1, v/v) and chloroform is added afterwards in order to obtain the ration 2:1 (v/v) [194,195].

*Procedure A*

The membrane preparation is homogenized in at least 10 vol methanol of chloroform–methanol (1:1, v/v) in a Potter-Elvehjem apparatus at room temperature or at 0°C. Methanol and/or chloroform (20–25°C) are added in order to obtain at least 20 vol chloroform–methanol (2:1, v/v) per vol of original preparation or aqueous homogenate and the suspension is further homogenized until the dispersion is maximal. The tube is closed and centrifuged after standing for 10 min with occasional resuspension if necessary. The procedure is repeated and the combined supernatants contain usually about 99% or more of the lipids. Occasionally it may be necessary to add methanol to the tube after the first extraction in order to be able to sediment protein material, which remains suspended in the chloroform–methanol–water mixture during centrifugation.

The combined supernatants are mixed with chloroform to a ratio 2:1 (v/v) if necessary and 0.2 vol of 0.1 M KCl (in water) per vol extract is added. After mixing, a biphasic system is obtained upon standing or centrifugation. The upper layer is removed by aspiration (see also Section 8.9.2). The lower layer may be washed with additional portions of 'Folch upper phase', consisting of chloroform–methanol–water (3:48:47, by vol). This is especially appropriate if radioactive non-lipids have to be removed. The lower phase is concentrated with a rotary vacuum evaporator and to prevent foaming during this process, small amounts of organic solvents may be added. Benzene, ethanol or propanol is suitable. It is often not necessary to evaporate to dryness, on the contrary this may be harmful for the lipids. In order to evaporate water, the lipid concentrate is mixed with 5 vol benzene-methanol (4:1, v/v) and concentrated further. The residual fluid (about 0.2 ml) is transferred to a small screw-cap bottle by means of several portions of chloroform–methanol (2:1, v/v) (the presence of the alcohol is necessary) and dried in a stream of nitrogen at 20–25°C. The lipids are stored in benzene–methanol or toluene–methanol (4:1, v/v) at – 20°C or lower.

To recover the gangliosides from the upper layer, the excess of organic solvent is evaporated and the residual fluid is dialysed against water and lyophilized afterwards. During dialysis the ganglioside concentration has to be more than 150 $\mu$g ml$^{-1}$, the critical micelle concentration, in order to prevent losses [196].

### *8.9.2 Bligh and Dyer extraction*

If large amounts of water are present it is often not suitable to concentrate the material to be extracted. Ultrafiltration may cause loss by adsorption to the filter membrane and during vacuum concentrating, autoxidation and enzymic degradation are difficult to rule out and the concentrated material is not always easily collected. The procedures based on the technique of Bligh and Dyer [9] do not use such large volumes of solvents as those described in Section 8.9.1 and they may be suitable in these cases. Filtration of the monophasic system has been introduced in the procedure in order to prevent the loss of acidic phospholipids [197]. The recovery is not as good as in Section 8.9.1.

#### *Procedure B*

A 1 g sample, containing 80% water, is homogenized with 3 ml chloroform–methanol (1:2, v/v). More chloroform–methanol should be added if the liquid is biphasic. The homogenate is filtered through a sintered glass funnel and the residue is re-extracted with 1 ml chloroform. After filtration, the combined filtrates are mixed with 1 ml 0.1 M KCl in water and centrifuged. The upper layer together with any interfacial material is removed by aspiration using a pipette with a suction bulb. Eventually, the aspirated lower layer will separate in the pipette and can flow back. If necessary, the chloroform layer is filtered and washed again with synthetic upper phase. The extract is evaporated and stored and the same precautions are taken as described in Section 8.9.1.

Many alterations of the procedure are permissible, but the authors state that it is imperative that the ratio of chloroform–methanol–water is 1:2:0.8 before dilution with chloroform and 0.1 M KCl, and is 2:2:1.8 afterwards. This enables the extraction of suspensions of membranes, organelles and cells, including micro-organisms [6].

### *8.9.3 Thin-layer chromatography*

#### *(i) Preparation and storage of the plates*

Clean, dry glass plates of 200 x 200 x 3 or 4 mm are coated with a slurry of the adsorbent. This slurry is made by intensively mixing a specific amount of silica gel, binder or other compound and water. Short (15–30 s) mixing in a Waring blendor improves the quality of the layer and is especially recommended if solid substances are added. Coating is accomplished immediately after mixing and the layers are air-dried for 0.5–1.5 h. They are activated for 30 min at 120°C and stored in a desiccator. If high reproducibility is an important factor, the desiccator should be evacuated, or the atmospheric composition in the desiccator has to be standardized. This can be done by filling it with nitrogen of known humidity.

#### *(ii) Application of the sample*

Lipid solutions have to be applied to the plate as compactly as possible. This can be done by means of commercially available capillary pipettes. The solvent must be volatile in order to allow rapid evaporation and as apolar as is suitable to dissolve the lipids. A row of very small spots, overlapping each other somewhat, is made on the plate, about 1.5–2 cm from the bottom. Quantitative application can be achieved by using a micro syringe with an air-tight plunger. Care should be taken, that the solution is immediately absorbed by the layer, without damaging it and without back-flow along the outside of the needle. The recovery of this technique should be measured if calculations have to be based on the applied volume [198]. During application the hygroscopic layer should be protected from humid air, because this affects the separation of lipids. A plastic box with an opening for application is suitable and is commercially available. If the air is relatively dry or the application occurs very rapidly, the decrease in activity may be neglected.

#### *(iii) Thin-layer composition and solvent systems*

*Hydrocarbons*

Silica gel G with heptane-benzene 9:1 (v/v); toluene and benzene.

*Neutral lipids*

Silica gel G with petroleum ether (b.p. 40–60°C)–diethylether–formic acid (80:20:2, by vol.) [23].

If BHT has to be separated from the front and from cholesterol esters, this solvent (II) is preceded by a development with petroleum ether (b.p. 40–60°C)–diethyl ether (96:4, v/v) to the top of the plate (I). Solvent II is then allowed to move to half-way [11] (Fig. 8.1). Also very effective is hexane or petroleum ether (b.p. 60–70°C)–ethyl ether–acetic acid (90:10:1, by vol) [199]. Generally, the resolved lipids are (from bottom to top): phospholipids (starting point), monoglycerides, diglycerides (double spot), cholesterol, free fatty acids, triglycerides, cholesterol esters, BHT, front with hydrocarbons.

*Mammalian phospholipids*

Silica gel H containing sodium carbonate (the slurry is made in 1 mM $Na_2CO_3$) with chloroform–methanol–acetic acid–water (25:15:4:2, by vol) [68].

The resolved lipids are (from bottom to top): lysophosphatidyl-choline, sphingomyelin, phosphatidylcholine, phosphatidylinositol, phosphatidylserine, phosphatidylethanolamine. Careful equilibration of the paper-lined tank with solvent is important. By increasing the amount of water or acetic acid the $R_F$ of phosphatidylethanolamine can be increased. The reverse is also true. The resolution of phosphatidylinositol and phosphatidylcholine can be influenced by varying the amount of methanol [68].

Two-dimensional thin-layer chromatography on silica gel H containing 3%

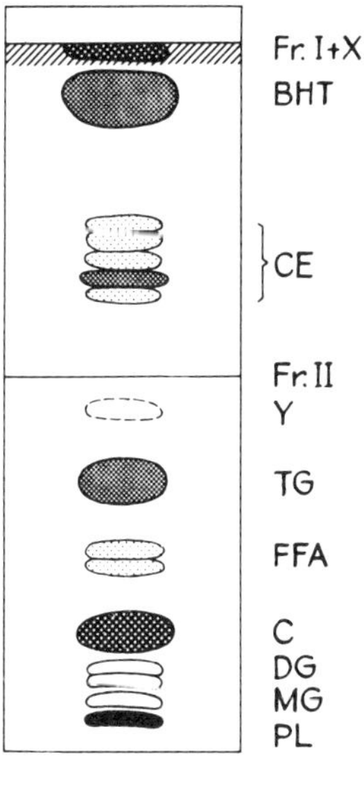


**Fig.8.1** Thin-layer chromatogram of neutral lipids from human sclera on silica gel G. Solvent I, light petroleum (b.p. 40–60°C)–diethyl ether, (96:4, v/v); Solvent II light petroleum (b.p. 40–60°C)–diethyl ether–formic acid (80:20:2, v/v), followed by charring. PL, phospholipids; MG, monoglycerides; DG, diglycerides; C, cholesterol; FFA, free fatty acids; TG, triglycerides; Y, unidentified; Fr.II, front of Solvent II; CE, cholesterol esters; BHT, butylated hydroxytoluene; Fr.I, front Solvent I; X, hydrocarbons, including squalene.

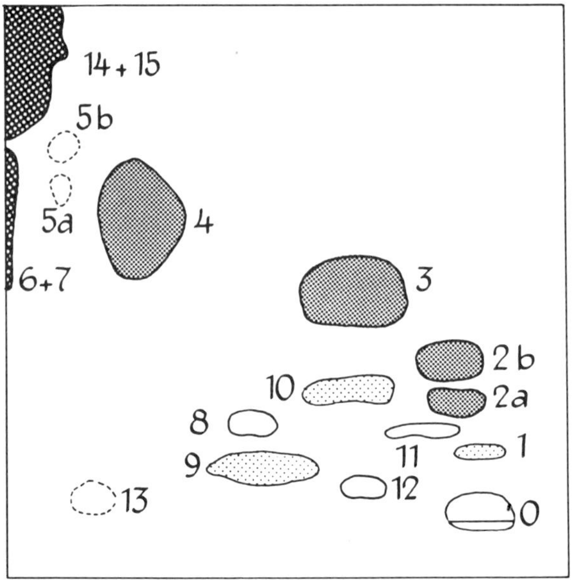


**Fig. 8.2** Thin-layer chromatogram of polar lipids from human sclera on silica gel containing 3% (w/w) alkaline magnesium silicate. Solvents: Y-direction, chloroform–methanol–7 M ammonia (90:54:11, v/v); X-direction, chloroform–methanol–acetic acid–water (90:40:12:2, v/v). 0, origin; 1, lysophosphatidylcholine; 2a and 2b, sphingomyelins; 3. phosphatidylcholine; 4. phosphatidylethanolamine; 5a and 5b, cerebrosides; 6. free fatty acids; 7. diphosphatidylglycerol; 8. phosphatidylinositol; 9. phosphatidylserine; 10. lysophosphatidylethanolamine; 11. unidentified; 12. glycolipid; 13. phosphatidic acid; 14. neutral lipids; 15. butylated hydroxytoluene.

alkaline magnesium silicate (from Woelm, Eschwege, Germany) or 3% magnesium hydroxy carbonate. A 10% (w/v) suspension of this mixture in water should have a pH of 8.0–8.4. Solvent systems: first direction, chloroform–methanol–7 M ammonia (90:54:11, by vol.), second direction chloroform–methanol–acetic acid–water (90:40:12:2, by vol). Between the two runs, the plates are dried for 15 min in air, or (if the relative humidity of the air exceeds 55%) for 0.5 h *in vacuo* above conc. $H_2SO_4$ [23,70] (Fig. 8.2). The $R_F$ values can be increased by lowering the amount of chloroform in the particular solvent systems.

Two-dimensional thin-layer chromatography on silica gel G or on silica gel H containing 10% (neutral) magnesium silicate as binder. The latter powder mixture is ground for 2 h on a ball mill at a slow speed for optimal results [9]. Solvent systems: first direction, chloroform–methanol–28% ammonia (65:25:5, by vol.), second direction, chloroform–acetone–methanol–acetic acid–water (3:4:1:1:0.5 , by vol) [9,57]. Each of the solvent systems for two-dimensional chromatography is also suitable for one dimension.

*Bacterial phospholipids*

Bacterial phospholipids can be separated and quantitatively determined by silica gel G thin-layer chromatography and radioactivity determination after growth of the organism on [$^{32}$P] phosphate-containing medium. Two-dimensional separation can be carried out best with a first development in chloroform–methanol–7 M ammonia (65:35:5, by vol) followed by a chloroform–methanol–acetic acid-water system (125:37:9.5:1.5, by vol for example) in the second direction [6,200].

*Glycolipids*

One-dimensional in all systems. On silica gel G with chloroform–methanol–water (65:25:4, by vol) if phospholipids are removed before. Four main classes are separated: ceramide trihexoside-*N*-acetyl-galactosamine, ceramide trihexosides, ceramide dihexosides and ceramide monohexosides [201]. A ganglioside fraction can be resolved on silica gel G with chloroform–methanol–2.5 M ammonia (60:40:9, v/v) [82] or with propanol–water (70:30, v/v) [202] in which water may be replaced by ammonia. The following classes are resolved: tri-, di- and mono-sialogangliosides. Acylated and non-acylated glycosyldiglycerides can be separated well in the following systems, diethyl ether saturated with water–isopropanol–water (100.4.5:3, by vol): monoglycosyldiglycerides [203] chloroform–methanol–7 M ammonia (60:35:5, by vol): di- and triglycosyldiglycerides, chloroform–methanol–7 M ammonia (70:20:2, by vol): all glycosyldiglycerides [200]. Other systems are reviewed by Sastry [204].

*Selected lipid classes*

The chromatography of selected classes of lipids is described in several review articles: acidic phospholipids, long-chain bases, alkyl– and alkenyl–ether lipids [3,6]; sterols, steroids [205], bile acids [206], prostaglandins [207]. All these classes are discussed by Christie [7].

*Molecular species*

Separation one-dimensional on silica gel G or H containing 2.5–25% $AgNO_3$, usually 5 or 10% (w/w). For fatty acid methyl esters hexane–diethyl ether (90:10, v/v) is used for the resolution of the saturated, monoenoic and dienoic from the other long-chain fatty acids as one group. Hexane–diethyl ether (40:60, v/v) or methanol–diethyl ether (1:9, v/v) resolves the saturated plus monoenoic plus dienoic acids from the trienoic, tetraenoic, pentaenoic and hexaenoic fatty acid methyl esters but not the saturated, monoenoic and dienoic analogues (see review by Morris [208]). With slight variations in the solvent system, many other simple lipid classes can be separated. Purified polar lipid classes can be resolved according to their degree of unsaturation as well [8,208, 209]. Various recent publications have been cited above. After detection with a non-destructive reagent, the lipids are eluted as soon as possible for further analysis, including gas chromatography.

### *(iv) Detection of lipids on chromatograms*

#### *General detection reagents*

Iodine vapour. The plate is placed in a container saturated with iodine vapour, all lipids stain brown; this stain is very sensitive.

Water. Plates are sprayed till they are translucent, the lipids appear as opaque spots or bands.

Rhodamine 6G. 0.01% solution in water, spray and view under ultraviolet light while wet; yellow or blue spots.

Sulphuric acid. 10% in water, spray and heat at 150–180°C in an oven (15–30 min), extremely sensitive. This reagent is also used for densitometric analysis.

#### *Specific detection reagents*

Ninhydrin. 0.25 g in 100 ml acetone–lutidine (9:1, v/v), fresh or stored at 4°C. Spray and heat 3–5 min at 100–120°C or leave the plate 2 h at 20–25°C. Purple spots appear; respraying may be necessary [210].

Phosphate stain. Solution A: 16 g ammonium molybdate in 120 ml water; solution B: 40 ml conc. HCl is shaken 30 min with 10 ml mercury and 80 ml solution A. The mixture is filtered. To prepare the spray, 200 ml conc. $H_2SO_4$ and all of solution B are added to the remainder of A, and diluted with water to 1 l. Upon spraying the phospholipids stain blue [211]. Recently, a modification of this reagent has been described [212]. A special method for detection of phospholipids in chromatograms is the use of 8–anilinonaphtalene–1–sulphonate (ANS) [213]. Since this reagent does not interfere with phosphorus determinations nor with liquid scintillation radioactive counting, it can be used readily to isolate samples for subsequent analysis.

Glycolipid stain. Spray with 1% sodium periodate until the layer is just saturated. After 10 min the plate is placed in a tank containing sulphur dioxide. When the reaction is terminated (colourless plate), the gel is lightly sprayed with Schiff reagent (1 g pararosaniline hydrochloride in 100 ml water and decolourized by a sulphur dioxide stream). Glycolipids stain blue, phosphatidylglycerol purple and after 40 min phosphatidylinositol stains yellow. The intensity of the glycolipid colour increases over several hours [214]. α-Naphthol is also used as a glycolipid stain [215].

Ganglioside stain. 2 g resorcinol in 100 ml water. Add 10 ml of this solution to 80 ml conc. HCl containing 0.5 ml. 0.1 M copper sulphate and dilute with water to 100 ml. Spray after at least 4 h, cover the plate with a glass plate and heat at 120°C. Sialic acid-containing substances stain blue-violet [216,217].

Cholesterol. Cholesterol can be recognized after spraying with sulphuric acid and heating at 100–120°C. It stains purple before charring. Most other lipids char more slowly. A modification, which allows staining of cholesterol, and bile acids has recently been described [218].

Other stains. Hydroxylamine for esters [219], Dragendorff [220] and *cis*-aconite [221] reagents for choline, orcinol–$FeCl_3$/Bial's reagent for glycolipids [222], Schiff reagent for plasmalogens [68].

### *(v) Quantification of thin-layer chromatograms*

#### *Spectrophotometric analysis*

Phospholipids. Lipids are stained by iodine vapour or by charring and the spots are circled and numbered, and the chromatogram is photographed (Polaroid camera). Spots containing not more than 500 $\mu$g lipid and blanks of corresponding dimensions are transferred to test tubes. To each tube 0.4 ml conc. $H_2SO_4$–72% $HClO_4$ (3:2, v/v) is added. Tubes are covered with glass beads and digestion is performed at 180°C for 30 min. After cooling 0.6 ml of a mixture of 22 ml amino–naphthol–sulphonic acid reagent and 500 ml 0.26% ammoniumheptamolybdate.4 $H_2O$ is added. The former reagent contains 3.43 g $Na_2S_2O_5$, 63 mg 1–amino–2–naphthol–4–sulphonic acid and 125 mg $Na_2SO_3$ per 25 ml water. After rapid mixing tubes are heated for 20 min at 90°C (or 10 min at 100°C) and afterwards cooled with tap water. Silica gel is removed by centrifugation and the extinction is measured at 820 nm. Reagent blanks and standard inorganic phosphate are treated similarly [69]. The blanks can be kept very low, if acid-washed glassware and reagents of the highest purity are used. Total phosphorus recovery from two-dimensional thin-layer chromatograms of phospholipids usually ranges from 96–100%.

Cholesterol and cholesterolesters. Standard cholesterol is run simultaneously in amounts varying from 10 to 50 $\mu$g; after chromatography water is used as detection spray. Spots of equal surface area are transferred to tubes and 1.3 ml *iso*propanol is added. Tubes are closed and heated for 7 min at about 65°C with occasional mixing. After cooling, 1 ml from the *iso*propanol extracts is pipetted and 2 ml ferric chloride reagent is added followed by 2 ml conc. $H_2SO_4$ with immediate mixing. The ferric chloride reagent is a solution of 50 mg $FeCl_3 \cdot 6H_2O$ in 100 ml glacial acetic acid. After 30 min the extinction is measured at 540 nm [223]. This method is based on the Zlatkis procedure [224].

Neutral lipids. After drying the chromatograms are sprayed with water or with 0.05% Rhodamine in ethanol, while the monoglycerides are masked. The spots are marked, if necessary under UV light, and the lipids are eluted. Triglycerides, free fatty acids and diglycerides are eluted with 3 portions of 5 ml diethyl ether (15 min 4°C), cholesterol, cholesterolesters and monoglycerides with 3 portions of 5 ml chloroform–methanol (4:1, v/v) at 40°C (10 min). Specific determinations are performed on the lipids [68] or they are analysed by gas chromatography.

## References

1. Rouser, G., Nelson, G. J., Fleischer, S. and Simon, G. (1968), In: *Biological Membranes: physical fact and function*, (Chapman, D. ed.), p. 5, Academic Press, London.
2. Van Hoeven, R. P. and Emmelot, P. (1972), *J. Memb. Biol.*, **9**, 105.
3. Marinetti, G. V. (1967), and (1969), *Lipid Chromatographic Analysis*, Vol. 1 and 2, Marcel Dekker, New York.
4. Lowenstein, J. M. (1969), *Methods in Enzymology*, vol. XIV, Academic Press, New York.
5. Johnson, A. R. and Davenport, J. B. (1971), *Biochemistry and Methodology of Lipids*, Wiley Interscience, New York.
6. Kates, M. (1972), *Techniques of Lipidology*, North-Holland Publishing Company, Amsterdam.
7. Christie, W. W. (1973), *Lipid Analysis*, Pergamon Press, Oxford.
8. Viswanathan, C. V. (1969), *Chromat. Rev.*, **11**, 153.
9. Rouser, G., Kritchevsky, G., Siakotos, A. N. and Yamamoto, A. (1970), In: *Neuropathology: methods and diagnosis,* (Tedeschi, C. G., ed.), p. 691, Little Brown, Boston.
10. Spanner, I. (1973), In: *Form and Function of Phospholipids*, (Ansell, G. B., Hawthorne, J. N. and Dawson, R. M. C., eds.), p. 43, Elsevier Publishing Company, Amsterdam.
11. Kawai, K., Fujita, M. and Nakao, M. (1974), *Biochim. biophys. Acta,* **369**, 222.
12. Weinstein, D. B., Marsh, J. B., Glick, M. C. and Warren, L. (1969), *J. biol. Chem.*, **244**, 4103.
13. Bretscher, M. S. (1972), *Nature New Biology,* **236**, 11.
14. Zwall, R. F. A., Roelofsen, B. and Colley, C. M. (1973), *Biochim. biophys. Acta,* **300**, 159.
15. Caspar, D. L. D. and Kirschner, D. A. (1971), *Nature New Biology,* **231**, 46.
16. Wallach, D. F. H. and Sun Lin, P. (1973), *Biochim. biophys. Acta,* **300**, 211.
17. Hanahan, D. J. (1973), *Biochim. biophys. Acta,* **300**, 319.
18. Schmid, P., Calvers, J. and Steiner, R. (1973), *Physiol. Chem. Phys.*, **5**, 157.
19. Radin, N. S. (1969), In: *Meth. Enzym.*, (Lowenstein, J. M. ed.), Vol. XIV, p. 245, Academic Press, New York.
20. Folch, J., Lees, M. and Stanley, G. H. S. (1957), *J. biol. Chem.*, **226**, 497.
21. Bligh, E. G. and Dyer, W. J. (1959), *Can. J. Biochem. Physiol.*, **37**, 911.
22. Broekhuyse, R. M. (1974), *Clinica chim. Acta,* **52**, 53.

23. Broekhuyse, R. M. (1972), *Biochim. biophys. Acta,* **280**, 637.
24. Hauser, G. and Eichberg, J. (1973), *Biochim. biophys. Acta,* **326**, 201.
25. Hauser, G., Eichberg, J. and Gonzalez-Sastre, F. (1971), *Biochim. biophys. Acta,* **248**, 87.
26. White, G. L. and Larrabee, M. G. (1973), *J. Neurochem.,* **20**, 783.
27. Bjerve, K. S., Daae, L. N. W. and Bremer, J. (1974), *Analyt. Biochem.,* **58**, 238.
28. Suzuki, K. (1965), *J. Neurochem.,* **12**, 629.
29. Saito, T. and Hakomori, S. (1971), *J. Lipid Res.,* **12**, 257.
30. Zwaal, R. F. A. (1970), Thesis, Utrecht, p. 11.
31. Zahler, H. P. and Wallach, D. F. H. (1967), *Biochim. biophys. Acta,* **135**, 371.
32. Roelofsen, B., De Gier, J. and Van Deenen, L. L. M. (1964), *J. cell. comp. Physiol.,* **63**, 233.
33. Houtsmuller, U. M. T. and Van Deenen, L. L. M. (1965), *Biochim. biophys. Acta,* **106**, 564.
34. Wells, M. A. and Dittmer, J. C. (1963), *Biochemistry,* **2**, 1259.
35. Wuthier, R. E. (1966), *J. Lipid Res.,* **7**, 558.
36. Williams, J. P. and Merrilees, P. A. (1970), *Lipids,* **5**, 367.
37. Dawson, R. M. C. (1960), *Biochem. J.,* **75**, 45.
38. Dawson, R. M. C., Hemington, N. and Davenport, J. B. (1962), *Biochem. J.,* **84**, 497.
39. Dawson, R. M. C. (1967), In: *Lipid Chromatographic Analysis*, (Marinetti, G. V. ed.), Vol. I, p. 171, Marcel Dekker Inc., New York.
40. Wells, M. A. and Dittmer, J. C. (1966), *Biochemistry,* **5**, 3405.
41. Brockerhoff, H. (1963), *J. Lipid Res.,* **4**, 96.
42. Short, S. A., White, D. C. and Aleem, M. I. H. (1969), *J. Bact.,* **99**, 142.
43. Lairon, D., Amic, J., Lafont, H., Nalbone, G., Domingo, N. and Hauton, J. (1974), *J. Chromat.,* **88**, 183.
44. Duncan, J. H., Lennarz, W. J. and Fenselou, C. C. (1971), *Biochemistry,* **10**, 927.
45. Cicero, T. J. and Sherman, W. R. (1971), *Biochem. biophys. Res. Commun.,* **43**, 451.
46. Brundish, D. E., Shaw, N. and Baddiley, J. (1967), *Biochem. J.,* **105**, 885.
47. Veerkamp, J. H. (1972), *Biochim. biophys. Acta,* **273**, 359.
48. Renkonen, O. (1965), *J. Am. Oil Chem. Soc.,* **42**, 298.
49. Shaw, N. and Stead, A. (1974), *Biochem. J.,* **143**, 461.
50. Veerkamp, J. H. and Van Schaik, F. W. (1974), *Biochim. biophys. Acta,* **348**, 370.
51. Snyder, F. (1973), *J. Chromat.,* **82**, 7.
52. Slotboom, A. J., De Haas, G. H., Bonsen, P. P. M., Burbach-Westerhuis, G. J. and Van Deenen, L. L. M. (1970), *Chem. Phys. Lipids,* **4**, 30.
53. Van Golde, L. M. G. and Van Deenen, L. L. M. (1966), *Biochim. biophys. Acta,* **125**, 496.
54. Brundish, D. E., Shaw, N. and Baddiley, J. (1966), *Biochem. J.,* **99**, 546.
55. Carter, H. E. and Weber, E. J. (1966), *Lipids,* **1**, 16.
56. Hirsch, J. and Ahrens, E. H. (1958), *J. biol. Chem.,* **233**, 311.
57. Rouser, G., Kritchevsky, G. and Yamamoto, A. (1967), In: *Lipid Chromatographic Analysis*, (Marinetti, G. V. ed.), Vol. I, p. 99, Marcel Dekker Inc., New York.
58. Sweeley, C. C. (1969), In: *Meth. Enzym.,* (Lowenstein, J. M., ed.), Vol. XIV, p. 255, Academic Press, New York.

59. Carroll, K. K. (1961), *J. Lipid Res.*, **2**, 135.
60. Radin, N. S. (1969), In: *Meth. Enzym.*, (Lowenstein, J. M. ed.), Vol. XIV, p. 268, Academic Press, New York.
61. Rouser, G., Kritchevsky, G., Yamamoto, A., Simon, G., Galli, C. and Bauman, A. J. (1969), In: *Meth. Enzym.*, (Lowenstein, J. M. ed.), Vol. XIV, p. 272, Academic Press, New York.
62. Ellingboe, J., Nystöm, E. and Sjövall, J. (1969), In: *Meth. Enzym.*, (Lowenstein, J. M. ed.), Vol. XIV, p. 317, Academic Press, New York.
63. Vorbeck, M. L. and Marinetti, G. V. (1965), *J. Lipid Res.*, **6**, 3.
64. Rouser, G., Kritchevsky, G., Simon, G. and Nelson, G. J. (1967), *Lipids*, **2**, 37.
65. Wuthier, R. E. (1966), *J. Lipid Res.*, **7**, 544.
66. Kates, M. (1967), In: *Lipid Chromatographic Analysis*, (Marinetti, G. V. ed.), Vol. I, p. 1, Marcel Dekker Inc., New York.
67. Fang, M. and Marinetti, G. (1969), In: *Lipid Chromatographic Analysis*, (Marinetti, G. V. ed.), Vol. I, p. 598, Marcel Dekker Inc., New York.
68. Skipsky, V. P. and Barclay, M. (1969), In: *Meth. Enzym.*, (Lowenstein, M. J. ed.), Vol. XIV, p. 530, Academic Press, New York.
69. Broekhuyse, R. M. (1968), *Biochim. biophys. Acta*, **152**, 307.
70. Broekhuyse, R. M. (1969), *Clinica chim. Acta*, **23**, 457.
71. Karlsson, K. A. and Pascher, I. (1971), *J. Lipid Res.*, **12**, 466.
72. Svennerholm, E. and Svennerholm L. (1963), *Biochim. biophys. Acta*, **70**, 432.
73. Arvidson, G. A. E. (1968), *Eur. J. Biochem.*, **4**, 478.
74. Sundler, R. and Åkesson, B. (1973), *J. Chromat.*, **80**, 233.
75. Luthra, M. G. and Sheltaway, A. (1972), *Biochem. J.*, **128**, 587.
76. Morris, L. J. (1966), *J. Lipid Res.*, **7**, 717.
77. Kuksis, A. and Marai, L. (1967), *Lipids*, **2**, 217.
78. Roehm, J. N. and Privett, O. S. (1970), *Lipids*, **5**, 353.
79. Akesson, B. (1969), *Eur. J. Biochem.*, **9**, 463.
80. Slawson, V. and Stein, R. A. (1970), *Lipids*, **5**, 713.
81. Perry, J. A. and Haag, K. W. (1973), *J. Chromat. Sci.*, **11**, 447.
82. Ledean, R., Salsman, K. and Cabzera, M. (1968), *J. Lipid Res.*, **9**, 129.
83. Freeman, C. P. and West, D. (1966), *J. Lipid Res.*, **7**, 324.
84. Nielsen, H. (1974), *J. Chromat.*, **89**, 275.
85. Renkonen, O. and Varo, P. (1967), In: *Lipid Chromatographic Analysis*, (Marinetti, G. V., ed.), Vol. I, p. 71, Marcel Dekker Inc., New York.
86. Cherayil, G. D. and Scaria, K. S. (1970), *J. Lipid Res.*, **11**, 378.
87. Hamberger, A. and Svennerholm, L. (1971), *J. Neurochem.*, **18**, 1821.
88. De Vries, G. H. and Norton, W. T. (1974), *J. Neurochem.*, **22**, 259.
89. Roughan, P. G. and Batt, R. D. (1968), *Analyt. Biochem.*, **22**, 74.
90. Rudel, L. L. and Morris, M. D. (1973), *J. Lipid Res.*, **14**, 364.
91. Kirchner, J. G. (1971), *J. Chromat.*, **63**, 73.
92. Smíd, F. and Reinisová, J. (1973), *J. Chromat.*, **86**, 200.
93. Nutter, L. J. and Privett, A. S. (1968), *J. Chromat.*, **35**, 519.
94. Siakotos, A. N. and Rouser, G. (1966), *Analyt. Biochem.*, **14**, 162.
95. Romas, J. R. and Palmer, I. S. (1972), *Analyt. Biochem.*, **49**, 585.
96. Gartzke, J. and Bolte, K. D. (1973), *J. Chromat.*, **84**, 109.
97. Schiefer, H. G. and Neuhoff, V. (1971), *Z. Physiol. Chem.*, **352**, 913.
98. Nicolosi, R. J., Smith, S. C. and Santerre, R. F. (1971), *J. Chromat.*, **60**, 111.
99. Chapman, L. R. (1972), *J. Chromat.*, **66**, 303.

100. Ackman, R. G. (1969), In: *Meth. Enzym.*, (Lowenstein, J. M., ed.), Vol. XIV, p. 329, Academic Press, New York.
101. Morrisson, W. R. and Smith, L. M. (1964), *J. Lipid Res.*, **5**, 600.
102. Fulk, W. K. and Shorb, M. S. (1970), *J. Lipid Res.*, **11**, 276.
103. Hansen, R. P. and Smith, J. F. (1966), *Lipids,* **1**, 316.
104. Morrison, W. R., Lawrie, T. D. V. and Blades, J. (1961), *Chemy. Ind.*, **1961**, 1534.
105. Farquhar, J. W., Insull, W., Rosen, P., Stoffel, W. and Ahrens, E. H. (1959), *Nutr. Abstr. Rev.*, **17**, supplement.
106. Burgher, R. D. and Ackman, R. G. (1964), *J. Lipid Res,* **5**, 130.
107. Appelqvist, L. A. (1972), *J. Lipid Res.*, **13**, 146.
108. Nelson, G. J. (1974), *Lipids,* **9**, 254.
109. Bayne, A. W. and Duncan, W. R. M. (1970), *J. Lipid Res.*, **11**, 293.
110. O'Brian, J. S. and Rouser, G. (1964), *Analyt. Biochem.*, **7**, 288.
111. Radin, N. S. (1965), *J. Am. Oil Chem. Soc.*, **42**, 569.
112. Thenot, J. P., Horning, E. C., Stafford, M. and Horning, M. G. (1972), *Analyt. Letters,* **5**, 217.
113. Thenot, J. P. and Horning, E. C. (1972), *Analyt. Letters,* **5**, 519.
114. Wayne-Moss, C. (1974), *Appl. Microbiol.*, **28**, 80.
115. Fulk, W. K. and Short, M. S. (1970), *J. Lipid Res.*, **11**, 279.
116. Klopfenstein, W. E. (1971), *J. Lipid Res.*, **12**, 773.
117. Minnikin, D. E. (1972), *Lipids,* **7**, 398.
118. Veerkamp, J. H. (1970), *Biochim. biophys. Acta,* **210**, 267.
119. Brian, B. L. and Gardner, E. W. (1968), *Appl. Microbiol.*, **16**, 549.
120. Wood, R. D., Raju, P. K. and Reiser, R. (1965), *J. Am. Oil Chem. Soc.*, **42**, 161.
121. Thomas, A. E., Sharaun, J. E. and Ralston, H. (1965), *J. Am. Oil Chem. Soc.*, **42**, 789.
122. O'Brian, J. F. and Klopfenstein, W. E. (1971), *Chem. Phys. Lipids,* **6**, 1.
123. Kuksis, A. (1972), *J. Chromat.*, **10**, 53.
124. Kuksis, A. (1972), *Can. J. Biochem,* **49**, 1245.
125. Litchfield, C. (1968), *Lipids,* **3**, 170.
126. Watts, R. B., Dils, R. and Wehr, H. (1972), *J. Chromat.*, **66**, 239.
127. Casparrini, G., Horning, M. G. and Horning, E. C. (1969), In: *Drugs affecting Lipid Metabolism*, Plenum Press, New York.
128. Cicero, T. J. and Sherman, W. R. (1971), *Biochem. biophys. Res. Commun.*, **42**, 428.
129. Cicero, T. J. and Sherman, W. R. (1973), *Analyt. Biochem.*, **54**, 32.
130. Karlsson, K. A., Samuelsson, B. E. and Steen, G. O. (1973), *Biochim. biophys. Acta,* **316**, 336.
131. Esselman, W. J., Ackermann, J. R. and Sweeley, C. C. (1973), *J. biol. Chem.*, **248**, 7310.
132. Henning, R. and Stoffel, W. (1973), *Z. Physiol. Chem.*, **354**, 760.
133. Thorpe, C. W., Pohland, L. and Firestone, D. (1969), *J. Ass. Off. Analyt. Chem.*, **52**, 774.
134. Fumagalli, R. (1969), In: *Lipid Chromatographic Analysis.*, (Marinetti, G. V., ed.), Vol. II, p. 187, Marcel Dekker Inc., New York.
135. Ishikawa, T. T., MacGee, J, Morrisson, J. A. and Glueck, C. J. (1974), *J. Lipid Res.*, **15**, 286.
136. Swell, L. S. (1966), *Proc. Soc. exp. Biol. Med.*, **121**, 1290.
137. Eto, Y. and Suzuki, K. (1972), *J. Neurochem.*, **19**, 109.

138. Sweeley, C. C. and Vance, D. (1967), In: *Lipid Chromatographic Analysis*, (Marinetti, G. V., ed.), Vol. I, p. 465, Marcel Dekker Inc., New York.
139. Clamp, J. R., Bhatti, T. and Chambers, R. E. (1971), *Meth. biochem. Analysis*, **19**, 229.
140. Zanetta, J. F. Breckenridge, W. C. and Vincendon, G. (1972), *J. Chromat.*, **69**, 291.
141. Ando, S. and Yamakawa, T. (1971), *Biochem.*, **70**, 335.
142. Gee, M. and Walker, H. G. (1962), *Analyt. Chem.*, **34**, 650.
143. Niedermeyer, W. (1971), *Analyt. Biochem.*, **40**, 465.
144. Holm, M., Mänsson, E., Vanier, L. and Svennerholm, L. (1972), *Biochim. biophys. Acta*, **280**, 356.
145. Kannan, R., Seng, P. N. and Delrich, H. (1974), *J. Chromat.*, **92**, 95.
146. Jamieson, G. R. and Reid, E. H. (1974), *J. Chromat.*, **101**, 185.
147. Karlsson, K. A., Pascher, I., Pimlott, W. and Samuelsson, B. E. (1974), *Biomed. Mass Spectr.*, **1**, 49.
148. Raheja, R. K., Kaur, C., Singh, A. and Bhatia, J. S. (1973), *J. Lipid Res.*, **14**, 695.
149. Wells, W. W., Pittman, T. A. and Wells, H. J. (1965), *Analyt. Biochem.*, **10**, 450.
150. Roberts, R. N., Johnston, J. A. and Fuhr, B. W. (1965), *Analyt. Biochem.*, **10**, 282.
151. Broekhuyse, R. M. (1968), *Biochim. biophys. Acta*, **163**, 269.
152. Roberts, R. N. (1967), In: *Lipid Chromatographic Analysis*, (Marinetti, G. V. ed.), Vol. I, p. 447, Marcel Dekker Inc., New York.
153. Agranoff, B. W., Bradley, R. M. and Brady, R. O., (1958), *J. biol. Chem.*, **233**, 1077.
154. Dittmer, J. C. and Wells, M. A. (1969), In: *Meth. Enzym.*, (Lowenstein, J. M. ed.), Vol. XIV, p. 482, Academic Press, New York.
155. Snyder, F. and Stephens, N. (1959), *Biochim. biophys. Acta*, **34**, 244.
156. Renkonen, O. (1961), *Biochim. biophys. Acta*, **54**, 361.
157. Renkonen, O. (1962), *Biochim. biophys. Acta*, **56**, 367.
158. Hanahan, D. J. and Olley, J. N. (1958), *J. biol. Chem.*, **231**, 813.
159. Garland, P. B. and Randle, P. J. (1962), *Nature*, **196**, 987.
160. Vaughan, M. J. (1962), *J. biol. Chem.*, **237**, 3354.
161. Chernick, S. S. (1969), In: *Meth. Enzym.*, (Lowenstein, J. M., ed.), Vol. XIV, p. 627, Academic Press, New York.
162. Holla, K. S. and Cornwell, D. G. (1965), *J. Lipid Res.*, **6**, 322.
163. Weigel, W. (1972), *Z. Physiol. Chem.*, **353**, 113.
164. Radin, N. S., Lavin, F. B. and Brown, J. R. (1955), *J. biol. Chem.*, **217**, 789.
165. Dubois, M. K., Gilles, K. A., Hamilton, J. K., Rebers, P. A. and Smith, F. (1956), *Analyt. Chem.*, **38**, 350.
166. Exterkate, F. A. and Veerkamp, J. H. (1971), *Biochim. biophys. Acta*, **231**, 545.
167. Maker, H. S., Lehrer, G. M., Weiss, C. and Levine, L. (1974), *Analyt. Biochem.*, **61**, 471.
168. Bergmeyer, H. U., Bernt, E., Schmidt, F. and Stork, H. (1970), In: *Methoden der Enzymatischen Analyse*, (Bergmeyer, H. U. ed.), Vol. 2, p. 1163, Verlag Chemie, Weinheim.
169. McLean, C., Werner, D. A. and Aminoff, D. (1973), *Analyt. Biochem.*, **55**, 72.
170. Dixon, J. S. and Lipkin, D. (1954), *Analyt. Chem.*, **26**, 1092.

171. Plackett, P. (1967), *Biochemistry,* **6**, 2746.
172. Kabat, E. A., and Mayer, M. M. (1961), *Experimental Immunochemistry,* p. 548, Thomas, C. C. Springfield, USA.
173. Böttcher, C. J. F., Van Gent, C. M. and Pries, C. (1961), *Recl. Trav. chim. Pays-Bas. Belg.,* **80**, 1157.
174. Sloane-Stanley, G. M. (1967), *Biochem. J.,* **104**, 293.
175. Lea, C. H. and Rhodes, D. N. (1955), *Biochim. biophys. Acta,* **17**, 416.
176. Wheeldon, SL. W. and Collins, F. D. (1957), *Biochem. J.,* **66**, 435.
177. Siakotos, A. N. (1967), *Lipids,* **2**, 67.
178. Böttcher, C. J. F., Pries, C. and Van Gent,C. M. (1961), *Recl. Trav. chim. Pays-Bas. Belg.,* **80**, 69.
179. Reid, W. D., Haubrich, D. R. and Krishna, G. (1971), *Analyt. Biochem.,* **42**, 390.
180. McKibbin, J. M. M. (1967), In: *Lipid Chromatographic Analysis,* (Marinetti, G. V. ed.), Vol. I, p. 497, Marcel Dekker Inc., New York.
181. De Koning, A. J. (1966), *Analyst,* **91**, 523.
182. Work, E. (1964), *Nature,* **201**, 1107.
183. Gatt, R. and Berman, E. R. (1966), *Analyt. Biochem.,* **15**, 167.
184. Slonekar, J. H. (1972), *Meth. Carbohyd. Chem.,* **6**, 20.
185. Kärkkäinen, J. and Vihko, R. (1969), *Carbohyd. Res.,* **10**, 113.
186. Varma, R. S., Varma, R., Allen, W. S. and Wardi, A. H. (1974), *J. Chromat.* **93**, 221.
187. Varma, R., Varma, R. S., Allen, W. S. and Wardi, A. H. (1973), *J. Chromat.,* **86**, 205.
188. Warren, L. (1959), *J. biol. Chem.,* **234**, 1971.
189. Svennerholm, L. (1958), *Biochim. biophys. Acta,* **24**, 604.
190. Craven, D. A. and Gehrke, C. W. (1968), *J. Chromat.,* **37**, 414.
191. Siakotos, A. N. and Rouser, G. (1971), *Lipids,* **6**, 254.
192. Coles, L. (1970), *J. Lipid Res.,* **11**, 164.
193. Naoi, M., Lee, C. and Roseman, S. (1974), *Analyt. Biochem,* **58**, 571.
194. Ways, P. and Hanahan, D. J. (1964), *J. Lipid Res.,* **5**, 318.
195. Broekhuyse, R. M. (1969), *Biochim. biophys. Acta,* **187**, 354.
196. Kanfer, J. N. and Spielvogel, C. (1973), *J. Neurochem.,* **20**, 1483.
197. Palmer, F. B. (1971), *Biochim. biophys. Acta,* **231**, 134.
198. Brain, K. R. and Turner, T. D. (1971), *J. Chromat.,* **61**, 157.
199. Mangold, H. K. (1969), In: *Thin Layer Chromatography,* (Stahl, E. ed.), p. 363, Springer-Verlag, New York.
200. Exterkate, F. A. and Veerkamp, J. H. (1969), *Biochim. biophys. Acta,* **176**, 65.
201. Svennerholm, E. and Svennerholm, L. (1963), *Biochim. biophys. Acta,* **70**, 432.
202. Kuhn, R. and Wiegandt, H. (1963), *Chem. Ber.,* **96**, 866.
203. Heinz, E. (1967), *Biochim. biophys. Acta,* **144**, 321.
204. Sastry, P. S. (1974), *Adv. Lipid Res.,* **12**, 251.
205. Lisboa, B. P. (1969), In: *Lipid Chromatographic Analysis,* (Marinetti, G. V., ed.), Vol. 2, p. 57, Marcel Dekker Inc., New York.
206. Eneroth, P. (1969), In: *Lipid Chromatographic Analysis,* (Marinetti, G. V., ed.), Vol. II, p. 149, Marcel Dekker Inc., New York.
207. Ramwell, P. W. and Daniels, E. G. (1969), In: *Lipid Chromatographic Analysis,* (Marinetti, G. V., ed.), Vol. 2, p. 313, Marcel Dekker Inc., New York.
208. Morris, L. J. (1966), *J. Lipid Res.,* **7**, 717.

209. Renkonen, D. (1971), In: *Progress in Thin-Layer Chromatography and Related Methods*, (Niederweiser, A. and Patalei, G., eds.), Vol. II, p. 143, Ann Arbor Publishers, Ann Arbor.
210. Marinetti, G. V. (1964), *New Biochemical Separations*, p. 339, Van Nostrand, Princeton, N. J.
211. Vaskovsky, V. E. and Kostetsky, E. Y. (1968), *J. Lipid Res.*, **9**, 396.
212. Vaskovsky, V. E. and Svetaskov, V. I. (1972), *J. Chromat.*, **65**, 451.
213. Gitler, C. (1972), *Analyt. Biochem.*, **50**, 324.
214. Shaw, N. (1968), *Biochim. biophys. Acta*, **164**, 435.
215. Siakotos, A. N. and Rouser, G. (1965), *J. Am. Oil Chem. Soc.*, **42**, 913.
216. Svennerholm, L. (1957), *Biochim. biophys. Acta*, **24**, 604.
217. Miettinen, T. and Takki-Luukainen, J. T. (1959), *Acta chem. scand.*, **13**, 856.
218. Goswani, S. K. and Fray, C. F. (1970), *J. Chromat.*, **53**, 389.
219. Skidmore, W. D. and Entenman, C. (1962), *J. Lipid Res.*, **3**, 471.
220. Mangold, H. K. (1961), *J. Am. Oil Chem. Soc.*, **38**, 708.
221. Vaskovsky, V. E. and Suppes, Z. S. (1971), *J. Chromat.*, **63**, 455.
222. Stahl, E., Ed. (1965), *Thin-layer Chromatography: a laboratory handbook*, Springer-Verlag Berlin.
223. Broekhuyse, R. M. (1976), *Clinica chim, Acta*, **66**, 88.
224. Zlatkis, A., Zak, B. and Boyle, A. J. (1963), *J. Lab. clin. Med.*, **41**, 486.

ERRATA (Chapter 9)

*Page* 324 line 30 should read
(2) Horse-radish peroxidase (Worthington or Sigma Type II) [17 units] is dissolved in

*Page* 351 reference 108, *Analyt. Biochem.* **12**, 509.

# 9 Techniques for the analysis of membrane carbohydrates

G. M. W. COOK



## 9.1 Introduction

The chemistry of the plasma membrane and the cell surface is a subject which has been of considerable interest to workers in many facets of biology for a number of years. This is understandable since a detailed knowledge of the biochemistry of this organelle is essential for explaining in molecular terms a wide variety of biological phenomena – the interaction of cells with one another or their environment. In this respect the carbohydrate constituents of animal cell membranes are now considered to be especially important, conferring upon the cell surface a degree of molecular variation not easily possible with other classes of molecules present in biological membranes. The large number of examples of biological recognition would suggest that a molecular architecture unique to the periphery of each particular cell type is indeed a fundamental requirement of cells.

Carbohydrates are found in significant quantities in a number of subcellular fractions though in terms of quantity, the amount of carbohydrate present in plasma membranes is quite small in comparison to the protein and lipid content, usually being of the order of 2–10% of the total membrane constituents. Indeed the carbohydrate content of plasma membranes is disproportionately small in comparison with the important biological properties they bestow on this organelle. The importance of studying membrane carbohydrates is to be able to draw comparisons between structural entities and biological function. In order to deduce the biological role of a particular monosaccharide it is not sufficient just to estimate its level in an isolated membrane, rather the purpose of studying membrane carbohydrates is to see how structural variations in the oligosaccharide moieties of plasma membranes influence changes in biological properties.

The methods for estimating sugars and solving oligosaccharide structures are well defined and a number of the fundamental chemical techniques are described in Section 9.5. The special problems associated with the study of membrane carbohydrates are basically those of obtaining enough material in a sufficiently purified form on which to perform the necessary chemical procedures. Two main approaches have been made towards elucidating the chemistry of membrane carbohydrates namely what Kraemer [1] terms 'non-disruptive' and 'disruptive' studies. In the former studies the cell, and whenever possible its permeability barrier, is kept intact and the constituent sugars of the cell periphery studied by solubilizing fragments of the plasma membrane by, for example, enzyme treatments, whereupon these fragments can be isolated from the supernatant fluid and subjected to conventional analytical and structural techniques. Alternatively the presence of a particular sugar residue may be inferred by the use of carbohydrate binding proteins as cell surface probes. On the other hand 'disruptive' studies cover those procedures which are performed on isolated plasma membrane fractions. It is these two lines of investigation, peculiar to membrane carbohydrates, which will be dealt with in the main body of this chapter. It is not intended to give a complete historical review of the subject but rather to concentrate on the principles on which the experimental design of these approaches are based, together with the problems of interpretation of the results. However, where applicable these points will be illustrated by reference to representative papers in the literature.

## 9.2 Structure of complex carbohydrates

The complex carbohydrates of the higher animals may be classified as glycoproteins, glycolipids and proteoglycans. The proteoglycans consist of glycosaminoglycans (or mucopolysaccharide) sidechains attached to a protein *backbone.* As regards cell membranes, it is the first two classes of macromolecules which will be the more usually encountered. In the case of the stroma of the human erythrocyte, which has so often been used as the classical model for plasma membranes, the carbohydrate content is approximately 8% by weight of the membrane. Only 7% of this carbohydrate is in glycosphingolipid, the bulk of the erythrocyte membrane carbohydrate being sialoglycoproteins [2].

Glycoproteins may be regarded as those proteins with covalently bound carbohydrate prosthetic groups. Whilst the glycoproteins are not distinguishable

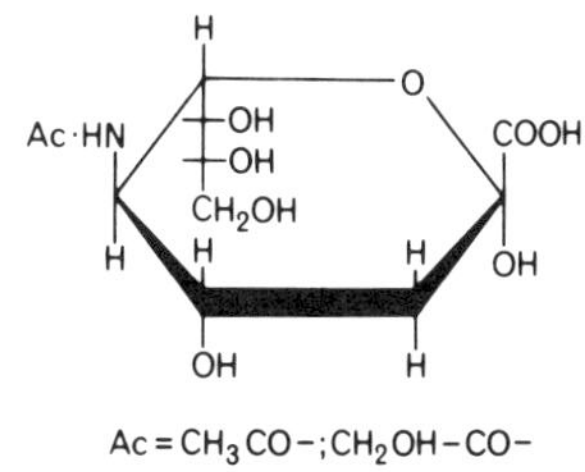


**Fig. 9.1** Sialic acid is the group name for all acylated neuraminic acids; the term neuraminic acid being reserved for the unsubstituted structure $C_9H_{17}O_8N$ common to all. This figure illustrates two of the commonly occuring sialic acids–*N*-acetylneuraminic acid (Ac = $CH_3CO$–) and *N*-glycolylneuraminic acid (Ac = $CH_2OH$–CO–). In addition to the *N*-acetyl group the molecule can, in addition, be acetylated at other positions (e.g. C–4, C–7) in *N*-acetyl-*O*-diacetylneuraminic acid and (at C–4) in *N,O*-diacetylneuraminic acid. In glycoproteins the sialic acid residue is linked α-ketosidically at C–2 usually to galactose or *N*-acetylgalactosamine; this is the bond which is sensitive to neuraminidase (EC 3.2.1.18). The sialic acid may either be attached at the C–3 or C–6 position on the partner sugar, the position of attachment of the sialic acid residue may influence the rate of hydrolysis by a neuraminidase (e.g. *Vibrio cholerae* neuraminidase cleaves both 2–3 and 2–6 bonds readily whilst the myxovirus neuraminidase cleaves the 2–3 linkage at a much faster rate). The carboxyl group (p$K_a$ 2.6 for *N*-acetylneuraminic acid) of the sialic acids is responsible for a significant portion of the net negative surface charge of animal cells.

by their amino acid composition, the carbohydrate moieties consist of a characteristic group of sugars. Of the neutral sugars, two stereoisomers of D-glucose, D-galactose and D-mannose are readily found, glucose itself is not a common constituent of glycoproteins and its presence is largely confined to collagen. In addition the carbohydrate portion includes the amino sugars D-glucosamine and D-galactosamine, invariably as *N*-acetyl derivatives, as well as L-fucose and the sialic acids. Though glycoproteins do not contain uronic acids and have a scarcity of sulphate they are readily distinguishable from the glycosaminoglycans by the size of their carbohydrate groups. The carbohydrate side chains of glycoproteins are relatively short and in the membrane glycoproteins examined so far are found to range from four to eight glycose residues. In contrast the proteoglycans possess much larger carbohydrate chains composed of a repeating disaccharide units. Generally glycosaminoglycans contain carbohydrate chains which are larger but structurally simpler. In the glycoproteins of membranes two types of linkage of carbohydrate to protein occur, namely an alkali-stable *N*-glycosidic bond involving the amide nitrogen of

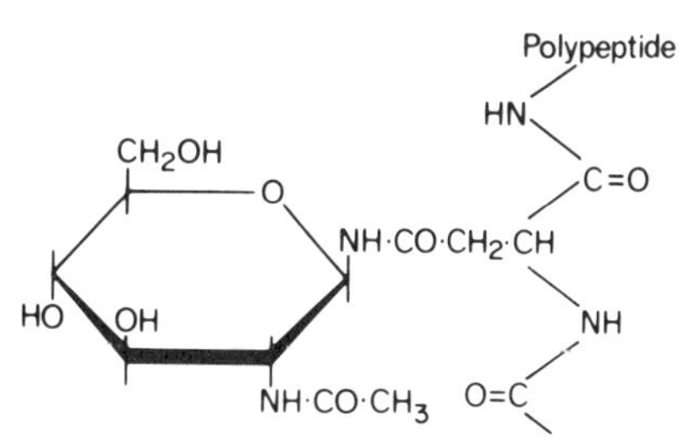


**Fig. 9.2** Diagram illustrating the *N*-glycosidic linkage of carbohydrate to protein. The *N*-acetylglucosamine-asparagine linkage is through a nitrogen atom, in all other types of carbohydrate-protein linkage the bond involves an oxygen atom. For the purposes of this diagram only the first sugar of the carbohydrate group of a glycoprotein is shown.

asparagine and the C-1 of an amino sugar residue and alkali-labile *O*-glycosidic bonds to serine or threonine. The former linkages have been regarded as characteristic of *serum type* (others have used *plasma type* in a synonymous manner) and the latter as *mucin* type glycoproteins. Kraemer [1] has persuasively argued that a determination of the carbohydrate-protein linkage in an unknown glycoprotein 'in itself, yields considerable predictive power for the types of structural patterns to be expected in the carbohydrate moiety'.

Earlier studies suggested that membrane glycoproteins were more like the *mucin* type for, as Eylar [3] observed, they have a much higher ratio of galactosamine to glucosamine than plasma proteins and a predominance of serine and threonine, as opposed to asparagine, linked to the carbohydrate. With the present state of knowledge it may, however, be premature to use this classification rigorously for membrane glycoproteins.

**Fig. 9.3** Diagram illustrating the *O*-glycosidic linkage between carbohydrate and protein commonly found in glycoproteins. In this linkage *N*-acetylgalactosamine can bind to serine, as shown in this figure, or threonine. As in the case of Fig. 9.2 for purposes of diagrammatic clarity only the first sugar of the carbohydrate group of the glycoprotein is shown.

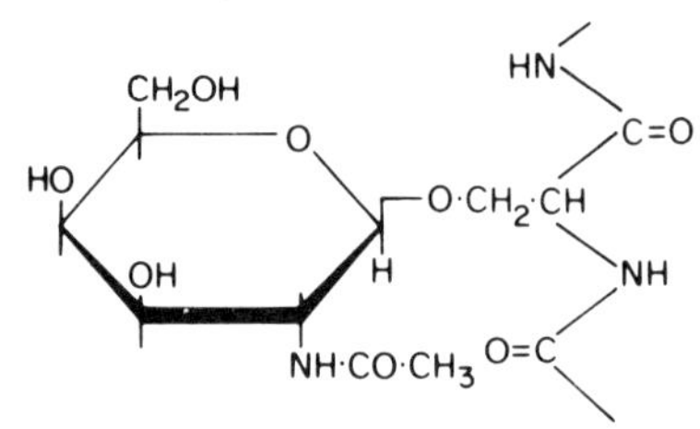


The other major group of complex carbohydrates commonly encountered in membranes are the glycolipids. For some years it was held that glycolipids were characteristic of such specialized tissue as brain but it is now clear that these materials are a common feature of plasma membranes. The glycolipids of the higher animals are glycosphingolipids, that is they are compounds composed of ceramide to which is attached a carbohydrate moiety. Ceramide is an *N*-acyl derivative of sphingosine. The carbohydrate fraction of the glycolipids contains, with the exception of mannose, the same sugars as are found in the glycoproteins. In addition, glucose is a common constituent of these compounds. Based on the nature of the composition of the carbohydrate moiety of these materials one can classify the glycolipids into cerebrosides and haematosides. The latter are glycolipids which contain sialic acid but no hexosamine whilst globosides are aminoglycolipids, containing hexosamine but no sialic acids. The term ganglioside covers all glycolipids containing sialic acids, although it is a term which some authors prefer to restrict to the sialoglycolipids of the central nervous system. Though these terms are useful for subdividing the glycolipids they are basically trivial terms which convey nothing of the structural relationship of one glycolipid to another.

## 9.3 Non-disruptive studies of membrane carbohydrates

The rationale behind this approach to the study of membrane carbohydrates is that by removing components from the intact cell by enzymatic means one is more likely to be examining material derived from the plasma membrane as opposed to the cytoplasm. If the integrity of the cell is maintained it may be possible to monitor any change in cellular behaviour brought about the removal of membrane components thus providing possible evidence for the biological function of a particular membrane constituent. In addition *non-disruptive* studies have the advantage over the *disruptive* method in that membrane artefacts produced as a result of the cell fractionation procedures are eliminated. This is not to say that the results of non-disruptive studies are free of difficulties of interpretation and the possible problems associated with this type of study will be discussed in this section.

### *9.3.1 Sialic acids*

In the early 1960's considerable interest focused upon the sialic acids in plasma membranes and their removal from the cell periphery by treating intact cells with neuraminidase (*N*-acetylneuraminate glycohydrolase EC3.2.1.18) provided a useful technique for studying the possible role of a particular sugar residue in this organelle. Sialic acids are a group of acylated neuraminic acids including *N*-acetylneuraminic acid, *N*-glycolylneuraminic acid, *N.O* diacetylneuraminic acid and *N*-acetyl-*O*-diacetylneuraminic acid, though only the first two are likely to be encountered in membrane studies. The *O*-acetyl forms are marked by an acid lability of their *O*-acetyl groups and particularly careful conditions are required in their isolation and characterization. In the case of glycoproteins with a low isoionic point it is possible to release sufficient amounts of the *O*-acetyl forms of sialic acid by heating the glycoprotein dissolved in distilled water at its own pH for 1–3 hours at 100°C for subsequent indentification. Following hydrolysis in boiling water, the hydrolysate may be dialysed against distilled water and the dialysate concentrated by lyophilization for chromatographic examination.

Because these sialic acids possess a relatively strong carboxyl group [p$K_a$ 2.6 – for *N*-acetylneuraminic acid] their presence at the cell periphery makes a significant contribution to the net negative surface charge of a number of cells. Indeed the charge properies of the sialic acids have been particularly useful in detecting their presence and consequently that of the glycoproteins at the cell periphery. The action of neuraminidase on intact and viable cells was readily monitored by means of the cell electrophoresis technique [4–7], and the concomitant presence of free sialic acids being detectable in the supernatant fluid. In cell electrophoresis, in common with other aspects of membrane study, the saline washed human erythrocyte was taken as an ideal model, so it is not surprising that the initial studies on sialic acids and cell surfaces should have been conducted on erythrocytes. Some details of the techniques used in analysing the supernatant fluids from neuraminidase treated cells will be covered. In addition the attendant problem of using neuraminidase for surface studies of intact cells will be discussed.

*(i) Estimation of sialic acids*

A number of colorimetric methods are available for estimating sialic acids. These methods may be classed as one-step or two-step procedures. In the former procedure the reagent is added directly to the sample under examination and at the conclusion of the reaction the colour-formation is measured in a spectrophotometer. The two-step method involves the release of the sialic acid from the substrate by hydrolysis in dilute acid or by treatment with neuraminidase before the assay can be performed. The Bial's orcinol [9] and Svennerholm resorcinol [10] procedures are examples of the one-step procedure whilst the use of 2-thiobarbituric acid [11,12] for the estimation of sialic acids is a two-step method. Both the resorcinol [10] and the 2-thiobarbituric acid [11] methods are described in detail in Section 9.5. Brown *et al.* [8] have described a factor (probably a protein-bound bivalent form of iron) in liver, kidney and intestinal mucosa of rabbit which causes the sialosyl residues in some glycoproteins to react as free sialic acid in the thiobarbituric acid assay [11].

The one-step resorcinol procedure [10] has been used to analyze the supernatant fluids of neuraminidase treated human erythrocytes [4]. Though this procedure is easy to perform it has been largely superseded by the considerably more sensitive methods [11,12] involving the use of 2-thiobarbituric acid. In this latter procedure once the sialic acids have been released by acid hydrolysis or neuraminidase they are then subjected to periodate oxidation in phosphoric acid [11]. Aminoff [12] makes the point that the high acidity of the periodate reagent as recommended by Warren [11] is not essential for colour formation and that a selective choice of conditions for oxidation can be more informative, for example oxidation at pH 1.2 and 7 enables one to distinguish between acid-labile compounds and free sialic acids. When oxidized at pH7 free *N*-acetylneuraminic acid still gives an intense chromophore whilst cytidine monophosphosialic acid though giving a colour reaction at pH 1.2 gives but a trace at neutral pH. Following oxidation excess periodate is destroyed with arsenite and a pink chromophore, with absorption maxima at 549 nm, formed by heating the mixture with 2-thiobarbituric acid. This chromophore is then extracted into cyclohexanone [11], or acid butanol (butan-1-ol containing 5% (v/v) 12 M HCl) [12], where the colour is far more intense and stable than in water, for spectrophotomeric evaluation. Chemical evidence indicates that $\beta$-formyl pyruric acid formed on periodate oxidation of sialic acids when coupled with 2-thiobarbituric acid forms the pink chromophore. A whole range of sugars likely to be encountered in complex carbohydrates of animal cells has been tested [11] in this reaction and with the exception of L-fucose none of them produced a chromophore or interfered with the colour formation. L-fucose decreases the optical density in the method by 35% and is considered to be due to the formation of acetaldehyde following periodate oxidation of this sugar: 1 $\mu$mol of this aldehyde inhibiting colour formation to a similar extent as 1 $\mu$mol of L-fucose. In the case of the neuraminidase treatment of intact cells the release of appreciable quantities of fucose is unlikely, though where dilute acid treatment is used, for example when examining isolated membrane fractions, the presence of this sugar cannot be discounted because of the location of fucosyl residues at the non-reducing terminus of carbohydrate groups and their ease of hydrolysis.

The great advantage of the 2-thiobarbituric acid procedure over the one-step methods is the increased sensitivity of the method (the molar extinction coefficient for *N*-acetyl neuraminic acid in the 2-thiobarbituric acid method [11] is some 12 times greater than in the resorcinol technique [10]). Tissue cells are invariably available in considerably smaller quantities than red blood cells so the increased sensitivity of the 2-thiobarbituric acid procedure, compared to resorcinol, is of particular value when examining the sialic acids present in the plasma membranes of cells derived from tissue cultures. Unfortunately the estimation by 2-thiobarbituric acid of sialic acids in the supernatant fluids derived from nucleated cells may be complicated by the presence of 2-deoxyribose, which gives a strong red colour in the 2-thiobarbituric acid reaction probably due to the formation of malonaldehyde following periodate oxidation of this sugar. The chromophore produced from this latter sugar has an absorption maximum at 532 nm, therefore dichromatic readings at 532 nm and 549 nm should be made so that an appropriate correction (see Section 9.5) can be made. A survey of the literature indicates that this procedure is largely followed by investigators working with membrane systems, though when very large amounts of 2-deoxyribose derivatives are present a differential extraction procedure may be useful for the estimation of the sialic acid chromophore. When an aqueous solution of the chromophores is extracted with two volumes of isoamyl alcohol 85% of the chromophore with maximum at 532 nm is removed as compared to only 10 to 15% of the 549 nm peak [11]. Subsequent extraction of the aqueous material with cyclohexanone affords a partially purified sialic acid chromophore. Alternatively whilst the chromophore derived from 2-deoxyribose is stable to the addition of strong base the chromophore due to sialic acid is destroyed under such conditions and therefore could be used as a means for determining the contribution of interfering deoxy sugars to the optical density at 549 nm.

A way of obviating the problems raised by the presence of contaminants would be to separate the sialic acid by column chromatography, using Dowex-1-formate resin. Unfortunately this is not always a practical solution to the problem when one is handling very limited quantities of material. This latter point is also illustrated by workers using conditions of acid hydrolysis to determine total cellular sialic acid levels and will be returned to later in this section. Such estimations of total sialic acid are necessary as one way in which to gauge whether this material is unique to the cell membrane; the results of such analysis being compared to the amount of sialic acid released by treating the intact cell with neuraminidase. Wallach and Eylar [13] using the 2-thiobarbituric assay [11] on acid hydrolysates of Ehrlich ascites-carcinoma cells found considerable amounts of material which could not be removed by chromatography on Dowex-1 resin or by extraction by isoamyl alcohol and therefore it was necessary for an alternative correction procedure suggested by Warren [11] to be followed, namely the absorbancy at 562 nm due to *N*-acetyl neuramic acid was corrected for absorbancy of 2-deoxyribose at 532 nm. Wallach and Esandi [14 ] found that the measurement of total sialic acid by the acid hydrolysis of cells was complicated by the presence of lipid breakdown products and these authors had to confine their measurements to total non-lipid sialic acid performed on lipid extracted material. McQuiddy and Lilien [15] also experienced difficulty assaying for total sialic acids when they examined acid hydrolysates (0.1 N $H_2SO_4$ at 80° for 1 h) of chick embryo neural retina cells. They comment that the recommended hydrolysis methods were originally developed for the analysis of relatively purified glycoprotein samples; this is a very useful point and one which the potential investigator in this field should not forget. Dilute homogenates gave a higher yield of sialic acid but then the amount of hydrolysate that can be suitably handled on a column is a limiting factor. Assuming optimal conditions are obtained, hydrolysis of total tissue homogenates in which the sialic acid level is extremely low when compared to other deoxy sugars also complicates the assay and again presents an entirely different problem to that experienced with isolated glycoproteins. Certainly, if this situation is not fully appreciated, erroneous interpretation of results may arise in the latter case from underestimation of sialic acid when correcting for chromophore due to contaminants. This point is nicely illustrated in experiments performed by Rosenberg and Einstein [16], these workers added varying amounts of DNA to a constant amount of a mucin preparation (as a source of sialoglycoprotein) after which the sialic acid present was determined, by use of the 2-thiobarbituric acid assay [11], following hydrolysis in dilute acid (0.1 N $H_2SO_4$ at 80°C for 1 h). Correcting the absorbancy at 549 nm by reference to the absorbancy at 532 nm was adequate for correcting for amounts of DNA less than 100 μg, however, Rosenberg and Einstein [16] were able to show experimentally that above this amount overcorrection resulted, that is to say the amount of sialic acid estimated was lower than that actually present.

In addition to the problem of estimating sialic acids in tissue hydrolysates the experimenter working with cells should consider the possibility that the suspending media (tissue culture media supplemented with serum) or the environment (body fluids in case of primary cultures) from which the cells under study are removed may well contribute to the analytical result when studying the action of neuraminidase on intact cells. Consequently, washing conditions for the cells under study should be established prior to enzymatic digestion.

*(ii) The use of neuraminidase*

In experiments designed to elucidate the function of sialic acids in plasma membranes by treating intact cells with neuraminidase, in addition to the problems associated with their chemical estimation, the purity of the enzyme preparation used is important. Neuraminidase is available from a number of sources both biological and commercial and it is essential that the experimenter check carefully the suitability of any particular preparation for the intended experiments. The work of Kraemer [17] on commercial preparations of neuraminidase derived from *Clostridium perfringens* illustrates this point well. As Kraemer [17] points out any experimenter intending to investigate the relationship between membrane bound calcium and sialoglycan structure would find *C. perfringens* neuraminidase a particularly useful *tool*, this enzyme being indifferent to calcium ions as regards activity and is indeed fully active in the presence of EDTA. However, Kraemer [17] found that a commercial sample of this enzyme contained cytotoxic, haemolytic and phospholipase activities and was not satisfactory for investigations of membrane-bound sialic acids unless it could be further purified. In addition the work of Emmelot and Bos [18–20], also serves as an example of the potential problem posed by contaminants in neuraminidase preparations. Though neuraminidase was shown to have a differential effect on the ATPases and 5′ nucleotidase, as well as on $K^+$ dependent *p*-nitrophenyl phosphatase [19] of rat liver plasma membranes, it was found [20] that the effects of the neuraminidase on the membrane-bound activities could not be due to the release of sialic acid. Notwithstanding the failure to detect proteolytic and phospholipase activity in the neuraminidase used in their earlier studies [17,18] Emmelot and Bos [20] observed that 'the present results force us to conclude that this preparation was contaminated by some other enzymes or inhibitor which must have been responsible for the enzyme inhibitions'. Any worker who had not performed the extensive studies which Emmelot and Bos [18–20] had undertaken might well have been misled into assigning a spurious role to sialic acids in membrane function. These latter authors also sound another cautionary note when they point out that the nature of the inhibitory agent in the preparation of neuraminidase used by themselves cannot be traced since the particular material 'is no longer on the market'; certainly now that the problems associated with this aspect of membrane chemistry are well documented the need to check the enzyme preparation used carefully is paramount and would appear to be an aspect of experimental technique which is so often overlooked.

Before leaving this aspect of the subject a recent paper of Yuhas *et al.* [21] on *Vibrio cholerae* neuraminidase and cell viability should be mentioned. It was found [21] that the viability of a line of murine malignant alveolar carcinoma cells was unaffected by neuraminidase treatment as judged by either eosin or trypan blue exclusion tests. On the other hand when viability *in vitro* of these same cells was assessed *in vitro* by either colony formation or $^3$H-thymidine incorporation neuraminidase treatment was found to bring about a reduction in viability. It remains to be seen whether this cell line is unique as regards the observed phenomena and whether the reduction in reproductive integrity which was measured was as a result of the removal of sialic acid residues or some additional enzymatic effects of the enzyme preparation. Certainly this work has important implications for those workers interested in designing experiments to test the possible role of membrane sialic acids in 'masking' tumour-specific antigens. No doubt investigators working in this area of membrane carbohydrates would be wise not to rely entirely on vital staining techniques and if suitable separation techniques can be devised the use of neuraminidase linked to a suitable support (neuraminidase conjugated to agarose is commercially available) might be a useful tool, by helping to confine the action of the enzyme to the cell periphery.

### *9.3.2 Use of proteolytic enzymes*

#### *(i) Enzyme digestion*

The use of proteolytic enzymes in elucidating glycoprotein structure is well documented. A general feature of structural studies on the carbohydrate groups of glycoproteins, as well as the elucidation of the nature of the carbohydrate-polypeptide linkage is the initial digestion of the macromolecule with a protease of broad specificity (e.g. pronase) such that the carbohydrate groups of the glycoprotein may be obtained with a minimum number of amino acid residues attached. A glycopeptide is usually more suceptible to the action of glycosidases than the intact glycoprotein; in addition the macromolecule may contain more than one type of carbohydrate group which would complicate structural analyses, a glycopeptide obviates this problem. In the early 1960's a number of papers on the effect of proteolytic enzymes on cells, principally erythrocytes, showed that carbohydrate containing fragments, glycopeptides, were cleaved from the membrane [22]. Therefore if the action of proteolytic enzymes on intact cells can be suitably controlled and the integrity of the cells preserved and thus the effects of the enzyme treatment on the cell monitored (i.e. by cell electrophoresis, immunological techniques), one is provided with the potential means not only of demonstrating the presence of a particular oligosaccharide moiety at the cell periphery but also of isolating it and examining its chemical structure. If these glycopeptides still possess biological activity one is provided with a unique source of material for direct structural examination of important features of the cell membrane. In addition to the release of glycopeptides from intact cells the proteolysis of isolated membranes and membrane glycoproteins is also equally important for the structural investigation of membrane complex carbohydrates.

As the methods for investigating the structure of glycopeptides, once isolated, are identical whether the material is derived from intact cells or isolated membranes or glycoproteins, the principles involved will be discussed here, though the study of glycopeptides derived from the latter two sources might well have been considered in the section of this chapter dealing with disruptive procedures. The isolation of glycopeptides from intact cells does present certain special problems not only of technique but also of interpretation, on the other hand characterization of glycopeptides from isolated membranes also presents questions peculiar to the source material. A glycopeptide derived from a well-characterized (by means of enzymatic and morphological studies) membrane preparation may mean that though the subcellular site of the particular residue under study can be stated with greater assurance, the problem of changes induced as a result of the isolation procedure cannot necessarily be ignored. Conversely, the availability of an isolated membrane glycoprotein may make it possible to obtain glycopeptides, for structural examination, which are not available to the action of the protease in the intact cell. An example of the latter material is the $\beta$-glycopeptide of Jackson *et al.* [23]. This glycopeptide was isolated following the digestion of the major sialoglycoprotein of the human erythrocyte with trypsin and as Jackson *et al.* [23] point out was not isolated by investigators working with intact cells, presumably because the particular trypsin-sensitive peptide linkages are not available when the molecule is *in situ*. The special problem of relocation of membrane constituents within the plane of the membrane following enzyme treatment should also be considered and will be dealt with in further detail in the section on lectins. Those facets of the subject peculiar to work with intact cells will be dealt with now, before describing the general chemical procedures used in elucidating their structure.

With the human erythrocyte the question of assigning isolated glycopeptides to the plasma membrane presents no problem but in the case of tissue cells the situation is somewhat more complicated. Tissue cells possessing a multimembrane system might well yield glycopeptides from intracellular membranes as well as a number of contaminating fragments derived from the cytoplasm. These are the points which the investigator contemplating the removal of surface constituents by proteolytic enzymes should consider. The work of Snow and Allen [24] on the release of materials by trypsin (EC3.4.4.4) from monolayers of BHK cells is a particularly useful guide to the problems associated with this field. The release of nucleic acids from these cells was monitored by labelling the cellular RNA with [5-$^{3}$H] uridine and cellular DNA with [2-$^{14}$C] thymidine. Incorporation of [1-$^{14}$C] glucosamine solely into the hexosamines and sialic acids of high weight macromolecules acted as a marker for complex carbohydrates. Snow and Allen [24] showed that crystalline trypsin released 24% of the total macromolecules containing amino sugars (as measured by radioactive marker) and 39% of the total sialic acid of the cells (measured with the thiobarbituric acid reaction [12] following anion exchange chromatography) but only 1% of the cellular nucleic acids and almost all (99%) of the cells were impermeable to trypan blue. On the other hand whilst a crude trypsin preparation released 33% of the total macromolecules containing amino sugars it also caused lysis of the cells and the release of similar quantities of RNA and DNA amounting to 11% of total cellular nuclei acid. This work would indicate that in experiments designed to remove surface glycopeptides it is essential to use crystalline trypsin, for although crude trypsin also removes surface glycopeptides the material is likely to be heavily contaminated with intracellular material.

Jett and Jamieson [25] have performed a related study. Conscious of the fact that the nature of the glycoprotein component of the cell membrane is frequently inferred from the nature of the glycopeptides obtained by proteolytic enzyme treatment of intact cells, these authors set out to test the validity of this assumption using RAJ1 (a human Burkitt lymphoma) lymphoid cells. Using [1-$^{14}$C] glucosamine as a marker, they were able to show that trypsin treatment of intact labelled viable cells resulted in the release of three radioactive products of molecular weights 15 000, 60 000 and over 200 000 (the materials were fractionated on a calibrated column of Sephadex G-200). However, these workers were only able to obtain the two components of lowest molecular weight by treating isolated plasma membranes, labelled with [1-$^{14}$C] glucosamine, with trypsin. The 200 000 molecular weight component was absent and could only be isolated, apart from intact cells, from the cytoplasm or from the culture medium. This finding led these authors to conclude that this high weight material may not be a typical glycopeptide but a secretory product, rather than a cell surface component, and that its release is accelerated by trypsin treatment. Jett and Jamieson [25] emphasize the importance of identifying the subcellular origin of the material released by the proteolysis of intact cells and particularly when incorporated radioactive sugars are used as markers.

It has to be admitted that the majority of the structural studies on membrane glycoproteins have been carried out on glycopeptides derived from erythrocytes and that in future more attention will have to be paid to tissue cells in order to obtain a more comprehensive knowledge of membrane carbohydrates.

In the examples quoted above of glycopeptides released from intact cells reference has been made to treatment with trypsin though other proteolytic enzymes have been used, for example papain treatment of tumour cells (Walborg *et al.* [26]). It is useful then to treat such glycopeptides with a protease of broad specificity, such as pronase from *Streptomyces griseus* in order to minimize variation in their peptide chain length. Ideally, one would wish to obtain the carbohydrate group with only one amino acid but this is not the case with membrane-derived material. Pronase digestion of glycopeptides released by papain treatment has been used by Wray and Walborg [27] whilst the direct use of pronase on intact cells has been performed by Cook and Eylar [28]. The latter work involved erythrocytes which retain their cellular integrity following treatment with pronase though this may not be the case for all cell types. Allen and Cook [29] found for example that pronase digestion caused extensive damage to murine macrophages.

### *(ii) The study of glycopeptides released by enzymes*

Following digestion of intact cells or glycopeptides with proteolytic enzymes, the initial stage in the isolation of the glycopeptides is to separate them from the contaminating peptides and amino acids and this is best accomplished by gel filtration, usually with Sephadex G-25 or G-50, when the glycopeptides can be resolved from the peptides and amino acids, which being usually of lower molecular weight penetrate the gel. Glycopeptides derived from the trypsinization of human erythrocytes have calculated molecular weights of 16 000; 6000; 2000 and 11 000 for $\alpha_1$, $\alpha_2$, $\alpha_3$ and $\beta$-glycopeptides respectively (Jackson *et al.* [23]). Jackson and Seaman [30] quote values of 12 000 and

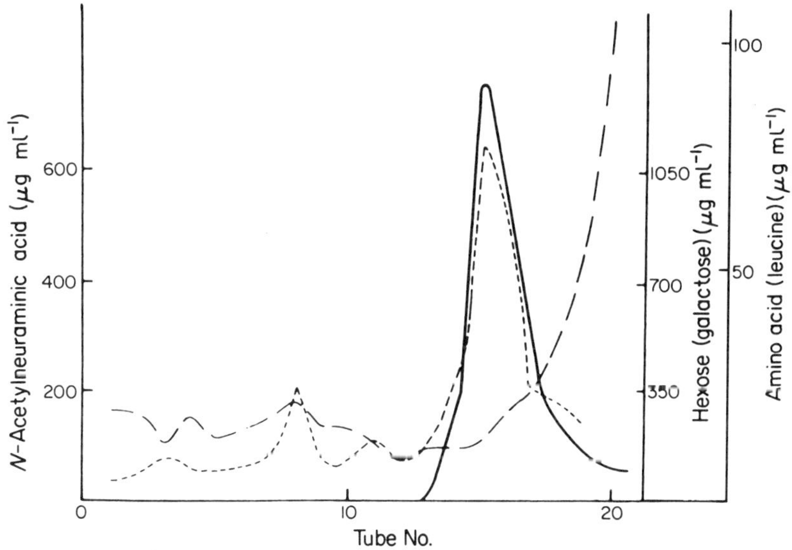


**Fig. 9.4** Separation of glycopeptides from contaminating peptides and amino acids by means of gel filtration through Sephadex G–25. Human erythrocyte membranes digested with pronase were passed through a column of Sephadex G–25 and the elution diagram is shown in this figure. Note that the glycopeptide as represented by sialic acid (*N*-acetylneuraminic acid ————) and hexose (galactose – – – – – –) elutes in water ahead of the contaminating peptides and amino acids (ninhydrin reaction —— —— ——).

6300 for two glycopeptides released by trypsin from erythrocyte stroma and measured by equilibrium sedimentation. Brief (30 min at 37°C) proteolytic enzyme digestion of platelets yields a macroglycopeptide of molecular weight 120 000 (Pepper and Jamieson, [31]). Pronase-digested glycopeptides derived from papain treatment of Novikoff ascites tumour cells yields two molecular weight fractions of 2000–3300 on gel filtration (Wray and Warborg [27]). Whilst water alone may be used as a convenient solvent for gel filtration a buffer, preferably a volatile buffer (such as 0.1 M pyridine acetate pH 5.0), may also be used. The eluates may be analyzed for sialic acids (by the use of the 2-thiobarbituric acid technique [11,12] see 9.5.1(iii)) hexose (using anthrone reagent – see 9.5.1(i)) and amino acids (using ninhydrin reagent). Fractions possessing carbohydrate may be pooled and if necessary the buffer removed by extensive dialysis, the glycopeptide fraction may then be lyophilized for further analyses or fractionation. If carbohydrate units of widely differing size are present it is possible that in addition to separating carbohydrate from non-carbohydrate containing material the glycopeptides may emerge from the column as more than one component. Wray and Walborg [27] were able to fractionate pronase-digested glycopeptides into two fractions by gel filtration on Sephadex G-50.

Whilst gel filtration will yield glycopeptides reasonably free of contaminating peptides and amino acids it is likely that within any one glycopeptide fraction there will be a family of glycopeptides especially if the glycoprotein from which they are derived has a number of different carbohydrate groups of similar molecular weight. Though ultracentrifugation studies might indicate homogeneity this is not sufficient proof to commence structural studies on the carbohydrate groups and it is essential to resolve the glycopeptides into as many different components as possible. In the case of glycopeptides derived from membranes many will be anionic because of the presence of sialic acid residues and as such they may be fractionated on columns of diethylaminoethyl (DEAE) cellulose (or DEAE Sephadex). Salt free samples of glycopeptide are placed on the column at a very low ionic strength (usually 0.0025 M sodium phosphate buffer) elution can then be achieved by increasing the ionic strength using a linear or other suitable gradient. The fractionation should normally be performed at a pH close to neutrality since it is essential not to remove or degrade the sialic acid. Pepper and Jamieson [32] using DEAE Sephadex A-50 ion exchange column chromatography performed preliminary experiments at pH values of 7.2, 6.2, 5.2 and 2.9 and obtained only partial resolution of human platelet glycopeptides. At pH 2.5 the low pH of the bound sialic acid enabled satisfactory adsorption to be effected though these authors are aware of the disadvantage 'that prolonged contact with solution of low pH may hydrolyse some NANA'. Again the eluate may be analysed for sialic acid and hexose (or hexosamine) by the use of the appropriate colorimetric technique. Columns of charcoal-celite developed with increasing concentrations of ethanol may be used for fractionating glycopeptides with substantially different carbohydrate groups, however, glycopeptides with a large sialic acid content cannot be removed from such columns even with high concentrations of ethanol.

Once purified glycopeptides have been obtained, a number of studies are possible which should yield important information about the carbohydrate groups of the parent glycoprotein, assuming a purified glycoprotein has been used (see Section 9.4.2 for the procedures available for isolating glycoproteins), or membrane, if intact cells or membrane fraction has been examined. Significant differences in the carbohydrate composition of the glycopeptides would indicate the occurrence of more than one type of carbohydrate unit. This is the most likely result when whole membranes, as opposed to isolated macromolecules have been used as starting material, though purified glycoproteins may have a number of structurally different carbohydrate groups.

In addition to providing a means of removing non-carbohydrate containing contaminants as well as fractionating glycopeptides of differing size gel filtration studies, as discussed further in Chapter 7, may also be used for estimating the molecular weight of the glycopeptides by comparison with the penetration into the gels of standard polymers of known molecular weight. However, Jackson and Seaman [30] using a column of Sephadex G-100 (calibrated with ribonuclease A, chymotrypsinogen A and ovalbumin) found that two glycopeptides obtained by treating human red cells with trypsin had molecular weights ≅51 000 and ≅3500, though when the molecular weight of the first of the two glycopeptides was also examined by equilibrium sedimentation it was shown to be 12 000. As Jackson and Seaman [30] point out, the difference from the apparent molecular weight obtained by gel filtration as opposed to equilibrium sedimentation probably reflects the effects of the branched and extended glycopeptide chains on the exclusion volume of the glycopeptides.

A number of techniques are available for determining the structure of the carbohydrate units of glycopeptides, namely graded acid hydrolysis, enzymatic determination of sequences with glycosidases, serial periodate oxidation (Smith degradation) and methylation techniques. At the present time no one membrane-derived glycopeptide has been subjected to the extensive series of structural techniques covered here. Methylation can provide important structural information about the linkages between sugar residues and degree of branching within the carbohydrate groups. The methylated sugars may be identified by paper chromatography or gas liquid chromatography, tetra-*O*-methyl ethers indicate a terminal non-reducing position, tri-*O*-methyl ethers are obtained from singly substituted sugars within the carbohydrate groups whilst di-*O*-methyl sugar derivates indicate that the sugar residue is present as a branch point. Based upon differences in the stability of the glycosidic bonds between various sugar residues in glycoproteins it is possible, using mild acid hydrolysis, to obtain data relating to sugar sequence by measuring the rate of release of the various sugar residues. Such graded hydrolysis (0.025 N to 0.1 N $H_2SO_4$ at 80°C for 1 h) has usually been confined to the release of sialic acids, for apart from some fucose other monosaccharides will not be released. Exo-glycosidases, enzymes which release monosaccharides from the terminal, non-reducing, positions in the carbohydrate groups, may also be used for sequence studies. Such treatment has advantages over graded acid hydrolysis, as graded acid hydrolysis can result in random hydrolysis of glycosidic linkages and the release of oligosaccharides. In addition, as long as the enzymes are purified the cleavage of a particular sugar residue will give some information as to its configuration. If one purified glycosidase is used at a time it will be necessary to recover the modified glycopeptide, for example by gel filtration, in order that the action of another glycosidase may be attempted. Probably from a practical point of view a 'cocktail' of purified glycosidases is preferable and can give valuable information about sequences from the order in which monosaccharides are released. When isolating glycosidases from plant sources care should be taken to see that lectins are not present. In this laboratory, preparations of glycosidases isolated by isoelectric focusing Jack bean extracts were found to contain haemagglutinating activity consistent with the presence of concanavalin A, which could only be conveniently removed by passage through columns of Sephadex G-50 (C. W. Lloyd personal communication). A difficulty associated with the use of glycosidases is obtaining sufficient amounts of purified material. Apart from neuraminidase, other glycosidases have not been widely used and it is only recently that they are becoming commercially available. Kornfeld and Kornfeld [33,34] have, however, made use of glycosidases obtained from Jack beans in studying the structure of the PHA receptor site of human erythrocytes. No doubt greater use of these materials will be made in the future though some reticence to experiment with these enzymes may exist because of the low activity they exhibit towards membrane-derived material and thus extended incubation periods are required in order to release measurable quanities of sugar. When performing these experiments involving long periods of incubation it is usual to add small quantities of toluene to the mixture to retard bacterial growth.

The foregoing methods require sizeable quantities of glycopeptide and this is not always possible with membrane-derived material. In addition when studying specific changes which take place in cell membranes, for example when comparing malignant and normal plasma membranes, these might be limited to a subtle alteration in just one oligosaccharide unit. In order to investigate oligosaccharide structure in these cases special techniques are required. A method which is finding increasing use in this area is that of double labelling. In this procedure a normal cell is incubated with a radiolabelled sugar (e.g. $^{3}H$ glucosamine) whilst its malignant counterpart is incubated with the same sugar intermediate but labelled with a different radioactive isotope (e.g. $^{14}C$ glucosamine). It is then possible to isolate radioactive labelled glycopeptides by treatment of the intact cells or isolated membranes with trypsin and the solublized material can then be extensively digested with pronase. The enzyme digests from both cell types are mixed with each other and co-chromatographed on Sephadex gels, the effluent fractions from the gel filtration being examined for both isotopes. Any differences in oligosaccharide structure may be reflected in alterations in the elution profile for one isotope as compared to the other. Various other sugar intermediates may be tested, fucose may be a particularly useful alternative to glucosamine, being less likely to be converted into other sugars. A necessary control is to see that the alteration in the elution profile is the same irrespective of the isotope used to label the cells. Warren *et al.* [35] have used this method to great advantage in exploring the differences in plasma membranes of normal and transformed fibroblasts. They observed that an early eluting fucose labelled glycopeptide from transformed cells could be made to resemble material from untransformed cells by neuraminidase treatment.

In the introductory paragraphs to this chapter the nature of the carbohydrate-polypeptide linkages in glycoproteins was discussed. As regards membrane glycoproteins two types of bond are to be found namely *O*-glycosidic and *N*-glycosidic linkages. In the former type of bond carbohydrate (usually an *N*-acetyl galactosamine residue) and residues of the α-amino-β-hydroxyl acids, serine and threonine are involved, whilst in the latter linkages (glycosylamine linkage) carbohydrate (via an *N*-acetyl glucosamine residue) is linked *N*-glycosidically to asparagine. The *O*-glycosidic linkage may be split by mild alkali by means of a β-elimination reaction:

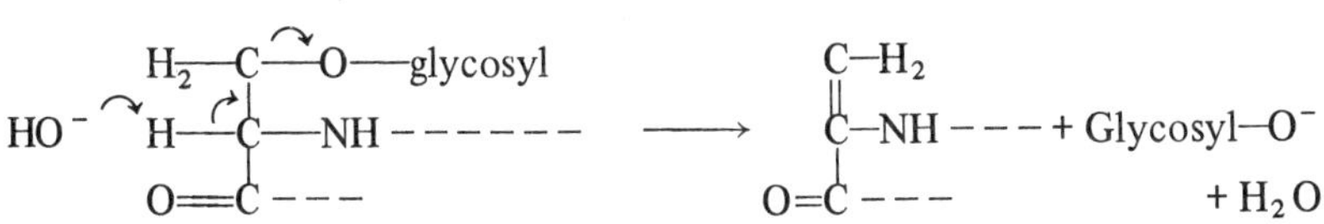


In this reaction serine and threonine are converted into dehydroalanine and α-aminocrotonic acid respectively. The reaction may be performed in the presence of sodium borohydride, when the sugar residue is reduced and on subsequent hydrolysis of the oligosaccharide may be identified as the sugar alcohol. The use of sodium borohydride also makes possible the identification of the amino acid involved for the unsaturated amino acids formed on β-elimination, dehydroalanine and α-aminocrotonic acid, are reduced to alanine and α-aminobutyric acid respectively. Whilst dehydroalanine is readily converted to alanine by sodium borhydride the reduction of α-aminocrotonic acid to α-aminobutyric acid requires the addition of palladium chloride [36] and in the case of both acids their formation, from the experimental data so far recorded, is found to be smaller than expected from the quantities of serine and threonine destroyed. Spiro [37] suggests that alkaline-sulphite treatment is much more reliable and specific than alkaline-borohydride palladium chloride treatment. He considers that preliminary evidence for the presence of an *O*-glycosidic bond in a glycoprotein exists if mild alkaline treatment [Spiro reports 0.05–0.5 N NaOH, at various temperatures (0–45°C) and lengths of incubation (15–216 h) have been used] results in a significant decrease in the content of serine and threonine but that final proof requires the isolation and characterization of glycopeptides containing the linkage. Cook and Eylar [28] finding appreciable quantities of serine in glycopeptides derived by pronase digestion of erythrocyte membranes, suggested that this was indicative of the predominant type of carbohydrate-protein linkage that may occur in these antigens. Certainly following treatment with alkali the serine content was significantly lowered (Cook and Eylar, unpublished observation). In 1967 Winzler *et al.* [38] showed the treatment of tryptic glycopeptides derived from the intact human red cell with tritium-labelled sodium borohydride in the presence of dilute alkali (0.2 N NaOH) resulted in a loss of serine, threonine and the selective destruction of galactosamine and the appearance of some alanine and α-amino-butyric acid, the latter amino acids being radioactive. Later studies by Thomas and Winzler [39] confirmed that residues of *N*-acetylgalactosamine were involved in the *O*-glycosidic linkage.

By way of contrast to the *O*-glycosidic linkage *N*-glycosidic bonds are fairly stable to alkali and the procedure adopted by several laboratories to prove the existence of this linkage has been the isolation of the linkage compound 2-acetamido-1-*N*-β-L-aspartyl-2-deoxy-β-D-glucopyranosylamine. This procedure is not practicable at the present time with the amounts of membrane-derived material available. The resistance of membrane-derived glycopeptides to the release of oligosaccharides by mild alkali borohydride treatment is usually taken to be indicative of the existence of glycosylamine-type linkage [34].

### *9.3.3 Lectins – their use with intact cells*

Over the last eight years there has been a growing interest in the use of lectins for examining the carbohydrate constituents of cell membranes. Lectins are a group of haemagglutins derived from plant sources which have an affinity for various carbohydrate residues. The term lectin as used in this chapter will be confined to carbohydrate-binding proteins of plants, though other authors have extended the term for animal and fungal agglutinins. Though this group of materials has been known since the end of the last century [40] and some of the first information on the carbohydrate nature of blood group specificity was obtained with lectins [41] it is only recently that the full potential of these materials in membrane research has been realized. No doubt this interest has been stimulated by the observation that certain lectins, in particular wheat germ agglutinin and concanavalin A, have the property of being able to preferentially agglutinate tumour cells. Naturally, such studies have been prompted by the desire to correlate the characteristic properties of malignant cells with an alteration in the structure of the plasma membrane. However, though the use of lectins for analysing the surface of cells is considerable, the difficulty in interpreting the results of agglutination assays should not be underestimated. Several mechanisms have been proposed for the apparent increased agglutinability of malignant cells by lectins and these data have been reviewed in detail elsewhere [42] and will therefore only be briefly dealt with here.

Initially it was suggested that the appropriate lectin receptor site on the normal cell might be present in a *cryptic state* [43] on the surface of the normal cell and that these sites may be exposed by gentle proteolytic enzyme digestion. An alternative mechanism envisages the receptor sites being present in a *semi-cryptic* state, that is the receptors are available for binding with the lectin but that other surface structures, labile to trypsin, prevent agglutination taking place. In view of the work of Inbar and Sachs [44] who, using $^{63}$Ni-labelled concanavalin A, were able to show a positive correlation between the ability of cells to be agglutinated by the lectin and their ability to bind the labelled lectin, a cryptic site mechanism proved attractive. However subsequent work by various groups [45–48] showed that there is no difference between the amount of lectin bound to transformed and normal cells including 3T3 fibroblasts, and the transformed variant. The reason for these two distinctly different sets of results can probably be attributed to the labelled form of lectin used. In the later work Ozanne and Sambrook [46] indicated that in their hands labelling of concanavalin A with $^{63}$Ni was unsatisfactory since varying results were obtained with different batches. Ozanne and Sambrook [46] found that the portion of $^{63}$Ni counts bound to the cells that was competed for by hapten was only 30–40% and they had difficulty in saturating the cells with radioactive lectin. Likewise Cline and Livingston [45] noted that the binding of $^{63}$Ni-labelled concanavalin A to cells showed little competition by an excess of unlabelled lectin and in common with the results of Ozanne and Sambrook [46] found that labelling was only slightly (15–30% as compared to 30–40%) inhibited by hapten. The main body of the work of Cline and Livingston [45] and Ozanne

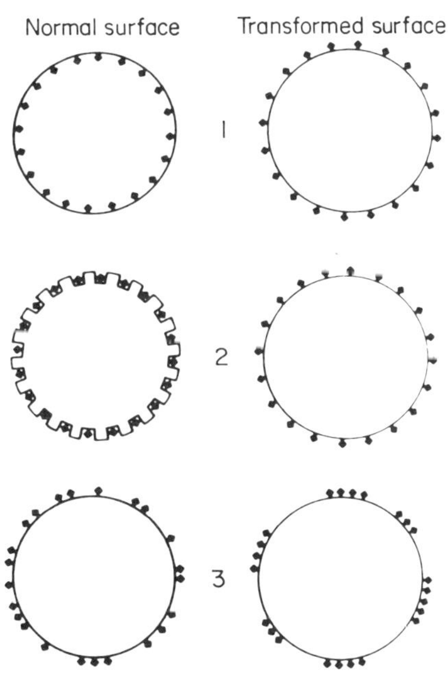


**Fig. 9.5** The possible explanations that have been advanced to explain the increased agglutinability of transformed cells by lectins are illustrated diagrammatically in this figure. 1. Cryptic site model 2. Semicryptic site model 3. The redistribution mechanism.

and Sambrook [46] was performed with lectins to which the label was covalently bound, $^{3}H$ (as acetyl residues – the authors state 2 to 3 per molecule) in the former case and with $^{125}I$ in the latter.

An alternative view of the mechanism of agglutination of cells by lectins is that in which an enhanced agglutinability is associated with the topographic distribution of the lectin-binding sites in the membrane. Nicolson [49] has produced electron microscope evidence which shows that concanavalin A binds in an essentially random distribution on mouse 3T3 cells but when 3T3 cells transformed by SV40 virus were examined the lectin bound in a more clustered distribution, which Nicolson [49] proposes is responsible for the increased agglutinability of these cells. In a subsequent paper Nicolson [50] showed, again using electron microscopic techniques, that in the case of murine 3T3 fibroblasts, where the topographic distribution of lectin is essentially random, the lectin sites become more clustered following treatment with proteolytic enzymes. In addition when Nicolson [50] examined thin sections of the cell-aggregates the concentration of lectin (measured as ferritin-labelled concanavalin A) in the region of cell-to-cell contact was very high and added additional weight to the idea that the agglutination occurs between regions in the cell membrane where there is a high density of agglutinin receptors. Subsequently, Rosenblith *et al.* [51] and Nicolson [52] showed by means of an electron microscopic examination of the distribution of lectin at the surface of cells which had been held at low temperature or after fixation, that the actual clustering seen on transformed or protease-treated cells is a secondary effect due to the lectin acting as a multivalent ligand, apparently crosslinking neighbouring lectin-binding sites. The inherent distribution of lectin-binding sites on 3T3 fibroblasts, protease-treated 3T3 and SV40 transformed cells is actually the same on all the cells and that neither transformation nor protease treatment leads directly to the formation of clusters of binding sites. It would seem that the enzyme treatment and transformation alters the membrane in such a way as to allow greater mobility of the lectin-binding sites in the treated and transformed cells as compared to the normal fibroblasts.

These mechanisms have been discussed here, albeit briefly, because it is hoped to convey the difficulties associated with this field and the care needed in preparing reagents and interpreting results. For the investigator intending to analyse the surface sugars of a line of cultured cells no doubt the procedures used in isolating the cells prior to lectin agglutination assay must be considered when interpreting the results. In addition the conditions used for culturing a cell should also be taken into account, especially if various chemical supplements have been added to the culture medium. The recent investigations of Horwitz *et al.* [53] on membrane fatty acid replacements and their effect on growth and lectin-induced agglutinability are particularly relevant in this context. By use of culturing techniques in which the cells are grown in a lipid depleted serum to which selected exogenous lipids, especially those which are not subject to metabolic conversion once incorporated into the cell, provides a neat technique for altering the lipid composition and hence for assessing the question of fluidity in membrane processes. Horwitz *et al.* [53] have used this particular technique on 3T3 cells and SV103 3T3 cells and shown that different fatty acids introduced as growth supplements have a striking effect on the agglutinability of the cells by wheat germ agglutinin and concanavalin A. Transition temperature, which these authors define as the temperature region in which agglutination is altered, was found to vary depending on the supplement used. For example by comparison with cells grown in regular serum elaidate supplementation increased the thermal transition temperature for concanavalin A-mediated agglutination whilst oleate lowered it. Of particular interest is the finding that the transition temperatures and the effects of fatty acyl replacements are the same for both normal and transformed cells suggesting that both types of cell have membrane regions of similar fluidity. If the clustering of lectin-binding sites described above is important then the increased agglutinability of transformed cells by lectin might be explained by an increased fluidity of mobility of membrane components. However, Horwitz *et al.* [53] find that in the case of untransformed cells where a very fluid membrane has been achieved by oleate enrichment, the level of agglutinability is not altered and they suggest various explanations to clarify this observation. That the size of the molecules containing the lectin receptor or binding site is important cannot be excluded though, as the authors point out, such differences would necessarily have to be quite large to explain the results in terms of lateral diffusion – the translational diffusion constant depending on the cube of the molecular weight. Certainly the mechanism of lectin-mediated agglutination is complex and many factors have to be considered.

The absence of an agglutination reaction cannot safely be taken to mean that a particular sugar residue is absent from the cell periphery and experiments with a labelled preparation of the lectin are an essential adjunct to such an investigation. Conversely, the finding that a certain lectin produces an agglutination reaction, which is inhibited by the appropriate hapten does not necessarily indicate that the monosaccharide residue is present in the cell membrane. An example of this practical problem is amplified by Jansons and Burger [54] who quote the work of Kornfeld and Kornfeld [34] on the structure of the carbohydrate moiety of the phytohaemagglutinin receptor of human erythrocytes which does not contain the haptenic inhibitor, *N*-acetylgalactosamine. Therefore, though agglutination and/or binding of lectin at a cell surface may give presumptive evidence for a particular carbohydrate residue being present at the cell surface rigorous proof requires the isolation and determination of the chemical structure of the particular receptor molecule from the membrane under study.

In using lectins for analysing surface carbohydrates it is essential that these agglutinins be obtained in a purified form, a factor which is perhaps not so essential when one is using these materials for blood typing. In those cases where one is working with an extract from a single source for blood group typing it is only rarely that such an extract will contain more than one specificity, hence the use of relatively crude extracts is quite practicable. The techniques used for purifying lectins are not difficult and follow very much the procedures used in conventional protein chemistry. Some plant extracts (e.g. castor bean extracts) are extremely toxic and in those cases great care should be exercised in isolating such materials. Extraction of a finely-ground meal of the plant material followed by ammonium sulphate fractionation and column chromatographic techniques on ion exchange resins or other adsorbents such as hydroxylapatite is the general pattern followed in a number of the published purification procedures. More recently, increasing attention has been paid to the potential provided by affinity chromatography for the rapid purification of milligram quantities of material and this technique has become widely utilized in the isolation of lectins. Using their property of binding to carbohydrate residues it is possible to obtain, in a one step procedure, a sizeable degree of purification by passing an extract of the plant material through a column of material possessing the appropriate sugar residues. The lectin will bind to the column and excess non-specific protein can be removed by washing the column extensively with buffer. The lectin can then be eluted in buffer containing the appropriate competitive inhibitor. The lectin may be freed of the appropriate hapten by extensive dialysis or by gel filtration, in the latter case columns of Sephadex are particularly useful as long as it is remembered that these same materials are particularly useful as affinity adsorbents and consequently a gel with sufficient degree of crosslinking (e.g. Sephadex G-25 does not bind concanavalin A but the Sephadex gels G-50 to 200 will bind the same lectin) must be used. It may be necessary, if gel filtration is being used for removing the hapten, to recycle the lectin preparation several times, though the actual presence of the competitive inhibitor may be monitored chemically. Regarding the actual affinity adsorbents used, concanavalin A (from Jack beans *Canavalia ensiformis*) and lectins from the common lentil (*Lens culinaris* also known as *Lens esculenta*) have been purified by use of crosslinked dextran gels (Sephadex: Pharmacia). In the case of the lectins of the castor bean (*Ricinus communis*) commercially available agarose gel (Agarose A 0.5 M – BioRad) has been shown by Nicolson and Blaustein [55] to be a suitable adsorbent for their isolation. In addition to these affinity adsorbents it is possible to prepare gels in which either a glycoprotein possessing the appropriate haptenic residues is linked (via the use of cyanogen bromide) to a suitable support (Sepharose 4-B: Pharmacia) or the requisite residue is synthesized and then also subsequently linked to a supporting material. The ovomucoid-Sepharose conjugate for the isolation of wheat germ agglutinin as described by Marchesi [56] is an example of the former type of adsorbent. The author and his colleagues have experienced difficulty preparing an ovomucoid-Sepharose conjugate which acts as an efficient affinity medium for the preparation of wheat germ agglutinin, this may be due to variations in the ovomucoid sample used. Wheat germ agglutinin has however been successfully prepared using chitin as a ligand matrix for affinity chromatography following the method of Bloch and Burger [57]. It was found necessary to wash the chitin thoroughly before use. The conjugates of agarose and *N*-aminocaproyl-D-galactopyranosylamine for the purification of soybean agglutinin (Gordon *et al.* [58]) and *N*-aminocaproylfucosamine [59] (Available commercially from Miles-Yeda Ltd., Rehovot Israel and marketed in the U.K. by Miles Laboratories, Stoke Poges.) are examples of the latter.

For those readers wishing to prepare their own samples of lectins excellent coverage with full practical details of the purification and properties of carbohydrate-binding proteins is given in *Methods in Enzymology* Vol. 28 Complex Carbohydrates Part B [60]. A number of commercial sources are listed in Table 9.1. The question of the suitability of any particular preparation will rest upon the actual application for which it is required. In the case of concanavalin A Uchida and Matsumoto [61] examined a number of commercial samples and demonstrated that they contained varying amounts of demetallized material, which would have an effect on the binding capacity of the preparation. The method adopted by Uchida and Matsumoto [61] was to examine the samples using Sephadex G-50 as an affinity adsorbent. Two fractions characterized as demetallized and partially metallized concanavalin A (the lectin is possibly metallized either with $Mn^{2+}$ or $Ca^{2+}$ or incompletely with both cations) were demonstrated and a third fraction which was only eluted from the column when glucose was incorporated into the medium was found. Of the three samples examined the proportion of material present as the third fraction varied as much as 25% to 94%, the latter figures being obtained on two different batches from the same supplier. Treatment of the commercial lectins with $Mn^{2+}$ and $Ca^{2+}$ (dialysis against 1.0 M NaCl containing 25 mM $MnCl_2$ and 25 mM $CaCl_2$, followed by dialysis against 1.0 M NaCl represent a suitable remetallization regimen) rendered the protein homogeneous with respect to carbohydrate-binding ability and Uchida and Matsumoto concluded that 'treatment with the essential bivalent cations of any commercial concanavalin A protein is necessary for providing a homogeneous preparation with respect to carbohydrate-binding ability'.

Reference has already been made to agglutination assays and the use of binding studies with radioactively labelled lectins. A frequently used technique for assaying agglutination is to estimate the degree of clumping on a serological scale (usually 0 – no agglutination to ++++ complete agglutination) [50,62]; by which the degree of agglutination can be observed microscopically [44]. In their assay system Inbar and Sachs [44] suspended washed cells ($1–3 \times 10^6$ cells ml) in a $Ca^{2+}$ and $Mg^{2+}$ free saline solution (buffered with phosphate), 0.5 ml of this suspension was mixed in a 35 mm Petri dish with an equal volume of lectin-containing solution. Greenberg and Jamieson [63] used Microteston tissue culture plates (Falcon Plastics, Oxnard, Calif.) for determining lectin-mediated agglutination of platelet membranes which was monitored by phase contrast microscopy (at a magnification x125). Burger and Goldberg [62] found that an aliquot (0.2 ml) of cell suspension ($5 \times 10^6$ cells $ml^{-1}$) mixed with a tenth of

TABLE 9.1
*Some properties of the more commonly used lectins in cell surface studies*

| Source | Molecular weight[a] | Sub units | Sugar specificity | Number of binding sites | Example of use in membrane studies |
|---|---|---|---|---|---|
| *Canavalia ensiformis* (Jack bean) | 55 000 | 2 | α-D-Man<br>α-D-Glc | 2 | Numerous comparative studies of normal and transformed cells. See text. Section 9.3.3. |
| *Glycine max* (soybean) | 110 000 | | D-GalNAc | 2 | As above |
| *Lens esculenta*, also known as *Lens culinaris* (common lentil | 42 000–69 000 | 2 | α-D-Man | 2 | Used as affinity adsorbent when conjugated to a suitable support medium. See Section 9.4.2 on preparation of Glycoproteins and Glycolipids. |
| *Phaseolus vulgaris* I<br>(red kidney bean) II | 138 000<br>98 000–138 000 | 8<br>4 | <br>D-GalNAc | | This PHA in conjunction with *Agaricus bisporus* PHA may be useful for probing carbohydrate-protein bonds at cell surfaces. The latter lectin binding with *O*-glycosidically linked oligosaccharide and E-PHA (see b below) with *N*-glycosidically linked oligosaccharide. |
| *Ricinus communis*[c] I<br>(castor bean) II | 120 000<br>60 000 | | β-D-Gal<br>β-D-Gal D-GalNAc | | Numerous comparative studies on normal and transformed cells. |
| *Triticum vulgaris*[c] (wheat) | 26 000 | | $(D\text{-}GlcNAc)_2$ | | Contemporary interest in lectins as tools for analysing cell membranes due largely to experiments with this lectin. |

[a]The molecular weights in this table follow closely those quoted by Sharon and Lis [40]; except in case of *Ricinus communis*, where Sharon and Lis [40] quote a value of 98 000. The values quoted here for *Ricinus communis* are those assigned by Nicolson and Blaustein [55] to the two fractions obtained by gel filtration in the final stages of purification.

[b]Sharon and Lis [40] designated *Phaseolus vulgaris* phytohaemagglutin (PHA) as I erythoagglutinin and II leukoagglutinin. Commercially available material PHA-P may be separated into two distinct entities which have been described E–PHA, an erthyroagglutinating PHA and L-PHA a leukoagglutinating PHA. Both possess mitogenic activity (see [60] for preparation of these lectins).

[c]With the exception of *Ricinus communis* lectins full laboratory details for the preparation of all the lectins mentioned in this table are given in Methods in Enzymology' Vol. 28. Complex Carbohydrates Part B. (ed. Ginsburg V.) [60].

The preparation of *Ricinus communis* lectins is described by Nicolson and Blaustein [55]. A novel affinity chromatography procedure for the preparation of wheat germ agglutinin is given by Bloch and Burger [57].

[d]These lectins are available commercially and though this list is not claimed to account for all the available preparations, at the time of writing the following sources have been noted.

Concanavalin A (from Jack beans)
BDH Chemicals Ltd., Calbiochem Ltd., Miles Laboratories Ltd.,* Pharmacia (Great Britain) Ltd.,* Sigma London Chemical Company Ltd.,
Soybean agglutinin
Miles Laboratories Ltd.,*
*Lens culinaris* agglutinin
BDH Chemicals Ltd.,
*Phaseolus vulgaris* PHA
Calbiochem Ltd.,
Wheat germ agglutinin
Calbiochem Ltd., Miles Laboratories Ltd., Sigma London Chemical Company Ltd.,

*Indicates that a preparation bound to a suitable support is available from this manufacturer for use in affinity chromotography. It will be appreciated by the reader that the specification for any particular lectin may vary from one supplier to another and the individual investigator would no doubt consult the supplier as regards the suitability of a particular product for his application.

[e]Concanavalin A [acetyl $-{}^{3}H$] 75–125 Ci m $mol^{-1}$ is marketed by New England Nuclear, Boston, Mass. U.S.A.

the volume of lectin-containing solution could be conveniently observed microscopically as a hanging drop preparation. Hollow ground microscope slides have also been found to make convenient vessels for direct visual assay of agglutination. Agglutination assays, based on quantitative spectrophotometric methods [64] as well as electronic particle counting [65] have been used. In general the agglutination assays are carried out at room temperature and the reactions observed over a period of 5 to 30 minutes.

The quantitation of lectin binding to cells with radioactively labelled material has been described by various workers. The need in such experiments to establish the degree of *non-specific* binding of lectin is important and may be deduced by conducting experiments in the presence and absence of appropriate haptens. The ability of an excess of native lectin to inhibit the binding of radioactive-labelled lectin is also an important control directed towards confirming the nature of the labelled lectin cell interaction. At the conclusion of the incubation period the cells can be collected, washed and processed for determining the level of radioactivity present. Precipitation of cellular protein with trichloroacetic acid solution (final concentration 5% w/v) followed by washing in the same solution, digestion in 1.0 N NaOH and drying of the digest onto glass filter paper for estimating radioactivity in the liquid scintillation spectrometer [66] is a suitable processing procedure. Some authors [45] have washed cells directly on glass filters before assaying for radioactivity. Certainly the use of filters for washing and assay of the cells *in situ* has many advantages, assuming the filter materials do not bind appreciable quantities of lectin, as efficient washing is rapidly achieved. The experimental details provided by Cuatrecasas [67], who examined the interaction of iodinated ($^{125}$I) wheat germ agglutinin and concanavalin A with isolated fat cells, are particularly useful in illustrating the advantages and problems of washing cells after interaction with radioactive-labelled lectins on filters. At the conclusion of Cuatrecasas's experiments, which were performed at 24°C, the samples were diluted with fifteen volumes of ice cold buffer and poured directly onto appropriate Millipore filters (25 mm) connected to a vacuum source. The filters were then washed with a further fifty volumes of ice cold buffer and Cuatrecasas points out that the entire procedure of dilution, filtration and washing consume 10–15 seconds. The processed filters may then be counted directly, in the example quoted, in a well-type counter. A problem associated with such a procedure is the effect of the filter on the assay. In his studies Cuatrecasas [67] tested a variety of filters of different chemical composition in order to select those which adsorbed the least amount of radioactivity in the absence of cells, a procedure which any intending worker using this method of washing and processing would need to undertake with his own materials. Cuatrecasas found that for wheat germ agglutinin a nylon filter (NRWP, Millipore Corporation), where less than 1% of the total $^{125}$I labelled agglutinin adsorbed to the filter, was the most suitable. In the case of concanavalin A Teflon (LSWP, Millipore Corporation) with about 1.5% of the labelled lectin adsorbing to the filter was the best material. Apparently the adsorption of the radioactivity to these filters was unaffected by high concentrations of the appropriate haptens (*N*-acetyl-D-glucosamine or α-methyl-D-mannopryanoside) or unlabelled native lectins. As might be expected filters composed of cellulose or of cellulose esters were unsuitable as they bound very large amounts of lectins, a binding which was sensitive to the presence of simple sugars or the native lectin.

The actual preparation of radioactive-labelled lectins has been described in detail by a number of authors ($^{3}$H labelled material – see Section 9.5.3; $^{125}$I and $^{131}$I labelled lectins using chloramine-T see Cuatrecasas [67], Kornfeld *et al.* [68]). A recent paper by Phillips *et al.* [69] details a novel method for producing $^{125}$I labelled concanavalin A. It is a common feature of a number of the methods described for the preparation of radioactive-labelled lectins to do the labelling in the presence of hapten in order to protect the carbohydrate binding site. Phillips *et al.* [69] have, instead of protecting the binding sites with soluble hapten, bound the lectin to beads of Sephadex G–75; following removal of inactive concanavalin A by washing, the bound lectin is iodinated on the beads by means of the lactoperoxidose procedure, after which the beads are washed and the lectin eluted using a glycine-HCl buffer (pH 2.0). The authors of this method claim three advantages for their procedure, firstly the lacto-peroxidase and free iodine are easily removed by washing thereby eliminating the need for extensive column chromatography and dialysis with attendant damages or loss of material. A second advantage is that the carbohydrate binding sites on the lectin are protected during the coupling reaction, and finally the Sephadex bead technique allows one to separate the inactive concanavalin A from the active material. Certainly the high specific activity of their product ($5–6 \times 10^4$ cpm $\mu g^{-1}$ concanavalin A) enables one to work with extremely small quantities of cells as well as allowing the study of the binding of low initial concentrations of the lectin. This methodology is certainly very interesting and though the work is recent no doubt the principles outlined in this study are applicable, with suitable modifications, to other lectins where suitable affinity adsorbents are available.

In addition to their contributions to the question of the preparation of labelled lectins Phillips *et al.* [69] have also described a procedure, which they term the 'microfuge method', for studying lectin binding with a minimum of non-specific binding to the reaction vessels. Basically the method involves centrifuging the cells through a barrier of 5% bovine serum albumin at 9000 g in a Beckman model 152 microfuge for 1 min. The bottom of each tube was cut off with a razor blade just above the cell pellet and removed directly for counting purposes. Phillips *et al.* were able to show that in the absence of cells no appreciable (65 cpm) radioactivity was recoverable at the base of the tube, in addition this method obviated the error produced by using glass reaction tubes for the washing procedure. It was demonstrated experimentally that the glass vessels bind a substantial amount of $^{125}$I-labelled concanavalin A which was removed when the washing procedure was performed in glass tubes, thereby indicating that washing methods involving the use of repeated washes in buffer may give falsely low values for binding.

## 9.4 Disruptive studies of membrane carbohydrates

Two aspects of *disruptive studies*, namely the analysis of isolated membrane fractions for component sugars, and the solubilization procedures designed to aid the isolation of the intact macromolecule for structural studies will be discussed here. In the latter case a number of the procedures used to solubilize glycosubstances function equally well for the solubilization of membrane proteins and as such these methods receive detailed treatment elsewhere in this volume. This chapter will be confined to the use of techniques which depend upon the glycose nature of the molecule to bring about a satisfactory fractionation.

### *9.4.1 Carbohydrate analysis of isolated cell membranes*

The results of the analysis for carbohydrates of a variety of plasma membrane preparations are given in Tables 9.2–9.5. In these tables the published data have, in some cases, been normalized to nmoles monosaccharide $mg^{-1}$ protein for ease of comparison. Some authors present only mean values for the quantity analysed, whilst others quote a standard deviation (or standard error). Often analytical data is published without a statement as to how many determinations have been performed and in those cases where the number of experiments performed is cited it is not always clear as to how many separate batches of membrane have been analysed. Obviously it is preferable that at least duplicate and if possible triplicate determinations be performed and wherever possible on more than one batch of membrane, so that at least some assessment can be made on the reproducibility of the method of analysis used, as well as the expected variations of components from preparation to preparation.

A number of papers express the amounts of carbohydrate present as a percentage of dry weight though few authors detail how measurement of this latter quantity is achieved. Bosmann *et al.* [70] give details of the determination of the dry weight of HeLa cell membrane preparations which may be generally useful. In their work membrane samples were dried over $P_2O_5$ and KOH under vacuum and weighed with a Sartorious balance until a constant weight was achieved. Similar techniques were used by Takeuchi and Terayama [71]. When lipid extractions are performed on membrane samples it is often convenient to evaporate the organic phase at 40°C in a current of air [82] the residue again being dried to constant weight *in vacuo*, except that a combination of $P_2O_5$ and paraffin wax are used.

The gross analyses of erythrocyte membranes together with more detailed results on the level of constituent sugars are given in Table 9.2. Invariably the

TABLE 9.2
*Carbohydrates in the human erythrocyte membrane*

Gross analysis

| Total carbohydrate (%) | Neutral sugar (%) | Hexosamine (%) | Sialic acid (%) | Reference |
|---|---|---|---|---|
| 7.2 | 4.0[a] | 2.0[b] | 1.2[c] | Rosenberg and Guidotti [72] |
| 10.3[d] | 4.3 | 4.1 | 1.9[c] | Tanner and Boxer [73] |

Some more detailed analysis of constituent sugars.

($n$ mol mg$^{-1}$ protein)

| Mannose | Galactose | Fucose | Hexosamine | Sialic acid. | Protein (%) | Reference |
|---|---|---|---|---|---|---|
| 450[a] | | | 230[b] | 79[c] | 49.2 | Rosenberg and Guidotti [72] |
| 222 | | | 260 | 86 | | Blumenfeld [74] |
| 62[e] | | | 112[e] | 97[f] | | Juliano and Rothstein [75] |
| 66 | 379 | 43 | 255(GlcNAc)<br>180(GalNAc) | 145 | 42.5 | Tanner and Boxer [73] |
| | | | 190[g] | 126[c] | | Pratt and Cook (in preparation) |

[a]Neutral sugar assayed as galactose by method of Dubois *et al.* [76]
[b]Hexosamine assayed as free amino sugars. Hydrolysis performed in 3 N HCl for 19 h at 110°C.
[c]Sialic acid measured by technique of Warren [11]
[d]Carbohydrates determined by gas liquid chromatography (Bhatti *et al.* [77] and by the Warren assay for sialic acid. See [c]above
[e]Hexose and hexosamine measured by methods described by Winzler [78]
[f]Sialic acid assayed by Aminoff technique [12] Both this measurement and [e]above performed on delipidated protein residue.
[g]Hexosamine assayed as glucosamine following hydrolysis in 4 N HCl for 4 h at 100°C.

method of choice for sialic acid analysis is the 2-thiobarbituric acid technique (either of Warren [11] or Aminoff [12]) and appropriate hydrolysis conditions are detailed in the Tables. Neutral sugar and hexosamine can be determined by the colorimetric techniques (described in Section 9.5) though more recently methods that separate and estimate neutral sugars that rely on automated ion exchange chromatography and gas-liquid chromatography have come into wider use. In the case of hexosamines it is usual to separate the amino sugars from neutral sugars and some amino acids present in the hydrolysate by adsorption onto a small column of Dowex-50 (H+ form) followed by elution in dilute acid and colorimetric estimation. When glucosamine and galactosamine are both present these two hexosamines may be separated and estimated by cation

TABLE 9.3
*Carbohydrates in plasma membranes of animal cells*

| Source | Neutral sugar (%) | Hexosamine (%) | Sialic acid (%) | References |
|---|---|---|---|---|
| Rat liver | | 1[a] | 0.1[b] | Takeuchi and Terayama [71] |
| HeLa cells | | 1.1[c] | 0.3[c] | Bosmann *et al.* [70] |
| Pig lymphocytes | 3.4[d] | | 0.6[e] | Allan and Crumpton [82] |
| Meth A tumour cells | 4.5[f] | | | McCollester [83] |
| Rat sarcolemma | 3.3[g,h] | 4.6[g,i] | 0.3[g] | Vandenburgh *et al.* [84] |

[a]Hexosamine and sialic acid determinations performed on two separate membrane preparations.
[b]Material dissolved in IN KOH and proteins precipitated with 5% perchloric acid. Precipitate hydrolysed in 0.1 N $H_2SO_4$ or 5% perchloric acid at 80°C for 15 min. Sialic acid (as *N*-acetylneuraminic acid) assayed by procedure of Warren [11].
[c]Material precipitated with 1% phosphotungstic acid in 0.5 N HCl, washed and lipid extracted ($CHCl_3$: $CH_3OH$ 2:1 v/v). The extracted material hydrolysed in 0.1 N $H_2SO_4$ at 85°C for h. Precipitation with TCA (final concentration 4%) followed by analysis of supernatant fluid by Warren (1959) procedure for sialic acid. Hexosamine (expressed as galactosamine) released by hydrolysis in 2 N HCl at 100°C for 16 h and assayed by method of Gatt and Berman [85].
[d]Neutral sugar (expressed by the authors as mg glucose per mg protein) determined by anthrone reagent.
[e]Sialic acid (expressed as *N*-acetyl derivative) by method of Warren following hydrolysis in 0.1 N $H_2SO_4$ at 80°C for 1 h.
[f]Represents 4% $CHCl_3$ insoluble and 0.5% $CHCl_3$ soluble carbohydrate (as glucose) estimated with sulphonated α-naphthol by method of Devor [86].
[g]Carbohydrates estimated as trimethylsilyl derivatives in a gas chromatogram.
[h]Includes values for fucose, mannose, galactose and glucose.
[i]Value represents *N*-acetylglucosamine 2.59% and *N*-acetylgalactosamine 2.03%.

exchange chromatography using the amino acid analyser. This technique has been used with red cell membranes by Rosenberg and Guidotti [72] and the method is described with full experimental details by Spiro [37]. This latter author recommends that hydrolysis of glycoproteins for the determination of hexosamine is best accomplished in sealed tubes at 100°C in 4 N HCl for 6 hours whilst in the case of human erythrocyte membranes Rosenberg and Guidotti [72] found experimentally that using 3 N HCl at 110°C optimal release of glucosamine and galactosamine occurs at about 19 h. These conditions were used by the authors though they point out that no correction was made for destruction of the sugars. Ion exchange chromatography may also be applied to the determination of neutral sugars and full practical details of borate-complex anion exchange chromatography of neutral sugars is given by Spiro [37]. Neutral sugars may also be analysed by gas-liquid chromatography.

Whilst sugars may not be introduced directly into a gas chromatograph because they dehydrate and char, as opposed to undergoing volatilization, several volatile derivatives have been successfully used. The technique of gas-liquid chromatography enables one to identify sugars by reference to the relative retention time as compared to standards, while quantitation is achieved by integration of the peak areas produced by the detector for each sugar. For sugar analysis two types of volatile derivatives are commonly used, namely, trimethylsilyl ethers and alditol acetates. The former derivatives are prepared by treating the liberated sugars with a mixture of trimethylchlorosilane and hexamethyldisilazane in anhydrous pyridine. This method is applicable to monosaccharides produced by hydrolysis or methyl glycosides formed by methanolysis. The formation of trimethylsilyl derivatives is rapid and the reaction mixture may be injected directly into the chromatograph, however because of solvent equilibrium mixtures of pyranosidic, furanosidic, and anomeric forms of the monosaccharides or methyl glycosides multiple peaks are obtained for each sugar. The gas chromatography of trimethylsilyl derivatives of monosaccharides has been described by Sweely *et al.* [79]. Practical details of the gas chromatography of methyl glycosides as trimethylsilyl derivatives are given by Laine *et al.* [80]. The work of McPherson *et al.* [81] describes the carbohydrate analysis of membrane derived glycoproteins by gas liquid chromatography using the trimethylsilyl ethers of the methyl glycosides of the constituent sugars. In addition to the trimethylsilyl derivatives alditol acetates have become popular sugar derivatives for use in gas-liquid chromatography, being prepared by the reduction of sugars, for instance by borohydride, followed by acetylation with acetic anhydride. The use of these derivatives has the advantage that only one peak is produced for each sugar, the reduction of the carbonyl group during the preparation eliminating the possibility of the formation of anomers or ring isomerization. Practical details on the preparation of alditol acetates of monosaccharides and their separation and estimation in the gas chromatograph are given by Spiro [37] and Laine *et al.* [80].

The gross analysis of some cell membrane preparations derived from animal cells apart from erythrocytes, is shown in Table 9.3. More detailed analyses are given in Table 9.4. Table 9.5 is confined to the results of sialic acid analyses. In most cases, as with the erythrocyte, the 2-thiobarbituric acid technique of either Warren [11] or Aminoff [12] is the preferred method for estimating sialic acids. As previously discussed, the danger of deoxyribose-containing compounds interfering in the above reaction is much greater when dealing with fractions derived from nucleated cells and it may be necessary to make the appropriate corrections. The use of a correction formula based on absorbency measurements at 549 nm and 532 nm was used by Takeuchi and Terayama [71].

It is hoped that the accompanying tables will give the reader an idea of the amounts of sugar one is likely to encounter in an isolated membrane preparation. Certainly, as mentioned in the introduction to this chapter, it is not intended to catalogue all the papers dealing with the analyses of cell membranes but rather to provide some representative examples which give an idea of the methodology currently employed in this field.

In addition to the sialic acids, hexosamines are now commonly estimated in membrane preparations and the colorimetric methods are often used for the

TABLE 9.4
*Detailed carbohydrate analyses of some plasma membranes*

| Source | (*n* mol mg$^{-1}$ protein) | | | | | | Reference |
|---|---|---|---|---|---|---|---|
| | Sialic acid | Fucose | Hexosamines | Galactose | | Mannose | |
| Rat liver | 33 ± 3 | | 50 ± 3 | | 65 ± 2[a] | | Emmelot *et al.* [87] |
| Ehrlich ascites carcinoma | 27 | | 100[b] | | | | Molnar [88] |
| Mouse fibroblasts (L-cells) | 15.0 ± 2.2[c]<br>24.3 ± 1.2 | 7.6 ± 7.1[c]<br>4.7 ± 3.6 | 53.5 ± 12.6[c]<br>35.7 ± 11.7 | 59.4 ± 8.2[c]<br>83.6 ± 23.3 | | 23.7 ± 6.6[c]<br>40.0 ± 12.3 | Glick *et al.* [89] |
| Rat ascites hepatoma | 28.8 | 33.6 | 52.2 (D-GalNH$_2$)<br>59.2 (D-GlcNH$_2$) | 97.8 | | 57.2 | Shimizu and Funakoshi [90] |
| (AH 7974F) | 20.7[d] | 16.1[d] | 19.9 (D-GalNH$_2$)[d]<br>53.7 (D-GlcNH$_2$)[d] | 43.2[d] | | 57.2[d] | |
| Pig lymphocytes | 36[e] | | | | | 328[f] | Allan and Crumpton [82] |
| Chick embryo fibroblasts | 106 ± 3 | | 180 ± 10[g] | | 1340 ± 40[h] | | Perdue *et al.* [91] |
| – transformed by avian sarcoma virus (RBA) | 79 ± 3 | | 190 ± 10[g] | | 1970 ± 70[h] | | |
| – infected by avian leukosis virus (RAV–49) | 114 ± 3 | | 180 ± 10[g] | | 1090 ± 100[h] | | |
| Rat skeletal sarcolemma | 12.8 | 21.5 | 117.8(D-GalNH$_2$)<br>150.2(D-GlcNH$_2$) | 67.5 | | 31.4 | Vandenburgh *et al.*[i] [84] |

[a] Value quoted by authors was for hexose per mg protein

[b] This value corresponds to hexosamine found in 'smooth' microsomes as opposed to 'plasma membranes'. The sialic acid content of 'smooth microsomes' is 30 *n* moles $mg^{-1}$ protein.

[c] These values correspond to plasma membranes prepared by the method of Warren *et al.* [92] using $Zn^{2+}$ to stablize the membranes whilst in the lower rank fluorescein mercuric acetate (FMA) was used.

[d] These values were obtained on 'lipid-free membranes' as opposed to the values for 'intact membranes' in the upper mark.

[e] In the original paper the authors reported plasma membrane sialic acid (expressed as *N*-acetylneuraminic acid, although the authors point out that by analogy with pig erythrocyte-stroma glycoprotein the sialic acid was *N*-glycolylneuraminic acid) as 11 ±1 $\mu$g per mg of protein, for the purposes of this table the mean value has been converted to *n* moles $mg^{-1}$ protein.

[f] The authors originally quoted 'carbohydrate' determined by anthrone reagent as 59 ± 8 $\mu$g glucose $mg^{-1}$ protein. In this table the mean value is shown as *n* moles $mg^{-1}$ protein.

[g] Values quoted by the authors for 'amino sugar'. Glucosamine was used as a standard.

[h] Values quoted by the authors for 'neutral sugar'. Neutral sugar content determined as glucose equivalent.

[i] Value of 112.2 *n* moles $mg^{-1}$ protein was quoted for glucose.

TABLE 9.5
*Sialic acid content of plasma membranes*

| Tissue* | Sialic acid (*n* moles $mg^{-1}$ protein) | Sialic acid released by neuraminidase (per cent) | Method of analysis | Reference |
|---|---|---|---|---|
| Rat liver | 33 ± 3 | 70 | A.1 | Emmelot *et al.* [87] |
| Rat liver | | | | |
| (bicarb) | 33 ± 2 | 69 ± 3 | A.1 | Emmelot & Bos [93] |
| (bicarb.$CaCl_2$) | 31 ± 3 | 71 ± 2 | A.1 | Emmelot & Bos [93] |
| (citric acid) | 19 ± 4 | 68 ± 1 | A.1 | Emmelot & Bos [93] |
| Hepatoma-484 | | | | |
| (bicarb.$CaCl_2$) | 45 ± 2 | 75 ± 3 | A.1 | Emmelot & Bos [93] |
| (citric acid) | 34 ± 3 | | A.1 | Emmelot & Bos [93] |
| Mouse liver (bicarb) | 30 ± 1 | 61 ± 1 | A.1 | Emmelot & Bos [93] |
| Hepatomas (bicarb) | | | | |
| T147042 | 36 ± 1 | 44 ± 3 | A.1 | Emmelot & Bos [93] |
| T4189 | 27 ± 0 | 55 ± 1 | A.1 | Emmelot & Bos [93] |
| T143066 | 28 ± 2 | 63 ± 3 | A.1 | Emmelot & Bos [93] |
| Mouse liver | | | | |
| (unfractionated) | 32 ± 2 | | B.1 | Evans [94] |
| (light subfraction) | 65 ± 9 | | B.1 | Evans [94] |
| (heavy subfraction) | 25.5 ± 2 | | B.1 | Evans [94] |
| Rat liver | | | | |
| (unfractionated) | 35 | | B.1 | Evans [94] |
| (light subfraction) | 73 | | B.1 | Evans [94] |
| (heavy subfraction) | 22 | | B.1 | Evans [94] |
| Rat ascites hepatoma AH7974F | 28.8 | | B.2 | Shimizu and Funakoshi [90] |
| Chick embryonic fibroblasts | | | | |
| (uninfected) | 106 ± 3 | | | Perdue *et al.* [91] |
| (RAV-49) | 114 ± 3 | | | Perdue *et al.* [91] |
| (morph$^t$ Fujinami) | 77 ± 6 | | | Perdue *et al.* [91] |
| (morph$^t$ Fujinami) | 67 ± 5 | | | Perdue *et al.* [91] |
| (RBA) | 79 ± 3 | | | Perdue *et al.* [91] |
| Murine fibroblasts (L cells) | | | | |
| $Zn^{2+}$ method | 15 ± 2.2 | 70–75 | A.3 | Glick *et al.* [89] |
| FMA method | 24.3 ± 1.2 | 80 | A.3 | Glick *et al.* [89] |

Method of Analysis A 2-thiobarbituric acid assay of Warren [11]
B 2-thiobarbituric acid assay of Aminoff [12]

Hydrolysis conditions 1 As used by Warren [11] 0.1 N $H_2SO_4$ at 80°C for 1 h.
2 0.1 N HCl at 80°C for 1 h.
3 0.05 N $H_2SO_4$ at 80°C for 1 h.

*The preparation buffer is given in parenthesis in those cases where the sialic acid content and the possible effect of the buffer has been investigated.

estimation of total hexosamine, often with glucosamine as standard [70,89]. Alternatively hexosamines may be separated and estimated, using ion exchange chromatography, on the amino acid analyser as employed by Shimizu and Funakoshi [90]. As with erythrocyte plasma membranes increasing use is being made of gas-liquid chromatography for estimating neutral sugars. Methanolysis and trimethylsilylation was used by Shimizu and Funakoshi [90] for the analysis of neutral sugars present in rat ascites hepatoma plasma membranes, whilst Glick *et al.* [89], working with L cells, analysed neutral carbohydrates as alditol acetates. More recently Vandenburgh *et al.* [84] used gas-liquid chromatography with a combination of methanolysis (1.0 N HCl in dry methanol at 100°C for 24 h under nitrogen) and trimethylsilylation for all the constituent sugars in their preparation, care being taken to re-*N*-acetylate the samples before derivatization.

Obviously analytical data on the carbohydrate content of cell membranes considerably strengthens the evidence that these materials are important in membrane structure. A lot of the earlier analytical data in this field was performed on material derived from liver cells. This is perhaps not surprising as this was one of the first tissues in which serious attempts were made to isolate highly purified plasma membranes. Emmelot *et al.* [87] and Takeuchi and Terayama [71], both working with rat liver, reported on the sialic acid and hexosamine content of their preparations. More recently Emmelot and Bos [93] have investigated the sialic acid content of cell membranes derived from rat and mouse livers as well as five hepatomas. It is interesting to note that these authors showed amongst rat liver and rat hepatomas that the level of sialic acid in the membrane depended on the homogenization medium used. In the case of rat liver and hepatoma citric acid apparently reduces the level of sialic acid as compared to other preparation conditions. The effect is less marked in the case of the rat hepatoma. Certainly, as illustrated by the above results, analytical data can be greatly influenced by the method used in isolating a membrane fraction and the effects may vary according to the tissue examined. These points should certainly be borne in mind when attempting comparative studies such as between malignant cells and their normal counterparts. For such comparisons it may be more profitable to isolate larger entities (such as glycopeptides) in order to seek a relationship between changes in biological behaviour and alteration in carbohydrate structure.

The influence of preparative techniques on analytical data is not confined to the liver but to surface membranes stabilized by chemical means e.g. Warren *et al.* [92]. In the latter technique the plasma membrane is sufficiently stable to allow the organelle to be isolated as whole membranes, which can be counted in a haemocytometer, rather than being reduced to vesicles with all the attendant difficulties of purification and identification. The plasma membrane isolation techniques of Warren *et al.* [92] are finding wide acceptance for the preparation of cell membranes so it is particularly important that this same group [89] have compared the carbohydrate composition of membranes prepared by their $Zn^{2+}$

procedure with those isolated following treatment with fluorescein mercuric acetate (FMA). They found that the average amounts of sialic acid, hexosamine, galactose and fucose were considerably less (when expressed as $\mu$moles x $10^{-10}$ per surface membrane) in plasma membranes prepared by the FMA procedure than in those prepared with the $Zn^{2+}$ technique. Membranes prepared by the $Zn^{2+}$ procedure contain more (10% as opposed to 4.7%) of the cell protein so that when carbohydrates are expressed per mg of protein the FMA membranes contain higher concentrations of all the carbohydrates with the exception of fucose and hexosamine. Glick *et al.* [89] suggest that as the membranes are not significantly contaminated with nuclei, mitochondria or endoplasmic reticulum the increased sugars in the $Zn^{2+}$ membranes are unlikely to come from these sources. These authors conclude that there may be a loss of glycoproteins from membranes prepared by the FMA procedure. Certainly these results amplify the point that when interpreting analytical data one must bear in mind that the results may well be dependent in the isolation procedures which have preceded the analyses. It would seem desirable for any membrane system to be studied by a variety of techniques so that correlations between biological function and membrane chemistry may be more safely drawn.

### *9.4.2 Isolation of glycoproteins and glycolipids*

No chapter on the carbohydrates of biological membranes would be complete without reference to the isolation of membrane glycoproteins and glycolipids. Some details of the isolation of glycoproteins have been reviewed recently [95–97] as have practical details on the solubilization, isolation and characterization of glycosphingolipids [98]. In addition the isolation of membrane proteins (Chapter 7) is covered in this present book and therefore attention will be paid in this chapter to those examples of glycoprotein isolation which make particular use of the carbohydrate of the macromolecule for their separation.

A major problem of fractionating membrane glycoproteins is the insolubility of biological membranes in various solvents. Detergents are, however, particularly good solubilizing agents. Recently use has been made of sodium deoxycholate, a detergent which has little effect on specific molecular interactions, to isolate membrane glycoproteins by means of affinity chromatography. Allan *et al.* [99] have made use of the lectin, concanavalin A attached to agarose beads as an affinity adsorbent for the isolation of glycoproteins from pig lymphocyte plasma membranes solublized in sodium deoxycholate. The lectin was attached to Sepharose 4B (Pharmacia) by means of cyanogen bromide reaction. (Insoluble forms of concanavalin A are available commercially including Glycosylex A$^{tm}$ which is the trademark of Miles Laboratories, Inc. and Con A Sepharose® which is the product of Pharmacia Fine Chemicals AB.) Pig lymphocyte plasma membrane were solubilized in 1% sodium deoxycholate and the soluble fraction eluted from a column of concanavalin A agarose with the same solvent until the optical density of the eluate at 280 nm reached base line. Elution of the glycoprotein fraction, which represented most of the membrane bound glycoprotein, was achieved by use of α-methyl glucopyranoside and isolated by precipitation with 2% acetic acid and the precipitated material freed of deoxycholate and sugar by washing with 95% ethanol. Subsequently Hayman and Crumpton [100] published an improved technique using a column of *Lens culinaris* phytohaemagglutinin Sepharose conjugate. They pointed out that the procedure using concanavalin-A Sepharose suffers from two major disadvantages, namely only 80% of the protein added to the column was recovered and that the yield of glycoproteins was low. The advantage of *Lens culinaris* lectin over concanavalin A is that it has the ability to bind the same sugar but with a 50-fold lower affinity, this decrease in strength of binding of sugar has obvious advantages regarding recovery of material. Indeed the use of this adsorbent resulted in a two-fold increase in yield of glycoproteins. Such studies have been recently extended to a large range of other lectins by Adair and Kornfeld [101] who coupled a number of lectins, including *Ricinus communis, Abrus precatorius, Agaricus bisporus* PHA, as well as *Phaseolus vulgaris* and *Lens culinaris* PHA and wheat germ agglutinin, to Sepharose 2B by means of the cyanogen bromide technique in order to examine the glycoproteins of the human erythrocyte membrane. In this investigation the non-ionic detergent, Triton X-100 (Octyl phenoxy polyethoxyethanol) was used to solubilize the membranes which subsequently eluted from the lectin-Sepharose column with the appropriate haptene sugar. Adair and Kornfeld [101] were able to separate erythrocyte membrane material into two distinct classes of glycoprotein, one of which were almost devoid of sialic acid and which interact with *Ricinus communis* and *Abrus precatorius* lectins and the major sialoglycoprotein of the human erythrocyte which contains receptors for wheat germ agglutinins and various PHA's. These techniques are extremely useful and will no doubt find wider application in the membrane field. The use of a non-ionic detergent makes possible the use of such fractionation techniques as iso-electric focusing and ion-exchange chromatography. A major need though is to develop techniques for freeing the samples of non-ionic detergent. Due to the presence of Triton X-100 in their samples Adair and Kornfeld [101] were forced to test their fraction for lectin receptor activity with erythrocyte ghosts as the whole cells were lysed by the detergent in the solubilized fractions. The binding of Triton X-100 to various resins has been exploited in the case of Bio-Rad SM 2 as a means of removing this detergent from samples of protein (see Chapter 6). In the case of membrane proteins Bio-Rad SM 2 has been shown to be very effective in removing Triton X-100 from erythrocyte membrane proteins fractionated by isoelectric focusing in the presence of the detergent (Pratt and Cook – unpublished data), but unfortunately removal of the detergent is accompanied by losses of the protein fractions in excess of 75%.

The application of non-ionic detergents to solubilizing membranes has mainly centred round the erythrocyte though the use of this detergent is potentially of much wider importance. Butters and Hughes [102] working with a plasma membrane fraction from KB cells, a cell line derived from a human carcinoma of the nasopharynx, studied the solubilization of the varous membrane components in Triton X-100. Glycoproteins were readily solubilized by low concentrations of Triton X-100 (optimal concentrations 0.5% v/v), glycolipids and some protein was also solubilized; solubility being defined as those components in supernatants obtained by centrifugation at 100 000 g for 1 hour. Apparently several non-glycosylated polypeptides of low molecular weight as well as sphingomyelin, cholesterol and triglycerides were resistant to extraction with Triton X-100 in common with the findings of Yu *et al.* [103] and Kirkpatrick *et al.* [104]. Though attempts were made to fractionate extracts of the membrane by chromatographic and isoelectric focusing procedures the degree of resolution was disappointing and Butters and Hughes [102] suggest that the formation of mixed micelles composed of detergent and several membrane components occurs therefore limiting the chances of obtaining homogenous fractions. Nevertheless, some separation of membrane glycoprotein was obtained by chromatography on a column of sepharose 6B. However, experiments by Zalik and Cook [130] do not point to mixed micelle formation. Working with various isolated embryonic cell membranes solubilized in Triton X-100 they found that the pattern of membrane proteins obtained in isoelectric focusing experiments was not altered by the addition of other membrane systems prior to solubilization; the resulting pattern in the latter case is a composite pattern of bands derived from the two membrane systems.

Butters and Hughes comment that the major glycoproteins of the KB cell membranes are relatively easy to solubilize and they suggest this is due to their exposed position in the membrane, though they make it clear that the detergent is probably effective because it is able to disrupt the hydrophobic interactions between the glycoprotein and other membrane constituents. In this laboratory (Pratt and Cook, in preparation) the question of the factors which favour the solubilization of membrane glycoproteins by Triton X-100 have been examined in the case of the human erythrocyte. In common with Yu *et al.* [104] (who worked with erythrocytes), and Butters and Hughes [102] (who used nucleated cells) Pratt and Cook found that the glycoproteins (measured as sialic acid by the Warren assay) were also readily solubilized as compared to proteins by low concentration of Triton X-100 (up to 1% v/v examined). At a concentration of 1.0% v/v close to 100% of the available sialic acid was solubilized and this was independent of pH over a range 6 to 7.5. It was particularly interesting to find that following the removal of sialic acid from the erythrocyte membrane by neuraminidase the solubilization by Triton X-100 of the asialoglycoprotein (measured as hexosamine) was suppressed, even at alkaline pH. The addition of borate ions, however, dramatically altered this situation, the glycoprotein becoming readily solubilized. From this data it is suggested that besides the

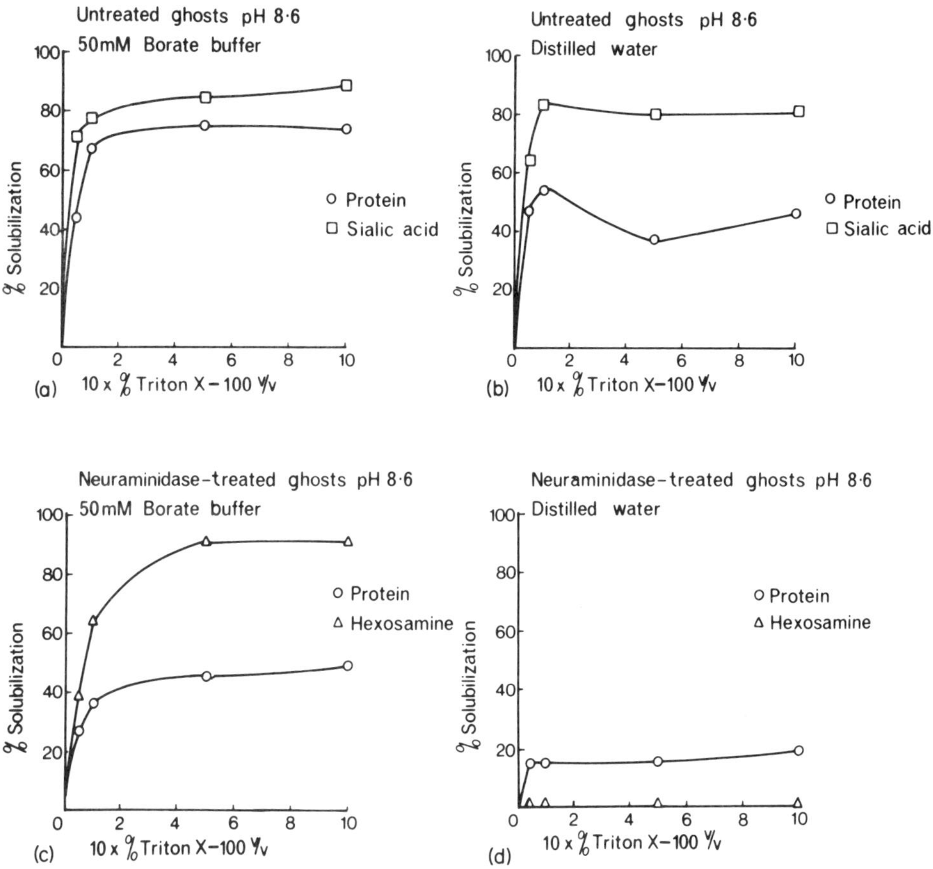


**Fig. 9.6** Solubilization of human (ORh + ve MN) erythrocyte ghost membranes in Triton X–100. (Pratt and Cook. In preparation).
(a) This graph shows that both glycoprotein (measured as sialic acid) and total protein are readily solubilized by low concentrations of Triton X–100 containing 50 mM borate buffer (pH 8.6).
(b) This graph shows the solubilization of glycoproteins (as sialic acid) and total protein by Triton X–100 in distilled water (brought to pH 8.6 with NaOH) and should be compared with (a), where borate buffer at the same pH is used. Note that although the glycoprotein is readily solubilized, total protein solubilization is depressed.
(c) In this graph hexosamine is determined to show the presence of asialoglycoprotein. Note that using borate buffer at pH 8.6 total protein is depressed but glycoprotein is readily solubilized.
(d) Note that on this diagram, as compared with (c), the absence of borate results in no solubilization of sialoglycoprotein (hexosamine). The proportion of total protein solubilized by Triton X–100 is also reduced compared with (c).

exposed position of the glycoprotein in the membrane the importance of charge on the glycosubstances should not be ignored, indeed the greatly enhanced solubilization of asialoglycoprotein by complexing with borate, a procedure which will preferentially confer anionic charge on the latter molecules, presents a novel method for the preferential solubilization of both sialylated, and providing there is sufficient glycosylation, asialoglycoproteins of membranes.

It is hoped the above paragraphs will give the investigator working with membrane systems an idea of the large number of approaches now available for investigating the carbohydrates of membrane systems. In a chapter of this size it is not possible to cover all facets of the subject, however, it is hoped that the methods discussed, which should be within the scope of any well-equipped biochemical laboratory, will stimulate greater interest in these membrane constituents.

## 9.5 Practical procedures for the study of membrane carbohydrates

### *9.5.1 Analytical methods for the investigation of the constituent sugars in supernatant fluids from enzyme-treated cells and isolated glycoprotein fractions*

A number of the problems associated with the application of a particular method to membrane analysis are dealt with in the main body of the chapter. The absence of a procedure from this section is in no way intended to indicate that such a method is inapplicable to the analysis of membranes or membrane-derived material.

#### *Hydrolysis*

In a number of the colorimetric methods described in this section it is necessary that the sugar being analysed is available in a free form. In the case of supernatant fluids derived from material treated with glycosidases this will not be a problem, however, in the case of glycopeptides and glycoprotein fractions the determination of optimal hydrolysis conditions for the release of the monosaccharide being examined is essential. The following conditions should be regarded as guide lines only. Free sialic acids may be obtained by treating the glycoprotein with 0.1 N $H_2SO_4$ at 80°C for 1 h; amino sugars by the use of 4 N HCl at 100°C for 4 h, preferably in sealed tubes under nitrogen. Hexoses may be released from glycoproteins by hydrolysis in 1 N to 2 N HCl or $H_2SO_4$ at 100°C for periods of 3–6 h. It is recommended that in these hydrolyses, the protein concentration not exceed 0.5% w/v. The hydrolysing acid may be removed subsequently by ion exchange resins or in the case of HCl by evaporation *in vacuo* over KOH.

### *(i) Hexoses*

#### *Estimation of total hexoses by anthrone reagent* [105]

This determination is based on the formation in concentrated sulphuric acid of furfural derivatives which react with anthrone to form a blue-green chromophore. The method may be applied directly to glycoproteins without prior hydrolysis. Hexosamines do not interfere in this reaction though sialic acid yields a chromophore equivalent to 8% of the colour given by galactose on a molar basis. Proteins, especially those rich in tryptophan may interfere in this reaction.

*Reagents*: Anthrone reagent. This reagent is prepared by adding to distilled water (280 ml) concentrated $H_2SO_4$ (720 ml), this mixture is allowed to cool to approx. 50°C. Anthrone (500 mg) and thiourea (10 g) are then added with stirring. The reagent is allowed to stand for 4 h before use. This reagent is best stored at 4°C and should be used within two weeks of manufacture.

*Standard*: A standard stock solution of glucose-free galactose (200 μg ml) is prepared in distilled water. This solution may conveniently be dispensed in small aliquots and stored frozen at –20°C; a fresh aliquot being thawed for each determination.

*Procedure*: Samples of hexose (20–200 μg) in 1 ml of distilled water are well mixed with the anthrone reagent (5 ml) in 16 x 150 mm Pyrex test tubes, capped with a glass marble and heated at 100°C for 15 m in (a boiling water bath or heating block of the type marketed by Scienco-Western Ltd. Trowbridge, Wilts. are suitable). A water blank and suitable diluted standards in 1 ml are examined at the same time. Tubes removed from the water bath (or heating block) are allowed to cool at room temperature for 20 min and their absorbancy determined at 620 nm.

#### *Determination of galactose by galactose oxidase*

Galactose is an important constituent of the cell surface. The enzymatic determination of galactose with galactose oxidase provides a relatively specific method for the estimation of this neutral sugar, however, it should be remembered that D-galactosamine and its more commonly encountered *N*-acetyl derivative show comparable reaction velocities to galactose as substrates in this assay.

Reagents for the enzymatic determination of galactose are available commercially in kit form, these kits are usually suitable for estimating samples of 25–100 μg galactose. A micro method capable of estimating galactose down to 2 μg has been described by Roth *et al.* [106] using 1 ml cuvettes and benzidine base as a chromophore.

The principle of this method depends on the formation of hydrogen peroxide through the oxidation, at C-6, of galactose by galactose oxidase and the coupled oxidation of the chromogen in the presence of peroxidase and the hydrogen peroxide. A note of caution should be sounded about the applicability of this method. The natural substrate for the oxidase is probably not galactose, for which the enzyme from *Polyporus circinatus* shows a low affinity ($K_m$ 240 mM for galactose). In the various assays galactose is present at concentrations far below those required to saturate the enzyme, such that the oxidase turns over far below its maximum capacity. Generally, galactosides and galactose containing polysaccharides show a very much lower $K_m$ for the enzyme and higher reaction velocity than does galactose under identical conditions [107]. For those oligosaccharides showing a higher affinity for the enzyme, activity will thus appear greater when compared to galactose standards than would be predicted on the basis of their galactose content alone. This consideration may well apply to the direct application of this procedure to the estimation of galactose in glycoproteins or glycopeptides possessing galactosyl residues at the non-reducing terminus of their carbohydrate groups. For this reason prior hydrolysis and separation of the component sugars (see below Preparation of a neutral sugar fraction) may provide a more accurate estimation, however, the loss of galactose in the hydrolysis step and its degree of recovery from the resin column must be checked experimentally. The direct application of the galactose oxidase method to glycoproteins may however be a useful guide to the effectiveness of a particular method (glycosidase treatment or Smith degradation) for removing galactosyl residues.

The procedure described below is derived from the method of Sempere *et al.* [108] and has been found to be very reliable (S. J. Kirtland, personal communication). Essentially the same method is available as a commercial kit (Galactostat kit – Worthington Biochemical Corporation, N.J., U.S.A.).

*Reagents*: (1) Galactose oxidase (*Dactylium dendroides* – Worthington or Calbiochem) [20 units] is dissolved in a small quantity of 50 mM phosphate (Na Salts) buffer [pH 7.0].

(2) Horse-radish peroxidase (Worthington or Sigma Type II) is dissolved in a small quantity of 50 mM phosphate (Na salt) buffer [pH 7.0].

(3) *o*-tolidine (Sigma) (2.5 mg) is dissolved in distilled water (2.0 ml). If this solution is not required immediately it should be stored in a dark bottle.

(4) The above reagents (1) (2) and (3) are combined and brought to a total volume of 50 ml with 50 mM phosphate (Na salt) buffer [pH 7.0]. These combined solutions are kept in a dark bottle.

(5) 0.5 M EDTA (Na salts) [pH 10.6].

*Procedure*: Test samples, standards (20–100 μg galactose contained in 1.5 ml, the standard galactose solution should be left to undergo mutarotation at room temperature for 2 h before use) and water blank (1.5 ml) are incubated at 37°C for 5 min. At time intervals reagent (4) (1.5 ml) is added and incubation continued in the dark at 37°C for 1 h. The reaction is terminated by the addition of reagent (5) (0.3 ml) and the optical density of the samples measured at 425 nm.

*Preparation of a neutral sugar fraction*

Hydrolysates designed for examining neutral sugars will also contain charged sugars, amino acids and peptides. The neutral sugars may be freed from these other molecules by passing the hydrolysate through a column of Dowex 50-X4 200–400 mesh ($H^+$ form) coupled directly to the top of a column of Dowex 1-X8 200–400 mesh (formate form). This latter resin may be prepared from Dowex 1-X8 ($Cl^-$ form) by passage of 3 N sodium formate through a bed of the resin on a Buchner funnel. The eluate can be tested for $Cl^-$ and the washing with sodium formate should be continued until the test is negative, whereupon the resin should be washed with copious amounts of distilled water. In this procedure the amount of resin used should contain approximately four times the number of equivalents of acid used in the hydrolysate [Dowex 50-X8 exchange capacity when dry = 5.1 (m. equiv $g^{-1}$) and wet = 1.7 (m. equiv $ml^{-1}$), Dowex 1-X8 has value of 3.2 (m. equiv $g^{-1}$) and 1.4 (m. equiv $ml^{-1}$) respectively]. The hydrolysate should be diluted with water to less than 0.4 N $H_2SO_4$ before application to the columns. Next the columns should be eluted with distilled water (equivalent to five column volumes) and the effluent reduced to dryness by lyophilization.

*Paper chromatography of neutral sugars*

Paper chromatography of the neutral sugar fraction may be used for identification purposes as well as quantitating the constituent monosaccharides. Whatman No. 1 paper is used and it is essential that the paper is washed for several days before use with chromatographic solvent, followed by distilled water. The washed paper is then air dried.

*Solvent*: butan-1-ol: ethanol: water (10:1:2 v/v) used in descending direction.

*Procedure*: Aliquots (containing 50–100 μg of each sugar) of the material under assay are spotted onto the chromatography paper by means of a capillary pipette and dried at the point of application with a domestic hair dryer, normally a 4 cm space is left between each spot. Samples of authentic sugars are also applied to the chromatography paper. The chromatogram should be developed in a descending direction for 3 to 4 days. At the completion of the run the chromatogram is air dried in a fume hood, with a good draught or at 60°C in a chromatography oven. The paper may then be treated with appropriate location reagents to detect the constituent sugars (see below).

In the above solvent Rgal values for hexoses are fucose, 2.21; mannose, 1.50; and glucose, 1.17.

*Detection Reagents*: Silver nitrate for reducing sugars [109].

The following solutions are prepared immediately before use.

(1) A saturated aqueous solution of $AgNO_3$ (0.5 ml) is mixed with acetone

[100 ml]. A white precipitate forms and enough distilled water is added dropwise to redissolve it.

(2) 0.5 N NaOH prepared in ethanol.

(3) 6 N $NH_4OH$.

The paper is dipped through solution (1) and allowed to dry in the air. Next the dried paper is dipped through solution (2), when the sugars appear as black spots. The chromatogram is immediately passed through solution (3) followed by repeated dipping in distilled water. The spots due to the sugars on the dry chromatogram turn dark brown and as long as the final washes have been performed immediately after treatment with solution (2) the background should remain a pale straw colour. It is essential that all traces of HCl present in hydrolysates be removed from the sample prior to chromatography.

For routine purposes the following aniline phthalate reagent [110] detection procedure is recommended.

The reagent is prepared before use by dissolving phthalic acid (1.66 g) and aniline (0.93 g) in water saturated butan-1-ol (100 ml). Chromatograms, free of solvent, are sprayed or dipped (as in the $AgNO_3$ procedure above – dipping prevents the formation of noxious aersols and enables one to achieve more easily an even background) through the reagent and then heated in a chromatography oven at 105°C for 10 min. Aldopentoses produce reddish-brown spots and other sugars yellow-brown spots. The procedure is relatively easy to use and the sensitivity may be increased by examining the developed chromatograms under U. V. light – when aldopentoses give a red fluorescence and other sugars a yellow fluorescence.

### *Quantitative paper chromatography*

Paper chromatography of the neutral sugar fraction is performed as described above. Strips of the paper corresponding to the region containing the standard sugars are removed and stained with one of the detection reagents, in order to locate the position of the sugars in the samples (unstained). Having located the regions occupied by the separated sugars the regions may be cut out and the sugars eluted and assayed for reducing sugar. Elution may be achieved by extracting the pieces of paper in a small volume of water (up to 5 ml) over a period of 2 h. Alternatively the pieces of paper may be eluted with water in a chromatography tank. Care should be taken to extract paper blanks, cut out from corresponding regions of the chromatogram, together with the extraction of a set of standards run on the same sheet, in order to determine the recovery of the method.

### *Estimation of reducing sugar by the Park–Johnson Ferricyanide Procedure* [111]

This method is based on the reduction of ferricyanide by sugar in neutral solution followed by the formation of Prussian blue by the addition of ferric ions.

The Prussian blue is kept in solution by the presence of sodium dodecyl sulphate. Because of differences in their reducing power separate standards must be used for each sugar estimated.

*Reagents*: (1) potassium ferricyanide (0.5 g) dissolved in distilled water to a total volume of 1 l).

(2) sodium carbonate (5.3 g) and potassium cyanide (0.65 g) are dissolved in distilled water (to a total volume of 1 l).

(3) ferric ammonium sulphate (1.5 g) and Duponol (sodium dodecyl sulphate – DuPont) are dissolved in 0.05 N $H_2SO_4$ (to a total volume of 1 l).

*Procedure*: Samples of reducing sugar (1–15 g) in distilled water (1.0 ml) are mixed with reagent (1) (0.5 ml) and reagent (2) (0.5 ml) and then heated at 100°C for 15 min. The mixture is cooled rapidly and reagent (3) (2.5 ml) added together with distilled water (0.5 ml). The reactants are allowed to stand at room temperature for 15 min to enable colour development to proceed and then the chromophore is estimated at 690 nm.

### *Thin layer chromatography of neutral sugars*

Thin layer chromatography has many advantages over paper chromatography giving separations in much faster times. Commercially available silica gel, layered on a flexible support, may, with chemical modification, be used for the separation of monosaccharides present in a neutral sugar fraction. The system described here is recommended by the manufacturers of 'Chromogram' sheets (Distillation Products Industries, Division of Eastman Kodak Co., Rochester, N.Y., U.S.A.) for the separation of various sugars and gives good separation of galactose and mannose, two of the neutral sugars most commonly found in membrane glycoproteins.

*Chromatography sheet*: Eastman Chromagram sheet 6061 (formerly type K301R2) – consisting of a 100 $\mu$m layer of silica gel coated on a support of solvent-resistent poly(ethylene terephthalate) sheet (200 $\mu$m thick). The gel used does not contain fluorescent indicator.

*Reagents*: Sodium bisulphite pretreatment solution. Sodium bisulphite (4 g) is dissolved in distilled water (80 ml) followed by the slow addition, with stirring, of ethanol (120 ml).

*Solvent*: Ethyl acetate: methanol: acetic acid: water (60:15:15:10 v/v).

*Procedure*: The sheet of chromatography gel is submerged in the bisulphite reagent for 1 min and then allowed to air dry. Treated sheets are then activated at 100°C for 30 min. Quantities of hexoses (1–10 $\mu$g) are dried onto the sheet and the chromatogram developed in an ascending direction in the above solvent. The developed and air-dried chromatograms are treated with the appropriate visualization reagent (see below) used as a spray and then treated at 85°C to produce spots corresponding to the various sugar zones.

*Visualization reagent*: A freshly prepared solution of 5% aniline hydrogen phthalate in glacial acetic acid is recommended.

Approx $R_F$ values in this system are, galactose: 0.33 mannose 0.41.

### (ii) Hexosamine

The methods for the determination of hexosamine require that the amino sugars be available in a free form. Amino sugars, unlike neutral sugars are not converted to furfural and similar chromogens by strong acids so that a convenient one step assay is not possible. Optimal hydrolysis conditions will vary depending on the nature of the material under assay and should be determined for each particular sample. Before subjecting hydrolysates to analysis it is preferable to free the samples of neutral sugars and non-basic substances, which may interfere with the hexosamine determination.

#### Preparation of hexosamine fraction

Hexosamines may be readily separated from neutral sugars and some amino acids and peptides by chromatography on Dowex 50 resin [112]. Dowex 50 x 4 (200–400 mesh) [$H^+$ form] is washed extensively, on a Buchner funnel, with 2 N NaOH, followed by distilled water, then 3 N HCl and a further wash in distilled water. Excess moisture is removed by continued suction at the pump. A 1:1 (w/v) suspension of the washed resin in distilled water is prepared and used for filling suitable chromatography columns. The chromatographic separation procedure adopted will depend on the state of the sample. Normally an acid-free hydrolysate (HCl can be readily removed by placing the sample, in a suitable container, *in vacuo* over NaOH, the pump being protected by a soda lime trap) redissolved in water may be transferred quantitatively to a small chromatography column (10 mm internal diameter – the body of a plastic disposable syringe is a suitable substitute) fitted with a small plug of clean Pyrex glass wool over which 5 ml of resin suspension has been allowed to settle. A column of this size is sufficient to handle the hydrolysate derived from up to 25 mg protein [113]. Once the sample has run into the resin the column is washed with four aliquots [5 ml each] of distilled water, the hexosamines may then be eluted with 2 N HCl (5–10 ml). The eluate may be freed of the HCl as described above and the residue either assayed directly for hexosamine by colorimetric methods or used for chromatographic identification.

Alternatively an aliquot (1 to 10 ml) of hydrolysate (containing 15–150 $\mu$g hexosamine) may be transferred directly to the above column, without prior removal of the HCl. The volume of hydrolysate which may be added to such a column will depend on the normality of the acid used, 10 ml of 0.25 N, 5 ml of 0.5 N and 2 ml of 1 N acid may be used and as long as these volumes are not exceeded the hexosamines should be completely adsorbed [112]. The column is washed with distilled water and the hexosamine eluted with 2 N HCl as described previously. The effluent may then be used for subsequent colorimetric and chromatographic procedures. Though quantitative recovery of hexosamine may be expected the accuracy of the resin separation should be checked in each case.

For the colorimetric determination of hexosamine it is not essential that the HCl in the column effluent be removed prior to analysis. In such cases the acid effluent can be treated in the following manner due to Boas [112]. Effluent (1–3 ml containing 10–30 μg hexosamine) in a 10 ml glass-stoppered volumetric flask is mixed with a drop of phenolphthalein (0.5% in ethanol) indicator, 4 N NaOH is added until the sample just turns red. The sample is back titrated with 0.5 N HCl until the colour just disappears and the solution is then made up to the final volume of the flask with distilled water. Standards and a blank are also prepared in 10 ml volumetric flasks, 4 N NaCl equal in volume to the NaOH used in the test sample being added to each flask before making up to volume with distilled water. This latter procedure is necessary to ensure that the salt concentration is equal in the unknown sample and standards, the condensation of hexosamines with acetylacetone in the colorimetric procedures is sensitive to salt concentration [112].

*Colorimetric estimation of hexosamine* [114]

Hexosamines may be determined by a number of methods most of which are based on the method of Elson and Morgan [115], in which free amino sugar is condensed with alkaline acetylacetone to form 2-methyl-3-acetyl pyrrole and 2-methylpyrrole. Following the formation of these chromogens, of which 2-methylpyrrole is the principal one on addition of Ehrlich reagent (4-dimethylamino-benzaldehyde) the mixtures give a reddish pink colour. The final colours developed by glucosamine and galactosamine in this type of reaction are identical spectrophotometrically and are of equal intensity for equal weights of the sugars. Though interactions between sugars and amino acids may cause interference in this method the prior separation of hexosamines by ion exchange chromatography obviates this problem.

*Reagents*: Alkaline acetylacetone reagent.

(1) 1 N $Na_2CO_3$ solution.

(2) 2,4-pentanedione (acetylacetone) – this material which should be colourless is freshly redistilled and kept at 4°C in a dark bottle. Immediately before use acetylacetone (2 ml) is dissolved in solution (1) (98 ml).

Ehrlich reagent 2.7% w/v 4-dimethylaminobenzaldehyde prepared in a mixture of ethanol-concentrated HCl (1:1 v/v). This reagent should also be prepared fresh each day. 95% ethanol.

*Procedure*: Aliquots of hexosamine containing solution (1 ml containing 5–30 μg hexosamine) are mixed with alkaline acetylacetone reagent (1 ml) in glass stopped test tubes. The samples are placed at 100°C for 45 min and then cooled under running tap water. Ethanol (4 ml 95% alcohol) is added, mixed thoroughly and Ehrlich reagent (1 ml) added and again mixed thoroughly. In the latter step care should be taken to allow release of $CO_2$ by loosening the stoppers carefully several times when mixing. The tubes are allowed to stand at room temperature for 1 h and read at 530 nm. Appropriate standards and blanks are subjected to the same procedure.

*Paper Chromatography of hexosamines*

Hexosamines and N-acetyl hexosamines may be separated on Whatman No. 1 paper using butan-1-ol: pyridine: water (6:4:3 by volume) in a descending direction. The sugars on the air-dried chromatograms may be detected using the $AgNO_3$ reagent (see above) or by the use of indirect Ehrlich reagent (in the indirect Ehrlich reaction mild alkali treatment precedes reaction with Ehrlich reagent).

Morgan–Elson reaction for detection of amino sugars and 2-acetamido derivatives [116]:

*Ehrlich reagent*: 4-dimethylaminobenzaldehyde (0.5 g) dissolved in a mixture of ethanol (15 ml), concentrated HCl (15 ml) and butan-1-ol (90 ml).

*Alkaline acetylacetone reagent*: Solution (1) acetylacetone (0.5 ml) in butan-1-ol (50 ml). Solution (2) 50% w/v aqueous solution of KOH (5 ml) mixed with ethanol (20 ml). Solution (2) (1 ml) was mixed with solution (1) (20 ml) immediately before use.

*Procedure*: The chromatogram is dipped through alkaline acetyl acetone reagent and heated at 105°C for 5 min before redipping the paper through Ehrlich reagent and heating at 90°C for 5 min. The amino sugars give dark red spots and their 2-acetamido derivatives give reddish purple spots, the latter sugars give a colour reaction without previous treatment with acetyl acetone. This is a useful means of distinguishing between hexosamines and *N*-acetyl hexosamines.

Using the above solvent and procedures $R_{glucosamine}$ values for galactosamine HCl = 0.91, 2-acetamido-D-galactose = 1.61 and 2-acetamido-D-glucose = 1.81.

*Identification of glucosamine and galactosamine by the identification of the pentose formed on oxidation with ninhydrin* [117]

The two hexosamines most commonly found in membrane complex carbohydrates are glucosamine and galactosamine. In addition to the paper chromatographic procedure described above the amino sugars, may for further identification purposes, be converted by oxidation with ninhydrin to pentoses. This procedure may be performed on small quantities of material and the pentoses identified by paper chromatography. Following ninhydrin oxidation glucosamine (and its epimer mannosamine) are converted to arabinose and galactosamine yields lyxose.

*Reagent*: ninhydrin (2 g) dissolved in distilled water (96 ml) and pyridine (4 ml).

*Procedure*: The hexosamine (20–50 μg) (which must be in a free form; if the hexosamine is present in HCl the acid must be removed) is dissolved in a minimum quantity of distilled water to which is added 10 volumes of the above reagent. If the free hexosamine has been dried down on a small square of plastic sheet the solubilization in water and mixing with reagent can be conveniently carried out on the same sheet. The mixture is then drawn into a thick walled capillary tube which is then sealed at both ends. This capillary and its contents is then heated at 100°C for 30 min. After heating, the contents of the tube may then be deposited directly, as a spot, on to a sheet of chromatography paper (Whatman No. 54) alongside marker quantities (20–50 μg) of arabinose and lyxose. Following development of the chromatogram in a descending direction with butan-1-ol:ethanol:water (4:1:1 v/v) the sugars are detected with alkaline silver nitrate reagent. Lyxose under these conditions migrates at 1.15 the rate of arabinose.

It is preferable that the hexosamine under examination be free of neutral sugars and amino acids. However, if these components are present a two-dimensional chromatographic procedure in which following migration in the first dimension the paper is treated with a 2% solution of ninhydrin in 95% ethanol containing 4% pyridine at 80°C followed by chromatography in the second direction is recommended [118].

### *(iii) Sialic acids*

Sialic acids freed by neuraminidase or acid hydrolysis may be separated from the remainder of the glycoprotein by ion-exchange chromatography on Dowex 1–8X (formate form). The preparation of this resin has been described under the section dealing with the preparation of a neutral sugar fraction. Hydrolysates should be diluted with water before being applied to a column of the resin, and the column washed copiously with distilled water followed by elution with 0.3 N formic acid (20 column volumes). The formic acid may be removed by lyophilization and the residue dissolved in a minimum quantity of distilled water for analysis.

Supernatant fluids from enzyme treated cells may be deproteinized by the addition (to a final concentration of 5% w/v) of trichloroacetic acid (TCA), this procedure may be particularly useful for removing coloured lysate material from erythrocyte preparations. Lichtman & Weed [119] have experienced difficulty with the use of trichloroacetic acid prior to the thiobarbituric acid assay [12] finding that 'low optical-density readings resulted unless the hydrogen ion concentration of the solution was reduced with sodium hydroxide'. In this laboratory we have found no significant difference in the formation of chromophore in the thiobarbituric acid reaction as described by Warren [11] when standards of *N*-acetylneuraminic acid freshly dissolved in 5% w/v TCA were compared to the equivalent amounts of standard dissolved in water. Molnar *et al.* [120] have used 5% w/v TCA at 80°C for 1 h as a suitable hydrolysis medium for releasing sialic acids, if the hydrolysates were extracted with diethyl ether and then adsorbed onto a column of Dowex-1-Cl the sialic acid could be quantitatively recovered on elution with 0.2 *N* HCl.

*Estimation of sialic acids by resorcinol reaction* [10]

This method is a one-step procedure and therefore can be employed to measure both free and glycosidically bound sialic acids. The resorcinol reagent is similar to Bial's reagent except that orcinol (3,5 dihydroxytoluene) is replaced by resorcinol (1,3 dihydroxybenzene). In this present method sialic acid containing glycoproteins on heating with the reagents yield a blue chromophore with an absorption maxima at 580 nm; the molecular extinction coefficient of *N*-acetyl neuraminic acid in the reaction (at 580 nm it is 8000 and 9500 for 15 min at 100°C and 110°C respectively) is higher than that quoted for the orcinol procedure (for *N*-acetylneuraminic acid the molecular extinction coefficient is 6000). *N*-glycolylneuraminic has a 30% greater molar absorbancy than the *N*-acetyl derivative. Aldohexoses and 6-deoxyhexoses do not appreciably interfere in the resorcinol reaction through the method is unsuitable for material containing ketohexoses, pentoses or 2-deoxyhexoses and in these cases it would be necessary to isolate the sialic acid from the biological material under examination (see above). The range measured in this reaction is 10–50 $\mu$g sialic acid per sample and this is well within the range encountered with experimental quantities of erythrocytes or when fractionating mg quantities of glycoproteins in gel filtration and other column chromatography procedures.

*Reagents*: Resorcinol (2 g) is dissolved in distilled water (100 ml). This solution may be kept for several weeks at 4°C. At least four hours before performing the estimation the above solution (10 ml) is added to a mixture of concentrated HCl (80 ml) and 0.1 M $CuSO_4$ (0.25 ml) and the volume made up to 100 ml with distilled water. This reagent may be used over a period of ten days if stored at 4°C.

Blank reagent containing HCl and $CuSO_4$ but not resorcinol is prepared as described above.

*Procedure*: Samples of sialic acid or sialoglycoprotein (containing 10–50 $\mu$g sialic acid) in water (2 ml) are mixed in glass-stoppered test tubes with resorcinol-HCl reagent (2 ml). The stoppered tubes are placed in a boiling water bath or a heating block at 100°C for 15 min., care being taken to loosen the stoppers to allow air to escape. The tubes are then cooled under running tap water. Next isoamyl alcohol (5 ml) is added and the chromophore extracted into the organic phase by vigorous shaking. The tubes are allowed to cool in ice for 15 min to allow the upper organic phase to separate out, this process can be accelerated by centrifugation at 500 g for 2–3 min. The organic phase is removed carefully with a pasteur pipette and the absorbancy measured at 580 nm. In addition to the samples an appropriate range of standards and a blank of distilled water should be examined at the same time. Blank reagent may be added to aliquots of the sample to control for any nonspecific colour development.

*Interfering substances*: Correction for the small contribution by neutral hexoses to the absorption at 580 nm may be made by dichromatic measurements at 450 and 580 nm of the samples and appropriate standards. (Galactose, glucose and mannose give a yellow chromophore with absorption maxima at 450 nm.) The following expression derived by Spiro [113] can be used to calculate the contribution of sialic acid to the absorption at 580 nm in a test solution containing both neutral sugars and sialic acids.

OD at 580 nm of sialic acid in test sample

$$\frac{(OD_{580} \times R_H - OD_{450})R_s}{R_s \times R_H - 1}$$

where $R_s$ = ratio of OD of sialic acid standards at 580 nm to that at 450 nm.
$R_H$ = ratio of OD of hexose standards at 450 nm to that at 580 nm.
$OD_{580}$ = Absorbancy of sample at 580 nm
and $OD_{450}$ = Absorbancy of sample at 450 nm

Using the above equation the calculated absorbancy of the sialic acid may then be used to determine the amount of sialic acid present in the test sample by reference to the standard curve for this sugar at 580 nm.

*Estimation of sialic acids by the use of 2-thiobarbituric acid* [11]

This method which only measures free sialic acids has been discussed in the main text of the chapter (see Section 9.3.1) and the various problems associated with its application to cellular systems detailed. Normally *N*-acetylneuraminic acid is used as a standard though *N*-glycolylneuraminic acid may also be estimated by this method but has a molar extinction coefficient in this reaction 19% lower than that for the *N*-acetyl derivative. Methoxyneuraminic acid (which is the methylglycoside of non-substituted neuraminic acid) does not react. Care should be taken to remove any sucrose from the samples. The presence of sucrose gives a dark yellow/brown chromophore.

*Reagents*: 0.2 M sodium metaperiodate in 9 M phosphoric acid. This solution should be stored in a brown bottle and is stable at room temperature for several months.

10% w/v sodium arsenite in a solution of 0.5 M sodium sulphate – 0.1 N $H_2SO_4$. This solution is also stable at room temperature for several months.

0.6% w/v 2-thiobarbituric acid in 0.5 M sodium sulphate solution. This solution may need warming during preparation. It has been found preferable to prepare fresh solution every two weeks, the presence of a slight white precipiate in the solution is of no consequence. Any reagent which turns bright yellow is discarded.

*Procedure*: Samples containing sialic acid (1–16 μg) in water (0.2 ml) are mixed with the periodate reagent (0.1 ml) and allowed to stand at room temperature for 20 min. After this period the periodate-treated samples are mixed thoroughly with the arsenite reagent (1 ml). Care should be taken to shake the tubes until the yellow brown colour disappears. At this stage the vessels and their contents may be stored at 4°C for up to 4 days without altering the production of the chromophore [11]. However, for most practical purposes the thiobarbituric acid reagent (3 ml) is added directly after the arsenite solution. Following the addition of thiobarbituric acid, the tubes are capped and heated at 100°C (a boiling water bath or heating block) for 15 min. The tubes are then cooled under running tap water for 5 min; at this stage the bright pink chromophore appears to fade and the solution may become cloudy, this does not affect the final result. The aqueous solution is then extracted with cyclohexanone (4.3 ml) [This solvent should be water white, if the solvent is yellow in colour it should be redistilled and the fraction of b.p. 154–156°C retained for extraction purposes] with vigorous shaking. A red organic layer forms on top of the aqueous layer on standing, though the separation may be accelerated by centrifugation at 500 g for 2 min. The organic layer is removed with a Pasteur pipette and the chromophore measured at 532 nm, 549 nm, and 562 nm using 1 cm cuvettes. In addition to the samples, quantities of authentic sialic acid as standards and water blanks are assayed at the same time.

*Interfering substances*: An inspection of the optical densities obtained at the above three wavelengths will immediately show whether any additional chromophore (with absorption maximum at 532 nm) to that due to sialic acids (absorption maximum at 549 nm) is present.

The following equations due to Warren [11] may be used. This first equation corrects for the optical density at 549 nm which does not derive from *N*-acetyl neuraminic acid and is based upon the molecular extinction coefficient for *N*-acetylneuraminic acid and 2-deoxyribose.

(1) μ Moles *N*-acetylneuraminic acid

$$= 0.090 \times OD_{549(nm)} - 0.033 \times OD_{532(nm)}.$$

An alternative correction, which may be useful for estimating sialic acids in the presence of large amounts of 2-deoxyribose makes use of the absorption values obtained at 532 and 562 nm.

(2) μ Moles *N*-acetylneuraminic acid

$$= 0.138 \times OD_{562(nm)} - 0.009 \times OD_{532(nm)}.$$

### *Paper chromatography*

The colorimetric methods described above do not characterize the type of sialic acid being analysed. Whilst paper chromatography will not provide rigorous proof of the identity of a sialic acid, in the case of membrane studies it may

TABLE 9.6
*Paper chromatography of sialic acids*

| | Solvents | | | | | | | Reference |
|---|---|---|---|---|---|---|---|---|
| | 1 | 2 | 3 | 4 | 5 | 6 | 7 | |
| $R_F$ value for *N*-acetyl neuraminic acid | 0.09 | 0.16 | 0.15 | 0.27 | 0.58 | 0.44 | 0.15 | Gottschalk [121] |
| | 0.11 | | 0.21* | | 0.75 | 0.63 | | Author's laboratory |
| $R_F$ value for *N*-glycolyl neuraminic acid | | | | | | 0.33 | 0.07 | Gottschalk [121] |
| | | | 0.12* | | | 0.52 | | Author's laboratory |

In the systems described above the solvents were used in a descending direction at room temperature. Whatman No. 1 paper was used.

Solvent system 1. butan-1-ol: acetic acid: water 4:1:5 (v/v)
2. butan-1-ol: pyridine: water 9:5:8 (v/v)
3. butan-1-ol: pyridine: water 6:4:3 (v/v)
4. butan-2-ol: acetone: acetic acid: water 6:6:3:5 (v/v)
5. butan-2-ol: acetic acid: water 4:1:5 (v/v)
6. butan-1-ol: propan-1-ol: 0.1 N HCl. 1:2:1 (v/v)
7. ethyl acetate: acetate acid: water 3:1:3 (v/v)

*These values were obtained with acid-washed paper (see text for details).

prove to be the only practicable technique. A number of solvent systems for use with Whatman No. 1 paper are given in Table 9.6. Solvent (3) may be particularly useful allowing the separation of sialic acids, amino sugars and acetamido sugars present in the same sample. It should be mentioned that in this solvent samples of purified *N*-acetylneuraminic acid migrate as two separate spots; this problem is not encountered with solvents (1), (5) and (6). Whether trace quantities of metal ions present in some samples of chromatography paper are responsible for this phenomenon is not entirely clear though it would appear that the use of solvents containing acids suppresses the formation of double spots. The problem of double spotting with other solvents may be overcome by washing the chromatography paper prior to use. A suitable procedure is to leave the paper in 0.1 N HCl for at least 48 h, followed by several changes of distilled water over a further 24 h. The washed paper is then allowed to dry at room temperature.

For the detection of sialic acids on the dried chromatograms two reagents are commonly used: namely Ehrlich reagent in which 4-dimethylamino-benzaldehyde is dissolved in a mixture of HCl, ethanol and butan-1-ol or the orcinol-trichloroacetic acid reagent of Klevstrand and Nordal [122]. Though Ehrlich reagent is the most specific, for routine purposes the Klevstrand and Nordal [122] procedure is more sensitive. Used as a dip reagent the latter method can be recommended for general use.

*Detection Reagent*: Orcinol [3,5-dihydroxytoluene] (0.5 g) and trichloroacetic acid (15 g) are dissolved in water saturated butan-1-ol (100 ml).

This reagent should preferably be prepared fresh before each experiment and must be discarded after one week.

The dry chromatograms are dipped through this reagent and then heated at 105°C for 10 min. Sialic acids appear as purple/brown spots on a white/yellow background.

*Thin-layer chromatography* [123]

Sialic acids may be separated on thin layers of Kieselgel G (E. Merck, Darmstadt. G denotes that the material contains calcium sulphate as a binder) activated at 120°C for 15 min. Recommended solvents are (1) propan-1-ol:water (7:3 w/v) and (2) propan-1-ol:1 N $NH_4OH$:water (6:2:1 by volume). In the latter solvent *N*-acetyl and *N*-glycolylneuraminic acids are resolvable ($R_F$ values = 0.36 and 0.30 respectively).

### *9.5.2 Structure of carbohydrate units*

*Release of monosaccharides by glycosidases*

The use of a mixture of glycosidases in order to obtain a sequential release of monosaccharides from a glycoprotein or glycopeptide is discussed in Section 9.3.2. An essential feature of such an investigation is the separation of the resulting monosaccharides from the enzymes and residual substrate. In the case of glycoproteins the released monosaccharides may readily be separated by the use of dialysis. Glycopeptides may be separated from released monosaccharides by means of charcoal–Celite chromatography.

*Charcoal–Celite chromatography*

Charcoal (Darco G-60-Kodak) and Celite 535 (Celite® is the registered trade mark of Johns–Manville for their diatomaceous silica filter aid – obtained in U.K. from Koch–Light Laboratories Ltd.) are mixed 1:1 by weight. This mixture is then mixed with a minimum quantity of distilled water and poured into a small glass chromatography column (a column 1 cm in diameter is ideal and should have a small plug of glass wool at the base of the column. Normally 1 ml of charcoal–Celite mixture is required for every 5 mg of substrate). The enzyme digest is added to the column, which is then washed with distilled water (1 column volume). These latter eluates should be retained for colorimetric analysis; this material should be free of any hexose or *N*-acetyl hexosamine. Next the column is developed with increasing concentrations of ethanol in water to eluate first the released monosaccharides, followed by the modified glycopeptide. The eluting conditions should be deduced for each system studied, normally 4–20% v/v ethanol is sufficient to elute the released monosaccharides whilst 50% v/v ethanol may be required to recover the glycopeptide. Effluents from the column may then be lyophilized and analysed for hexoses and *N*-acetyl hexosamines. The latter sugars as distinct from free hexosamines may be determined by the formation of a chromogen, which reacts with Ehrlich's reagent, on heating with sodium carbonate in the absence of acetylacetone.

### *9.5.3 Lectins – preparation of tritium ($^3$H-acetyl) labelled agglutinins*

A number of procedures for the radioactive labelling of lectins is described in the main body of this chapter (Section 9.3.3). The method of Ostrowski *et al.* [124] for the preparation of tritium-labelled antibodies may be employed in a similar manner for labelling lectins rapidly and with a minimum of manipulative procedures.

*Solutions*: Stock solution of phosphate buffered saline. This solution is prepared by dissolving NaCl (80 g) KCl (2.0 g) $KH_2PO_4$ (11.5 g anhydrous material $Na_2HPO_4$ (11.5 g anhydrous material or 14.4 g for 2 $H_2O$ or 29 g for 12 $H_2O$) in distilled water to make a total of 2 litre solution. This stock solution is diluted 1:5 (v/v) before use and for the initial labelling steps made 50 mM with respect to sodium acetate: pH = 7.4. Diluted stock solution without added sodium acetate is also required.

Acetic anhydride-$^3$H (a benzene solution of acetic anhydride $^3$H 500 mCi-mmol obtained from the Radiochemical Centre, Amersham, U.K. has been used).

*Procedure*: The lectin to be labelled is freshly dissolved (20–50 mg/ml$^{-1}$) in phosphate buffered saline (containing 50 mM sodium acetate). An aliquot (1 ml) of this solution is placed in a glass stoppered test tube which is cooled in ice and then mixed in a vortex shaker with $^3$H acetic anhydride in benzene (5 $\mu$l).

It is essential that the mixing be carried out rapidly, the whole procedure being performed in a fume cupboard designed for use with radioactive materials. The mixture is allowed to stand on ice for 1 h, after which time the contents of the tube are carefully transferred to a dialysis sac. Exhaustive dialysis against phosphate-buffered saline (containing sodium acetate) is performed at 4°C, followed by a further period of dialysis against phosphate-buffered saline (without sodium acetate). At this stage the lectin is essentially free of any radioactivity not associated with the protein, however, before being used with cells the material present in the dialysis sac should be passed down a suitable column (2.54 x 38.0 cm) of G-25 Sephadex equilibriated with phosphate buffered saline (without sodium acetate). The column is developed with the same solvent and the absorbancy of the effluent monitored at 280 nm. Fractions of 8 ml may be collected; small quantities of the fractions may be removed and assayed for radioactivity. The bulk of the radioactivity is normally found to coincide with the protein peak (absorbancy at 280 nm) emerging from the column ahead of a small quantity of radioactivity associated with low weight contaminants. The protein peak is pooled and assayed for radioactivity and lectin content (the Lowry procedure for the estimation of protein may be used together with a standard solution of unlabelled lectin). It has been found convenient to freeze (at –20°C) the labelled lectin in small aliquots; by limiting freezing and thawing to a minimum the aggregation of the lectin is retarded.

The above conditions of acetylation have been chosen so as to obtain a sufficient degree of labelling for experimentation without altering the biological properties of the lectins.

### *9.5.4 Electrofocusing of membrane proteins and glycoproteins*

In this section an analytical technique using a thin layer of polyacrylamide gel is described. The pH gradient is set up using Ampholines® (LKB; these are a series of commercial ampholytes – ampholytes are zwitterions which will act as buffers about their isolectric point). Care has been taken to design a gel which keeps the molecular sieving effect to a minimum. The technique is discussed in general terms in Chapter 7.

#### *Analytical electrofocusing in thin polyacrylamide gels*

*Apparatus*: The 2117 Multiphor apparatus of LKB-Produkter AB., Sweden was used entirely.

*Gel Stock Solutions*: (1) acrylamide (28 g) and *N N'*methylene-bis acrylamide (0.7 g) were made up to 100 ml in distilled water. This solution is prepared from laboratory grade acrylamide which has been recrystallized from chloroform before use. Kept in a dark bottle at 4°C the solution is stable for at least a month.

(2) Catalyst solution: riboflavin (0.007 g) *N N N' N'*tetramethylethylene-diamine (0.7 ml) and 10% v/v Triton solution (see 3 below) (10 ml) are made up to 100 ml with distilled water. This solution is kept in a dark bottle at 4°C and has been used for periods of one month.

(3) 10% v/v Triton X-100. Solutions of Triton X-100 (13–15% v/v) may be deionized by passage through mixed bed ion exchange resins (Amberlite 1R-120 (H form) and IR-45 (OH form)). The actual concentration of Triton X-100 in the effluent may be obtained by direct spectrophotometric measurements (at 277 nm) or by chemical means.

[Determination of Triton X-100. This determination is based on a modification of the method of Stevenson [125] by Wright and Plummer [126].

Samples (2 ml) were thoroughly mixed in turn with 2 M HCl (75 $\mu$l), 10% (w/v) $BaCl_2$ (50 $\mu$l) and 10% (w/v) phosphomolybdic acid. The white/yellow precipitate was pelleted by centrifugation at 1000 g for 10 min and the supernatant fluid carefully removed; excess moisture may be removed by allowing the inverted tubes to drain on filter paper. This precipitate is then dissolved in concentrated $H_2SO_4$ (3 ml) with thorough mixing. After 40 min the absorbancy is measured at 520 nm. Wright and Plummer [126] point out that it is necessary to dilute the sample under test as a linear relationship between extinction and detergent concentration is only obtainable up to 0.05% (w/v).]

*Preparation of gel layer*: Solution (1) (14.7 ml) is mixed in a Buchner flask with 10% v/v Triton X-100 (5.8 ml) and appropriate pH range ampholine [LKB Produkter AB., Sweden] (usually 1.8 ml 40% w/v 3–10 range used) and sufficient distilled water to make 57.8 ml of solution. The flask is stoppered and evacuated in the side arm in order to degas the mixture. Catalyst solution (4.2 ml) is added to the degassed acrylamide solution. This complete mixture is then carefully introduced between two glass plates (12.5 x 26 x 0.1 cm) held apart by a 0.2 cm spacer gasket (the equipment for the casting of gel layers is provided with the Multiphor apparatus) and photopolymerized. The glass plates used in this process are cleaned in a solution of Decon 90 detergent and then washed in distilled water followed by aqueous ethanol (approx. 20% v/v) and allowed to air dry. After photopolymerization the gel slab may be separated from one of the glass plates by introducing a little air between the gel surface and the plate by means of a small round ended spatula, at the same time 'levering' the plate off with the blunt end of another spatula. The polyacrylamide slab (characterized according to the definition of Hjerten [127] has $T = 6.8\%$ and $C = 2.4\%$) may be stored at 4°C in a damp chamber (a plastic sandwich box containing damp cotton wool plugs makes a suitable storage container) for up to a month before use.

*Preparation of samples:* Soluble protein and glycoproteins (100–200 μg) may be directly applied to the slab gel by soaking small pieces of chromatography paper (Whatman 3MM approx. 1.0 cm x 0.5 cm) with the sample and applying the paper to the gel surface. This method is not entirely satisfactory with detergent-solubilized membrane fractions and results in a noticeable amount of trailing of material close to the point of sample application. The following procedure is recommended.

Membranes may be solubilized in 1% v/v Triton X-100 (prepared by diluting deionized stock solution) at 4°C for 30 min. Following centrifugation at 100 000 g for 1 h the clear supernatant fluid is mixed with the appropriate quantities of gel stock solutions, etc., to give a mixture of the same composition as that used for casting the slab gel. This mixture is then cast by photopolymerization into blocks 1.5 x 1.0 x 0.3 cm. A suitable 'cell' for preparing these sample blocks is made from a gasket prepared from a silicone rubber insert from a universal bottle cap sandwiched between two microscope slides. The gasket may be lightly greased with silicone stop-cock grease and the glass slides held in place with spring paper clips.

*Separation procedure*: Using the template supplied by the manufacturer anode and cathode wicks (soaked in 1 M $H_3PO_4$ and 1 M KOH respectively) are placed in position on the gel slab together with the sample gels. Normally focusing is performed over the length (26 cm) of the slab, the sample gels being placed 3 cm from the cathode wick. Initially a voltage of 600 V is applied to the slab gel and raised over a period of 1 h to 1200 V; care is taken never to let the power exceed 30 W. Cooling may be achieved by tap water applied via the flow

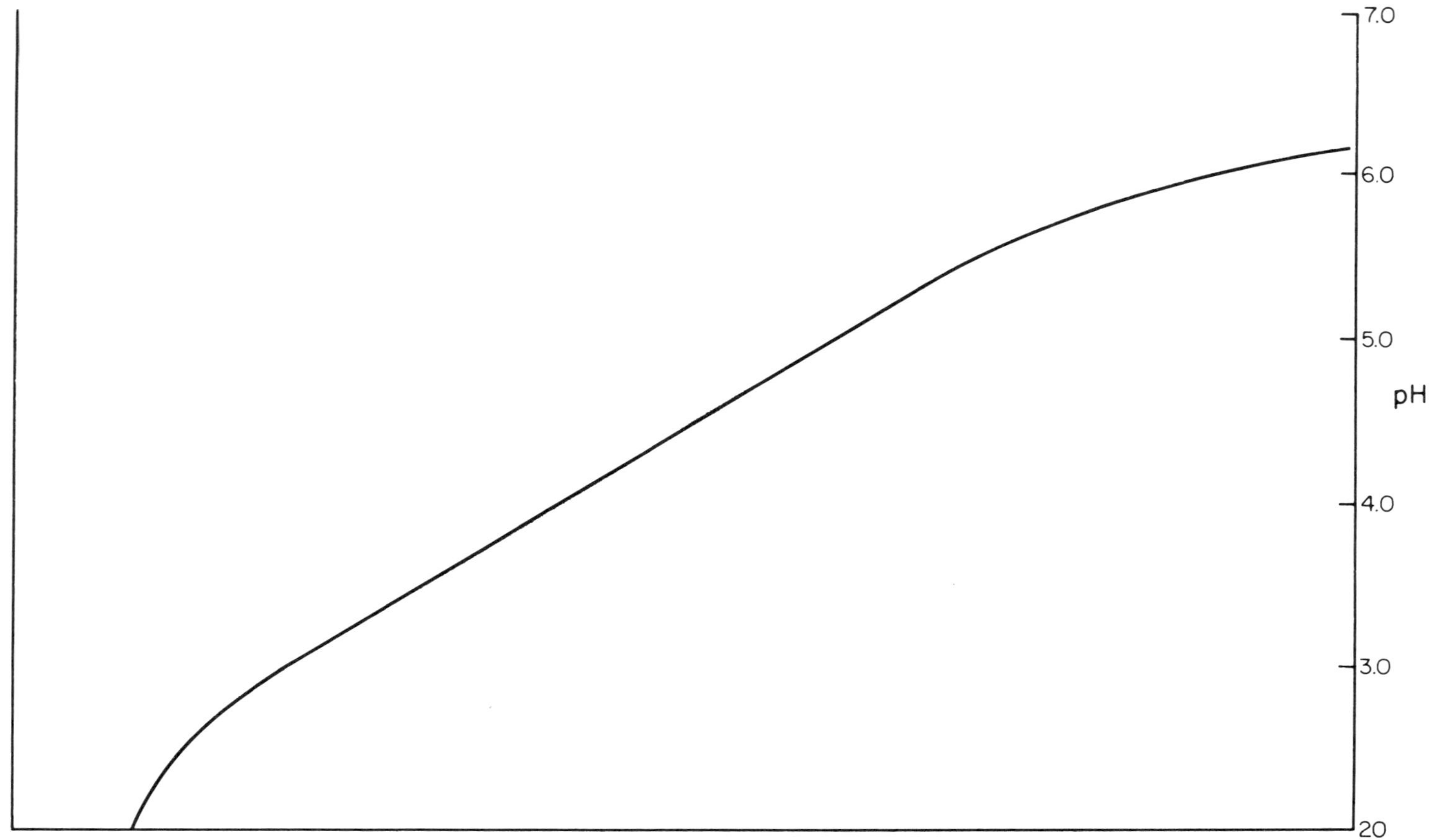
7.0
6.0
5.0
pH
4.0
3.0
20

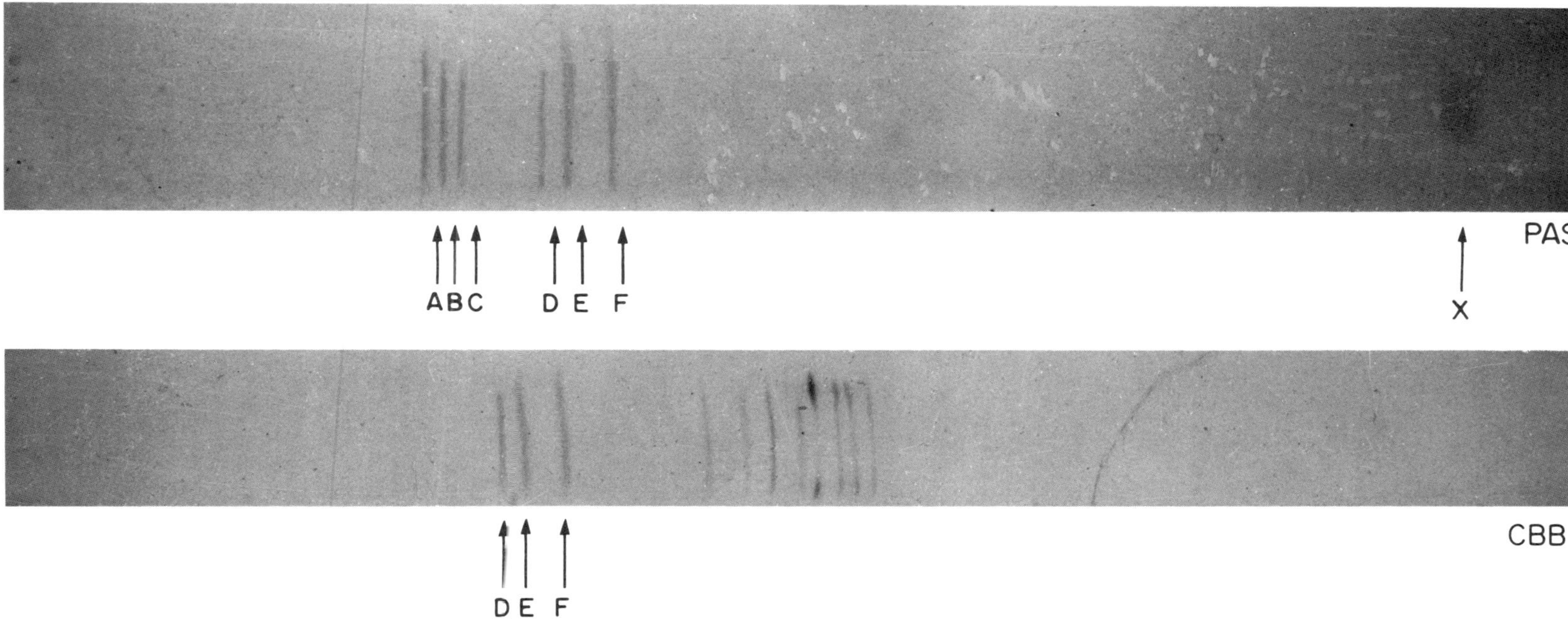


**Fig. 9.7** Analytical thin layer gel isoelectric focusing in polyacrylamide of human erythrocyte glycoprotein. A glycoprotein fraction prepared by pyridine extraction [96] was analysed by electrofocusing over a pH gradient 2.5–6.0 (the gel and associated staining techniques are described in full in the text. In this gel LKB Ampholines pH 2.5–4.0 (1.2 ml) 4.0–6.0 (1.2 ml) and 5.0–7.0 (0.6 ml) were used in the described gel mixture). The top of the figure shows the pH gradient produced in the gel; the PAS positive components (A,B,C,D,E and F) are shown in the photograph immediately under this graph. In this gel the point of application of the sample is marked 'X'. The bottom photograph shows the gel stained with Comassie brilliant blue for protein. Because of differences in the expansion and contraction of the gel slabs during the staining procedures the two gels, though originally from the same slab of acrylamide, are not directly comparable, however, the PAS positive components D, E and F show a staining reaction with Comassie brilliant blue as indicated by the respective arrows. Note the presence of material staining for protein but negative in the PAS reaction.

indicator supplied by the manufacturer, routinely a stable cooling temperature between 14–15°C has been achieved by this means. Focusing is allowed to proceed for 6–7 h at this stage the current will be approx. 3.5 mA.

*Determination of pH gradient*: The pH gradient may be determined by removing 1 $cm^2$ pieces of gel from along the edge of the gel, homogenizing them in a small volume of de-aerated distilled water. The pH of the extract can be determined with a glass electrode (Pye Unicam Ingold Type 405). Alternatively strips of narrow range pH-paper (Duotest: Machery–Nagel, Germany; Lyphan®–Dr. Gerhard Kloz, West Berlin) may be damped with distilled water and laid along the length of the gel to give a rapid indication of the pH gradient.

*Staining*: The slab gels are fixed in 12.5% w/v TCA solution at room temperature for 18 h. This solution is then removed and the now opaque gel soaked in distilled water (1–2 l) to remove the detergent and ampholytes. Usually the gel becomes transparent in 2–3 h. It is found that the gel usually becomes detached from the glass plate, however, as long as the various staining solutions are carefully aspirated and the plate left in position (for subsequent handling) under the gel no problems should be encountered. Enamel photographic dishes are suitable for performing the fixing and staining procedures.

Gels may be stained for protein using Coomassie Brilliant Blue or for carbohydrate using the periodic acid Schiff (PAS) reaction as described in Chapter 7.

---

**Fig. 9.8** Isoelectric focusing of 12 day chick embryonic liver plasma membranes (Zalik & Cook [130]), using the method described in Section 9.5. This photograph shows the gel following staining with Coomassie Brilliant Blue. On the right-hand side of the gel are shown the bands when a preparation of liver plasma membrane solubilized in Triton X–100 is examined. The series of bands immediately to the left represent those glycoproteins which interact with the lectin, wheat germ agglutinin. In the latter sample isolated whole membranes solubilized in non-ionic detergent were passed through a column of wheat germ agglutinin conjugated to agarose beads, following elution with detergent containing buffer to remove 'non-specific' membrane components the column was eluted with *N*-acetyl glucosamine (0.1 M) containing buffer to yield the fraction examined. The 'non-specific' fraction was examined in two separate portions and the results are shown on the two tracks at the left-hand-side of the gel. The track on the extreme left represents the pattern obtained when the sample was applied 3 cm from the cathode and the next 'track' over shows the result when the sample was applied 9 cm from the cathode; note that the resulting band patterns are identical showing that minimal 'molecular sieving' is taking place. The anode is at the top of the photograph and the cathode at the bottom. Some distortion of the band pattern on the right-hand 'track' may be seen, this is probably due to uneven heat dissipation from the sample gel being placed in such a manner that the right-hand edge of the sample block was positioned over one of the walls which form the channels in the cooling block. For optimal results it has been found preferable to position the sample blocks directly over the centre of one of the six channels in the cooling plate, avoiding the wall area where possible.

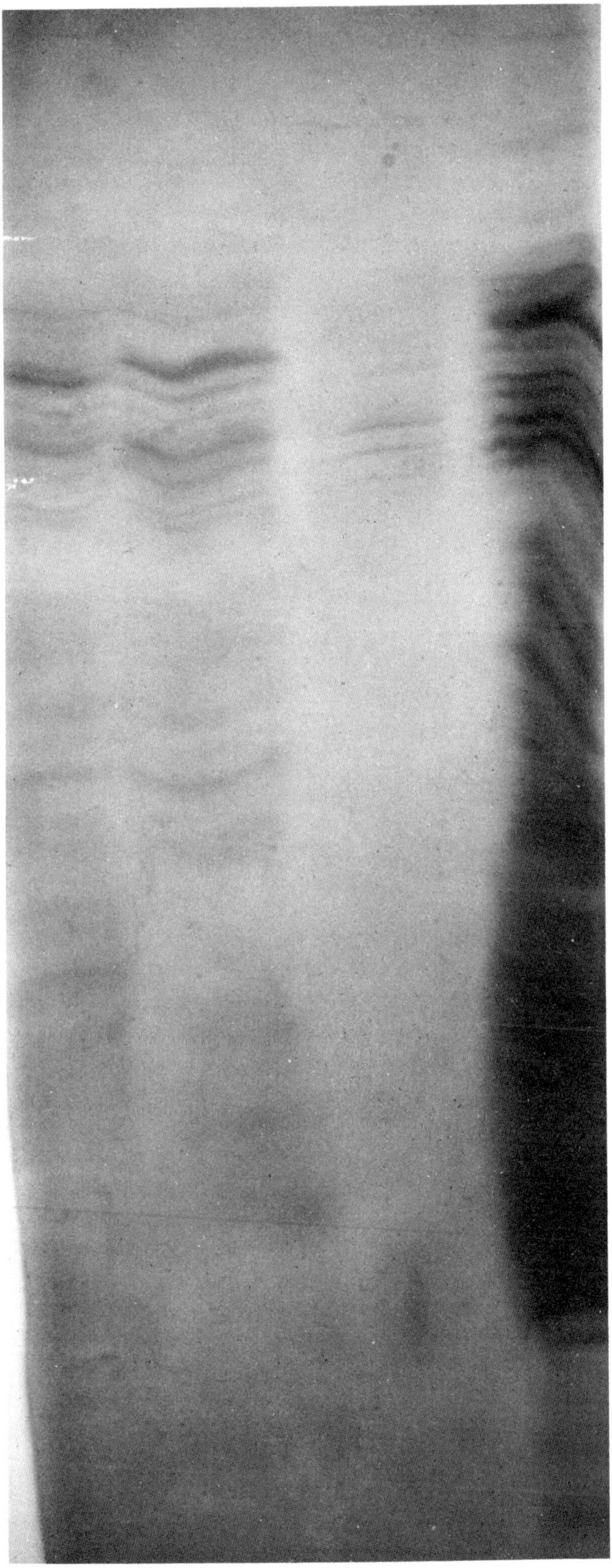

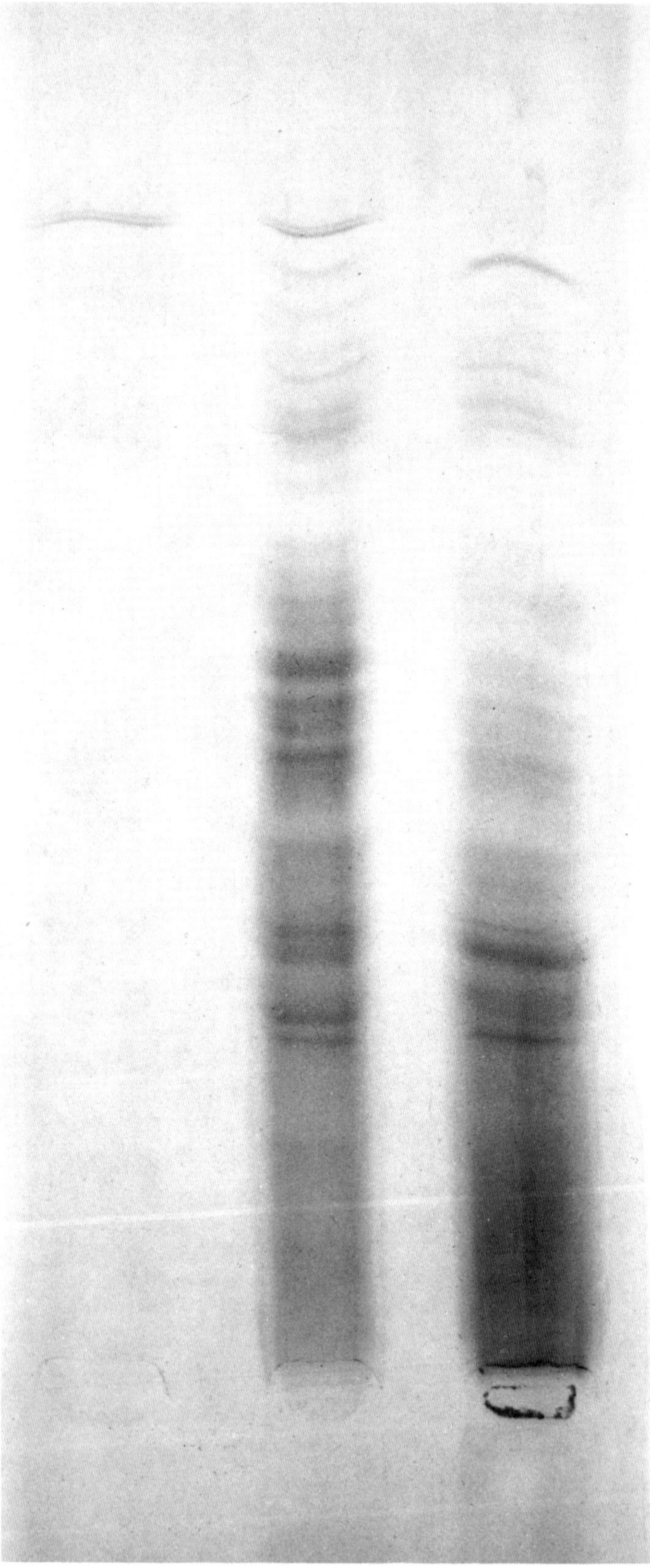

*Comments on the applicability of the described procedure for examining biological membranes*

The procedure described has the advantage of allowing between three and four different membrane samples to be examined at a time under identical conditions thus allowing a direct comparative analytical examination of relatively small quantities of complex systems. No doubt the individual worker may wish to adapt (use alternative non-ionic detergents, gel concentrations, etc) the given procedure to suit his particular problem. however, the above method has been found by the author and his colleagues to represent a general practical procedure for the isoelectric focusing of membrane proteins and glycoproteins. The method complements the hitherto more widely used sodium dodecyl sulphate–polyacrylamide gel electrophoresis technique (see chapter by Maddy & Dunn) which separates the components present in a mixture on the basis of apparent molecular weight.

In the procedure detailed above the gel has been designed to exert only a minimal sieving effect so that the constituent proteins and glycoproteins are reasonably mobile when migrating in the pH gradient and achieve their pI in the shortest possible time. As a direct test of this condition identical samples may be applied to different positions on the gel slab (see Fig. 9.8), which on staining should yield an identical band pattern irrespective of point of application. Such a test also serves to show that the components of the mixture are not being modified as a result of pH-induced changes at the point of application of the sample.

From the point of view of examining biological membranes it is particularly important to establish that following solubilization there is no reassociation of the membrane components. This latter problem has been discussed in Section

---

**Fig. 9.9** Isoelectric focusing of a mixture of solubilized preparations of human erythrocyte stroma and plasma membranes obtained from 12 day chick embryo liver (Zalik & Cook [130]), using the methods described in Section 9.5. Plasma membranes as well as frozen dried erythrocyte stroma were suspended each in 150 μl distilled water. The 'track' on the right-hand side of the photograph represents a sample of 100 μl of embryo liver membrane suspension mixed with 100 μl of 2% v/v Triton X–100. While on the left-hand-side of the gel the sample consisted of 100 μl erythrocyte stroma suspension mixed with 100 μl of 2% v/v Triton X–100. The sample in he centre consisted of 50 μl of the original aqueous suspension of embryonic liver mixed with an equal volume of erythrocyte stroma suspension, to this mixture was then added 100 μl of 2% v/v Triton X–100. It can be seen that this 'central track' consists of the bands present in both the liver and erythrocyte samples; the doublet seen at the top left-hand-side of the photograph is still clearly present when erythrocyte stroma are mixed with liver membrane prior to the addition of detergent. The gel shown in the photograph was stained with Coomassie Brilliant Blue and as in the preceding figure the anode is at the top and the cathode at the bottom of the figure.

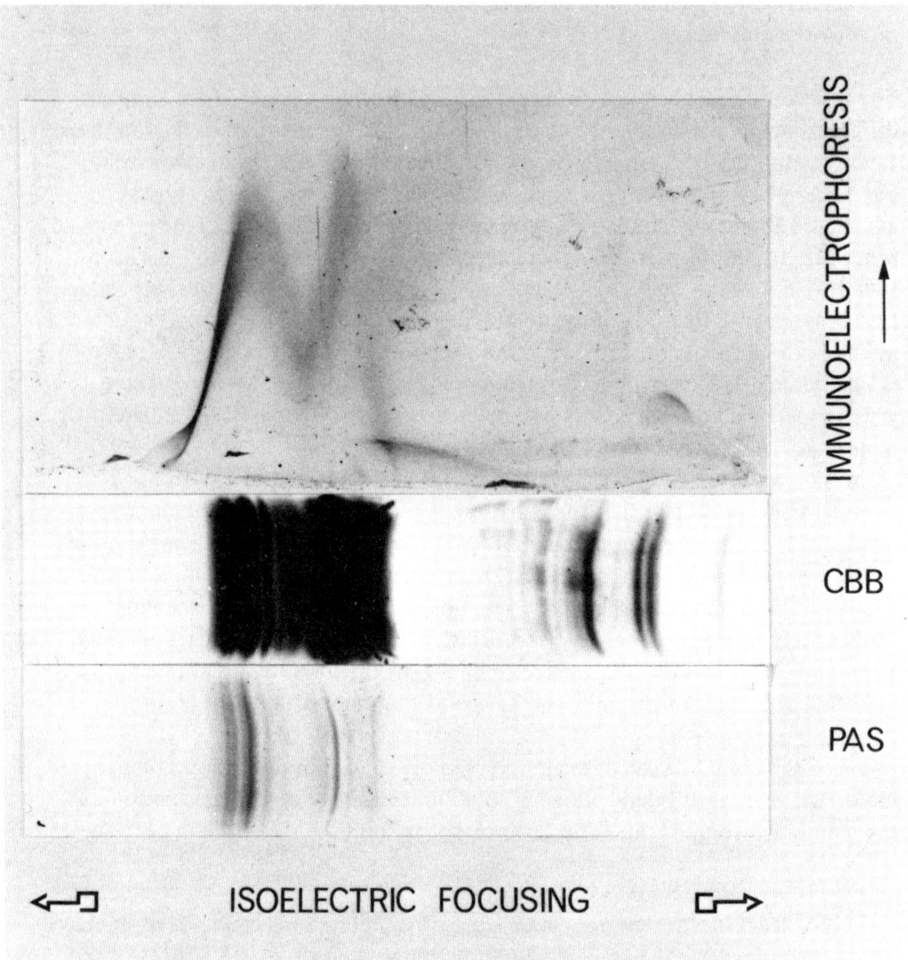


**Fig. 9.10** Combined analytical gel isoelectric focusing and immunoelectrophoresis (modified two dimensional Laurell technique) (Pratt & Cook, unpublished).

A strip of unstained acrylamide gel following isoelectric focusing of rabbit alveolar lavage (cell-free) in 1% v/v Triton X–100 was subjected to electrophoresis into an agarose gel containing guinea pig anti-lavage serum and non-ionic detergent. The resulting preciptin lines were stained with Coomassie Brilliant Blue. At the bottom of the figure corresponding isoelectric focusing strips (anode on the left and cathode on the right pH range 3.5–10) stained with Coomassie brilliant blue (top strip) and periodic acid–Schiff reagent (bottom strip) are shown. It can be seen that the antibody is directed mainly towards the acidic glycoproteins and that there is a reaction of identity between thc two groups of glycoproteins.

9.4.2. When mixtures of human erythrocyte stroma and plasma membranes prepared from 12 day chick embryonic liver were examined (Zalik & Cook [130]) in the isoelectric focusing system detailed in this chapter, the final band pattern was found to correspond to a combination of the banding pattern for each sample when solubilized and analysed separately (see Fig. 9.9). This experiment would suggest that artefacts, due to reassociation of membrane constituents are unlikely to be formed in this system, however, this type of experiment should not be taken as making a claim for the homogeneity of each stained band. Proteins of near pI would not necessarily be resolved and an additional criterion (e.g. SDS gel electrophoresis) should be used by an investigator wishing to establish the homogeneity of bands obtained in this procedure. Conversely, an apparently homogeneous sample in SDS gel electrophoresis should be tested by isoelectric focusing; by using a sample of the same batch of material completely devoid of residual SDS.

*Use of the technique in combination with immunoelectrophoresis*

Immunological techniques represent an extremely powerful tool (see chapter by Bjerrum & Bøg-Hansen) for the investigation of membrane proteins and glycoproteins. The isoelectric focusing method may be combined with immunoelectrophoresis (see Fig. 9.10) [Pratt & Cook, in preparation]. At the completion of the isoelectric focusing procedure a strip of the unfixed gel [$\simeq$ 2 cm in width] is removed and transferred carefully to one side of a glass plate (10 cm square). Agarose–antibody gel mixture (containing 1% v/v Triton X-100) is then cast directly onto the remaining glass (10 cm x 8 cm) surface and immunoelectrophoresis (Laurell, etc) performed as described in detail elsewhere in this volume. Conventional staining procedures may be used though it is essential to remove the acrylamide gel strip away from the bed of agarose immediately after electrophoresis and before drying; unless this is done a marked distortion by splitting of the agarose occurs due to the differential expansion and contraction of the acrylamide and agarose gels. The combination of isoelectricfocusing and immunoelectrophoresis enables one to characterize a membrane glycoprotein (or protein) as regards its pI and immunolgical properties. With the recent [128] successful combination of SDS gel electrophoresis with immunoelectrophoresis it should should now be relatively easy to obtain the apparent molecular weight, pI and immunological properties of membrane glycoproteins with analytical quantities of membrane preparations. In view of a recent report by Green, Dunn & Maddy [129] the appropriate controls to climinate nonspecific, nonimmune precipitation reaction in SDS should be performed.

**Acknowledgements**

The author wishes to thank Mr. Michael Abercrombie, F.R.S., for his interest in this work. In addition the author is grateful to his colleagues, in particular, Dr C. W. Lloyd, Dr M. F. Heath, Dr S. J. Kirtland and Mr R. S. Pratt who have read and commented on various aspects of the draft manuscript. Special thanks are due to Mrs E. Broad and Miss Clare Reid for painstaking secretarial help in the preparation of the manuscript.

Figs 9.1, 9.2, 9.3 and 9.5 are reproduced from Kemp, Lloyd and Cook (1973) Progress in Surface and Membrane Science by permission of Academic Press and Fig 9.4 from Cook and Eylar [1965) [28] by permission of Biochim. Biophys. Acta. The author is grateful to Mr Peter Lancaster for the production of Figs 9.6, 9.7, 9.8, 9.9 and 9.10.

**References**

1. Kraemer, P. M. (1971), In *Biomembranes*, (Manson, L. P. ed), Vol. 1, pp. 67, Plenum Press, New York.
2. Steck, T. L. (1974), *J. Cell Biol.*, **62**, 1.
3. Eylar, E. H. (1965), *J. theoret. Biol.*, **10**, 89.
4. Cook, G. M. W., Heard, D. H. and Seaman, G. V. F. (1961), *Nature*, **191**, 44.
5. Eylar, E. H., Madoff, M. A., Brody, O. V. and Oncley, J. L. (1962), *J. biol. Chem.*, **237**, 1992.
6. Cook, G. M. W. (1968), *Biol. Rev. Cambridge Phil. Soc.*, **43**, 363.
7. Weiss, L. (1973), *J. natn Cancer Inst.*, **50**, 3.
8. Brown, C. R., Srivastava, P. N. and Hartree, E. F. (1970), *Biochem. J.*, **118**, 123.
9. Klenk, E. (1941), *Hoppe-Seyler's Z. physiol. Chem.*, **268**, 50.
10. Svennerholm, L. (1957), *Biochim. biophys. Acta*, **24**, 604.
11. Warren, L. (1959), *J. biol. Chem.*, **234**, 1971.
12. Aminoff, D. (1961), *Biochem. J.*, **81**, 384.
13. Wallach, D. F. H. and Eylar, E. H. (1961), *Biochim. biophys. Acta*, **52**, 594.
14. Wallach, D. F. H. and Esandi, M. V. De P. (1964), *Biochim. biophys. Acta*, **83**, 363.
15. McQuiddy, P. and Lilien, J. (1971), *J. Cell Sci.*, **9**, 823.
16. Rosenberg, S. A. and Einstein, Jr., A. B. (1972), *J. Cell Biol.*, **53**, 466.
17. Kraemer, P. M. (1968), *Biochim. biophys. Acta*, **167**, 205.
18. Emmelot, P. and Bos, C. J. (1965), *Biochim. biophys. Acta*, **121**, 354.
19. Emmelot, P. and Bos, C. J. (1966), *Biochim. biophys. Acta*, **99**, 578.
20. Emmelot, P. and Bos, C. J. (1968), *Biochim. biophys. Acta*, **150**, 341.
21. Yuhas, J. M., Toya, R. E. and Pazmiño, N. H. (1974), *J. natn Cancer Inst.*, **53**, 465.
22. Cook, G. M. W., Heard, D. H. and Seaman, G. V. F. (1960), *Nature*, **188**, 1011.
23. Jackson, R. L., Segrest, J. P., Kahane, I. and Marchesi, V. T. (1973), *Biochemistry*, **12**, 3131.
24. Snow, C. and Allen, A. (1970), *Biochem. J.*, **119**, 707.

25. Jett, M. and Jamieson, G. A. (1973), *Biochem. biophys. Res. Commun.*, **55**, 1225.
26. Walborg, E. F., Jr., Lantz, R. S. and Wray, V. P. (1969), *Cancer Res.*, **29**, 2034.
27. Wray, V. P. and Walborg, E. F., Jr., (1971), *Cancer Res.*, **31**, 2072.
28. Cook, G. M. W. and Eylar, E. H. (1965), *Biochim. biophys. Acta*, **101**, 57.
29. Allen, J. M. and Cook, G. M. W. (1970), *Expl Cell Res.*, **59**, 105.
30. Jackson, L. J. and Seaman, G. V. F. (1972), *Biochemistry*, **11**, 44.
31. Pepper, D. S. and Jamieson, G. A. (1970), *Biochemistry*, **9**, 3706.
32. Pepper, D. S. and Jamieson, G. A. (1969), *Biochemistry*, **8**, 3362.
33. Kornfeld, S. and Kornfeld, R. (1969), *Proc. nat. Acad. Sci. U.S.A.*, **63**, 1439.
34. Kornfeld, R. and Kornfeld, S. (1970), *J. biol. Chem.*, **245**, 2536.
35. Warren, L., Fuhrer, J. P. and Buck, C. A. (1972), *Proc. natn Acad. Sci. U.S.A.*, **69**, 1838.
36. Tanaka, K. and Pigman, W. (1965), *J. biol. Chem.*, **240**, 1487PC.
37. Spiro, R. G. (1972), In Methods in Enzymology (Ginsburg, V., ed.) Vol. 28, pp. 3. Academic Press, New York and London.
38. Winzler, R. J., Harris, E. D., Pekas, D. J., Johnson, C. A. and Weber, P. (1967), *Biochemistry*, **6**, 2195.
39. Thomas, D. B. and Winzler, R. J. (1969), *J. Biol. Chem.*, **244**, 5943.
40. Sharon, N. and Lis, H. (1972), *Science*, **177**, 949.
41. Watkins, W. M. and Morgan, W. T. J. (1952), *Nature*, **169**, 825.
42. Cook, G. M. W. and Stoddart, R. W. (1973), *Surface Carbohydrates of the Eukaryotic Cell*, Academic Press, London and New York.
43. Burger, M. M. (1969), *Proc. natn Acad. Sci. U.S.A.*, **62**, 994.
44. Inbar, M. and Sachs, L. (1969), *Nature*, **223**, 710.
45. Cline, M. J. and Livingston, D. C. (1971), *Nature New Biology*, **232**, 155.
46. Ozanne, B. and Sambrook, J. (1971), *Nature New Biology*, **232**, 156.
47. Arndt-Jovin, D. J. and Berg, P. (1971), *J. Virol.*, **8**, 716.
48. Sela, B. A., Lis, H., Sharon, N. and Sachs, L. (1971), *Biochim. biophys. Acta*, **249**, 564.
49. Nicolson, G. L. (1971), *Nature New Biology*, **233**, 244.
50. Nicolson, G. L. (1972), *Nature New Biology*, **239**, 193.
51. Rosenblith, J. Z., Ukena, T. E., Yin, H. H., Berlin, R. D. and Karnovsky, M. J. (1973), *Proc. natn Acad. Sci. U.S.A.*, **70**, 1625.
52. Nicolson, G. L. (1973), *Nature New Biology*, **243**, 218.
53. Horwitz, A. F., Hatten, M. E. and Burger, M. M. (1974), *Proc. natn Acad. Sci. U.S.A.*, **71**, 3115.
54. Jansons, V. K. and Burger, M. M. (1971), In *Glycoproteins of Blood Cells and Plasma*, (Jamieson, G. A. and Greenwalt, T. J, eds.) 267 Lippincott, Philadelphia.
55. Nicolson, G. L. and Blaustein, J. (1972), *Biochim. biophys. Acta*, **266**, 543.
56. Marchesi, V. T. (1972), In *Methods in Enzymology* (Ginsburg, V., ed.) Vol 28, pp. 354, Academic Press, New York and London.
57. Bloch, R. and Burger, M. M. (1974), *Biochem. biophys. Res. Commun.*, **58**, 13.
58. Gordon, J. A., Blumberg, S., Lis, H. and Sharon, N. (1972), *FEBS Letters*, **24**, 193.
59. Inbar, M., Vlodavsky, I. and Sachs, L. (1972), *Biochim. biophys. Acta*, **255**, 703.

60. Ginsburg, V. (1972), *Methods in Enzymology,* Vol. 28, Academic Press, New York and London.
61. Uchida, T. and Matsumoto, T. (1972), *Biochim. biophys. Acta,* **257**, 230.
62. Burger, M. M. and Goldberg, A. R. (1967). *Proc. natn Acad. Sci. U.S.A.,* **57**, 359.
63. Greenberg, J. H. and Jamieson, G. A. (1974), *Biochim. biophys. Acta,* **345**, 231.
64. Gordon, J. A., Sharon, N. and Lis, H. (1972), *Biochim. biophys. Acta,* **264**, 387.
65. Lloyd, C. W. and Cook, G. M. W. (1974), *J. Cell. Sci.,* **15**, 575.
66. Stoddart, R. W. and Northcote, D. H. (1968), *Biochem. J.,* **105**, 61.
67. Cuatrecasas, P. (1973), *Biochemistry,* **12**, 1312.
68. Kornfeld, R., Gregory, W. T. and Kornfeld, S. A. (1972), In *Methods in Enzymology* (Ginsburg, V. ed.) Vol. 28, pp. 347. Academic Press, New York and London.
69. Phillips, P. G., Furmanski, P. and Lubin, M. (1974), *Expl Cell Res.,* **86**, 301.
70. Bosmann, H. B., Hagopian, A. and Eylar, E. H. (1968), *Archs Biochem., Biophys.* **128**, 51.
71. Takeuchi, M. and Terayama, H. (1965), *Expl Cell Res.,* **40**, 32.
72. Rosenberg, S. A. and Guidotti, G. (1968), *J. biol. Chem.,* **243**, 1985.
73. Tanner, M. J. A. and Boxer, D. H. (1972), *Biochem. J.,* **129**, 333.
74. Blumenfeld, O. O. (1968), *Biochem. biophys. Res. Commun.,* **30**, 200.
75. Juliano, R. L. and Rothstein, A. (1971), *Biochim. biophys. Acta,* **249**, 227.
76. Dubois, M., Gilles, K. A., Hamilton, J. K., Rebers, P. A. and Smith, F. (1956), *Analyt Chem.,* **28**, 350.
77. Bhatti, T., Chambers, R. and Clamp, J. R. (1970), *Biochim. biophys. Acta,* **222**, 339.
78. Winzler, R. (1955), In *Methods in Biochemical Analysis* (Glick, D. ed.) pp. 279, Interscience, New York.
79. Sweeley, C. C., Bentley, R., Makita, M. and Wells, W. W. (1963), *J. Am. chem. Soc.,* **85**, 2497.
80. Laine, R. A., Esselman, W. J. and Sweeley, C. C. (1972), In *Methods in Enzymology,* (Ginsburg, V. ed.), Vol. 28, pp. 159, Academic Press, New York and London.
81. McPherson, J. C., Clamp, J. R. and Manstone, A. J. (1971), *Immunochemistry,* **8**, 225.
82. Allan, D. and Crumpton, M. J. (1970), *Biochem J.,* **120**, 133.
83. McCollester, D. L. (1970), *Cancer Res.,* **30**, 2832.
84. Vandenburgh, H. H., Sheff, M. F. and Zacks, S. I. (1974), *J. Memb. Biol.,* **17**, 1.
85. Gatt, R. and Berman, E. R. (1965), *Analyt. Biochem.,* **15**, 167.
86. Devor, A. W. (1950), *J. Am. chem. Soc.,* **72**, 2008.
87. Emmelot, P., Bos, C. J., Benedetti, E. L. and Rümke, Ph. (1964), *Biochim. biophys. Acta,* **90**, 126.
88. Molnar, J. (1967), *Biochemistry,* **6**, 3064.
89. Glick, M. C., Comstock, C. and Warren, L. (1970), *Biochim. biophys. Acta,* **219**, 290.
90. Shimizu, S. and Funakoshi, I. (1970), *Biochim. biophys. Acta,* **203**, 167.
91. Perdue, J. F., Kletzien, R. and Wray, V. L. (1972), *Biochim. biophys. Acta,* **266**, 505.
92. Warren, L., Glick, M. C. and Nass, M. K. (1966), *J. Cell Physiol.,* **68**, 269.
93. Emmelot, P. and Bos, C. J. (1972), *J. Memb. Biol.,* **9**, 83.

94. Evans, W. H. (1970), *Biochem. J.*, **116**, 833.
95. Howe, C., Lloyd, K. O. and Lee, L. T. (1972), In *Methods in Enzymology* (Ginsberg, V. ed.), Vol. 28, pp. 236, Academic Press, New York and London.
96. Blumenfeld, O. O. and Zvilichovsky, B. (1972), *Ibid*, pp. 245.
97. Marchesi, V. T. (1972), *Ibid*, pp. 252.
98. Essel, W. J., Laine, R. A. and Sweeley, C. C. (1972), In *Methods in Enzymology*, (Ginsberg, V. ed.), Vol. 28, p. 140, Academic Press, New York and London.
99. Allan, D., Auger, J. and Crumpton, M. J. (1972), *Nature New Biology*, **236**, 23.
100. Hayman, M. J. and Crumpton, M. J. (1972), *Biochem. biophys. Res. Commun.*, **47**, 923.
101. Adair, W. L. and Kornfeld, S. (1974), *J. biol. Chem.*, **249**, 4696.
102. Butters, T. D. and Hughes, R. C. (1974), *Biochem. J.*, **140**, 469.
103. Yu, J., Fischman, D. A. and Steck, T. L. (1973), *J. supramolec. struct.*, **1**, 233.
104. Kirkpatrick, F. H., Gordesky, S. F. and Marinetti, G. V. (1974), *Biochim. biophys. Acta*, **345**, 154.
105. Roe, J. H. (1955), *J. biol. Chem.*, **212**, 335.
106. Roth, H., Segal, S. and Bertoli, D. (1965), *Analyt. Biochem.*, **10**, 32.
107. Avigad, G., Amaral, D., Arsensio, C. and Horecker, B. L. (1962), *J. biol. Chem.*, **237**, 2736.
108. Sempere, J. M., Gancedo, C. and Asensio, C. (1965), **12** , 509.
109. Trevelyan, W. E., Proctor, D. P. and Harrison, J. S. (1950), *Nature*, **166**, 444.
110. Partridge, S. M. (1949), *Biochem. J.*, **164**, 443.
111. Park, J. T. and Johnson, M. J. (1949), *J. biol. Chem.*, **181**, 149.
112. Boas, N. F. (1953), *J. biol. Chem.*, **204**, 553.
113. Spiro, R. G. (1966), In *Methods in Enzymology*, (Neufeld, E. F. and Ginsburg, V., eds.), Vol. 8, pp. 3, Academic Press, New York and London.
114. Davidson, E. A. (1966), *Ibid.*, pp. 52.
115. Elson, L. A. and Morgan, W. T. J. (1933), *Biochem. J.*, **27**, 1824.
116. Cramer, F. (1954), *Paper Chromatography*, Macmillan, London.
117. Stoffyn, P. J. and Jeanloz, R. W. (1954), *Archs Biochem. Biophys.*, **52**, 373.
118. Kabat, E. A. (1961), *Experimental Immunochemistry*, Charles C. Thomas, Springfield.
119. Lichtman, M. A. and Weed, R. I. (1970), *Blood*, **35**, 12.
120. Molnar, J., Robinson, G. B. and Winzler, R. J. (1965), *J. biol. Chem.*, **240**, 1882.
121. Gottschalk, A. (1960), *The Chemistry and Biology of Sialic Acids and Related Substances*, Cambridge University Press.
122. Klevstrad, R. and Nordal, A. (1950), *Acta chem. scand*, **4**, 1320.
123. Granzer, E. (1962), *Hoppe-Seyler's Z. physiol. Chem.*, **328**, 277.
124. Ostrowski, K., Barnard, E. A., Sawicki, W., Chorzelski, Lagner, A. and Mikulski, A. (1970), *J. Histochem. Cytochem.*, **18**, 490.
125. Stevenson, D. G. (1954), *Analyst*, **79**, 504.
126. Wright, D. L. and Plummer, D. T. (1973), *Biochem. J.*, **133**, 521.
127. Hjerten, S. (1962), *Archs Biochem. Biophys. Suppl.*, **1**, 147.
128. Converse, C. A. and Papermaster, D. S. (1975), *Science*, **189**, 469.
129. Green, J. R., Dunn, M. J. and Maddy, A. H. (1975), *Biochim. biophys. Acta*, **382**, 457.
130. Zalik, S. E. and Cook, G. M. W. (1976), *Biochim biophys. Acta*, **419**, 119.

# 10 Applications of pure phospholipases in membrane studies

ROBERT F. A. ZWAAL and BEN ROELOFSEN



## 10.1 General introduction

Detailed knowledge about the topographic arrangement of components in the membrane is an essential requisite to gain an insight into the molecular organization and function of biological membranes. A variety of techniques have been developed in the last decade which provide specific information on the localization of membrane constituents. Various investigators have shown that some proteins are only present at one side of the membrane, whereas other proteins extend through the membrane core (for recent reviews see [1] and [2] ). A non-random distribution of phospholipids between the outside and the inside of the erythrocyte membrane was first proposed by Bretscher [3,4] based on the observation that the relatively non-permeant label formylmethionylsulphone methylphosphate (FMMP) fails to react with phosphatidylethanolamine and phosphatidylserine in intact cells. These observations were essentially confirmed by Gordesky and Marinetti [5] using the non-penetrating probe 2,4,6-trinitrobenzenesulphonate, although they could label some of the phosphatidylethanolamine with intact cells. Since these reagents are intrinsically unable to label the choline-containing phospholipids, these results should be

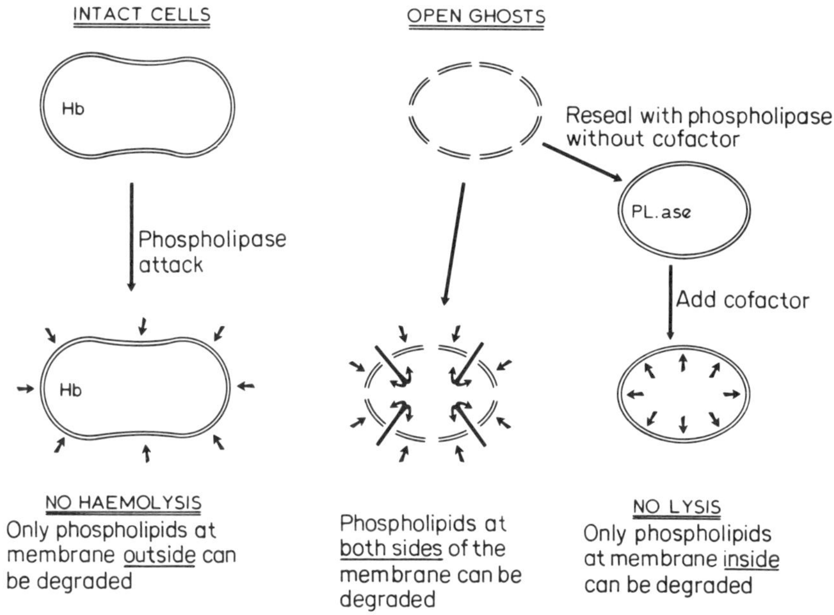


**Fig. 10.1** Scheme of phospholipase attack on intact cells and ghosts. Abbreviations: Hb = haemoglobin. PL.ase = phospholipase. Arrows indicate site of action of the phospholipases.

taken as indirect evidence that lecithin and sphingomyelin form the outer monolayer of the erythrocyte membrane.

Studies using highly purified phospholipases also favour the view that an asymmetric phospholipid distribution occurs in the erythrocyte membrane [6–9]. In order to obtain information about the localization of the phospholipids in red cell membranes, the degradation of the phospholipids by lipolytic enzymes is studied in a number of systems (Fig. 10.1). It can be argued that non-haemolytic breakdown of phospholipids by the action of phospholipases on intact cells, is restricted to those phospholipids which are located at the membrane exterior. Enzymatic action on non-sealed ghosts will lead to degradation of phospholipids on both sides of the membrane. Moreover, trapping of phospholipases (without cofactor) inside resealed ghosts followed by addition of cofactor, will lead to enzymatic degradation of those phospholipids which are present on the inside of the membrane, provided that no lysis of the resealed ghosts occurs.

Studies in which phospholipases have been used to investigate structural and functional aspects of various cell membranes are often confusing, since certain fundamental aspects of the use of lipolytic enzymes in membrane studies have been ignored. In many investigations impure enzymes have been used, introducing the possibility that contaminations in the phospholipase preparation have influenced the observed effects. Furthermore, it is not always obvious whether partial degradation of a certain phospholipid class can be attributed to the existence of two or more different physical states of the particular phospholipid, since in many cases no indications have been given as to whether the phospholipase treatment has been carried out under optimal conditions. In this respect, it should be emphasized that the percentage degradation of a certain phospholipid in a membrane (or liposome) is strongly dependent, not only upon the incubation time, but also on the amount of enzyme present. If both these aspects are not considered properly, the observed partial degradation of certain phospholipid classes does not allow any conclusion with respect to differences in physical state of that particular lipid.

In general, the accessibility of membrane phospholipids *in situ* to phospholipases is governed by at least four underlying principles.

(a) Substrate specificity of the phospholipases.
(b) Sidedness of phospholipids in membranes when only one side of the membrane is exposed to the phospholipase action.
(c) Compactness of the membranous lipid layer.
(d) Degree of shielding by membrane proteins or other membrane constituents.

(a): When certain phospholipases are selected as tools for membrane studies, the variability in substrate specificity of the various phospholipases should be taken into account. Phospholipases of the same type but from different sources may exhibit variations, not only with respect to the question which phospholipid classes are proper substrates for the individual enzymes, but also in the rate with which each of the individual substrates is hydrolysed. For example, *Bacillus cereus* phospholipase C fails to hydrolyse sphingomyelin while this phospholipid is rapidly degraded by *Clostricium welchii* phospholipase C, and pancreatic phospholipase $A_2$ exhibits its largest degradation rate on negatively charged glycerophospholipids like phosphatidylglycerol and phosphatidylserine, while *Naja naja* phospholipase $A_2$ is more active towards phosphatidylethanolamine. A second point to be considered concerns the degradation products formed by the phospholipase action. In this respect, it is not only important to know which split products remain in the membrane, but also what possible effects these degradation products might have on the structural and functional aspects of the membrane. For example, the action of phospholipase D produces an accumulation of phosphatidic acid in the membrane with a concomitant increase in negative charge.

(b): Assuming an arrangement of the membrane phospholipids in a bimolecular leaflet, it is clear that this will influence the percentage phospholipid degradation when only one side of the membrane can be attacked by the phospholipase action, particularly in those cases where the membranes are not grossly disintegrated by the enzymatic action. In this respect, it is important to work with either completely sealed or completely unsealed preparations.

Mixtures of these two will in general lead to confusing results.

(c): The phospholipase activity towards monomolecular films of phospholipids at the air-water interface is strongly contingent upon the initial surface pressure [9]. In general, each phospholipase has its characteristic surface pressure optimum (similar to pH and temperature optimum) where it exhibits its maximal activity, while at both higher and lower surface pressures the activity declines. A typical example of the influence of the compression state of a lecithin monolayer on the rate of hydrolysis by pancreatic phospholipase $A_2$ is given by Zografi *et al.* [10]. It will be clear from the foregoing, that the degree of packing of the lipids in the biological membrane will also strongly influence the enzyme activity towards the membrane. Since the number of enzyme units always refers to optimal conditions, the apparent enzyme activity towards biological membranes is in general lower than expected, since it is likely that the compactness of the lipid layer in the membrane is not optimal for the particular phospholipase. This also means that different numbers of units are required for different phospholipases to produce maximal breakdown of phospholipids in biological membranes. For example, it requires at least 8 IU of *N. naja* phospholipase $A_2$ per 0.25 ml of red cells to produce maximal breakdown of lecithin, whereas only 1 IU of sphingomyelinase per 0.25 ml of red cells is required to cause a maximal attack of sphingomyelin.

(d): It is conceivable, that shielding of the phospholipids by membrane proteins may influence the rate of phospholipid hydrolysis, or even completely prevent phospholipid breakdown. In this respect, it might be useful to distinguish between two different possibilities of shielding by membrane proteins: (i) accessibility of phospholipids to phospholipases is influenced by membrane proteins which are in direct interaction with the lipid substrates, (ii) the topographic arrangement of certain membrane proteins, which are not in direct interaction with the lipid substrates, might be such that the migration of phospholipases to the lipids is prevented. In the first case not only masking by proteins should be considered resulting in a complete blocking of the substrate, but also a competitive inhibition by membrane proteins on the enzymic activity might occur, especially in those cases where the lipid-membrane protein complex is of the same type as the enzyme-substrate complex between phospholipases and phospholipids. It will be clear, that isolation of such a lipo-protein complex from the membrane will in general not influence the phospholipid accessiblity by phospholipases. This is in contrast to the second type of shielding, which can occur only in a highly organized structure like a biological membrane. In this case, disintegration of the membrane structure will lead to exposure of the phospholipids to the phospholipases.

The first section of this chapter will consider extensively the study of phospholipid localization in red cell membranes. The second section will deal with the use of phospholipases in studying the lipid dependence of some membrane functions.

## 10.2 Localization of phospholipids in erythrocyte membranes as detected by phospholipases

### *10.2.1 Phospholipase action on non-sealed ghosts*

Treatment of non-sealed red cell ghosts with phospholipases is a necessary part of a 'sidedness experiment', in order to find out whether the phospholipids at both sides of the membrane are completely available to the enzyme. The approximate composition of the red cell ghost is 50 wt% proteins, 40 wt% lipids and 10 wt% carbohydrates. The lipids are composed of equimolar amounts of phospholipids and cholesterol. The formulae of the major phospholipid classes of the erythrocyte membrane are shown in Fig. 10.2, which also shows the site of action of phospholipase $A_2$, C and D and of sphingomyelinase on phosphoglycerides and sphingomyelin, respectively. In a wide range of mammalian erythrocytes, the sum of the choline containing phospholipids (lecithin and sphingomyelin) comprises 46–60% of the total membrane phospholipids [11]. Whereas the ratio of phosphatidylethanolamine to phosphatidylserine is relatively constant in all these species, the ratio of lecithin to sphingomyelin

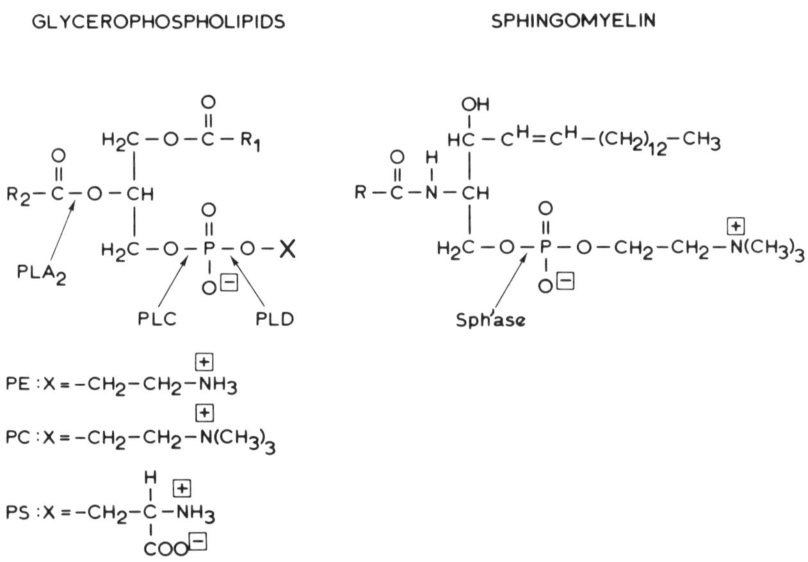


**Fig. 10.2** Major phospholipid classes of red cell membranes and sites of action of various phospholipases. Abbreviations: PE = phosphatidylethanolamine. PC = phosphatidylcholine. PS = phosphatidylserine. R, $R_1$, $R_2$ = alkyl residues. $PLA_2$ = phospholipase $A_2$ (phosphatide acyl-hydrolase, EC 3.1.1.4). PLC = phospholipase C (phosphatidylcholine cholinephosphohydrolase, EC 3.1.4.3). PLD = phospholipase D (phosphatidylcholine phosphatidohydrolase, EC 3.1.4.4). Sph'ase = sphingomyelinase (sphingomyelin cholinephosphohydrolase).

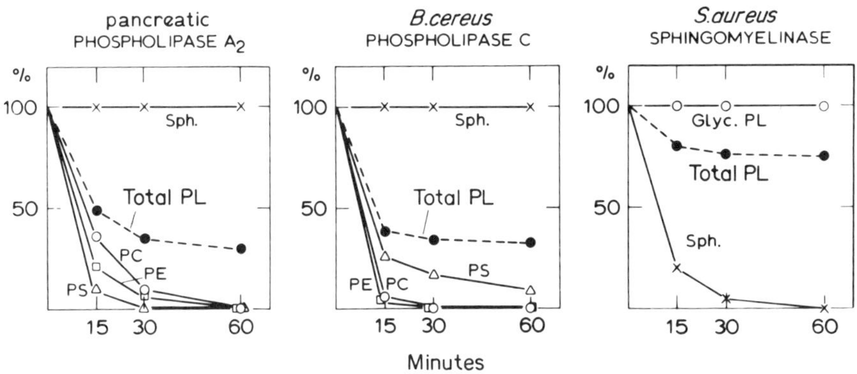


**Fig. 10.3** Percentage retention of phospholipids in human erythrocyte ghosts after incubation with phospholipases. Abbreviations: Total PL = initially present total phospholipids. Glyc. PL = initially present glycerophospholipids. PC = lecithin. PE = phosphatidylethanolamine. PS = phosphatidylserine. Sph = sphingomyelin.

varies markedly. The most dramatic deviation is found in ruminant red cells where lecithin is nearly completely absent and seems to be replaced by sphingomyelin.

The result of treatment of ghosts with both phospholipases $A_2$ and C is a complete digestion of lecithin, phosphatidylethanolamine and phosphatidylserine, exhibiting only minor differences with respect to the velocity with which these phospholipid classes are hydrolysed (Fig. 10.3). However, in both cases no detectable breakdown of sphingomyelin can be demonstrated. These observations are responsible for the fact that only 70% of the initially present phospholipids are attacked, as is indicated by the two dashed lines in Fig. 10.3, levelling off at 30% retention. Sphingomyelinase produces a complete breakdown of sphingomyelin without affecting the glycerophospholipids. As expected from the results shown in Fig. 10.3, a mixture of phospholipase C and sphingomyelinase is able to produce a complete degradation of the total phospholipid complement of the membrane.

The major conclusion which can be drawn from phospholipase attack on red cell ghosts, is that all phospholipids are readily available to the enzymes, provided that the lipids are proper substrates for the enzymes involved. Although in Fig. 10.3 the results are shown for human erythrocyte ghosts, there is in principle no difference when ghosts from other mammalian species are used.

When phospholipases from different sources are used, one should be aware of variations in substrate specificity. When pancreatic phospholipase $A_2$ is replaced by *N. naja* phospholipase $A_2$, a complete degradation of the glycerophospholipids again occurs although in this case phosphatidylethanolamine appears to be the most rapidly degraded phospholipid. A somewhat different picture is obtained when phospholipase C from *C. welchii* is used instead of

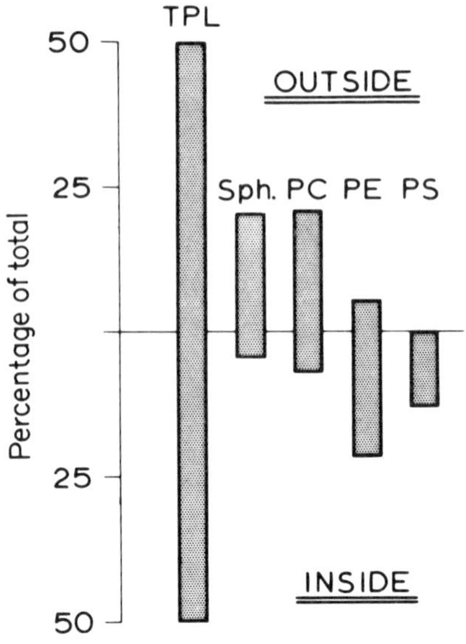


**Fig. 10.4** Proposed distribution of phospholipids between inner and outer layer of the human erythrocyte membrane. Abbreviations: TPL = total phospholipid. Sph = sphingomyelin. PC = lecithin. PE = phosphatidylethanolamine. PS = phosphatidylserine.

phospholipase C from *B. cereus:* both choline containing phospholipids – lecithin and sphingomyelin – are rapidly degraded, whereas phosphatidylethanolamine is hydrolysed at a much slower rate and phosphatidylserine is not degraded at all [19].

It should be noted that the lyso-compounds and fatty acids produced by the action of phospholipase $A_2$ remain in the membrane. The split products produced by phospholipase C, i.e. ethanolamine-, serine-, and choline-phosphate (the latter also produced by sphingomyelinase) are readily water soluble and hence are removed from the membrane. On the other hand, diglycerides and ceramides produced by phospholipase C and sphingomyelinase, respectively, remain in the membrane and are supposed to migrate into discrete pools [6,19].

### *10.2.2 Phospholipase action on intact cells*

In contrast to the action of lipolytic enzymes on non-sealed erythrocyte ghosts, the enzymes can only act on the exterior of the membrane of an intact cell. In principle, one should distinguish three possible results of incubation of erythrocytes with pure phospholipases, either alone or in combination:

(a) No phospholipid hydrolysis and no haemolysis.
(b) Hydrolysis of phospholipid without haemolysis.
(c) Both phospholipid hydrolysis and haemolysis.

In the last case, non-sealed ghosts will be produced due to haemolysis, and degradation of the lipid at both sides of the ghost membrane will occur. Only the second result provides direct information about those phospholipids which are present at the outside of the membrane. Therefore, secondary haemolytic activities in crude or partially purified phospholipase preparations must be completely removed by further purification. In general, nearly all commercially available phospholipase preparations are contaminated with haemolytic factors, whereas until now no pure phospholipase (from any source) has been shown to be haemolytic; phospholipase C from *C. welchii* may be a possible exception.

Although no haemolysis is produced by the separate action of the pure phospholipases on intact cells, it cannot necessarily be concluded that no phospholipid degradation occurs. Non-haemolytic breakdown of phospholipids has also been observed with phospholipase $A_2$ from sea-snake venom [20] and phospholipase $A_2$ from bee venom [8], and as shown in Table 10.1, both *N. naja* phospholipase $A_2$ and *S. aureus* sphingomyelinase also accomplish a non-lytic breakdown of phospholipids in intact human erythrocytes. However, both pancreatic phospholipase $A_2$ and *B. cereus* phospholipase C appear to be ineffective in this respect. Phospholipase $A_2$ from *N. naja* only attacks lecithin, converting over two-thirds of this phospholipid into its lyso-derivative. This corresponds with a degradation of 20% of the total phospholipid fraction. Treatment of intact cells with sphingomyelinase results in the degradation of 85% of the sphingomyelin, representing another 20% of the membrane phospholipids.

It comes within the framework of determining which phospholipids are on the membrane exterior, to combine the non-lytic degradative action of *N. naja* phospholipase $A_2$ and sphingomyelinase from *S. aureus.* While phospholipase $A_2$ alone exclusively attacks lecithin in intact cells, addition of sphingomyelinase

TABLE 10.1
*Non-lytic degradation of phospholipids in intact human erythrocytes by phospholipases*

| | % degradation of phospholipids | | | | |
|---|---|---|---|---|---|
| | Total phospholipid | Sphingomyelin | Phosphatidylcholine | Phosphatidylethanolamine | Phosphatidylserine |
| Phospholipase $A_2$ (pancreas) | 0 | 0 | 0 | 0 | 0 |
| Phospholipase C (*B. cereus*) | 0 | 0 | 0 | 0 | 0 |
| Phospholipase $A_2$ (*N. naja*) | 20 | 0 | 68 | 0 | 0 |
| Sphingomyelinase (*S. aureus*) | 20 | 85 | 0 | 0 | 0 |

TABLE 10.2
*Non-lytic degradation of phospholipids by phospholipases in human erythrocytes. Incubations were started in the presence of phospholipase $A_2$ (*N. naja*), followed by addition of sphingomyelinase (*S. aureus*) after 1 h.*

| Incubation time (min) | Enzyme (% degradation) | | | | | | |
|---|---|---|---|---|---|---|---|
| | Phospholipase $A_2$ (*N. naja*) | Sphingomyelinase (*S. aureus*) | Total phospholipid | Sphingomyelin | Phosphatidylcholine | Phosphatidyl-ethanolamine | Phosphatidylserine |
| 0 | ↓ | | 0 | 0 | 0 | 0 | 0 |
| 60 | ↓ | ↓ | 20 | 0 | 68 | 0 | 0 |
| 120 | ↓ | ↓ | 48 | 82 | 76 | 20 | 0 |

TABLE 10.3
*Percentage haemolysis and phospholipid hydrolysis by combined action of phospholipases in human erythrocytes. All incubations were carried out for 3 h. Enzymes were added to the incubation mixtures after time intervals indicated.* $PLA_2$ *= phospholipase* $A_2$ *(*N. naja*); PLC = phospholipase C (*B. cereus*); Sph'ase = sphingomyelinase (*S. aureus*).*

| Addition of enzyme at: | | | % haemolysis at: | Phospholipid degradation by the action of: | | |
|---|---|---|---|---|---|---|
| $t$ = 0 min. | $t$ = 60 min. | $t$ = 120 min. | $t$ = 180 min. | $PLA_2$ | Sph'ase | PLC |
| – | – | – | 2 | | | |
| $PLA_2$ | – | – | 1 | + | | |
| Sph'ase | – | – | 2 | | + | |
| PLC | – | – | 1 | | | – |
| Sph'ase | PLC | – | 100 | | + | + |
| $PLA_2$ | Sph'ase | PLC | 5 | + | + | – |
| $PLA_2$ | PLC | – | 2 | + | | – |
| $PLA_2$ | Sph'ase | – | 3 | + | + | |

not only results in the hydrolysis of more than 80% of the sphingomyelin, but also enables phospholipase $A_2$ to degrade 20% of the phosphatidylethanolamine and another 8% of the phosphatidylcholine (Table 10.2). Prolonged incubation times fail to produce more phospholipid breakdown. This successive action of the two enzymes, though not producing lysis of the cells, degrades approximately half of the total phospholipid complement of the membrane. It is noteworthy that no phosphatidylserine is attacked, although this phospholipid is rapidly degraded when ghosts are subjected to phospholipase $A_2$ treatment.

Although phospholipase $A_2$ treated cells are not lysed upon subsequent addition of sphingomyelinase, the reverse order of addition of these two enzymes has been shown to have a lytic effect [7]. Whether or not haemolysis is produced by the combined action of phospholipases appears to be strongly dependent upon the sequence of addition of enzymes to the incubation mixtures. A typical example is shown in Table 10.3. In contrast to *N. naja* phospholipase $A_2$ and *S. aureus* sphingomyelinase, phospholipase C from *B. cereus* is not able to produce non-lytic phospholipid breakdown by individual action. On the other hand, cells treated first with sphingomyelinase are completely lysed by addition of *B. cereus* phospholipase C. Apparently phospholipase C is able to produce phospholipid breakdown after sphingomyelinase treatment, which results in lysis of the cells. However, when the cells are first treated with phospholipase $A_2$ and then incubated with sphingomyelinase and phospholipase C, no significant haemolysis is produced. In this case, the possible substrates for phospholipase C at the membrane exterior (mainly lecithin) are already converted by the phospholipase $A_2$ treatment into lyso-compounds, which can hardly be considered as proper substrates for *B. cereus* phospholipase C.

TABLE 10.4
*Percentage haemolysis and phospholipid hydrolysis by phospholipase C and sphingomyelinase in mammalian erythrocytes. The following erythrocytes from the two categories have been tested,* Non-ruminants: *human, pig, rabbit, dog, rat;* ruminants: *ox and sheep. PLC = phospholipase C (*B. cereus*); Sph'ase = sphingomyelinase (*S. aureus*).*

| | non-ruminants | | | ruminants | | |
|---|---|---|---|---|---|---|
| | | Phospholipid degradation by the action of: | | | Phospholipid degradation by the action of: | |
| | % haemolysis | PLC | Sph'ase | % haemolysis | PLC | Sph'ase |
| Blank | 1 | | | 1 | | |
| PLC | 1 | – | | 2 | – | |
| Sph'ase | 2–3 | | + | 2–4 | | + |
| PLC + Sph'ase | 100 | + | + | 1–4 | – | + |

The combined action of sphingomyelinase and phospholipase C appears to be lytic for red cells of a variety of mammalian species. However, no lysis is produced by the combined action of the two enzymes on ruminant erythrocytes (Table 10.4). These cells are known to contain very little lecithin. Although sphingomyelinase is able to accomplish a non-lytic degradation of sphingomyelin in these cells, phospholipase C fails to produce further breakdown of glycerophospholipids and this inactivity is taken to be the reason for the absence of haemolysis.

The most extensive degradation of phospholipids in intact cells is achieved by a combined action of *N. naja* phospholipase $A_2$ and sphingomyelinase from *S. aureus* resulting in a breakdown of half of the membrane phospholipids. On the other hand, this enzyme combination produces a complete degradation of the membrane phospholipids of non-sealed ghosts. Starting from the principle that the membrane lipids are oriented in a bilayer structure, the most simple interpretation of these results is that the phospholipid fraction which can be degraded in intact cells (compare Table 10.2) comprises the outer monolayer of the membrane. This implies that the outer layer of the human erythrocyte membrane consists mainly of the two choline-containing phospholipids, e.g. lecithin and sphingomyelin, while in addition one-fifth of the phosphatidylethanolamine is present (Fig. 10.4). The inner layer is mainly composed of phosphatidylethanolamine and phosphatidylserine, while a minor fraction of the choline-containing phospholipids may also be present. A similar arrangement of the phospholipids has been proposed by Bretscher [3,4] and also by Gordesky and Marinetti [5] based on indirect evidence obtained by labelling the cells with relatively non-permeant reagents. With respect to other mammalian red cells, it is recalled that a mixture of phospholipase C and sphingomyelinase is lytic towards a wide variety of mammalian erythrocytes, but fails to produce lysis in ruminant red cells. These results can be taken to indicate that choline-containing phospholipids are generally present in the outer layer of mammalian red cells, including those from ruminants in which case sphingomyelin will be the major phospholipid at the outside of the membrane.

It should be noted that non-lytic breakdown of phospholipids by phospholipases has a dramatic influence on the osmotic resistance of the cell. When such cells are subjected to a decrease in osmotic strength in the order of 10–20 mosmol, spontaneous lysis occurs. On the other hand, no lysis is observed when the osmolarity is kept constant or increased. This may be of vital importance when phospholipase treated cells are used for further experiments in which lysis should be avoided. In addition, lysis also occurs when lyso-compounds, produced by the action of phospholipase $A_2$ on intact cells, are removed from the membrane by incubating the cells in the presence of fat-free albumin [6,21].

The failure of pancreatic phospholipase $A_2$ and *B. cereus* phospholipase C to degrade phospholipids in intact red cells, while phospholipids are accessible to sphingomyelinase and *N. naja* phospholipase $A_2$, does suggest that the substrate requirements are not fulfilled for the first group of enzymes, but that a proper enzyme-substrate complex can be formed with the latter group of enzymes. Two major possibilities should be considered in this context:

(a) The phospholipids are not readily available at the membrane exterior but shielded by proteins, and some phospholipases may be intrinsically able to disturb this shielding leading to exposure of the phospholipids.

(b) The phospholipids are directly available on the outside of the cell, but the ability of the different phospholipases to exert their action depends strongly on the alignment and the packing of the phospholipid molecules in the native membrane.

The first possibility may be less likely since pre-treatment of erythrocytes with proteases does not make the lipids accessible to degradation by pancreatic phospholipase $A_2$ or *B. cereus* phospholipase C [22]. The second possibility, however, seems to find some support from studies dealing with the activity of phospholipases towards monomolecular films of phospholipids at various initial surface pressures [9]. Sphingomyelinase (*S. aureus*) and phospholipase $A_2$ (*N. naja*) when injected under a monolayer of sphingomyelin or lecithin, respectively, are still able to produce phospholipid breakdown at surface pressures above 300 $\mu$N cm$^{-1}$. On the other hand, pancreatic phospholipase $A_2$ and *B. cereus* phospholipase C can only exert their action at initial surface pressures below below 300 $\mu$N cm$^{-1}$. From these observations it might be concluded that the surface pressure in the intact red cell membrane is above 300 $\mu$N cm$^{-1}$. Furthermore, phospholipase C is able to exert its action to a mixed monolayer of lecithin and sphingomyelin of 420 $\mu$N cm$^{-1}$ provided that sphingomyelin is degraded first or simultaneously by sphingomyelinase. This may explain why the combined action of phospholipase C and sphingomyelinase is lytic to a variety of red cells, except those from ruminants.

In this respect, the action of *C. welchii* phospholipase C may be interesting. Although controversial observations exist in the literature, the most extensively purified preparation of this enzyme has been shown to attack both choline-containing phospholipids, lecithin and sphingomyelin [23]. Since partially purified preparations of this enzyme from our laboratory have been found to attack monolayers of lecithin at an initial surface pressure above 350 $\mu$N cm$^{-1}$, it is anticipated that this enzyme will have the same haemolytic effect towards red cells as a mixture of *B. cereus* phospholipase C and *S. aureus* sphingomyelinase. Preliminary experiments from our laboratory show that a partially purified *C. welchii* phospholipase C preparation is strongly lytic under normal conditions, whereas no haemolysis is produced when the enzyme activity is inhibited with a mixture of EDTA and *o*-phenanthroline, prior to addition to the cells. Moreover, red cells treated with *N. naja* phospholipase $A_2$ followed by treatment with *C. welchii* phospholipase C are lysed at a much slower rate than control cells with this phospholipase C alone, in spite of the strongly increased osmotic fragility of the phospholipase $A_2$ treated cells. The most obvious explanation is that with control cells, *C. welchii* phospholipase C is able to degrade both lecithin and sphingomyelin (which produces lysis similar to the combined action of *B. cereus* phospholipase C and *S. aureus* sphingomyelinase), while with phospholipase $A_2$ treated cells, *C. welchii* phospholipase C has the opportunity to degrade lyso-lecithin and sphingomyelin. The conversion of lyso-lecithin by *C. welchii* phospholipase C occurs at a much slower rate than the conversion of lecithin [24], resulting in a much lower rate of the haemolysis process. The observed effects with *C. welchii* phospholipase C, both on monolayers and intact cells, make it likely that the particular enzymatic properties of this phospholipase are responsible for causing haemolysis. At present, no other pure phospholipase has been shown to have haemolytic properties.

### *10.2.3 Phospholipase action on resealed ghosts*

Further evidence for the existence of a non-random phospholipid distribution can be obtained by directly attacking the inside of the red cell membrane, after trapping a phospholipase without its cofactor inside resealed ghosts. Starting the enzymic reaction by addition of cofactor will lead to phospholipid hydrolysis which is restricted to those phospholipid classes which are present on the inside of the membrane, as long as no lysis of the resealed cells occur. For this purpose, it is essential that the enzyme is trapped within the greatest possible number of resealed cells, and that no hydrolysis of phospholipids occurs during the resealing process.

It was first recognized by Hoffman [25,26] that a ghost suspension cannot be considered as a homogeneous population. A certain fraction of the ghost population is capable of resealing, whereas the rest of the ghosts remains leaky. Bodemann and Passow [27] and Schwoch and Passow [28] have distinguished between three types of ghosts: ghosts which reseal directly after haemolysis (type I); ghosts which reseal after reversion of the haemolysis procedure by addition of alkali ions (type II); ghosts which remain leaky even after reversion (type III). The ratio of the three ghost fractions appeared to be dependent on a variety of parameters. For details, the reader is referred to the above-mentioned articles from Passow's group.

It needs no argument that for a sidedness experiment with phospholipases, the fraction of type II ghosts should be as high as possible. While haemolysis at 0°C for 5 min followed by resealing at 37°C eliminates virtually all type I ghosts, the fraction of type III ghosts can be drastically minimized by haemolysing at pH 6.0. Moreover, the tonicity of the haemolysate should be 40 mosmol or more, in order to prevent irreversible damage to the ghost membrane. During the resealing process at 37°C, it is absolutely essential that the phospholipase does not exert its action. Resealing in the presence of EDTA seems to be the obvious way. Phenanthroline should be avoided since this may bind to iron from haemoglobin as well. However, many phospholipases appear not to be completely blocked by addition of EDTA, pancreatic phospholipase $A_2$ forming a main exception. Although this enzyme fails to attack phospholipids at the cell exterior, it appears to be able to hydrolyse the phospholipids at the inside of the resealed cell membrane.

From these experiments with resealed ghosts it is found that pancreatic phospholipase $A_2$, when trapped within resealed ghosts, is able to produce substantial breakdown of glycerophospholipids before the cells begin to lyse (Fig. 10.5). At this point, approximately 25% of the lecithin, 50% of the phosphatidylethanolamine, and 65% of the phosphatidylserine are degraded. This comprises 30–35% of the total phospholipid complement of the membrane. It is remarkable that, before the onset of lysis, the degradation of phosphatidylcholine tends to cease, whereas hydrolysis of the other two glycerophospholipids appears to be in full swing. As expected, breakdown of all the glycerophospholipids goes to completion during lysis of the resealed cells.

It is likely, that non-lytic breakdown of phospholipids by phospholipases which are trapped inside resealed ghosts, is restricted to those phospholipids which are present at the membrane interior. The observation that lecithin degradation tends to level off around 25% breakdown (corresponding to 75% retention in Fig. 10.5), does suggest that this fraction is present at the inner side of the human erythroctye membrane. This fraction may correspond to that lecithin fraction which cannot be hydrolysed by *N. naja* phospholipase $A_2$ treatment of intact cells (compare Table 10.2 and Fig. 10.4). On the other hand, the nonlytic degradation of phosphatidylethanolamine and phosphatidylserine from the inside is not completely complementary to the observations using intact erythrocytes, described in the previous section. This is mainly due to the fact that the resealed cells apparently start to lyse when 30–35% of the total

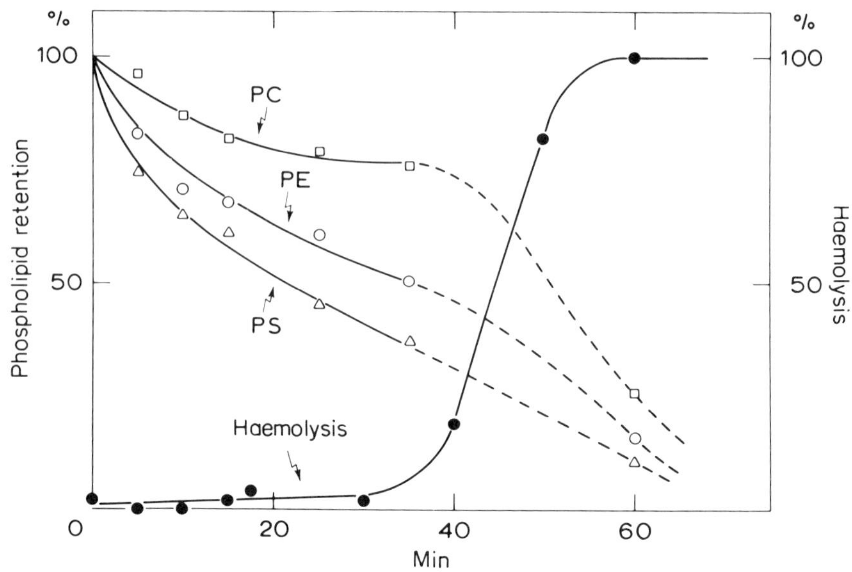


**Fig. 10.5** Percentage retention of glycerophospholipids after trapping of pancreatic phospholipase $A_2$ (without $Ca^{2+}$) inside resealed human erythrocyte ghosts followed by addition of $Ca^{2+}$ to start enzymatic breakdown at $t = 0$. Abbreviations: PC = lecithin. PE = phosphatidylethanolamine. PS = phosphatidylserine.

phospholipids of the membrane have been converted into their respective lyso-derivatives. The relatively rapid attack of phosphatidylethanolamine and phosphatidylserine does suggest that a major fraction of these two phospholipids is present at the membrane interior, although no quantitative estimations as to the extent can be made.

Due to its absolute requirement for $Ca^{2+}$ [29], pancreatic phospholipase $A_2$ can be completely inhibited by EDTA. This is an essential premise for experiments with resealed ghosts, when no breakdown of phospholipids can be allowed to occur during the resealing procedure. So far, other phospholipases have been found to be less satisfactory in this respect. For example, the ghosts are difficult to seal in the presence of *N. naja* phospholipase $A_2$; they often remain leaky and some phospholipid degradation may occur during the resealing process, even in the presence of high EDTA concentrations. Although in some cases the *N. naja* phospholipase $A_2$ action inside resealed ghosts produces results similar to the action of pancreatic phospholipase $A_2$, the reproducibility leaves much to be desired.

Finally, pancreatic phospholipase $A_2$ is unable to attack the membrane outside of intact erythrocytes, but is capable of attacking the membrane inside of resealed ghosts. Similar observations have been made with *B. cereus* phospholipase C [30]. Although the reason for this phenomenon remains obscure, it might be speculated, that either the compression state (surface pressure) of the phospholipids at the membrane inside is smaller than at the outside, or that *B. cereus* phospholipase C and pancreatic phospholipase $A_2$ are able to attack the negatively charged phospholipids (present on the inside) at higher initial surface pressures than phosphatidylcholine (present on the outside).

## 10.3 Applications of lipolytic enzymes in studying functional aspects of membranes

So far, the discussion above has been limited mainly to methodological aspects of using phospholipases in determining sidedness of phospholipids in red cell membranes. In principal, the same methodology can also be used in studies concerning functional aspects of biomembranes. It should be emphasized that in these studies it is even more important to use pure lipolytic enzymes, since any contamination in the enzyme preparation may influence the outcome of the experiment, and may therefore lead to incorrect interpretations.

Three different possibilities for applying phospholipases in the study of membrane functions can be distinguished.

### *10.3.1 Involvement of phospholipids as a juncture for membrane proteins*

When enzymatic hydrolysis of membrane phospholipids results in liberation of certain proteins (including enzymes) from the membrane, the conclusion seems to be justified that these proteins are anchored in the membrane via direct lipid-protein interactions. For example, treatment of human red cell ghosts with pure *B. cereus* phospholipase C, though releasing less than 1% of the ghost protein into the medium, liberates up to 20% of the ghost acetylcholinesterase as a consequence of phospholipid breakdown [8]. Since all the main glycero-phospholipid classes are degraded at similar rates, no conclusion as to the involvement of a particular phospholipid class can be drawn. However, the role of a certain phospholipid class has been established for anchoring NADH-dehydrogenase in mitochondrial cristae [31,32]. At relatively low concentrations of *N. naja* phospholipase $A_2$, only phosphatidylethanolamine and phosphatidylcholine are degraded, without liberating NADH-dehydrogenase from the membrane. On the other hand, prolonged incubation periods on higher enzyme concentrations also result in a nearly complete breakdown of cardiolipin with a concomitant release of NADH-dehydrogenase. This might imply that this enzyme is anchored in the electron transport particle by direct interaction with cardiolipin, without influencing the accessibility of this phospholipid to phospholipase $A_2$.

Extraction of phospholipase-treated erythrocyte ghosts with acetic acid, hypertonic NaCl, potassium thiocyanate or water, does not release any significant additional amount of protein, compared with that released from non-digested ghosts [8]. This may imply that either there is association between the proteins in the normal ghost (which is maintained after degradation of the lipids), or that on breakdown of the lipids the normal hydrophobic lipid-protein interactions are replaced by hydrophobic protein-protein interactions. The last possibility might explain the observed shrinkage of ghosts after phospholipase C treatment, while the vesicular structure is maintained [6,19].

### *10.3.2 Localization of phospholipid metabolizing processes in the membrane*

It will be clear that, when phospholipases can be used for determining phospholipid distributions in membranes, these enzymes may also be useful tools in localizing phospholipid turnover and renewal processes in biological membranes. For the red cell membrane, two mechanisms have been described which play a major role in phospholipid renewal: (i) total exchange of phospholipids between serum lipoproteins and the erythrocyte membrane [33–35]; (ii) incorporation of fatty acids into lysolecithins, which are either produced in the membrane or supplied by exchange with the surrounding serum [36,37]. It has recently been shown that treatment of intact human erythrocytes with $^{14}C$-labelled fatty acids in plasma, resulted in an incorporation of the label in that lecithin fraction which cannot be degraded with *N. naja* phospholipase $A_2$ [38]. Since the treatment of intact cells with phospholipase $A_2$ leads to an exclusive degradation of that lecithin fraction (more than two thirds of the total) which is located in the outer monolayer, it can be concluded that *in vitro* incorporation of fatty acids into lecithin in human red blood cells occur predominantly at the inner part of the membrane. Consequently, it was speculated that the acyltransferases involved are located almost exclusively at the inside of the membrane. On the other hand, there is a good possibility that renewal of lecithin by exchange of the whole molecule between plasma lipoproteins and the erythrocyte membrane takes place in the outer monolayer only.

Although ruminant red cells differ strongly from those of other mammals in that they contain very little lecithin, no such differences have been observed between various mammalian plasma's, which all contain relatively high amounts of lecithin. It has recently been discovered that a membrane-bound phospholipase $A_2$, which is highly specific for lecithin, is present in ruminant erythrocyte membranes and cannot be found in other mammalian red cells [39]. Moreover, this lecithinase is localized at the exterior of the cell membrane, since it can be nearly completely inactivated by pronase treatment of intact cells. Preliminary experiments may suggest that this enzyme plays a role (for unknown reasons) in keeping lecithin out of the red cell membrane of ruminants, this in spite of the high lecithin concentrations in the plasma.

### *10.3.3 Determination of a phospholipid requirement of membrane processes*

A wide variety of membrane-processes are known to require lipids for proper functioning. Phospholipases have been extensively used in studies on the lipid-requirement of various membrane-bound enzymes. When the decrease in phospholipid content, due to enzymatic hydrolysis, runs simultaneously with lowering of activity of some membrane-bound enzyme(s), it may seem justified to conclude that the enzyme needs phospholipid for activity. However, it might always be possible that any contamination (no matter how small) in the phospholipase preparation produces the observed inactivation of the enzyme under study. Therefore, the observed results should be confirmed by reactivation experiments, in which the phospholipase-inactivated membrane-bound enzyme activity is restored by addition of phospholipids. In this case, one should be aware that the split products, produced by the phospholipase action, might either occupy possible binding sides of the re-added phospholipids, or intrinsically be inhibitory towards the membrane-bound enzyme activity.

Until now, the most extensively studied membrane-bound lipid-requiring enzyme system is the ($Na^+ + K^+$)-stimulated $Mg^{2+}$-dependent ATPase. However, the right nature of the lipid involved has continuously been subject to controversial opinions. From studies using various pure phospholipases, it has been concluded that at least one of the glycerophospholipids plays a specific role in the ($Na^+ + K^+$)-ATPase of the human red cell membrane, since this enzyme system is inactivated by phospholipases $A_2$ and C and not by sphingo-myelinase [40] (Table 10.5). After removal of the diglycerides produced by *B. cereus* phospholipase C, the ATPase activity could be completely restored by addition of phosphatidylserine (and to a lower extent also by phosphatidic acid), whereas both lecithin and phosphatidylethanolamine had no effect. The requirement of the ($Na^+ + K^+$)-ATPase for phosphatidylserine was further confirmed by the observation that this enzyme system was inactivated when phosphatidylserine was completely converted into phosphatidylethanolamine by treatment with phosphatidylserine–decarboxylase (*E. coli*).

Substantial evidence exists which points into the direction that the ($Na^+ + K^+$)-stimulated ATPase occupies an asymmetric alignment in the red cell membrane, and that at least the active centre is localized on the inside of the membrane [41–43]. The localization of both the phosphatidylserine molecules and the active centre of the ($Na^+ + K^+$)-ATPase at the inside of the human erythrocyte membrane is probably an essential prerequisite for active transport of these cations through this membrane.

## 10.4 Practical procedures

### *10.4.1 Preparation of phospholipases*

Phospholipase $A_2$ from porcine pancreas is prepared as described by de Haas *et al.* [13]. Phospholipase $A_2$ from *N. naja* venom is purified as described by Cremona and Kearney [14], with the exception that gel filtration on Sephadex G–75 is replaced by gel filtration on Sephadex G–100 (column: 4 cm x 180 cm) at room temperature, using as elution buffer: 0.05 M Tris-HCl

TABLE 10.5

*Phospholipid breakdown and $(Na^+ + K^+)$-ATPase activity after treatment of human erythrocyte ghosts with pure phospholipases*

| Enzyme | Number of experiments | Sphingomyelin | Phosphatidyl-choline | Phosphatidyl-ethanolamine | Phosphatidyl-serine | ATPase activity |
|---|---|---|---|---|---|---|
| | | % Control | | | | |
| No phospholipase | – | 100 | 100 | 100 | 100 | 100 |
| Phospholipase $A_2$ (pancreas) | 12 | 100 | 0 | 0 | 0 | 0 |
| Phospholipase C (*B. cereus*) | 19 | 100 | 0 | 0 | 0–5 | 0 |
| Sphingomyelinase (*S. aureus*) | 3 | 0 | 100 | 100 | 100 | 120–140 |

(pH 7.4), 5 mM Ca, 50 vol. % glycerol. Phospholipase C from *B. cereus* is prepared by the method of Zwaal *et al.* [15], with the exception that the bacteria are grown in a fermentor (5 l culture in a 7 l vessel) under continuous aeration. The foam produced during growth, which appears to contain virtually all the phospholipase C activity, is collected from the air-outlet pipe on the fermentor vessel. After condensing the foam by cooling to 4°C, the mixture is centrifuged and the supernatant is used as the starting material for purification. The main advantage of the foaming technique appears to be a much higher production of phospholipase C by the bacteria, combined with a three to four-fold increase in specific activity in the starting material. Sphingomyelinase from *Staphylococcus aureus* is purified as described by Colley *et al.* [6].

The phospholipases are routinely stored in 50 vol. % glycerol at – 20°C (solution does not freeze), in order to retain enzymatic activity over a prolonged period of time (1–2 years).

### *10.4.2 Preparation of erythrocytes and ghosts*

#### *(i) Isolation of red cells*

Erythrocytes are isolated from freshly collected blood (defibrinated with acid–citrate–dextrose or heparin) by centrifugation for 10 min at 3000 x *g*. The cells are washed four times with 0.87% NaCl, 0.25 mM $CaCl_2$, 0.25 mM $MgCl_2$, pH 7.4 with 0.05 M Tris. It is strongly advised to carry out the subsequent incubations with phospholipases (as described below) at the same day on which the blood has been collected.

When cells are prepared for the resealing experiments the washing is carried out with 0.9% NaCl.

#### *(ii) Preparation of ghost*

In principle any ghost preparation can be used, provided that the ghost is of the non-sealed type. The various modifications of the procedure according to Dodge *et al.* [12] have been found to be most useful. In case EDTA is used in the buffers, it is advised to omit this compound in the final washings of the ghost in order to avoid any inhibitory effect of this chelating agent on phospholipases.

### *10.4.3 Treatment of ghost with phospholipases*

Suspensions of 45 mg of erythrocyte ghosts in 20 ml 0.1 M Tris buffer (pH 7.4; 10 mM $CaCl_2$, 0.25 mM $MgCl_2$) are incubated at 37°C under stirring with 5–15 IU of either phospholipase $A_2$, phospholipase C or sphingomyelinase. At different incubation times 4 ml aliquots are taken and enzyme activity is immediately blocked by addition of *o*-phenanthroline and EDTA (final concentration: 12 mM of each). (The enzymatic activity must be stopped by the addition of appropriate inhibitors prior to lipid extraction and thin-layer chromatography, since it has been shown that some phospholipases exert their action in organic solvents, particularly when silicic acid is present [16].) In contrast to the phospholipase $A_2$ treated ghost samples, which can be directly subjected to the lipid extraction procedure, the ghost samples from the incubations with phospholipase C or sphingomyelinase are washed twice with 40 ml 0.1 M Tris buffer (pH 7.4) to remove the phosphorus containing split products. The washed ghosts are resuspended in 4 ml Tris buffer.

### *(i) Lipid extraction procedure*

Methanol A.R. (20 ml) is added to the ghost samples (4 ml) and the mixture is stirred for 5 min. Subsequently 20 ml of chloroform A.R. is added, followed by another 5 min of stirring. After centrifugation for 10 min at 3000 x *g* the supernatant is collected and the extraction procedure is repeated twice. The pooled extracts are collected in a pear-shaped flask and taken to dryness under reduced pressure at 37°C. A repeated evaporation with absolute ethanol may be necessary in order to obtain a water-free residue. The residue is dissolved in 8 ml chloroform–methanol (1:1, v/v).

### *(ii) Determination of phospholipid breakdown*

For determination of total phospholipid degradation, duplicate 1 ml samples are taken from the lipid solution in chloroform–methanol and transferred to a test tube. The tubes are placed in an oven at 80–100°C to allow the solvents to evaporate completely. Subsequently, the lipids are digested by adding 1 ml 70% $HClO_4$ followed by heating at 190°C until the solution is colourless (the top of each tube is closed with a marble). Phosphorous determinations on the digests are carried out by a modification of the procedure of Fiske-Subbarow as described by Böttcher *et al.* [17].

The individual phospholipid composition of each lipid extract is determined after evaporating the remaining 6 ml of the original sample under reduced pressure, and dissolving the residue in approximately 200 μl chloroform–methanol (1:1, v/v). Duplicate samples ($\hat{=}$ 100 μl each) are separated by two-dimensional thin-layer chromatography on silica gel HR (Merck), using the solvent systems described by Broekhuyse [18] (see also Fig. 10.6). The spots are located by staining with iodine vapour. All spots are scraped from the plate and transferred to test tubes. 1 ml of 70% $HClO_4$ is added to each of the samples followed by destruction at 190°C and subsequent determination of phosphorous as described above. Prior to determination of the optical density, the silica should be removed by centrifugation. The phospholipid composition can be directly calculated from the ratios of the optical densities at 830 nm of the individual samples. From these data and the total phospholipd content of the

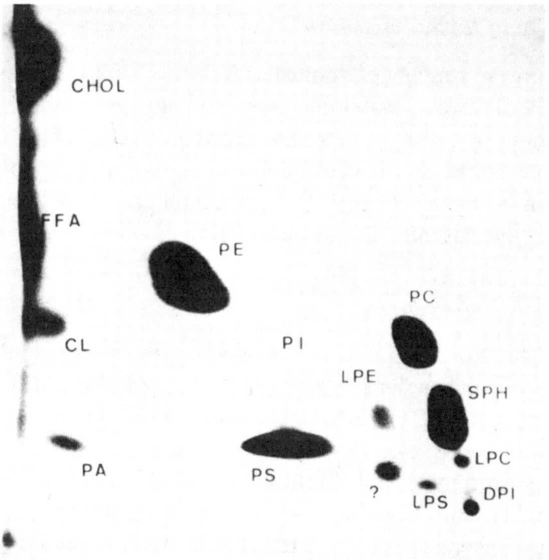


**Fig. 10.6** Two-dimensional thin-layer chromatogram of lipids frequently found in red cell membranes. Glass plates (20 x 20 cm) are covered with silicagel H.R. (Merck) containing 2 wt % Mg-silicate (Woelm). The lipid mixture is applied at the right bottom corner (3 cm from the edges). Thin layer plates are developed in the first direction (in figure from bottom to top) with $CHCl_3$ : $CH_3OH$ : $NH_3$ (25%) : $H_2O$ (90:54:5.5:5.5 v/v/v/v). For the second direction (in figure from right to left) $CHCl_3$ : $CH_3OH$ : $CH_3COOH$ : $H_2O$ (90:40:12:2, v/v/v/v) is used as developer. Abbreviations: CHOL = cholesterol. FFA = free fatty acids. CL = cardiolipin. PA = phosphatidic acid. PE = phosphatidylethanolamine. PI = phosphatidylinositol. PS = phosphatidylserine. PC = lecithin. SPH = sphingomyelin. LPE = lyso-phosphatidylethanolamine. LPC = lyso-lecithin. LPS = lyso-phosphatidylserine. DPI = diphosphoinositides.

corresponding sample, the absolute quantity of each of the individual phospholipid classes can be calculated.

### *10.4.4 Treatment of intact cells with phospholipases*

Unless otherwise stated, the incubations are carried out as follows: Aliquots of enzyme are mixed with 5 ml of the isotonic saline solution (used to wash the cells) followed by addition of 0.25 ml of packed cells. The amounts of enzyme added are routinely: 10 IU of phospholipase $A_2$, 10 IU of phospholipase C and 5 IU of sphingomyelinase. Incubations with *N. naja* phospholipase $A_2$ contain $CaCl_2$ in a concentration of 10 mM instead of 0.25 mM. The mixtures are incubated for 1 h at 37°C with gentle stirring, followed by centrifugation for 5 min at 3000 x *g*. Blanks are run in an identical manner except that no enzymes are present.

*(i) Determination of haemolysis*

Percentage haemolysis can be determined as follows: 100 μl of the supernatant is diluted with 5 ml of distilled water and the absorbance is measured at 418 nm, using distilled water as a blank. The absorbance corresponding to 100% haemolysis is determined by analogous dilution of 100 μl of supernatant obtained from an incubation (1 h, 37°C) of 0.25 ml packed cells in 5 ml distilled water (instead of the isotonic incubation buffer).

*(ii) Lipid extraction procedure*

After taking aliquots for haemolysis determination, the supernatants are removed by suction, and enzymatic activity is inhibited by addition of a mixture of *o*-phenanthroline and EDTA (final concentration: 12 mM of each, in a final volume of 0.5 ml). The lipids are extracted by addition of 5 ml methanol A.R. and, after stirring for 5 min, 5 ml chloroform A.R.. After stirring for another 5 min, the mixture is centrifuged for 5 min at 3000 x *g* and the supernatant is collected in a pear-shaped flask. This extraction procedure should be repeated twice and the pooled extracts are taken to dryness under reduced pressure at 37°C. The residue is taken up in 150 μl chloroform–methanol (1:1, v/v).

*(iii) Determination of phospholipid breakdown*

The phospholipids are separated by two-dimensional thin-layer chromatography and determined as phosphorous after destruction with 70% $HClO_4$ at 190°C, as described above (Section 10.4.3). Percentage degradation of glycerophospholipids after phospholipase $A_2$ or C attack is determined using sphingomyelin (which is not degraded) as an internal standard. For determination of sphingomyelin breakdown by sphingomyelinase, one of the glycerophospholipids (usually lecithin) is used as an internal standard.

In the case of phospholipid degradation by a combined action of sphingomyelinase with either phospholipase $A_2$ or C, none of the phospholipids can be used as an internal standard since in principle each of them can be degraded. Therefore, lipids extracted from incubated cells and corresponding blanks are quantitatively transferred to the thin-layer plates, allowing direct comparison of the quantity of each of the individual phospholipid classes before and after phospholipase treatment of the cells. An alternative method can be followed in case phospholipase $A_2$ is involved in a combined action. The two-dimensional thin-layer chromatography technique easily allows the determination of lysoglycerophospholipids (compare Fig. 10.6). Since the lyso-compounds produced by the action of phospholipase $A_2$ on erythrocytes remain in the membrane, percentage conversion of glycerophospholipids by this enzyme can also be detected by directly comparing the remaining phospholipid with the corresponding lyso-derivativc formed. Within the limits of accuracy, both methods give the same result.

### *10.4.5 Treatment of resealed ghosts with phospholipases*

*(i) Trapping of phospholipase $A_2$ inside resealed ghosts*

Washed human red cells (25 ml) are lysed by addition of 155 ml 10 mM NaCl, 1 mM EDTA (final osmolarity = 60 mosmol) and stirred for 5 min at 0°C. Prior to resealing, the haemolysate is centrifuged (10 min, 12 000 x *g*, 0°C) and two-third of the supernatant is removed, in order to reduce the volume of the ghost suspension to be resealed. The ghosts are resuspended in the remaining one-third of the supernatant, and 750 μl, pancreatic phospholipase $A_2$(1 IU/$\mu l^{-1}$) is added, followed by stirring for 5 min at 0°C. The ghosts are subsequently resealed by addition of 8.4 ml 1 M NaCl while stirring at 0°C. After 5 min, the mixtures are brought to 37°C and stirred for one hour. The resealed cells are washed four times with 160 mM NaCl (centrifuge 10 min, 12 000 x *g*, 0°C) to remove external phospholipase.

*(ii) Incubation of resealed cells with trapped phospholipase A*

The hydrolytic action of pancreatic phospholipase $A_2$ from the inside of the resealed cell produces lysis at a certain stage of the incubation. Since the moment at which lysis occurs is strongly dependent upon the detailed experimental conditions, it is essential that for each experiment a haemolysis versus time curve is determined first, in order to know beforehand after which time periods samples should be drawn for determination of phospholipid breakdown. Therefore, 1 ml of packed resealed ghost is suspended in 5 ml 150 mM NaCl, 5 mM $CaCl_2$ at 37°C under stirring, in order to start the enzymatic reaction from the inside. After different time intervals, 200 μl of the suspension is transferred into 5 ml 160 mM NaCl (or 5 ml water for 100% haemolysis), and directly centrifuged at 0°C for 5 min at 12 000 x *g*. The supernatant is monitored (without further dilution) at 418 nm, using 150 mM NaCl as a blank. The corresponding haemolysis versus time curve, for the experimental conditions used here, is shown in Fig. 10.5. Since significant lysis starts after 35 min, samples for determination of phospholipid breakdown should be taken within this time period. To prepare these samples, 4 ml of packed resealed ghost suspension is mixed with 20 ml 150 mM NaCl, 5 mM $CaCl_2$, and incubated at 37°C while stirring. After different time periods (not exceeding 35 min), 6 ml samples are taken and transferred into test tubes containing 20 mg of EDTA in order to stop the enzymic reaction. To check for absence of enzymatic breakdown when $Ca^{2+}$ is omitted during the resealing and subsequent incubation procedure, 2 ml of packed resealed ghost is suspended in 10 ml 160 mM NaCl (without Ca) and a 6 ml sample (corresponding to the longest time period) is taken and transferred to a test tube containing 20 mg of EDTA.

*(iii) Lipid extraction procedure*

The ghost samples (6 ml) are extracted with 30 ml methanol A.R. and 30 ml chloroform A.R. as described in Section 10.4.3. The final residue is taken up in 150 μl chloroform–methanol (1:1, v/v).

*(iv) Determination of phospholipid breakdown*

The phospholipids are separated by two-dimensional thin-layer chromatography and determined as phosphorous after destruction with 70% $HClO_4$ at 190°C, as described earlier above (Section 10.4.3). Percentage degradation of glycerophospholipids is determined, using sphingomyelin (which is not degraded) as an internal standard.

## References


1. Wallach, D. F. H. (1972), *Biochim. biophys. Acta,* **265**, 61.
2. Juliano, R. L. (1973), *Biochim. biophys. Acta,* **300**, 341.
3. Bretscher, M. S. (1972), *Nature New Biology,* **236**, 11.
4. Bretscher, M. S. (1973), *Science,* **181**, 622.
5. Gordesky, S. E. and Marinetti, G. V. (1973), *Biochim. biophys. Res. Commun.,* **50**, 1027.
6. Colley, C. M., Zwaal, R. F. A., Roelofsen, B. and Deenen, L. L. M. van (1973), *Biochim. biophys. Acta,* **307**, 74.
7. Verkleij, A. J., Zwaal, R. F. A., Roelofsen, B., Comfurius, P., Kastelijn, B. and Deenen, L. L. M. van (1973), *Biochim. biophys. Acta,* **323**, 178.
8. Zwaal, R. F. A., Roelofsen, B. and Colley, C. M. (1973), *Biochim. biophys. Acta,* **300**, 159.
9. Zwaal, R. F. A. (1974), *Biochem. Soc. Trans.,* **2**, 821.
10. Zografi, G., Verger, R. and Haas, G. H. de (1971), *Chem. Phys. Lipids,* **7**, 185.
11. Deenen, L. L. M. van and Gier, J. de (1974), *The Red Blood Cell,* (Surgenor, S. M. ed.), pp. 147, Academic Press, New York and London.
12. Dodge, J. T., Mitchell, C. and Hanahan, D. J. (1963), *Archs Biochem. Biophys.,* **100**, 119.
13. Haas, G. H. de, Postema, N. M., Nieuwenhuizen, W. and Deenen, L. L. M. van (1968), *Biochim. biophys. Acta,* **159**, 103.
14. Cremona, T. and Kearney, E. B. (1964), *J. biol. Chem.,* **239**, 2328.
15. Zwaal, R. F. A., Roelofsen, B., Comfurius, P. and Deenen, L. L. M. van (1971), *Biochim. biophys. Acta,* **233**, 474.
16. Goerke, J., Gier, J. de and Bonsen, P. P. M. (1971), *Biochim. biophys. Acta,* **248**, 245.
17. Böttcher, C. J. F., Gent, C. M. van and Pries, C. (1961), *Analytica chim. Acta,* **24**, 203.
18. Broeckhuyse, R. M. (1969), *Clinica chim. Acta,* **23**, 457.
19. Coleman, R., Finean, J. B., Knutton, S. and Limbrick, A. R. (1970), *Biochim. biophys. Acta,* **219**, 81.
20. Ibrahim, S. A. and Thompson, R. H. S. (1965), *Biochim. biophys. Acta,* **99**, 331.
21. Gul, S. and Smith, A. D. (1972), *Biochim. biophys. Acta,* **288**, 237.

22. Roelofsen, B., Zwaal, R. F. A., Comfurius, P., Woodward, C. B. and Deenen, L. L. M. van (1971), *Biochim. biophys. Acta*, **241**, 925.
23. Takahashi, T., Sugahara, T. and Ohsaka, A. (1974), *Biochim. biophys. Acta*, **351**, 155.
24. Stahl, W. L. (1973), *Archs Biochem. Biophys.*, **154**, 47.
25. Hoffman, J. F. (1958), *J. gen. Physiol.*, **42**, 9.
26. Hoffman, J. F. (1962), *J. gen. Physiol.*, **45**, 837.
27. Bodemann, H. and Passow, H. (1972), *J. Memb. Biol.*, **8**, 1.
28. Schwoch, G. and Passow, H. (1973), *Molec. Cell. Biochem.*, **2**, 197.
29. Haas, G. H. de, Bonsen, P. P. M., Pieterson, W. A. and Deenen, L. L. M. van (1971), *Biochim. biophys. Acta*, **239**, 252.
30. Low, M. G., Limbrick, A. R. and Finean, J. B. (1973), *FEBS Letters*, **34**, 1.
31. Awasthi, Y. C., Berezney, R. Ruzicka, F. J. and Crane, F. L. (1969), *Biochim. biophys. Acta*, **189**, 457.
32. Awasthi, Y. C., Ruzicka, F. J. and Crane, F. L. (1970), *Biochim. biophys. Acta*, **203**, 233.
33. Lovelock, J. E., James, A. T. and Rowe, C. E. (1960), *Biochem. J.*, **74**, 137.
34. Mulder, E. and Deenen, L. L. M. van (1965), *Biochim. biophys. Acta*, **106**, 348.
35. Sakagami, T., Minarii, O. and Orii, T. (1965), *Biochim. biophys. Acta*, **98**, 356.
36. Mulder, E. and Deenen, L. L. M. van (1965), *Biochim. biophys. Acta*, **106**, 106.
37. Robertson, A. F. and Lands, W. E. M. (1964), *J. Lipid Res.*, **5**, 88.
38. Renooij, W., Golde, L. M. G. van, Zwaal, R. F. A., Roelofsen, B. and Deenen, L. L. M. van (1974), *Biochim. biophys. Acta*, **363**, 287.
39. Zwaal, R. F. A., Flückiger, R., Moser, S. and Zahler, P. (1974), *Biochim. biophys. Acta*, **373**, 416.
40. Roelofsen, B. and Deenen, L. L. M. van (1973), *Eur. J. Biochem.*, **40**, 245.
41. Marchesi, V. T. and Palade, G. E. (1967), *J. Cell Biol.*, **35**, 385.
42. Martin, K. (1970), *Biochim. biophys. Acta*, **203**, 182.
43. Fortes, P. A. G., Ellory, J. C. and Lew, V. L. (1973), *Biochim. biophys. Acta*, **318**, 262.

# 11 Immunochemical gel precipitation techniques in membrane studies

O. J. BJERRUM and T. C. BØG-HANSEN



## 11.1 Introduction

Serological methods utilizing the specificity of antibodies for recognition of molecular structures have been used for a long time in differentiation and typing of blood group markers on red cell surfaces by the haemagglutination technique [1]. Due to the clinical significance of such typing much effort has been devoted to the elucidation of the nature of these antigens. This has been facilitated by the presence of individual iso-antibodies and of soluble blood group substances in secretions. A number of the blood group-determining glycoproteins have been isolated and partially characterized [2–4], and the carbohydrate chains of some of these have been sequenced [2,5]. In the cytotoxity test, specific antibodies have been used in the investigation of other types of tissue antigens (histo-compatibility antigens) present on leucocytes and thrombocytes [6]. Due to the great importance in organ transplantation much effort has also been devoted to the isolation and characterization of these antigens [2]. In experimental membrane research the effect on membrane function produced by binding specific antibodies to the membranes has been utilized [7–11]. Antibodies labelled with certain marker substances have been used for determining the localization of different antigens in the cell membrane [12–14].

Less work has been performed utilizing the specificity of antibodies in the direct analysis of solubilized membrane constituents by means of gel precipitation techniques. This is possibly a consequence of the early solubilization procedures, which destroyed the antigenic determinants of the solubilized membrane proteins or of certain glycolipids. However, the use of milder solubilization procedures with non-ionic detergents has [15,16] made such immunochemical analysis possible and provided the possibility of establishing a more direct correlation between the protein constituents of the membranes and the results obtained with the blood group and tissue antigens.

In this chapter we will show how modern immunochemical precipitation techniques, either alone or in combination with other physiocochemical methods, can be used for (1) identification, (2) quantification, (3) characterization, (4) preparation and (5) comparative analysis of membrane proteins. Such analyses have become reliable following the recent developments in immunochemical gel precipitation techniques which began with the double immunodiffusion techniques of Ouchterlony [171]. An increase in resolution was obtained by separation of the antigens by electrophoresis before the immunodiffusion step, described by Grabar and Williams [18] as immunoelectrophoretic analysis. With this method analysis of complicated protein mixtures were possible but difficult. Later Resler [19] and Laurell [20] independently introduced crossed antigen-antibody electrophoresis which in the following is called crossed immunoelectrophoresis.* In this method the antigens are first electrophoretically separated in one dimension. The next step is a second electrophoresis of the separated antigens, perpendicular to the first electrophoresis, into an antibody containing gel. At pH 8.6 the antibodies do not migrate in the electric field and formation of immunoprecipitates takes place when the antigen-antibody complexes reach such a size that they cannot move in

*The naming of the various quantitative immunoelectrophoresis techniques described in this chapter follows the nomenclature proposed by Axelsen, Krøll and Weeke [21].

the agarose gel [22]. The advantages are an even higher resolving power than the classical immunoelectrophoresis and a much more precise identification of the different immunoprecipitates, characterized by their position, individual shape and staining intensity. Clarke and Freeman made the method quantitative [23]. Many different modifications of the method have subsequently been developed to improve the applicability of the method [21,24–26]. Some of these modifications will be described in the following. Up-to-date technical and review papers concerning the mentioned immunochemical gel precipitation techniques can be found in the following references [17,21,24,172].

Immunochemical gel precipitation techniques were first employed in the analysis of membrane proteins by Howe *et al.* [27] and Vulpis [28], who examined water and butanol extracts of human erythrocyte membranes by means of double immunodiffusion and immunoelectrophoresis. Furthmayr and Timpl [29] used the immunodiffusion technique for monitoring gel filtration experiments on water soluble proteins of human erythrocyte membranes, and Hamaguchi and Cleve [30] applied immunoelectro-focusing to the same proteins. The use of gels containing non-ionic detergents for the immunochemical analysis of membrane proteins was first described by Demus and Mehl [15] and such gels were later used in the crossed immunoelectrophoresis of erythrocyte membrane proteins by Bjerrum and Lundahl [16]. Lundahl and Liljas [31] combined polyacrylamide gel electrophoresis with subsequent immunochemical analysis by crossed immunoelectrophoresis. Analysis by quantitative immunoelectrophoresis of isolated protein bands from sodium dodecylsulphate has also been performed [32,52].

The general problem of membrane solubilization is discussed in Chapter 6. As is described below, several of the different solubilization techniques have been used for preparing membrane antigens for immunochemical analysis as water extracts [9,27–30,33,34], as extracts with organic solvents [27,35,37–43], or by solubilization with ionic detergents [7–9,15,32,45–53], or by solubilization with non-ionic detergents [7–9,15,46,48,50,54–60].

A variety of membranes have been investigated by immunochemical precipitation techniques. Thus, erythrocyte membranes have been investigated by double immunodiffusion and immunoelectrophoresis [15,27–30,33,36–42,45–47,49,55,61,166] and by quantitative immunoelectrophoresis [31,32,34,56]. Similar investigations have also been performed with liver cell plasma membranes [48,54 and 9,48,57,132], cancer cell membranes [54,62 and 50,60] and bacterial membranes [7,8,44,46,51, and 59]. In all three cases the later references are to investigations which make use of quantitative immunoelectrophoresis. Milk fat globule membranes [52,63,64], rabbit thymocyte membranes [65,133], rat brain synaptosome membranes [58,66] and herpesvirus proteins [67–69] have only been examined by quantitative immunoelectrophoresis. Our own experience in the field is restricted to erythrocyte membranes, milk fat globule membranes and proteins of *Herpesvirus hominis*, and therefore in the following discussion these membrane systems will be used to illustrate the use of the different immunochemical techniques, and it is our hope that this presentation may serve as a model and guideline along which other membrane systems may be studied.

## 11.2 Preparation of membrane material suitable for immunochemical analysis

### *11.2.1 Membrane preparation*

The preparative isolation of different membrane systems is dealt with in the chapters in Part I of this book. However, it must be realized that the composition of the isolated membranes depends to a considerable extent on the isolation procedure employed [70]. For immunization purposes the membrane preparation does not have to be of high purity but should be isolated under as mild conditions as possible to be certain that none of the membrane specific proteins are lost during the isolation procedure (See Section 11.11.1). Once the antibodies are obtained they can be used for monitoring the concentrations of the proteins in the membrane preparation (see Sections 11.5 and 11.6).

### *11.2.2 Solubilization with non-ionic detergents*

Many methods are available for the solubilization of membrane proteins [71–73] but not all are suitable for subsequent immunochemical analysis. The solubilizing agent should disintegrate the membrane and keep the proteins in solution but on the other hand it should not destroy the antigenic structure of the proteins. Accordingly, the solubilizing agent must not affect the antibodies in the electrophoresis. Non-ionic detergents fulfil these requirements. In aqueous solutions such detergents dissolve many membranes to a high degree [9,48,56,58,59,67,74] and do not affect the antigen-antibody reaction to any appreciable extent [16,52,75]. Conformational changes of the solubilized proteins have not been detected [76] and enzymatic activity is often retained [8,15,55]. However, we recently found that 1% (w/v) non-ionic detergent in the gels prevents the formation of immunoprecipitates of human serum lipoproteins [52].

From our experience with erythrocyte membranes [56], milk fat globule membranes [52] and *Herpesvirus hominis* proteins [77] we have found that the optimum conditions for solubilization with the non-ionic detergents Triton X-100 and Berol EMU-043 include low ionic strength (less than 0.01), alkaline pH (pH 8–9), a detergent–protein ratio higher than 1:1 and a final protein concentration below 5 mg $ml^{-1}$ (otherwise aggregation will occur) [see also 74,78,79]. Repetition of the solubilization procedure on the unsolubilized residue often increases the yield of solubilized material [56]. Sonication of aggregated membrane preparations for short periods of time (5–30 s at 20 000 Hz) may often be applied with success even after addition of detergent [52]. Sonication in this manner is reported not to affect the conformation of membrane proteins [80] but sonication for more than 3 min changes the conformation as monitored by crossed immunoelectrophoresis; flying precipitates appear and the area below the immunoprecipitates increases [81]. The degree of solubilization changes only slightly with increasing temperature and we therefore prefer 0–4°C to avoid undesirable enzymatic degradation of the solubilized proteins as far as possible (see Section 11.2.2.(ii)). The concise experimental conditions for the solubilization of the membrane preparations in question can be found in Section 11.11.1. Solubilized material is defined for our purposes as that which remains in solution after ultra-centrifugation at 100 000 x *g* for 1 h [73]. This centrifugation was also found necessary for obtaining unhindered migration in the agarose gels containing 1% (w/v) of non-ionic detergent.

A general guide to the type of non-ionic detergent to be chosen for a particular solubilization experiment cannot be given, since membrane proteins are solubilized to various degrees by different detergents [78,79,82] and since also individual precipitation patterns are observed with proteins solubilized and immunoelectrophoresed in different detergents [56]. By varying the ionic strength, pH and also the detergent concentration selective solubilization can be achieved [78,79,82]. Table 11.I gives a survey of some commonly used non-ionic detergents. (Concerning physico-chemical data of these detergents [169] is the book of choice.) The purity, molecular weight distribution and U.V. absorption of the membrane proteins may play a role in

TABLE 11.1
*Some non-ionic detergents commonly used for solubilization of membrane proteins*

| Trade name | Chemical composition | Manufacturer | $A_{280}$(1 cm) for 1% (w/v) solutions in water | References |
|---|---|---|---|---|
| Berol EMU-043 | Ethylene oxide adduct of fatty alcohols | MoDoKemi, Stennungsund, Sweden | 1.3 | [56] |
| Brij 58 | Polyoxyethylene cetyl ether | Atlas Chemical Industries, Wilmington, U.S.A. | n.d. | [7,75] |
| Lubrol W | Cetylpolyoxyethylene condensate | I.C.I., Millbank, London, U.K. | n.d. | [54] |
| Lubrol WX | | | 0.05 | [56] |
| Nonidet P 40 | Alkylphenyl ethoxylate | Shell Chemicals, London, U.K. | 20 | [56, 75] |
| Triton X-100 | *p*-acetylphenoxy-polyoxyethylene | Rohm and Haas, Philadelphia, U.S.A. | 23 | [56, 75] |
| Tween 80 | Polyoxyethylene sorbitane mono-oleate | Atlas Chemie, Essen, F.D.R. | n.d. | [75] |
| Tween 20 | | | 0.43 | [56, 59] |

n.d. not determined

the choice of non-ionic detergent. U.V. absorption is important for protein determinations (see section 11.2.2.(i)) and for monitoring fractionation experiments. Use of immunochemical elution profiles obtained by fused rocket immunoelectrophoesis [83] may, however, overcome the latter diffuculty when U.V. absorbing detergents are used (see Section 11.8).

*(i) Determination of protein concentration*

Non-ionic detergents interfere with most protein determination and U.V. absorption can only be used if the detergents have low extinction coefficients at 280 nm (see Table 11.1). Protein determination according to Lowry *et al.* [84], with the modification of Bonsall and Hunt [85], allows determination of relative concentrations; the absolute concentrations tend to be over-estimated [56], see Fig. 7.1, Chapter 7. Determination of proteins by use of the Biuret reaction requires relatively large amounts of protein but the method gives fairly reliable results if used with the modification of Futterman and Rollins [86]. Amino acid analysis is laborious but most accurate [87]. For micro-organisms, protein determination can be performed in the presence of detergent by use of radioactive counting if the organisms have been grown in radioactive media [8].

*(ii) Stability*

Since the non-ionic detergents do not inactivate contaminating hydrolytic enzymes, the stability of the solubilized material should be tested. Crossed immunoelectrophoresis is in itself a very sensitive technique for the detection of proteolytic degradation, which gives rise to certain characteristic immunoprecipitation patterns [88] (see Section 11.5). Otherwise, the stability of the solubilized material can be tested by performing crossed immunoelectrophoresis after a period of storage [56]. In the event of active enzymes being present the method of preparation of the membranes must be changed or suitable inhibitors must be found. For example, human erythrocyte membrane preparations solubilized in non-ionic detergents were found to undergo proteolytic degradation. Protease inhibitors such as aprotinin phenyl methyl-1-sulfonyl fluoride and soya bean trypsin inhibitor were not effective, but they did not disturb the precipitation pattern either. However, the degradation was avoided by drastic reduction of the number of leucocytes contaminating the erythrocytes. The latter was obtained by repeated washings in autologous plasma of the erythrocytes. The higher density of the plasma lead to a better separation of the erythrocytes and leucocytes, so that the latter could be removed from the top of the erythrocytes [89] (For experimental details see Section 11.11.1). The solubilized material is best stored at –90°C or below. For shorter periods of time (about 1 month) –20°C is sufficient. It should be noted that aggregation of the solubilized proteins can occur upon freezing. Dilution of the samples often reduces the aggregation. Repeated thawing and freezing of the solubilized material can be avoided by dividing it into small aliquots before storage.

### *11.2.3 Solubilization with ionic detergents*

Sodium dodecylsulphate and sodium deoxycholate are the most commonly used ionic detergents for solubilization of membrane proteins [73]. They are generally considered to be more denaturing than non-ionic detergents, a fact which can be explained by a higher critical micellar concentration for these detergents, allowing a higher free detergent concentration [169]. The denaturing effect of detergents is proportional to the concentration of the free detergent [90–92], but the use of suitably low concentrations of these detergents permits a high degree of solubilization of the membrane proteins with good retention of antigenecity. However, a titration must be performed to find the optimum detergent concentration. This detergent concentration is much more critical in the case of sodium dodecylsulphate than deoxycholate, but immunochemical analysis of membrane proteins solubilized with ionic detergents is nevertheless possible and has been performed [9,29,32,38,49,52,53]. For erythrocyte membrane proteins the total number of precipitates observed in quantitative immunoelectrophoresis using the above two ionic detergents is reduced compared with non-ionic detergents [53,56]. This is due to decreased resolution of the precipitation pattern, resulting from the solubilized proteins having comparable migration velocities in the first dimension electrophoresis because of binding of the charged detergents to the proteins. Furthermore, binding of ionic detergents to the antibodies causes them to migrate anodically thereby decreasing the resolution, so that incorporation of ionic detergent in the agarose gels is inexpedient [56]. These problems are not present when double immunodiffusion is used, but the latter method does not permit reliable identification of the immunoprecipitates. Because of the facts that ionic and non-ionic detergents compete for the same binding sites of the solubilized protein [90,92] and the critical micellar concentration of the ionic detergent is reduced to the level of the non-ionic detergent by formation of mixed micelles [168] transfer of sodium dodecylsulphate solubilized proteins to non-ionic detergent is possible [52,53]. Quantitative immunoelectrophoretic analysis of membrane proteins solubilized in ionic detergents is then performed with better resolution as side effects of the ionic detergent are avoided (see 11.6.2).

### *11.2.4 Solubilization by other methods*

Membrane proteins liberated in media of high or low ionic strength may be subjected to immunochemical analysis [28,29,33]. Some antigenecity is also retained even after solubilization with denaturing agents such as urea [15,29,37], pyridine [29,32], ethanol [40], phenol [36,39,43], butanol [27,35,37], formic acid [37]. The resistant antigens mostly comprise glycoproteins but spectrin is also very resistant [93]. Solubilization by chemical modification of membrane proteins [87,94] has only little effect on antigenicity [81,95].

## 11.3 Production of antibodies

### *11.3.1 Immunization*

Membrane proteins should in general be good antigens due to their content of hydrophobic amino acids [96]. The antibodies used in our membrane studies were purified, concentrated immunoglobulins from rabbits immunized over a long period (18 months) with small amounts (100–500 μg) of membrane material. The immunization scheme described by Harboe and Ingild [97] was followed. When initiating production of antibodies against a new membrane system, at least 3 to 6 rabbits had to be immunized (to eliminate individual differences in the antibody response). It should be remembered that the antibody titre increases during the course of immunization. In the first approach, unfractioned membrane material should be used and 100–500 μg should be administered per injection (for experimental details see Section 11.11.2). After establishing a pool of antibodies and a reference pattern by crossed immunoelectrophoresis (see Section 11.6) it is easy to determine whether immunization with extracts or with solubilized membrane material gives the better antibody response.

Immunization with membrane material solubilized by non-ionic detergents does not, in our experience, give rise to antibodies directed against additional antigens but sometimes it results in a higher titre, possibly due to an adjuvant effect of the detergent. Reducing the antigenic burden on the rabbits by immunizing with a membrane fraction, for example an extract, will give rise to the formation of antibodies against the minor membrane components which have been enriched in such a fraction [56] (see Section 11.11.2. (i)).

### *11.3.2 Purification of the antiserum*

The surplus of protein present in unfractionated antiserum often gives rise to background staining of the immunoplates diminishing the resolution of the precipitation pattern. Furthermore, for certain experiments serum proteins may cause artefacts by their interaction with the antigen under examination, e.g. proteolytic enzymes with serum inhibitors, membrane proteins with complement factors and lectins with serum glycoproteins. However, the immunoglobulin fraction of the antiserum can be isolated by precipitation with 1.75 M ammonium sulphate and subsequent ion-exchange chromatography on QAE–Sephadex at pH 5.0 as described by Harboe and Ingild [97]. In addition, purification of the immunoglobulins permits the use of a higher antibody concentration in the gels. However, plasmin (a proteolytic enzyme present in serum) follows the immunoglobulins in the isolation procedure and is present in the antibody preparation in an active form because the natural serum inhibitors have been removed [98]. In some cases this enzyme gives rise to artefacts in the immunoelectrophoresis, since it causes degradation of certain proteins during their electrophoresis in the antibody-containing gel [98]. Plasmin is, however, totally inhibited by addition of the polyvalent protease inhibitor aprotinin (Trasylol R Bayer, Leverkusen 500 u per ml of antibody solution). To avoid bacterial growth 15 mM sodium azide should be added to the antibody preparation [97]. For experimental details see Section 11.11.2.

### *11.3.3 Evaluation of the antibody preparation*

As the antibody preparation is to be used for immunochemical analysis of all subsequently solubilized membrane proteins the question arises as to whether the antibodies precipitate all solubilized protein material during the immunoelectrophoresis. This can be ascertained directly by performing zone electrophoresis of solubilized material through a gel strip containing a very high concentration of the immunoglobulin fraction of the antibodies. If the antibodies precipitate all the proteins in the gel strip, subsequent protein staining will show that the gel beyond the antibody-containing region does not contain any protein bands. This was found to be the case for antibodies raised against human erythrocyte membranes [56].

It should be pointed out that an antibody pool corresponding to about 500 ml of antiserum is necessary before a systematic investigation of a membrane system can be started. This is because the antibody response changes during the course of immunization. Thus with subsequent bleedings the precipitation pattern will differ from the reference precipitation pattern established earlier. The problems associated with changing reference precipitation patterns have been discussed by Axelsen [99].

It is unnecessary to absorb the antibodies since quantitative immunoelectrophoretic techniques allow identification of non-membrane-specific precipitates by crossed immunoelectrophoresis with an intermediate gel (Section 11.11.7). In Section 11.6 this method is exemplified.

### *11.3.4 Production of monospecific antibodies*

Monospecific antibodies can be produced with the help of the polyvalent antibody preparation without the use of purified antigens. It is possible to cut single precipitates out of the gel after crossed or line immunoelectrophoresis and to use the precipitates for direct immunization of rabbits [100]. This has been successfully performed with one of the precipitates of *Herpesvirus hominis* [68] and for the spectrin protein precipitates of human erythrocyte membranes [93]. For experimental details see Section 11.11.2 (v). As indicated by

Vestergaard [68], the immunoprecipitates used for immunization must preferably be cut out of gel regions where the precipitates are 'free', i.e. not overlapped by other precipitates.

## 11.4 Establishing a reference precipitation pattern by crossed immuno-electrophoresis

Assuming that a solubilization procedure for the membrane material has been devised (Section 11.2), and that a satisfactory pool of antibodies is available (Section 11.3), the first step in an immunochemical characterization of the antigenic components of solubilized membrane material is the establishment of a reference immunoprecipitation pattern by crossed immunoelectrophoresis (see Section 11.11.4). A general description of crossed immunoelectrophoresis has been given by Weeke [101] and a review giving 505 references has been published recently [167]. A prerequisite for obtaining immunoprecipitates by quantitative immunoelectrophorsis is that the migration velocity of the antigens differs from that of the antibodies. The electrophoreses are normally performed at pH 8.6, where the mean migration velocity of the antibodies is zero and most antigens migrate towards the anode. Each membrane protein gives a distinct immunoprecipitate. This, in spite of the fact that membrane proteins solubilized with non-ionic detergents in gel filtration experiments seems to be present in large micelles [90]. Either these micelles are only composed of one protein species or the electric field disrupts the micellar structure.

Cathodically migrating proteins can be detected by performing agarose gel electrophoresis [102] or rocket immunoelectrophoresis [103], but by using agarose gels of differing electroendosomotic properties [63] and by varying the nature and pH of the electrophoresis buffer [22] it is possible to induce nearly all proteins to migrate anodically. However, it may be necessary to modify chemically the charged groups of the antibodies so as to endow the antibodies with a new isoelectric pH, at which the immunoelectrophoresis can be performed [104,105]. Thus quantitative immunoelectrophoresis can be carried out in the pH range 4.8 to 10.3 [104,105]. The antibody titre is only slightly affected by such chemical modifications of the antibody molecules [106].

For determination of the maximum number of immunoprecipitates, the crossed immunoelectrophoresis should be performed with a variety of different antigen-antibody ratios. Enumeration of the precipitates is possible as they can be recognized by their position on the immunoplates, as measured for example, by the distance from the application well to the middle of the peak under standardized electrophoretic conditions. However, determination of the migration relative to the middle of the precipitate of an added protein (e.g. albumin, $\alpha_1$-antitrypsin or carbamylated transferrin is much more precise [67,95,99] (see Fig. 11.7 in Section 11.6). Furthermore, the individual shape and staining intensity of the precipitate will help in the recognition. The numbers can be set on a drawing of an enlarged immunoplate [22]. Examination

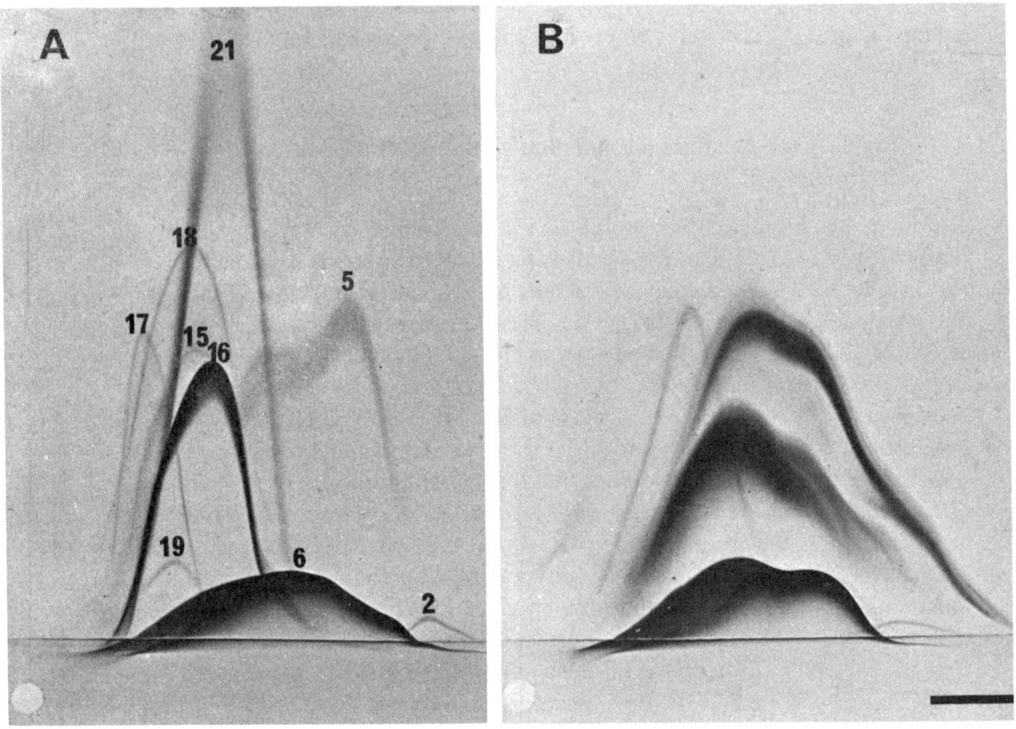


**Fig. 11.1** Crossed immunielectrophoresis of 10 μl (20 μg) of Berol-EMU-043-solubilized human erythrocyte membrane proteins **A** in agarose gels containing 1% (w/v) of Berol-EMU-043 and **B** in detergent-free gels. Note the decrease in number of precipitates and their change in position and morphology in the latter case. The first and second dimension electrophoreses were performed at 10 V $cm^{-1}$ for 60 min and 2 V $cm^{-1}$ for 18 h, respectively. The amount of antibody was 0.13 mg $cm^{-2}$. 500 u of aprotinin (Trasylol R) had been added per ml antibody solution. Numbers on the precipitation lines denote proteins as follows: No. 2 albumin, No. 6 spectrins, and No. 21 major glycoprotein. The bar represents 1 cm.

of different batches of membrane preparations is generally necessary to determine the reproducibility, in qualitative and quantitative terms, of the procedures for the preparation and solubilization of the membranes [99].

The results obtained by performing crossed immunoelectrophoresis of human erythrocyte membrane proteins solubilized and analysed with the non-ionic detergent Berol EMU–043 is shown in Fig. 11.1A. The antibodies employed were rabbit antibodies raised against the same membrane material. The precipitation pattern differs to some extent from earlier published patterns [56]. This is partly due to the use of a new pool of antibodies with changed titre against thc different proteins, but is also due to the inhibition of proteolytic activity in the antibody preparation by addition of aprotinin

(Section 11.3). A total of 12 precipitates are seen on the original plate. The reproducibility of the precipitation pattern is very good with the same membrane preparation. If another batch of erythrocytes was used the precipitation pattern was slightly different, but also reproducible. An impression of this can be obtained from comparison of Fig. 11.1 with Figs. 11.3 – 11.10, which represent another membrane preparation. The major precipitates are enumerated according to earlier published patterns [56,89]. Precipitate No. 2 corresponds to albumin. An additional precipitate appears occasionally in the same area. No. 6 corresponds to spectrin. (The earlier mentioned precipitates No. 8–12, correspond to cleavage products of spectrin [98]). Precipitate No. 21 corresponds to the major glycoprotein, also known as the MN-glycoprotein or glycophorin [89]. Precipitation patterns obtained by crossed immunoelectrophoresis of human erythrocyte membrane proteins using other non-ionic detergents and the ionic detergent sodium dodecylsulphate have been published elsewhere [53,56].

Analysis in detergent-free gels gives rise to a drastically changed precipitation pattern (Fig. 11.1**B**). Fewer and more blurred precipitates, consisting of several splitting precipitation lines, are seen. The numbering of the precipitates corresponds to that of Fig. 11.1**A**. Increased migration velocity for some of the proteins (No. 16, 21) is observed together with the formation of irregular shapes of the precipitates. Thus the presence of non-ionic detergent in the gel has the following effects: (1) it prevents unspecific precipitation in the gel, (2) it increases the number of precipitates, (3) it sharpens the outline of the precipitates (a similar effect is seen with dextran and polyethylenglycol [107]) and (4) it shields some charges on the proteins.

The quantitative immunoelectrophoretic techniques can detect as little as 10–100 ng of protein, depending on the antigenecity of the protein and the staining procedure used [108] (for the latter Coomassie Brilliant Blue is generally best [22]. However, the sensitivity can be increased 50–100 times by employing autoradiographic procedures involving the use of radioactive antigens or antibodies [109,110]. Recently, a relatively easy method of general applicability has been worked out. After the normal immunoelectrophoresis a second electrophoresis at pH 5 is performed. $^{125}$I-labelled swine anti-rabbit IgG antibodies are than electrophoresed through the already formed precipitates. The rabbit IgG in the precipitates binds the radioactive swine antibodies and can then be detected by autoradiography [111].

As mentioned in Section 11.2 the presence of non-ionic detergent does not inhibit proteolytic activity in the erythrocyte membrane preparations [112–114]. Fig. 11.2 shows an example of a crossed immunoelectrophoresis pattern for Berol-solubilized human erythrocyte membrane proteins which were extensively degraded by such enzymes. No effort was made to remove contaminating leucocytes. The appearence of multiple peaks (a-arrows), 'split' precipitates (b-arrows) in combination with 'flying' and skew precipitates indicates that degradation has taken place [88]. However,

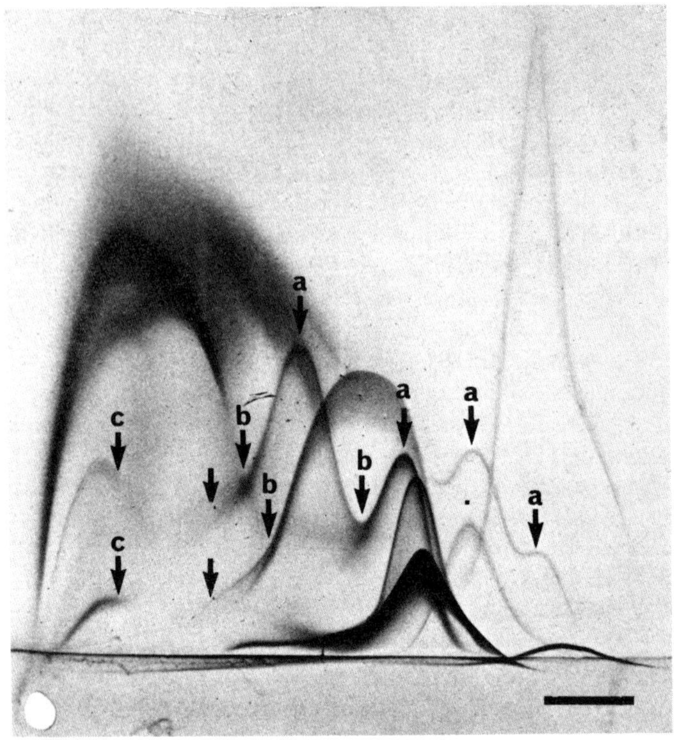


**Fig. 11.2** Crossed immunoelectrophoresis of 50 μg of Berol EMU-043-solubilized human erythrocyte membrane proteins, showing clear evidence of proteolytic degradation. Leucocytes had not been properly removed from the erythrocytes and the solubilized sample had been kept at room temperature for 16 h. The precipitation pattern is completely different from that of Fig. 11.1**A**. The amount of anti-erythrocyte membrane antibodies was 0.08 mg $cm^{-2}$. (a) indicates precipitates showing shoulders or camels, (b) indicates split precipitates. Conditions were otherwise as for the experiment illustrated in Fig. 11.1**A** except for addition of Trasylol. The bar represents 1 cm.

some of the observed effects could also be caused by the presence of certain glucosidases or neuraminidase in the erythrocyte membrane preparation [115,116]. The effect of the latter enzyme in quantitative immunoelectrophoresis has been investigated in detail with model systems [117].

## 11.5 Crossed immunoelectrophoresis for quantification of membrane proteins

Crossed immunoelectrophoresis is a quantitative method. The area between a given precipitation line and the base line is proportional to the amount of protein applied and inversely proportional to the antibody concentration in the

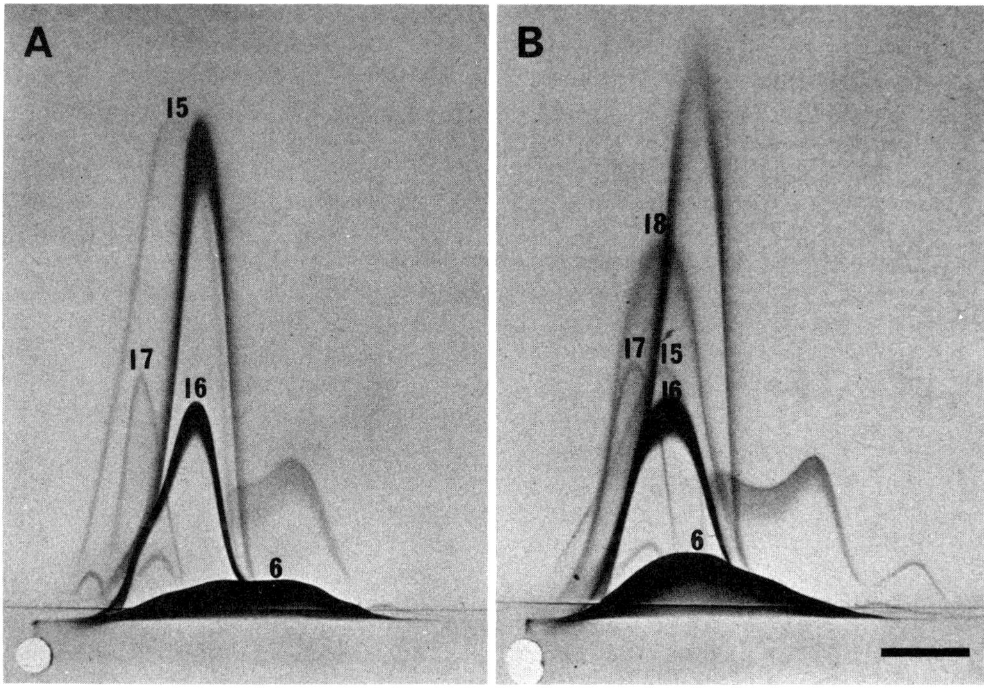


**Fig. 11.3** Crossed immunoelectrophoresis of about 20 μg of Berol EMU-043-solubilized human erythrocyte membranes. The membranes were washed **A** 5 times in 0.008 M phosphate buffer (pH 7.4) before solubilization, and **B** 6 times in the same buffer. Note the decrease in area below precipitate No. 15, due to dissociation of the corresponding protein from the membrane during the washing. The amount of antibody was 0.160 mg $cm^{-2}$. The conditions were otherwise as for the experiment illustrated in Fig. 11.1**A**. The bar represents 1 cm.

gel [23,118]. The area of a precipitate can be measured by planimetry, and when a pure protein is available, it is possible to determine immunochemically the absolute concentration of the protein in different mixtures once a calibration curve has been established [101]. The albumin content of different human erythrocyte membrane preparations has been determined in this way [89]. Depending on the extent of washing before lysis, the amount of albumin was found to vary between 0.02 and 0.2% of the membrane protein [89]. Binding to lysed ghosts of many other serum proteins has been observed [119].

If monospecific antibodies against a given membrane protein are available the relative concentrations of the proteins can easily be determined by rocket immunoelectrophoresis (Section 11.11.5) [103]. Crossed immunoelectrophoresis can be used directly for determination of relative protein concentrations. This may be illustrated by considering Fig. 11.3**A** and 11.3**B** which show the crossed immunoelectrophoresis patterns of human erythrocyte membrane proteins washed 5 and 6 times, respectively, with 0.008 M phosphate buffer (pH 7.4) after lysis of the erythrocytes (see Section 11.11.1). Comparison of the areas below the precipitates, using the area of precipitate No. 16 as reference, shows that the proteins corresponding to precipitates Nos. 6 and 15 were present in relatively small amounts in Fig. 11.3**B**. Furthermore, precipitate No. 18 is lacking in Fig. 11.3**A**. There are two reasons for these findings. Firstly, the degree of protein solubilization differed for the two membrane solutions, which explains the appearance of precipitate No. 18 in Fig. 11.3**A**. Nearly 100% of the membrane protein was solubilized in the experiment of Fig. 11.3**B** compared to about 90% in that of Fig. 11.3**A**. Secondly, some proteins were lost during the washing procedure, which explain the decrease in area of precipitate 6

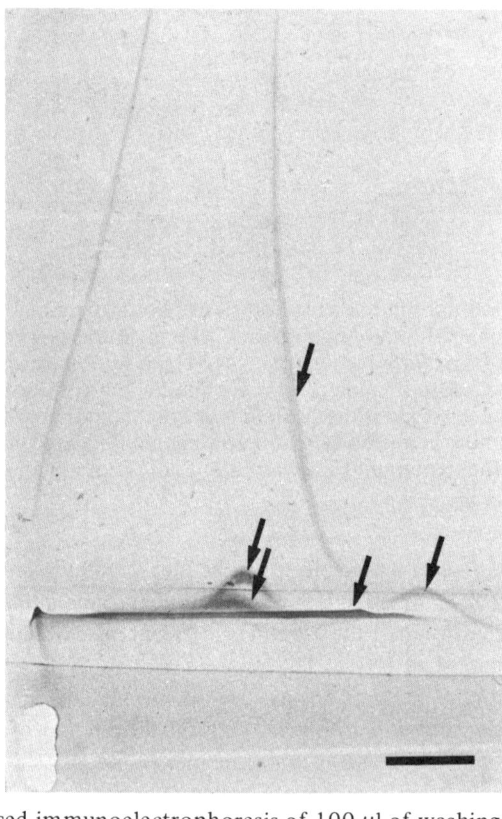

**Fig. 11.4** Crossed immunoelectrophoresis of 100 $\mu$l of washing buffer from the 5th washing of the human erythrocyte membranes used in the experiment of Fig. 11.3. The amount of the anti-erythrocyte membrane antibodies was 0.06 mg cm$^{-1}$. 5 precipitates are seen (arrows). The intermediate gel is blank. The conditions were otherwise as for the experiment illustrated in Fig. 11.3. The bar represents 1 cm.

and 15. This was verified by crossed immunoelectrophoresis of the washing buffer (Fig. 11.4). A total of 5 different precipitates were observed for the unconcentrated washing buffer. How these precipitates can be identified will be described in Section 11.6. Furthermore, determination of the volume of washing buffer allows us to calculate the relative percentage of protein lost during the washing procedure [93].

These experiments clearly show that some membrane protein is washed out of the membrane during the isolation procedure. This fact had been demonstrated earlier [30] but there was no technique available at that time for quantitative analysis. The extent of dissociation of proteins from the membranes is highly dependent on the ionic strength of the washing buffer. For example, at an ionic strength of 0.15 glyceraldehyde-3-phosphate dehydrogenase would have been dissociated from the erythrocyte membrane [55,120].

## 11.6 Quantitative immunoelectrophoresis for identification of membrane proteins

### *11.6.1 Membrane specificity of the precipitates*

Do the immunoprecipitates obtained in crossed immunoelectrophoresis correspond only to membrane proteins? In other words, are the precipitates obtained in this way membrane specific? This question can be answered by examining the proteins in the fluid compartments which the membrane separates.

#### *(i) Examination with antigens*

Immunoelectrophoresis with antibody preparations absorbed with the contents of the fluid compartments (depleted for membranes) which delimit the membrane should reveal only the membrane specific precipitates. However, in quantitative immunoelectrophoresis such absorbed antibodies give a heavily stained background and much quantitative information is lost by this approach. Absorption is not at all a prerequisite for the use of quantitative immunoelectrophoretic methods as they allow an identification of the different precipitates by use of crossed immunoelectrophoresis with an intermediate gel (Section 11.11.7). The use of these methods for such an identification is described below for human erythrocyte membrane proteins. In this case the two fluid compartments which have to be examined are serum and erythrocyte lysate. Fig. 11.5 shows the crossed-line immunoelectrophoresis patterns [121] (see Section 11.11.7) of human erythrocyte membrane proteins with (1) membrane-depleted lysate (Fig. 11.5**A**) and (2) human serum (Fig. 11.5**B**) in the intermediate gel. Fig. 11.5**C** shows a control experiment. In Fig. 11.5**A** precipitate No. 17 was elevated and fused with a line-immunoprecipitate, a reaction indicating identity of the precipitates [122]. Thus this precipitate corresponds to a protein

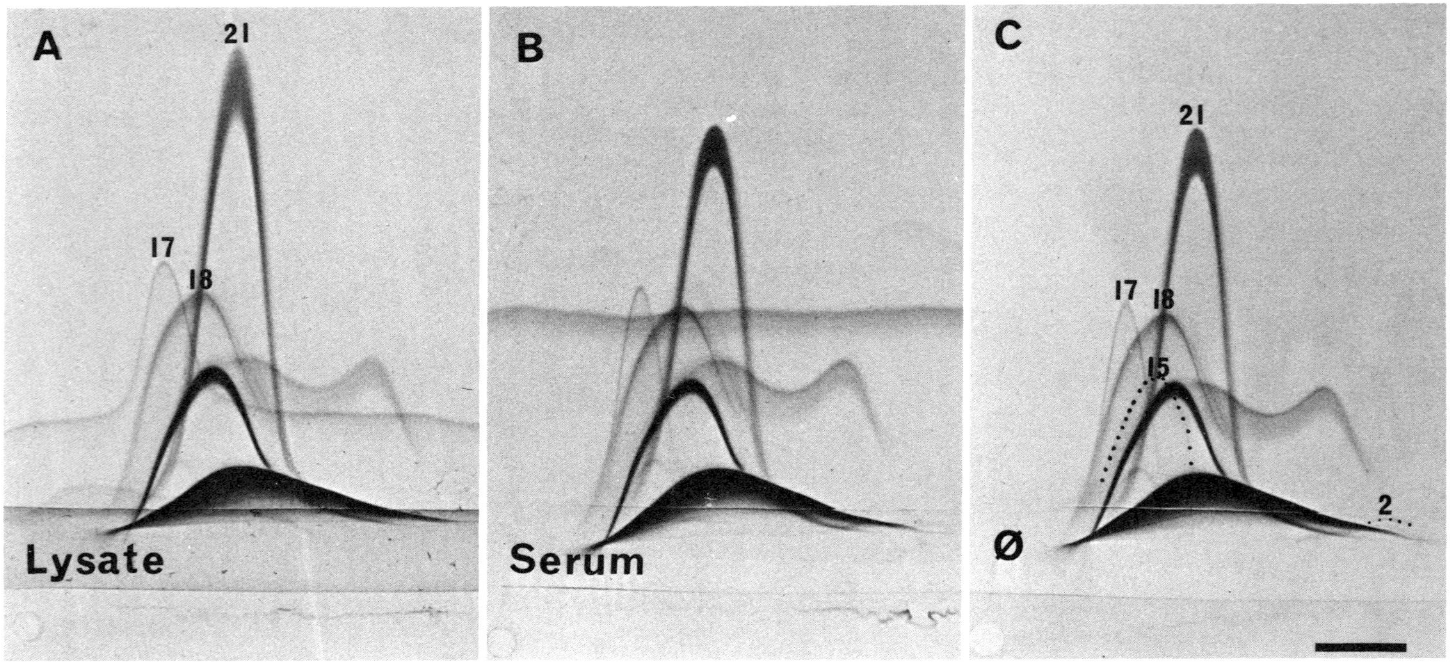


**Fig. 11.5** Crossed-line immunoelectrophoresis of 20 $\mu$g of Berol EMU-043-solubilized human erythrocyte membrane proteins using an intermediate gel containing **A** 50 $\mu$l of diluted (4×) membrane-depleted lysate of the same cells and 50 u of aprotinin, **B** 15 $\mu$l of human serum and **C** 50 $\mu$l of electrophoresis buffer as control ($\phi$). Compared to plate **C**, precipitates Nos 2, 15, 17 and 21 are affected in **A** and precipitates No. 2 and 15 in **B**. For explanation see text. Conditions were otherwise as for the experiment illustrated in Fig. 11.3. The bar represents 1 cm.

present in the lysate. Precipitates Nos 2 and 15 disappeared upon incorporation of the lysate in the intermediate gel, owing to absorption *in situ.* This means that the surplus of the corresponding antigens was so high that a line-precipitate was not formed [122]. Furthermore, an increase in the area below precipitate No. 21 can be seen relative to the control experiment (Fig. 11.5**C**). The latter finding indicates that non-precipitating cleavage products which absorb some antibodies are present in the lysate (since line-precipitates were not formed).

A comparison of Fig. 11.5**B** and 11.5**C** shows that two precipitates (Nos 2 and 15) disappeared upon incorporation of serum in the intermediate gel as a result of *in situ* absorption. A line-precipitate was observed which did not affect any of the bell-shaped precipitates, demonstrating that the antibodies reacted with a serum protein which was not present in the examined membrane preparation. To summarize, most of the precipitates were unaffected, indicating that these precipitates were membrane specific.

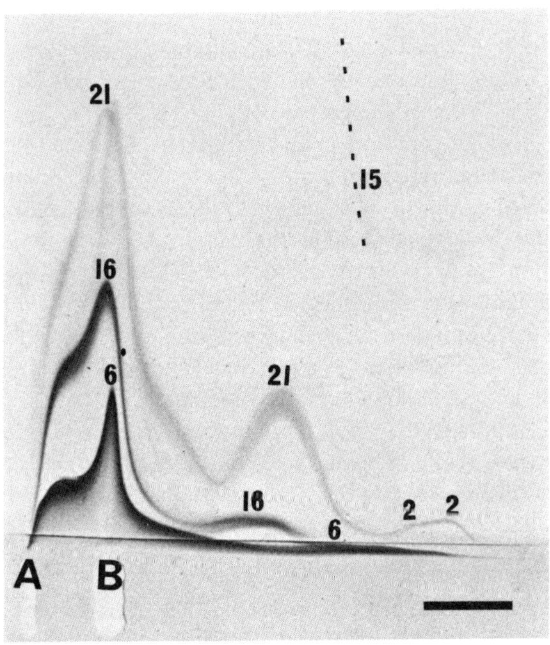


**Fig. 11.6** Tandem-crossed immunoelectrophoresis of 10 μg of Berol EMU-043-solubilized human erythrocyte membrane proteins (placed in well **A**) and 25 μl of concentrated ( x 5) washing buffer from the 5th washing of the erythrocyte membranes (placed in well **B**) (cf. Fig. 11.4). Proteins corresponding to precipitates Nos 2, 6, 16 and 21 were identified in the washing buffer by formation of double peaks. Antibody concentration was 0.04 mg $cm^{-2}$. Conditions were otherwise as for the experiment illustrated in Fig. 11.3. The bar represents 1 cm.

As shown in Fig. 11.4 (Section 11.5) some of the proteins in the washing buffer reacted with the anti-membrane antibodies. By performing tandem-crossed immunoelectrophoresis [123] of the concentrated washing buffer and the membrane preparation these precipitates were identified by the formation of double peaks (showing reaction of identity) to be Nos 2, 6, 16 and 21 (Fig. 11.6). The high precipitate was identified in other experiments as precipitate No. 15. This method gives a much clearer demonstration of the dissociation of proteins from the membrane than does comparison of the membrane preparations at different stages of washing (Fig. 11.3), since in the latter case the degree of solubilization also plays a role.

### *(ii) Examination with antibodies*

The analysis of the membrane specificity of the immunoprecipitates can also be performed with antibodies, using the technique employing an intermediate gel. Fig. 11.7**A** and 11.7**B** show the results of crossed immunoelectrophoresis of Berol-solubilized human erythrocyte membrane proteins with anti-human erythrocyte lysate (membrane depleted) and anti-human serum antibodies (Fig. 11.7**B**) incorporated in the intermediate gel, respectively. Fig. 11.7**C** shows a control with blank intermediate gel (see Section 11.11.7). In this series of experiments small amounts of human $\alpha_1$-antitrypsin were added to the membrane preparation as reference proteins. A corresponding amount of monospecific anti-$\alpha_1$-antitrypsin was added to the membrane antibodies. The resulting precipitate is designated AT in Fig. 11.7.

Incorporation in the intermediate gel of the antibodies raised against lysate (Fig. 11.7**A**) resulted in a complete depression of precipitate No. 15. This indicates the presence of high-titre antibodies against this protein in the anti-lysate preparation. Furthermore, a reaction of 'inward feet' [124], often in combination with a slight depression of the precipitate, was seen for precipitates Nos 2, 6, 16, 21, indicating the presence of antibodies of low titre. Antibodies against these proteins were anticipated since they were also demonstrated to be present in the washing buffer (see Fig. 11.6). Precipitate No. 6 (spectrin) was

---

**Fig. 11.7** Crossed immunoelectrophoresis of 20 $\mu$g of Berol EMU-043-solubilized human erythrocyte membrane proteins using an intermediate gel containing, **A**, 0.37 mg of rabbit antibodies against human erythrocyte lysate (membrane-depleted) with 50 u aprotinin, **B**, 0.18 mg of antibodies against human serum proteins (code 100 SF, Dakopatts A/S, Copenhagen, Denmark), with 205 u of aprotinin and **C** 100 $\mu$l of electrophoresis buffer as control ($\phi$). Compared to plate **C**, precipitates No. 6 and 15 are affected in **A** and precipitate No. 2 and AT in **B**. For explanation see text. $\alpha$-1-Antitrypsin and the corresponding specific antibody were added to the solubilized membranes and antibodies, respectively, to give a reference precipitate (AT). Conditions were otherwise as for the experiment illustrated in Fig. 11.3. The bar represents 1 cm.

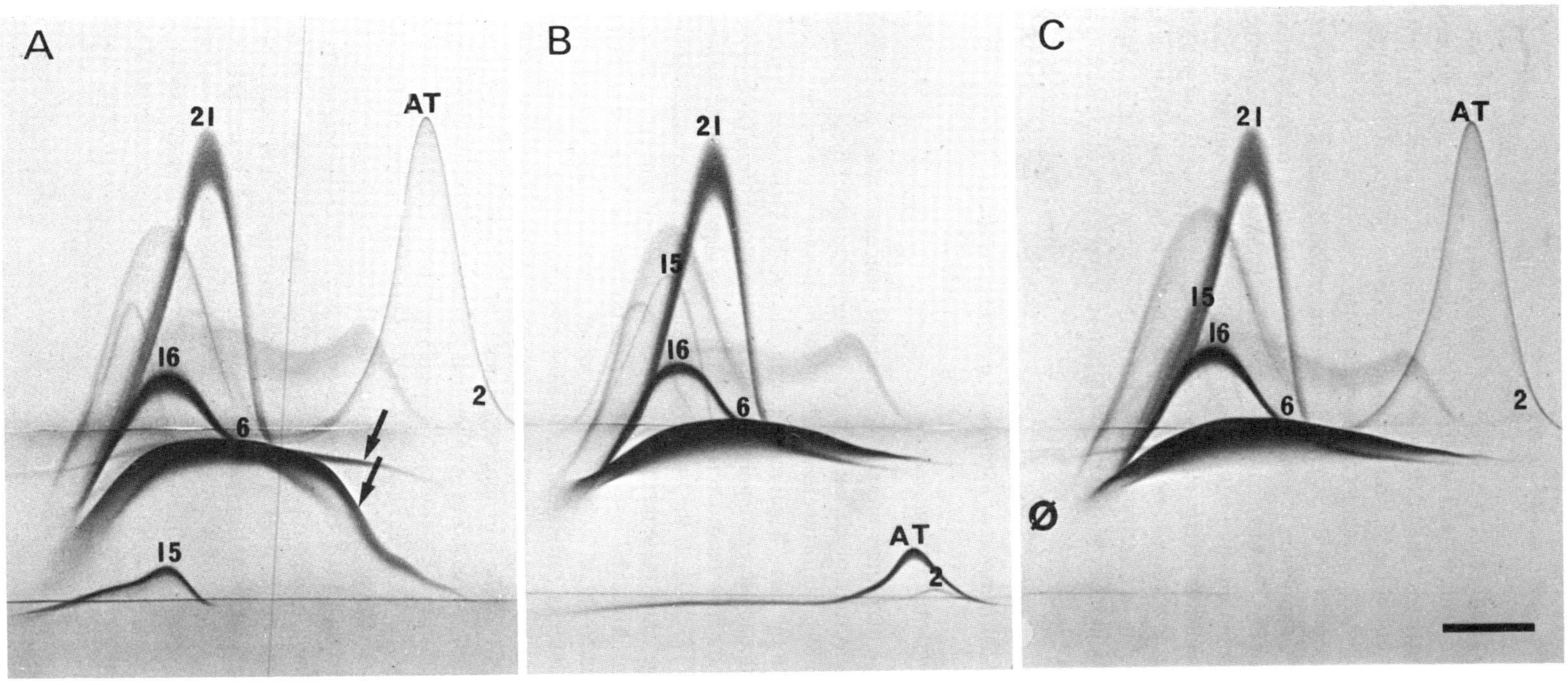
A
21
AT
16
2
6
15
B
21
15
16
6
AT
2
C
21
AT
15
16
6
2
Ø

split into two distinct precipitates (arrows) and precipitate No. 17 was unaffected. This was not expected from the results of Fig. 11.5 where lysate incorporated into the intermediate gel gave rise to elevation of precipitate No. 17, but not No. 6. This result may be a consequence of different immunogenicity in rabbits of the two proteins.

Incorporation in the intermediate gel of antibodies against serum proteins resulted in depression of precipitate No. 2 and the $\alpha_1$-antitrypsin precipitate. Precipitate No. 15 was slightly elevated. This finding contrasts with the experiments using serum in the intermediate gel, which resulted in the disappearance of precipitate No. 15. However, if we assume that protein No. 15 which we now know to be present in the lysate in high concentration, is haemoglobin or a derivative thereof this observation is explicable, since haemoglobin is known to bind to certain serum proteins (haptoglobin and hemopexin).

### *11.6.2 Identification of membrane specific precipitates*

#### *(i) Examination with antigens*

If purified membrane proteins are available, tandem-crossed and crossed-line immunoelectrophoresis can be used for identification of precipitation lines [66,89]. Unfortunately, only a few membrane proteins have been purified so far. On the other hand, many membrane proteins have been characterized by sodium dodecylsulphate polyacrylamide gel electrophoresis (SDS–PAGE), which is the method of choice for analysis and identification of membrane proteins [73,113]. If the two methods could be correlated it would be an advantage, since these methods make use of different kinds of molecular properties. The experience obtained with the two methods could then be combined to obtain a better description of the membrane proteins. In principle, two methods of correlation are feasible: (1) quantitative immunoelectrophoretic analysis of the protein bands from preparative SDS–PAGE and (2) analysis by SDS–PAGE of the immunoprecipitates.

The first approach has been applied to human erythrocyte membrane proteins, where the major protein bands have been isolated by preparative SDS–PAGE [32,53,125]. Fig. 11.8 shows the pattern obtained by rocket immunoelectrophoresis (Section 11.5) of these bands using antibodies against human erythrocyte membranes. 1% (w/v) Berol EMU-043 was added to the samples and the electrophoresis gels for the purpose of reversing the denaturation of the sodium dodecylsulphate-treated proteins and to avoid side effects of the charged detergent in quantitative immunoelectrophoresis [52]. The two detergents form mixed micelles which reduce the free SDS concentration [168]. The homogeneity of the isolated bands was checked by re-electrophoresis by SDS–PAGE. It appears that all the bands (designated according to [113]) contain antigenic material. Furthermore, the material from some of the protein

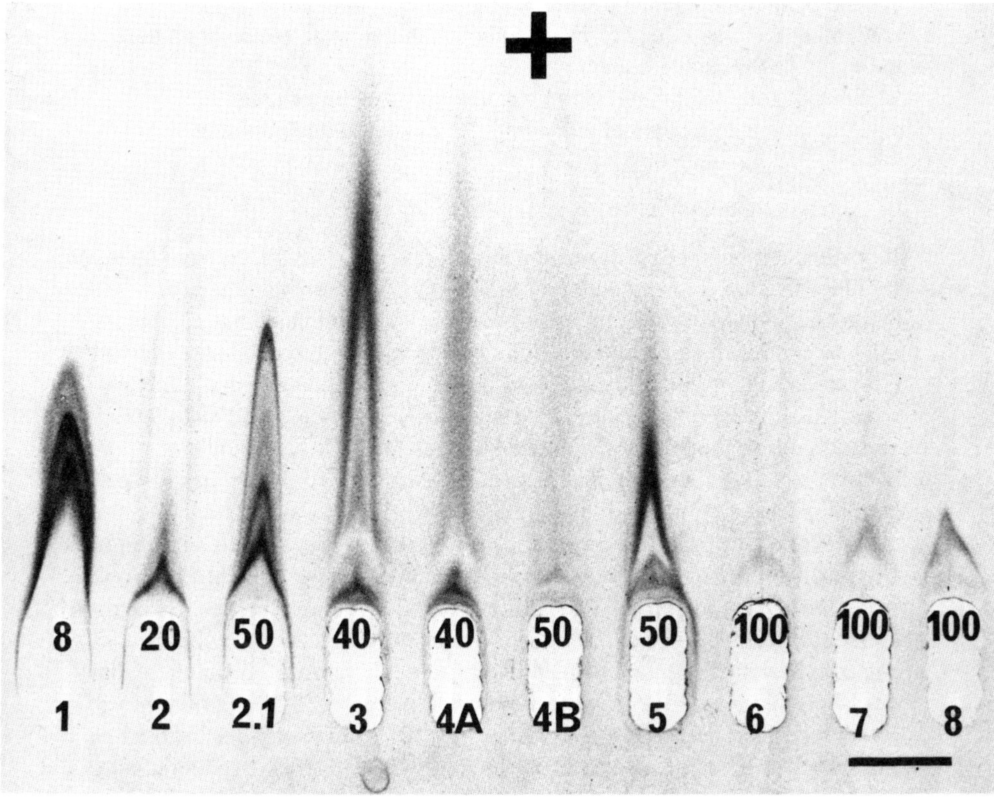


**Fig. 11.8** Rocket immunoelectrophoresis of isolated bands from preparative sodium dodecylsulphate-polyacrylamide gel electrophoresis of human erythrocyte membrane proteins. The applied volume in μl of each band is stated on the Figure. The protein concentration was 0.1 mg–0.4 mg ml$^{-1}$. The bands contained approximately 0.05% of sodium dodecylsulphate and were made 1% with respect to Berol. The electrophoresis was performed in the presence of 1% (w/v) of Berol EMU-043 with 3 V cm$^{-1}$ for 18 h. The antibody content of the gel was 0.3 mg cm$^{-2}$. The bar represents 1 cm.

bands gives rise to several precipitates, indicating the presence of several immunochemical components in each band. An investigation of the nature of these precipitates and a comparison with the pattern obtained by crossed immunoelectrophoresis is currently in progress [53]. The immunogenicity (i.e. the ability of an antigen to promote antibody response by immunization) of the sodium dodecylsulphate-treated membrane proteins also seems to be retained in some cases [51]. Because of this we are also currently investigating the antibody response in rabbits to the material in the various bands.

As far as the second possibility for obtaining a correlation between immunoelectrophoresis and SDS–PAGE is concerned, one approach has been to cut the immunoprecipitates out of the gel, extract them with a solution containing sodium dodecylsulphate and mercaptoethanol, concentrate the extracts and then subject them to analysis by SDS–PAGE [126]. Radiocactive labelling of the antigens seems to be a prerequisite due to the low amount of antigen present in an immunoprecipitate compared to the immunoglobulins.

*(ii) Examination with antibodies*

If purified antibodies against membrane proteins are available they can be used for identification of membrane proteins. Fig. 11.9 shows the pattern obtained by crossed immunoelectrophoresis of human erythrocyte membrane proteins using an intermediate gel containing mono-specific antibodies against spectrin (see Section 11.11.2. (v)). It can be seen that a precipitate appears in the intermediate gel, probably composed of two precipitation lines [93]. This precipitate is derived from precipitate No. 6, which thus seems to be composed of two precipitates. This finding is consistent with the results obtained using antibodies against lysate in the intermediate gel (Fig. 11.7).

The use of such specific antibodies provides a very easy way to be sure that two purified protein preparations are identical [43] and is of great value in the search for membrane proteins present in other fluid compartments. (Since these antibodies are stable enough to be sent through the mail they may be used to compare protein fractions obtained in different laboratories. Antibodies from our laboratory are available for such purposes on request.) Thus an erythrocyte glycoprotein has been identified in urine [61,127] and membrane proteins from fibroblasts have been found in serum [128,129].

## 11.7 Quantitative immunoelectrophoresis for characterization of membrane proteins

It is possible to characterize the proteins giving rise to the precipitates in the crossed immunoelectrophoresis reference pattern of a given membrane system according to: (1) possible enzymatic activity, (2) binding to lectins, (3) sialic acid content and (4) 'inside-outside' localization in the membrane.

### *11.7.1 Enzymatic activity*

Because of the mildness of the solubilization procedure involving non-ionic detergents, considerable enzymatic activity is retained after solubilization. Furthermore, precipitation in the gel does not seriously affect the activity of most enzymes [130,131]. These findings have been exploited in the characterization of membrane-bound enzymes in liver cells [48,132], erythrocytes [15,36,89] and lymphocytes [133]. Fig. 11.10 shows the result of esterase staining of a crossed immunoelectrophoresis plate for Berol-solubilized human erythrocyte membrane proteins (see Section 11.11.9). No. 20 indicates a

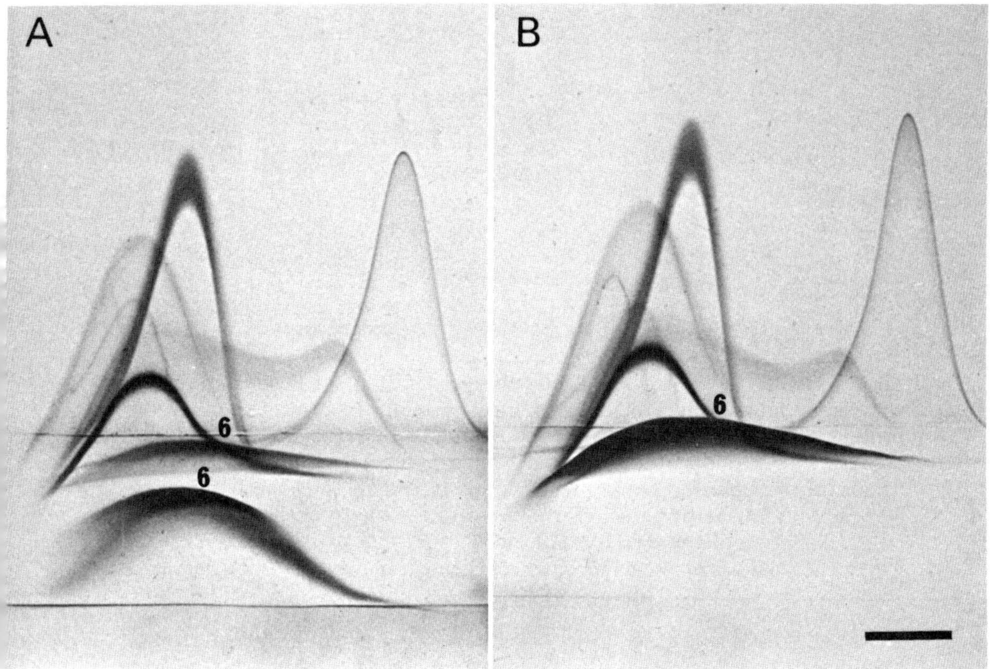


**Fig. 11.9** Crossed immunoelectrophoresis of 20 μg of Berol EMU-043-solubilized membrane proteins using an intermediate gel containing, **A**, 1.3 mg of specific antibodies against spectrin and 15 u of aprotinin (see Section 11.2(v) and **B** 300 μl of electrophoresis buffer as control (ϕ). Compared to **B**, only one precipitate was depessed in plate **A**, the latter composed of two precipitation lines. Thus the heavily stained precipitate No. 6 corresponds to at least two components. The conditions were otherwise as for the experiment illustrated in Fig. 11.7. The bar represents 1 cm.

blurred brown red precipitate which is the esterase-active precipitate. The other faintly stained precipitates (arrows) result from unspecific absorption [131] of the esterase to these precipitates. No Coomassie Brilliant Blue stainable precipitate was found in the region of the esterase active precipitate. The esterase in question is cholinesterase, since it is inhibited by eserine [89,134].

### *11.7.2 Detection and characterization of membrane proteins which interact with lectins*

We find that the immunoelectrophoretic methods allow partial characterization of membrane proteins with respect to certain recognizable molecular structures like carbohydrate parts. Moreover, identification of glycoproteins in gel precipitation tests is necessary in order to correlate these with other serological

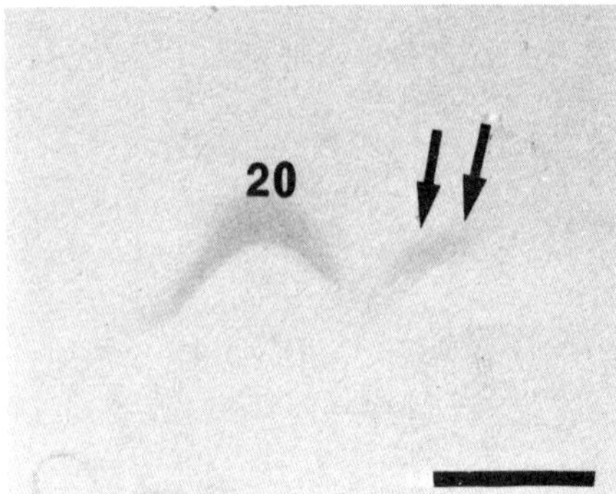


**Fig. 11.10** Crossed immunoelectrophoresis of 20 $\mu$g of Berol EMU-043-solubilized human erythrocyte membrane proteins. Instead of staining for protein with Coomassie Brilliant Blue R the plate was stained for esterase activity with 5-chloro-*o*-toluidin and 1-naphthylacetate (see Section 11.1, .9). The brown-red blurred broad precipitate corresponds to cholinesterase (No. 20). The fainter and sharper precipitates (arrows) are due to non-specific co-precipitation of the esterase molecules with other precipitates. The conditions were otherwise as for the experiment illustrated in Fig. 11.3. The bar represents 1 cm.

tests, and would be an important step towards the ultimate purification and characterization of these antigens.

Recently we have developed methods by which it is possible to detect – in a crude protein mixture – those proteins which interact with lectin. For our model studies we used concanavalin A (con A), a tetrameric lectin which interacts with carbohydrate structures of α-D-mannopyranoside configuration (for reviews of the properties of con A, see [135] and [136]). Our experimental approach was to include lectin in the electrophoretic system and to detect changes in the immunoprecipitation pattern. By including con A as immobilized con A (as con A–Sepharose) in an intermediate gel in crossed immunoelectrophoresis we were able to detect glycoproteins which bind to con A, since they would be missing in the reference precipitation pattern (cf. below). The method was named crossed affino-immunoelectrophoresis and may also be used to predict the result of experiments with affinity chromatography because an analytical scale affinity separation is actually performed during the electrophoresis [25,137].

By including lectin in the first dimension gel it is possible to study closely the interacting components, to determine binding specificities, and to detect minor interacting components, including hetereogeneous molecular forms [138]. In contrast to what we found for the carbohydrate-binding lectin con A, we found in this manner that poke weed mitogen did not interact with serum glycoproteins but interacted specifically with serum lipoproteins. This indicated that the association of poke weed mitogen to lymphocyte cell membranes leading to stimulation, may be explained by an interaction with membrane lipids rather than with membrane carbohydrates [138].

We have been interested in characterizing virus-specific membrane antigens in mammalian cells. Presence of 1% (w/v) non-ionic detergent did not interfere with the con A binding and earlier we reported the detection of four con A–binding *Herpes simplex* virus ( *Herpesvirus hominis*) proteins [69]. We did this by crossed affino-immunoelectrophoresis with immobilized con A. Further characterization of these glycoproteins is possible, since the method allows a distinction of glycoproteins with respect to the number of binding sites to lectin (Fig. 11.11). Three immunoelectrophoresis experiments must be performed as shown in Fig. 11.11. One reference experiment with an empty intermediate gel, **A**, one experiment with immobilized lectin, **B**, and one experiment with free lectin in the intermediate gel, **C**, [139]. All molecules with binding sites to the lectin will interact with the immobilized lectin and be retained in the intermediate gel during electrophoresis [25] provided the gel is not overloaded. In comparison only molecules with two or more binding sites will be able to form a precipitate with free lectin. (Since the lectin has four binding sites it acts as an antibody by joining several molecules in a complex, and before a precipitate is formed, gradually a macromolecular lattice is build up by lectin and glycoprotein molecules, each with at least two binding sites per molecule.) If crude mixtures of proteins are analysed without primary separation, one may find that glycoproteins with one binding site will stick to an already formed precipitate of glycoproteins with two or more binding sites or that glycoproteins of the latter type will cross-link small unsaturated complexes of con A and glycoproteins with only one binding site. We anticipate that clear-cut examples will be obtained with pure proteins, but often sufficient separation may be achieved conveniently during the first dimension electrophoresis in crossed electrophoresis [139].

The three experiments shown in Fig. 11.11 were made with the same amount of an antigen extract of *Herpesvirus hominis* type 1 (*Herpes simples* virus type 1) kindly provided by B. F. Vestergaard (Copenhagen). The proteins that appeared as immunoprecipitates of unchanged size and location in the pattern were proteins that did not interact with con A, i.e. they did not contain any binding sites to con A. Those proteins that disappeared as immunoprecipitates in experiments with con A–Sepharose were proteins retained totally on the con A–Sepharose particles in the intermediate gel (notably Nos 4, 6′, and 11, Fig. 11.11**B**). Proteins that were partially retained (as for example Nos 3 and 8) were proteins of which some molecules did not contain binding sites to con A – in other words, antigens Nos 3 and 8 exhibited heterogeneity with respect to the con A–binding site.

As accounted for above, proteins which were not retained in the intermediate gel with free con A (Nos 3, 4, 6′, and 8, Fig. 11.11**C**) were proteins containing only one binding site per molecule. Some of the precipitates (Nos 6′ and 8) appeared somewhat reduced in size in Fig. 11.11**C** compared to the reference experiment in Fig. 11.11A. We presumed that these proteins consisted of molecules which were heterogeneous with respect to the number of binding sites to con A per molecule, or that molecules with only one binding site were bound in the affinity precipitate (a, Fig. 11.11C) formed by glycoproteins with two or more binding sites to con A. (Con A had a slight cathodical migration in this experiment.)

We also performed crossed affino-immunoelectrophoresis of erythrocyte membrane proteins. As has been found by Fukuda *et al.* [140] in another experimental set up, neither did we find any binding to con A of the major glycoprotein, glycophorin. However, for some membrane preparations we found that the protein designated as No. 16 (cf. Fig. 11.1) was bound to con A, whereas for other preparations binding could not be found. This heterogeneity will be investigated further.

In experiments similar to those shown here for virus glycoproteins in collaboration with R. Schmidt-Ullrich (Boston) we have identified con A–binding proteins from rabbit thymocyte plasma membranes [133,65].

### *11.7.3 Sialoglycoproteins*

In most sialoglycoproteins the negative charge of the sialic acid groups contributes greatly to the mobility at pH 8.6 [141] and the removal of sialic acid residues from the proteins by neuraminidase treatment will therefore reduce the migration velocity of the sialoglycoproteins [141]. In immunoelectrophoresis a.m. Grabar such a change in migration velocity is very difficult to detect but it is easily detected in crossed immunoelectrophoresis [65,117] even in complex mixtures. To facilitate identification of the sialoglycoproteins, a sample consisting of a mixture of normal membrane proteins and neuraminidase-treated membrane proteins can be analysed [117]. Because of the retention of antigenecity of the asialo-proteins [142,143] and because they have the lower migration velocity, the result will be the appearance of 'camel' precipitates, and they are easily detected even in complex precipitation patterns. Although neuraminidase-treated samples with enzyme still present may be mixed with untreated samples without any danger of further degradation (since the enzyme is inactive at pH 8.7), the use of immobilized enzyme (for instance

---

**Fig. 11.11** Crossed affino-immunoelectrophoresis of 15 $\mu$l Triton X–100-solubilized *Herpesvirus hominis* type 1 antigens with intermediate gels containing **A**, electrophoresis buffer ($\phi$), **B**, Sepharose-bound con A (0.1 mg cm$^{-2}$), and **C**, free con A (0.1 mg cm$^{-2}$), where (a) indicates an affinity precipitate. In the upper gel was antiserum against herpesvirus hominis (6 $\mu$l cm$^{-2}$). Precipitates Nos 3, 4, 6, 8 and 11 are affected (for details see the text). Buffer: Tris-barbital pH 8.6, ionic strength 0.02, containing 1% (v/v) Triton X–100. Otherwise as for Fig. 11.3. The bar represents 1 cm.

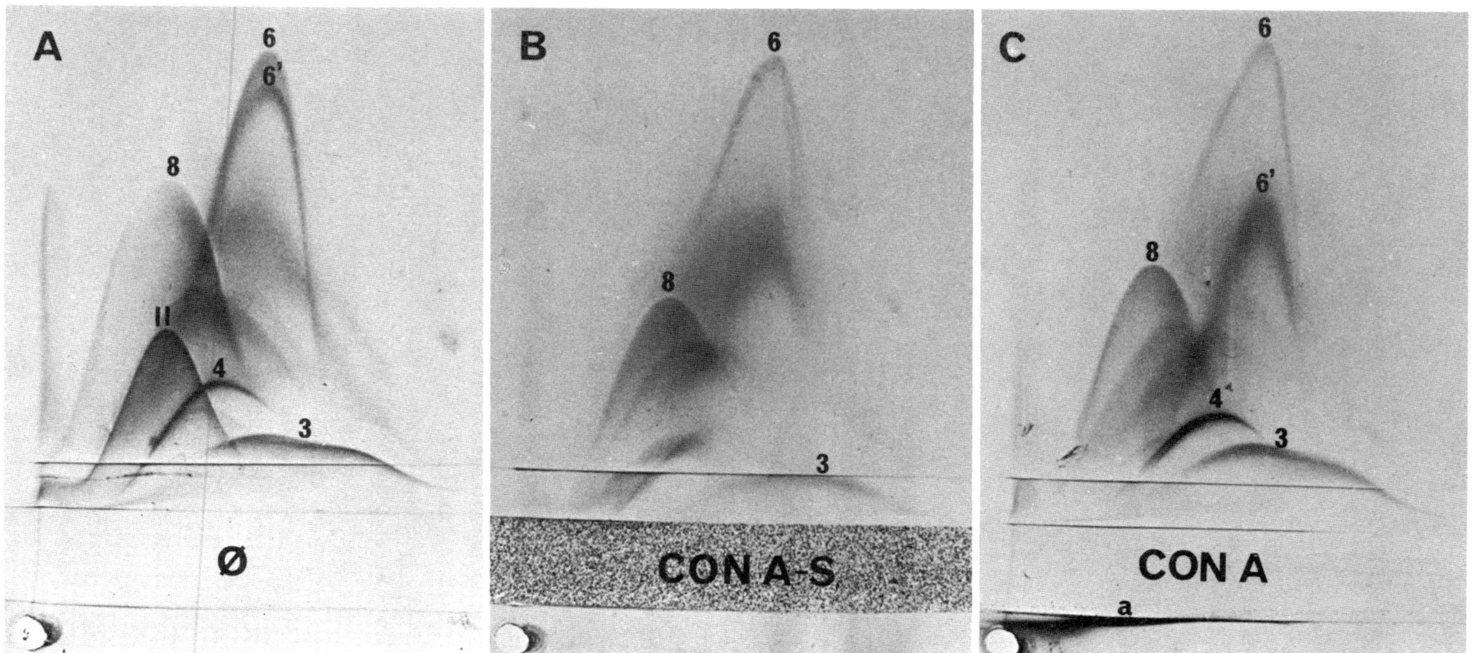
A
6
6'
8
II
4
3
Ø
B
6
8
3
CON A-S
C
6
6'
8
4'
3
CON A
a

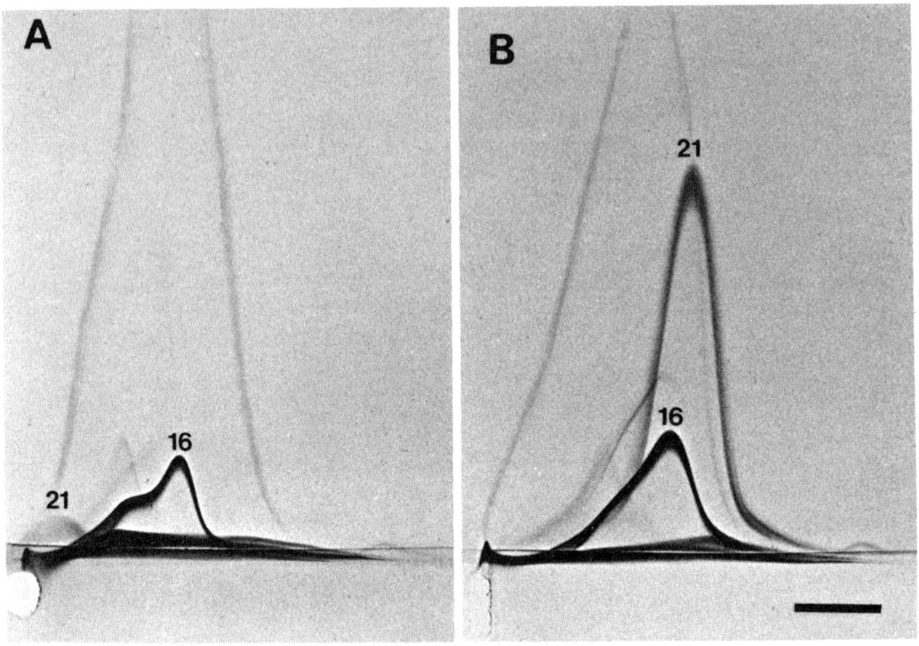


**Fig. 11.12** Crossed immunoelectrophoresis of Berol EMU-043-solubilized human erythrocyte membrane proteins of **A**, erythrocytes which had been treated with neuraminidase before lysis and **B**, an erythrocyte control which had been incubated under similar conditions (see Section 11.11.11). Compared to **B**, precipitates Nos 16 and 21 changed position and shape upon the neuraminidase treatment. Amount of anti-membrane antibody: 0.08 mg $cm^{-2}$. The conditions were otherwise as for the experiment illustrated in Fig. 11.3. The bar represents 1 cm.

neuraminidase coupled to agarose) is recommended, since this is very easy to remove after incubation [117].

The influence of neuraminidase treatment of intact erythrocytes (see Section 11.11.11) on the precipitation pattern of Berol-solubilized membrane proteins can be seen in Fig. 11.12, as a preliminary experiment for demonstration of the method. The pattern for the neuraminidase-treated sample (Fig. 11.12**A**) shows changes in two precipitates relative to that for the control sample (Fig. 11.12**B**) (Nos 21 and 16). One of these precipitates (No. 21) is that for the major glycoprotein which is known to contain sialic acid groups [36]. All sialoglycoproteins of human erythrocyte membranes are located on the outside of the cell-membrane [144] so the results obtained should be representative for all the sialoglycoproteins of the membranes. Thus this method also provides a way of identifying precipitates for proteins located on the outer side of the cell-membrane (see Section 11.7.4).

### *11.7.4 'Inside-outside' localization*

Several methods have been devised for the determination of the location of proteins within the membranes [145,146]. Of these, radioactive labelling of intact cells has been found suitable for use in combination with immunochemical techniques (see Section 11.6.4. (i)). The immunochemical approach *per se* also offers possibilities for such identification (see Section 11.7.4 (ii)).

#### *(i) Autoradiography following $^{125}$I-labelling of intact cells*

A variety of procedures are available for the labelling of exposed membrane proteins of intact isolated cells [145,146]. We describe here an example involving the lactoperoxidase-catalysed $^{125}$I-iodination of exposed proteins on the surface of human erythrocytes (Fig. 11.13) (see Section 11.11.12). The

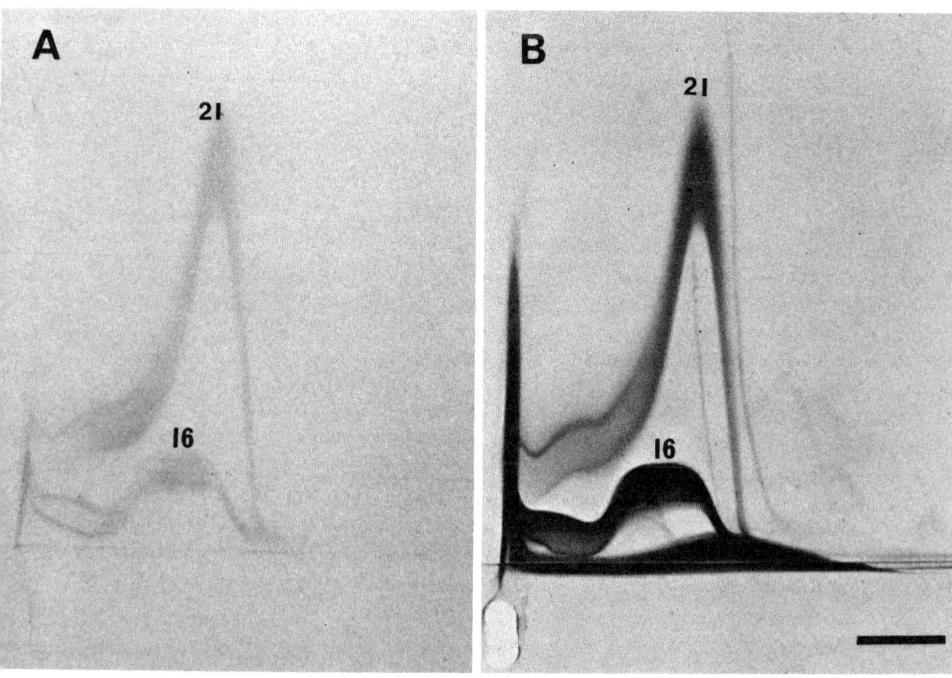


**Fig. 11.13** A, Autoradiography of the crossed immunoelectrophoresis plate for Berol EMU-043-solubilized human erythrocyte membrane proteins on which lactoperoxidase-catalysed $^{125}$I-iodination had been performed on the exposed proteins on the outer surface (see Section 11.11.12). **B**. The same plate after Coomassie Brilliant Blue staining. Two precipitates were radioactive (nos 16 and 21). The amount of antibodies was 0.12 mg cm$^{-2}$. The bar represents 1 cm.

autoradiography of a crossed immunoelectrophoresis plate for Berol-solubilized $^{125}$I-iodinated membranes (Fig. 11.13**A**) shows that only two precipitates are radioactive. Comparison with the Coomassie Brilliant Blue stained plate (Fig. 11.13**B**) enables identification of these precipitates as Nos 16 and 21. Thus, the proteins corresponding to these precipitates are exposed on the outer side of the membranes, a conclusion which agrees with the findings from neuraminidase treatment (see Section 11.7.3). The experiments shown here are preliminary but are presented as an illustration of the applicability of immunochemical methods.

*(ii) Absorption of the antibodies with intact isolated cells*

If a sample of antibodies raised against whole membrane is treated with a surplus of isolated intact cells, the cells will remove the antibodies directed against

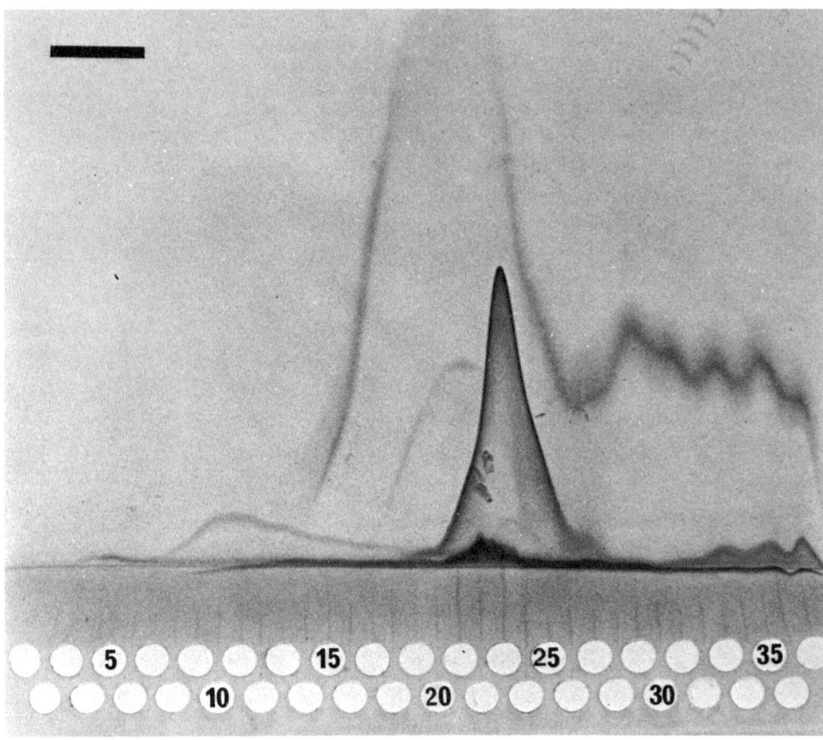


**Fig. 11.14** A fused rocket immunoelectrophoresis experiment in which 25 mg of Berol EMU-043-solubilized human erythrocyte membrane proteins, which have been EDTA-extracted prior to the solubilization, were fractionated by isotacophoresis in 1% (w/v) Berol EMU-043. The fractionation numbers of the 10 $\mu$l aliquots are indicated on the figure. The amount of anti-erythrocyte membrane antibodies: 0.14 mg cm$^{-2}$. The electrophoresis was performed at 2 v cm$^{-1}$ for 16 h. The bar represents 1 cm.

proteins exposed on the outer cell surface. Use of the depleted antibody sample in crossed immunoelectrophoresis of the solubilized membrane material will result in a pattern where these antigens are lacking. If only some of the antigenic determinants are exposed, the result will be an increase in the area below the corresponding precipitate. However, the experiment involving absorption with intact erythrocytes does not give information on whether the rest of the antigens are located within the membrane or on its inner surface. Differentiation between the latter two possibilities can be achieved by absorption experiments with isolated membranes, where the detection of any remaining antibody activity would indicate antigenic determinants located within the membranes. From results obtained to date it appears that all the antigenic determinants of membrane proteins are exposed on one or other of the two membrane surfaces [59,89].

In principle the technique was first applied by Fukui *et al.* [46] to the antigens from *E. coli* bacteria using Grabar immunoelectrophoresis. Using double immunodiffusion the technique has been applied to erythrocyte membranes [29,37,127] and with quantitative immunoelectrophoresis it has been used with success on *Acholeplasma laidlawii* membrane proteins [59] and human erythrocyte proteins [89]. In the case of the last mentioned system, acetylcholinesterase (precipitate No. 20) and the proteins corresponding to precipitates Nos 19 and 21 were found to be exposed on the outer side of the cell membrane. The experiments are described in detail in [89]. The finding that the antigens associated with precipitate No. 21 are on the outside of the membrane agrees with the results obtained from the experiments involving neuraminidase treatment (Section 11.7.3) and those involving $^{125}$I-labelling. However, the latter two methods also indicate the protein corresponding to precipitate No.16 to be exposed on the outside.

## 11.8 Quantitative immunoelectrophoresis for monitoring fractionation procedures

Fused rocket immunoelectrophoresis was developed by Harboe and Svendsen for monitoring fractionation procedures and gives an immunochemical elution profile of each individual protein fractionated [83]. The method has been applied to fractionation of membrane proteins from human erythrocyte membranes [34,64,147] and bovine fat globule membrane proteins [64].

Fig. 11.14 shows an example of fused rocket immunoelectrophoresis (with detergent-containing gels; see Section 11.11.8) of the eluate from a fractionation of human erythrocyte membrane proteins solubilized in Berol. The proteins were separated by preparative isotachophoresis in polyacrylamide gel containing 0.5% (w/v) Berol (for further details, see [64,147]). The number of the fractions is indicated on the figure. It appears that much information about the distribution of the proteins may be gained from this type of analysis compared to the information gained from ordinary UV-absorption monitoring. Moreover, several detergents commonly used in membrane studies absorb UV-light (see Section 11.2.2 and Table 11.1), and also Ampholine used as 'spacers' in the shown experiment results in difficulties in analysis of the elution profile by traditional methods for protein quantitation. Therefore a curve showing the UV-absorption of the effluent does not give a true picture of the separated proteins. Often biologically interesting components are analysed by their biological activity as receptors, as enzymes, etc., but contaminating constituents are usually difficult to analyse for. However, by means of this method, fused rocket immunoelectrophoresis, the immunochemical elution profile of each individual protein is provided, i.e. each individual protein appears as a continuous precipitate, indicating the fractions in which the protein is eluted, the relative distribution of the protein in the fractions, and the distribution of contaminating proteins. Identification of proteins appearing in the different fractions may be performed by crossed immunoelectrophoresis, by tandem-crossed immunoelectrophoresis etc. as described above (Section 11.11.6).

To us, this method offers much more detail about a fractionation experiment than traditional methods for analysis of protein separation, and consequently fractions corresponding to a given membrane protein may be pooled for further fractionation with maximum yield and minimum contamination on the basis of such analyses.

## 11.9 Quantitative immunoelectrophoresis for comparative studies of membrane proteins

Antibodies raised against certain proteins of one animal species often crossreact with the analogous proteins of another species. For example, rabbit antibodies raised against human erythrocyte membrane proteins have been shown by double immunodiffusion to cross-react with membrane proteins of erythrocytes from other mammalian species [33,41,148]. Thus there exists the possibility of establishing a relationship between membrane proteins of analogous cells from different animals by the use of a single antibody preparation if the different precipitates can be recognized. This is not possible with double immunodiffusion except with purified proteins but can be achieved by quantitative immunoelectrophoresis [149]. Fig. 11.15 shows a crossed immunoelectrophoresis pattern for solubilized human erythrocyte membrane proteins, analysed with rabbit antibodies raised against bovine erythrocyte membrane material. Four precipitates are observed and judged from their relative position they can be identified tentatively as Nos 5, 6, 16, and 21. A precise identification can be achieved by crossed-line immunoelectrohoresis with erythrocyte membrane proteins in the intermediate gel [149]. Such a molecular relationship cannot be found by physico-chemical methods such as sodium dodecylsulphate

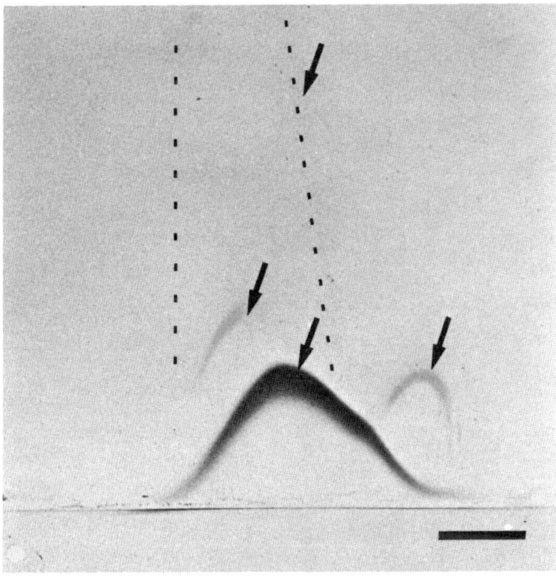

**Fig. 11.15** Crossed immunoelectrophoresis of 10 $\mu$g of Berol EMU-043-solubilized human erythrocyte membranes, using anti-bovine erythrocyte membrane antibodies at a concentration of 0.03 mg cm$^{-2}$. Four faint and blurred perecipitates were seen on the original plate (arrows). Conditions were otherwise as for the experiment illustrated on Fig. 11.3. The bar represents 1 cm.

polyacrylamide gel electrophoresis since the patterns for proteins from different animals only reveal size distribution [150,151].

Fig. 11.16 demonstrates how antibodies may be used for verifying that the same molecular structures (antigenic determinants) exist in the membranes of different cell types of the same animal. Antibodies raised in rabbits against bovine milk fat globule membrane proteins (Section 11.11.2 (iv)) cross-react with some of the major proteins of bovine erythrocyte membranes. This can be seen from crossed immunoelectrophoresis of Berol-solubilized membrane proteins of bovine erythrocytes, using rabbit antibodies against the same material (Fig. 11.16**B**). Some of the precipitates (Nos 6 and 16) exhibit the reaction of 'inward feet' [124] if antibodies against bovine milk fat globule membrane proteins are incorporated in the intermediate gel (Fig. 11.16). This shows that the antibody preparation against milk fat globules contains antibodies directed against the erythrocyte membrane proteins. The precipitates of the bovine precipitation pattern were identified by crossed-line immuno-electrophoresis and numbered according to the human reference pattern (Fig. 11.1) [149]. Since the milk fat globule membrane is believed to be derived from the lactating cell plasma membrane this experiment indicates a much closer immunochemical relationship between membrane proteins of different cell types

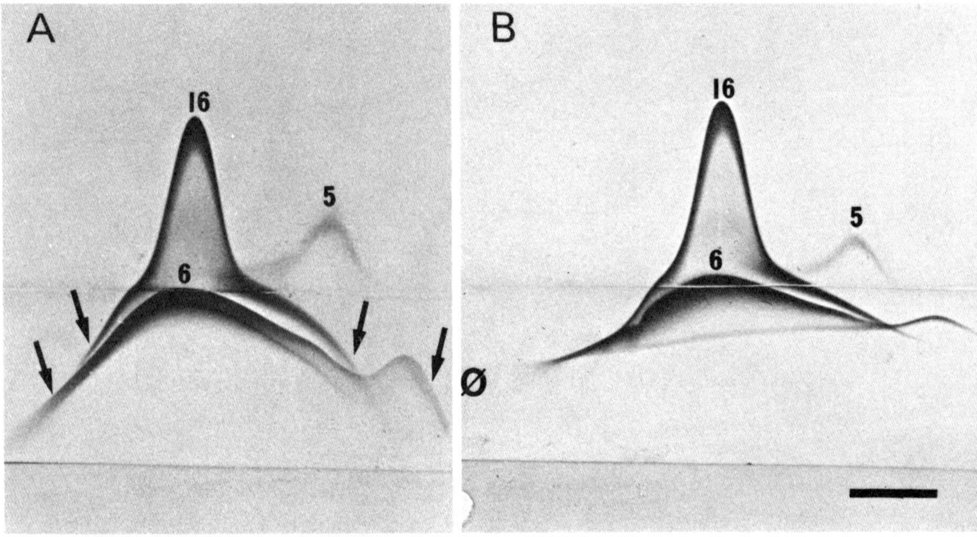


**Fig. 11.16** Crossed immunoelectrophoresis of 19 μg of Berol EMU-043-solubilized bovine erythrocyte membrane proteins using an intermediate gel containing **A**, 4.4 mg of rabbit anti-bovine fat globule membrane protein antibodies and 400 u of aprotinin and **B**, 800 μl electrophoresis buffer containing 400 u aprotinin as control (ϕ). The amount of rabbit anti-bovine erythrocyte membrane antibodies in the upper gel was 0.015 mg $cm^{-2}$. The precipitates are numbered according to Fig. 11.1**A** and ref. [149]. Compared to the control, the presence of anitibodies in he intermediate gel which cross-react with proteins Nos 16 and 6 (spectrin) is observed by he reaction of inward feet (arrow) [124]. Conditions were otherwise as for the experiment illustrated in Fig. 11.3. The bar represents 1 cm.

than hitherto suggested. However, in comparative studies of this kind the limitation of the immunochemical approach should be kept in mind. An animal will normally only form antibodies against an antigen which is recognized as 'not-self'. Therefore the species under investigation should not be too closely related to the animal chosen for immunization, since the antibody response will reveal only antigenic differences but fail to reveal antigenic similarities.

## 11.10 Concluding remarks

One of the main advantages of quantitative immunoelectrophoretic methods over classical Grabar immunoelectrophoresis for analysis of membrane proteins is that the former offer better possibilities for the identification of proteins from the point of migration velocity and the shape and staining intensity of the precipitates. For the further identification and characterization of proteins in crude mixtures, several approaches may be used, some of which have been described in previous sections. Reasonably straightforward methods include addition of purified proteins or absorption with specific antibodies [118], tandem-crossed immunoelectrohoresis with purified proteins [123], or incorporation of purified proteins or specific antibodies in an intermediate gel [124], as reviewed by Axelsen *et al.* [122]. Furthermore, precise identification may be achieved by adding specific antibodies to the first dimension gel so that the first dimension electrophoresis is performed like rocket immunoelectrophoresis, as suggested by Platt *et al.* [153]. Identification and partial characterization may also be achieved by exploiting the specific binding properties of some proteins [118], or by making use of the specific affinity of some proteins towards certain dyes [130]. Enzymes may be identified by histochemical staining for enzyme activity [130,131] and other molecular properties may be studied through the interaction of specifically reacting components other than antibodies, as has been described for lectins [25,138,139].

In addition to the various immunochemical methods described in this chapter several other modifications have been developed for specific studies, such as comparison of antibodies and examination of the partial identity of proteins [154]. Reversed immunoelectrophoresis [155], electrofocusing-immunoelectrophoresis [156] and the direct combination of thin-layer gel filtration and immunoelectrophoresis [157] have also been developed to solve problems of molecular characterization (molecular weight, heterogeneity etc.). As mentioned in Section 11.11.2 (v), line-immunoelectrophoresis may provide a shortcut to the production of specific antibodies.

The fact that all the above mentioned techniques can now be applied to membrane proteins after solubilization with non-ionic detergents makes it possible to examine the more functional properties of the membrane proteins. Thus, it has been possible, for example, to demonstrate the presence of complement components on erythrocyte membranes by crossed immunoelectrophoresis with an intermediate gel after solubilization of the complement-lysed membranes [119]. Immunoelectrophoretic analysis of receptor molecules should also be possible with the use of radioactive substrates and autoradiography.

An additional advantage of immunochemical methods is that antibodies are highly stable proteins and can thus readily be exchanged between different laboratories, thereby facilitating the identification of important membrane proteins.

Moreover, we feel that it is very important for the complete analysis of membrane proteins to have alternatives to sodium dodecylsulphate polyacrylamide gel electrophoresis, alternatives which are based on molecular properties other than molecular size.

The available methods are, however, somewhat limited, the main drawbacks being that:

(1) Not all membrane proteins are soluble in non-ionic detergents,
(2) Not all membrane proteins are antigenic, and detergents may disturb the antigenecity of certain proteins (as is the case for the lipoproteins), and
(3) The rabbits' immunological response is quite individual, which makes comparison between different laboratories difficult when different antisera are used (see Section 11.6.2(ii)).

## 11.11 Practical procedures

### *11.11.1 Solubilization procedures*

#### *(i) Erythrocyte membranes*

Erythrocytes from outdated citrated blood were washed repeatedly by centrifugation in autologous plasma (4 times at 600 x *g* for 20 min). The erythrocytes were contaminated with less than 5 leucocytes per million erythrocytes, as shown by lysis, staining with 0.02% (w/v) Gentian Violet and counting in a Bürker-Türk counting chamber (Carl Zeiss, Jena, Germany). The erythrocytes were then washed 6 times in *iso*osmotic phosphate buffer (pH 7.4). This washing is important since serum proteins absorbed to the lysed membranes will otherwise appear, in concentrated form, in the solubilized membrane material [89,119]. The membranes were then prepared according to Dodge *et al.* [158]. The content of haemoglobin in the washing buffer was measured by spectrophotometry and $A_{540}$(1 cm) was less than 0.035 for the supernatant of the final washing buffer. The freshly prepared membranes, or membranes stored at liquid nitrogen temperature, were stirred for 10–30 min at 4°C in a buffer consisting of 0.010 M glycine, 0.0038 M Tris (pH 9.2 at 4°C) and 1% (w/v) of detergent, at a membrane protein concentration of 3 mg $ml^{-1}$. The latter value corresponds to $A_{280}(1\ cm) - A_{310}(1\ cm) = 3.9$, measured for a sample aliquot which had been completely solubilized with 0.10 M sodium dodecyl-sulphate [79]. After centrifugation for 1 h at 4°C and 100 000 x *g* the supernatant was divided into aliquots of about 200 μl and stored frozen at – 20°C or lower.

#### *(ii) Herpesvirus hominis*

*Herpesvirus hominis* (*Herpes simplex* virus) was cultured on monolayers of Hep-2 cells for 24 h. The cells were then scraped off the glass and compacted by centrifuging at 2000 x *g* for 10 min [67]. The 0.25 g (wet weight) pellet was mixed with 1 ml of 5% Triton X-100 in 0.010 M glycine, 0.0038 M Tris (pH 9.2, 4°C) and sonicated on ice three times for 5 s at 20 000 Hz (Branson Sonifier B 12 with microtip, Donburg, Conn.). The supernatant obtained after ultracentrifugation at 100 000 x *g* for 1 h was used for the immunochemical analysis.

### *11.11.2 Production of antibodies*

*(i) Erythrocyte membrane proteins*

Three different types of antisera were produced by immunizing 2–4 rabbits with (a) human erythrocyte membranes, (b) EDTA-extracts of membranes according to Marchesi and co-workers [159] and (c) the membrane residue remaining after the EDTA extractions. For immunization, 0.1 ml of membrane material extract with a protein concentration 2–5 mg $ml^{-1}$ was mixed with equal volume of Freund's incomplete adjuvant (Statens Seruminstitut, Copenhagen) and injected subcutaneously in the back of the rabbits. The second injection was given after a period of 2 weeks and the subsequent injections at intervals of 4 weeks. Starting with the fourth injection the animals were bled from the ear vein (about 30–40 ml blood was drawn) one week after each injection. The immunoglobulin fraction of the immunoserum was isolated by salting out with 1.75 M $(NH_4)_2SO_4$ and ion-exchange chromatography on DEAE–Sephadex at pH 5.0 [97]. The antibody solution used was a mixture containing 14, 7, and 7 mg $ml^{-1}$ of immunoglobulins from the antisera produced against the various materials (a), (b) and (c), respectively. The different antibody preparations gave rise to different antibody specificities but produced approximately the same number of precipitation lines in crossed immunoelectrophoresis. An antibody preparation produced according to this procedure is now available from Dakopatts A/S, Frederiksberg, Copenhagen, Denmark.

Antibodies against bovine erythrocyte membrane proteins were produced by immunizing two rabbits with non-fractionated membranes. The concentration of the purified immunoglobulin fraction of the antiserum was 5.5 mg $ml^{-1}$.

*(ii) Erythrocyte lysate proteins*

One volume of packed human erythrocytes, washed as in Section 11.11.1 (i) was hypotonically lysed in 4 volumes of 0.008 M phosphate buffer (pH 7.4). The membranes were removed by centrifugation at 100 000 x *g* for 2 h. The supernatant was used directly for the immunization of rabbits, as described for the erythrocyte membranes. 0.3 ml of lysate mixed with an equal volume of Freund's incomplete adjuvant was given per injection [160]. Concentration of purified immunoglobulin fraction of the antiserum was 5.7 mg $ml^{-1}$.

*(iii) Herpesvirus hominis*

The virus used for immunization of rabbits was produced in a continuous rabbit cornea cell line, without the inclusion of serum in the maintenance medium.

Compacted 24 h infected cells were disintegrated by ultrasonic treatment (18 000 Hz for 30 s), mixed with an equal volume of Freund's incomplete adjuvant and injected (2 x 0.1 ml) intracutaneously into rabbits once a month over a period of 12 months [67]. (Similar antibodies are available from Dakopatts A/S, Frederiksberg, Copenhagen, Denmark)

*(iv) Milk fat globule membrane proteins*

Bovine milk fat globule membranes (obtained from butter milk after churning of fresh cream from a single cow), were washed 5 times in a volume of isotonic phosphate buffer (pH 6.5) equal to 15 times the cream volume and isolated by centrifugation at 100 000 x *g* for 1 h. After a further two washings, ca. 0.2 mg of membrane protein material mixed with Freund's incomplete adjuvant were injected in each of four rabbits, as described for the erythrocytes [52]. The concentration of the purified immunoglobulin fraction of the antiserum was 5.7 mg $ml^{-1}$.

*(v) Monospecific antibodies by use of immunoprecipitates*

This type of immunization was performed for the human erythrocyte membrane protein spectrin. Line immunoelectrophoresis [100] (Section 11.11.7) was performed on an EDTA-extract of human erythrocyte membrane [98] on 10 x 10 cm glass plate in the presence of 1% (w/v) of Berol EMU-043, using the polyvalent antibody preparation against human erythrocyte membrane proteins. 50 μg of protein was uniformly distributed in the intermediate gel on each of 8 plates and line immunoelectrophoresis was performed. The antibody concentration was 0.06 mg $cm^{-2}$. The plates were then washed for 24 h in 0.1 M NaCl, with several changes of buffer and pressed under filter paper (see Section 11.11.3 (v)). A protein-stained control plate showed that the different precipitation lines were sufficiently spaced to allow the major precipitation line to be cut out of the gel. The latter line was directly visible in the unstained gel and the gel strip containing the precipitate was removed with a narrow spatula. The 8 gel strips were mixed with 2 ml of 4 M KI solution and the suspension pressed several times through a syringe needle. The resulting suspension was mixed with 2 ml of Freund's incomplete adjuvant and a single rabbit was then immunized with 0.3 ml of the mixture, as described for the erythrocyte membranes (Section 11.11.2 (i)). It should be noted that a similar experiment with precipitation line No. 16 did not give rise to any antibody production in two different rabbits. Concentration of purified immunoglobulin fraction of the antiserum was 3.9 mg $ml^{-1}$.

### 11.11.3 Equipment for quantitative immunoelectrophoresis

Weeke [22] has described in detail the necessary equipment for performing immunoelectrophoresis. Only a short survey is given below.

*(i) Electrophoresis apparatus*

An apparatus with a cooled surface of glass or perspex is necessary to maintain a constant temperature (10–16°C). The smallest usable cooled area is 11 x 22 cm (for 2 plates of 10 x 10 cm or 3 plates of 10 x 7 cm). The connections to the electrode vessels are best made using moulded agarose wicks, which for each electrophoresis are connected to the gel plates via other agarose wicks placed on dialysis film (the film makes the handling of the gel easier and prevents protein contamination from one run to another [22]. Filterpaper wicks which connect the electrode vessels directly to the gel plates are also suitable (e.g. 5 sheets of Whatman No. 1, which are discarded after the run). Some kind of cover over the electrophoresis apparatus is necessary to avoid drying-out of the gels. Suitable electrophoresis apparatus is available from following firms: Holm-Nielsin APS, Analyseudstyr, Søborg, Denmark and Dansk Laboratorieudstyr A/S, Copenhagen, Denmark; Analysteknik, Wallentuna, Sweden; LKB, Bromma, Sweden. Additional firms are listed by Verbruggen [167].

*(ii) Power supply*

Rectified current with stabilized voltage should be used. The voltage output must be at least 300–500 V with a current capacity of 150–300 mA.

*(iii) Voltmeter*

A suitable voltmeter may be fitted with a pair of platinum electrodes to check the voltage-gradient in the gel directly.

*(iv) Equipment for preparing the immunoplates*

Glass plates are easily obtained from a glazier. The dimensions of the plate used for the first dimension electrophoresis should be a little greater than those of the immunoplate. A plate of 90 x 110 x 1.5 mm is suitable and allows the first dimension gel slab to be cut out with sharp edges for use on an immunoplate of 100 x 100 x 1.5 mm or 100 x 70 x 1.5 mm. A puncher has been devised to make uniform circular holes in the agarose gel [161]. For rocket (Section 11.11.5) and fused-rocket immunoelectrophoresis (Section 11.11.8) a template must be used for punching uniform holes of regular separation. Volumes of 1–15 μl are most easily applied in the gel holes or slits by means of double constriction pipettes or Hamilton syringes. For cutting out the gel and transferring it from plate to plate a 2 x 12 cm razor blade is used.

*(v) Equipment for rinsing, drying and staining*

Perspex plate holders with a capacity of 8 plates of 10 cm width are suitable. The holders are constructed in such a way that the plates can only be inserted from one side. Covered plastic vessels with a capacity of 2 l are convenient for rinsing and staining the plates. A hair drier can be used to dry rapidly the gel to a fine film.

### *11.11.4 Crossed immunoelectrophoresis in detergent-containing gels*

Crossed immunoelectrophoresis was performed according to Clarke and Freeman [23]. A 1.5 mm thick gel of 1%(w/v) agarose (Litex, batch AGS 105 Dx, Glostrup, Denmark) containing 0.10 M glycine, 0.038 M Tris and 1% (w/v) of Berol EMU-043 was poured onto a glass plate (90 x 110 x 1.5 mm). (Four types of agarose differing in electroendosmotic properties are available from Litex. The electroendosomosis of batch 105 Dx corresponds to the highest but one. It is important that the gel not has been kept in the water bath (55°C) for longer than 1 h after the melting of the agarose (100°C) as the gel then becomes gritty. 10 μl of sample, containing 10–40 μg protein, was applied in each of 4 wells. The gel plate was placed in an electrophoresis apparatus cooled to 16°C by thermostated water, and connected to the electrode vessels by agarose gel bridges. The detergent was included in the gel bridges but not in the electrode buffer, due to the slight migration of the uncharged detergent (1–3 cm) because of the electroendosmotic flow in the gel. Electrophoresis was performed at 10 V $cm^{-1}$ for 1.0 or 1.2 h. Afterwards, the gel was cut into 20 x 70 mm or 20 x 100 mm sections, each containing the separated antigens from one sample. Each gel section was transferred with a razor blade to a separate glass plate (70 x 100 x 1.5 mm or 100 x 100 x 1.5 mm). On the free part of the plate was poured another agarose gel, 1.0 mm thick, containing the antibodies, detergent and buffer as above. The second dimension electrophoresis was performed perpendicularly to the first at 2 V $cm^{-1}$ for 18 h. It is important that the electric field strength in the second dimension electrophoresis is not too high (maximum 3 V $cm^{-1}$), otherwise the precipitates become blurred. The plates were pressed under filter paper (about 5 g $cm^{-2}$), washed in 0.1 M NaCl for 30 min, pressed again, dried in the hot air stream from a hairdrier and finally stained with 0.5% (w/v) Coomassie Brilliant Blue R 250 (Searle, High Wycombe, Bucks., England) in ethanol–water–acetic acid solution (45:45:10). Destaining was performed with the same solvent mixture.

### *11.11.5 Rocket immunoelectrophoresis*

Rocket immunoelectrophoresis according to Laurell,[162] was performed as described by Weeke [95]. The electrophoreses were carried out on 100 x 100 x 1.5 mm or 100 x 200 x 1.5 mm glass plates using the same gel-buffer-detergent medium as for the crossed immunoelectrophoresis (Section 11.11.4). A row of wells were punched on the cathodic side of a 1.5 mm thick gel by the use of a template [22]. To avoid diffusion, the samples should be applied after the plate has been placed in the electrophoresis apparatus and the voltage (3 V $cm^{-1}$) applied. After the electrophoresis (about 16–18 h) the plates were treated as the crossed immunoelectrophoresis.

### *11.11.6 Tandem-crossed immunoelectrophoresis*

Tandem-crossed immunoelectrophoresis according to Krøll [123] was performed on 100 x 100 x 1.5 mm glass plates in the same way as crossed immunoelectrophoresis (Section 11.11.4) except for the punching of an additional well 10–15 mm in front of the normal well.

### *11.11.7 Crossed immunoelectrophoresis with an intermediate gel*

Line-immunoelectrophoresis [100] and crossed-line immunoelectrophoresis according to Krøll [121], and crossed immunoelectrophoresis with an intermediate gel according to Svendsen and Axelsen [163], were performed as described by Axelsen [124]. The first dimension electrophoresis was performed as for crossed immunoelectrophoresis (Section 11.11.4). After the transfer of the first dimension gel to the second dimension glass plate, the intermediate gel containing the antibodies or antigens was moulded to a thickness of 1.2 mm between a brass bar [124] placed parallel to the free edge of the first dimension gel and 1.2 or 2.2 mm apart. After the gel had congealed it was cut at a distance of 2 mm from the brass bar and the surplus gel strip removed together with the bar. The antibody-containing gel was poured onto the free part of the plate to a thickness of 1.0 mm. The conditions for the electrophoresis were otherwise as for the crossed immunoelectrophoresis.

### *11.11.8 Fused-rocket immunoelectrophoresis*

Fused-rocket immunoelectrophoresis was developed for the immunochemical monitoring of a fractionation procedure [83]. Rows of sample wells (with a diameter of 2.5 to 4 mm) were punched in a 2 mm thick agarose gel strip of width 20 mm on the cathodic side of the glass plate (100 x 100 mm or 100 x 200 mm) using a template (Section 11.11.3). As can be seen from Fig. 11.14, two rows of wells displaced from each other, were punched, thereby allowing more fractions to be applied. The antibody-containing gel was poured onto the free part of the plate to a thickness of 1.5 mm. Samples of equal volume from the fractionation experiment were then transferred to the sample wells in the same order as they were collected. Diffusion was allowed to take place for 30–60 min before the electrophoresis was started (2 V $cm^{-1}$ for 18 h). Rinsing, pressing, drying and final staining were performed as described in Section 11.11.4.

### *11.11.9 Crossed affino-immunoelectrophoresis*

This method [25,139] is a modification of the crossed immunoelectrophoresis with intermediate gel (cf. above) as described by Svendsen and Axelsen [163] who originally used antibodies in the intermediate gel. The principle of the present modification is evident from Fig. 11.11. Concanavalin A (con A) coupled to Sepharose (con A–Sepharose, batch no. TO–3352, Pharmacia Fine Chemicals) or free con A (batch no. 7623, Pharmacia Fine Chemicals) were added to the first intermediate gel to retain glycoproteins containing mannose in their carbohydrate part.

First dimension electrophoresis was performed as normally and the first intermediate gel with con A–Sepharose or free con A was cast between the first dimension gel and a brass bar placed parallel to this gel at a distance of 3 cm. The final concentration of con A was 0.1 mg $cm^{-2}$. After congelation the bar was carefully removed and 2 cm of the free edge of the first intermediate gel was cut off. The second intermediate gel (without con A or antibodies) was cast similarly. Finally the antibody-containing reference gel was cast. The thickness of the four gels were: first dimension gel, 1.7 mm; second intermediate gel, 1.25 mm; and reference gel containing antibodies, 1.0 mm.

We have found it important for this type of analysis to use an agarose gel with low electroendosmotic properties as for instance batch 091A, Litex (the gel type with the lowest electroendosmosis but one, cf. Section 11.11.4) and to separate the con A-containing gel and the antibody-containing gel with an extra intermediate gel because the cathodic backward flow of antibodies. Also we find that free con A from Pharmacia Fine Chemicals is suited for this kind of analysis by intermediate gel since this preparation has a slight cathodical migration at pH 8.9 in contrast to at least one other commercial preparation of con A [65].

### *11.11.10 Staining for esterase activity*

A staining solution [131], modified after Stern and Lewis [164], was prepared by mixing 25 mg of 5-chloro-*o*-toluidine (Fast Red TR) with 1 ml of 1% (w/v) 1-naphthyl acetate in acetone (both reagents from Sigma, St. Louis, Mo., USA), and 50 ml of 0.2 M sodium phosphate buffer (pH 7.4). The agarose gels were washed, pressed and dried at room temperature (20°C) and then kept in the above solution for 1 h at 30°C. Esterase activity gives rise to a brown-red precipitate. After rinsing the gel in water for 1 h and subsequent drying, protein staining can finally be performed in the normal manner (see Section 11.11.4). The technical details of enzymatic staining procedures can be found in [130] and in a recent survey by Brogren and Bøg-Hansen [131]. The latter authors also discuss the interpretation of staining patterns obtained in quantitative immunoelectrophoresis.

### *11.11.11 Neuraminidase treatment*

One volume of freshly obtained human erythrocytes was washed 5 times in 10 volumes of saline, and then once in a buffer of 0.125 M NaCl, 0.01 M $CaCl_2$ and 0.03 M sodium acetate (pH 5.5). After centrifuging (250 x *g* for 15 min), a 0.5 ml aliquot of the compacted erythrocytes was incubated with 350 u of purified neuraminidase (EC 3.2.1.1.8) from *Vibrio comma* (cholerae) (Behringwerke, Marburg, Germany), in a total volume of 3 ml of the acetate buffer at 37°C for 3 h with gentle shaking. A second 0.5 ml aliquot of the compacted erythrocytes incubated without neuraminidase served as control. After washing once with phosphate buffer (pH 7.4), the erythrocytes were lysed in 0.008 M phosphate buffer and then further washed and solubilized as described in Section 11.11.1 (i). The content of haemoglobin in the ghost preparation was higher than normal. Attempts at neuraminidase treatment of the solubilized material were unsuccessful, probably because the membrane proteins aggregated as a result of the low pH and the presence of $Ca^{2+}$ ions. Experiments with human serum proteins indicated that the detergent *per se* did not inhibit the enzyme.

### *11.11.12 Autoradiography of $^{125}$I-labelled proteins*

$^{125}$I-labelling was performed according to Reichstein and Blostein [165]. 5 ml of compacted erythrocytes, prepared as described in Section 11.11.11, were mixed with 5 ml of phosphate-buffered saline (1:9) (pH 7.4) containing 6 μl of $Na^{125}I$ solution (1.5 mCi) (thiosulphate-free, carrier-free, Studsvik AB, Sweden), 150 μl of lactoperoxidase solution (1 mg ml$^{-1}$) (Sigma, St. Louis, Mo., USA) and 100 μl of glucose solution (200 mg ml$^{-1}$). 10 μl of glucose oxidase solution (10 mg ml$^{-1}$) (Type II, Sigma, St. Louis, Mo., USA) was then added and the mixture incubated at 37°C for 30 min. After 3 washings of the erythrocytes with phosphate-buffered sodium iodide (1 + 9), the membranes were then prepared according to Dodge *et al.* [158]. After solubilization with Berol EMU-043 (see Section 11.11.1) followed by crossed immunoelectrophoresis (see Section 11.11.4) the autoradiography of the immunoplates was performed with a RPX-omat film (Kodak, Copenhagen) and the film processed after 3 weeks' exposure in RPX-omat processor, Model M6-N, Kodak.

**Acknowledgements**

Miss A. Möllnitz is thanked for her skilful technical assistance. Thanks are also due to the following other members of our group for valuable discussions: C.-H. Brogren, C. S. Nielsen, J. Ramlau, P. J. Svendsen and B. F. Vestergaard. The English text was revised by Dr. M. P. Hancock and typed by Mrs. G. Bardenfleth and Mrs. L. Muir.

Financial support for this work has been provided by The Danish Medical Science Research Council and Kong Christian den Tiendes Fond.

## References

1. Race, R. R. and Sanger, R. (1968), *Blood Groups in Man,* Blackwell Scientific Publications, Oxford.
2. Aminoff, D. (ed.) , (1970), *Blood and Tissue Antigens,* Academic Press, New York and London.
3. Cleve, H., Hamaguchi, H. and Hütterroth, T. (1972) *J. exp. Med.,* **136**, 1140.
4. Marchesi, V. T., Tillack, T. W., Jackson, R. L., Segrest, J. P. and Scoott, R. E. (1972), *Proc. natn. Acad. Sci. U.S.A.,* **69**, 1445.
5. Ginsburg, V., Kobota, A., Hickey, C. and Sawicka, T. (1971), in *Glycoproteins of Blood Cells and Plasma,* (Jamieson, G. A. and Greenwald, T. J., eds.), Lippincott, Philadelphia.
6. Ceppellini, R. and van Rood, J. J. (1974), *Seminars in Hematology,* **11**, 233.
7. Kahane, I. and Razin, S. (1969), *J. Bact.,* **100**, 187.
8. Néeman, Z., Kahane, I., Kovartovsky, I. and Razin, S. (1972), *Biochim. biophys. Acta,* **266**, 255.
9. Gurd, J. W. and Evans, W. H. (1974), *Arch. Biochem. Biophys.,* **164**, 305.
10. Lauf, P. K., Rasmussen, B. A., Hoffman, P. G., Dunham, P. B., Cook, P. H., Parmelee, M. L. and Tosteson, D. D. (1970), *J. Memb. Biol.,* **3**, 1.
11. Jϕrgensen, P. L., Hansen, O., Glyn, I. M. and Cavieres, J. D. (1973), *Biochim. biophys. Acta,* **291**, 795.
12. Nicolson, G. L., Marchesi, V. T. and Singer, S. J. (1971), *J. Cell Biol.,* **51**, 265.
13. Nicolson, G. L., Masouredis, S. P. and Singer, S. J. (1971), *Proc. natn. Acad. Sci. U.S.A.,* **68**, 1416.
14. Oppenheim, J. D. and Salton, M. R. (1973), *Biochim. biophys. Acta,* **298**, 297.
15. Demus, H. and Mehl, E. (1970), *Biochim. biophys. Acta,* **211**, 148.
16. Bjerrum, O. J. and Lundahl, P. (1973), *Scand. J. Immunol.,* 2, Suppl. 1, 139.
17. Ouchterlony, Ö. (1967), In: *Handbook of Experimental Immunology,* (Weir, D. M., ed.), pp. 655, Blackwell Scientific Publications, Oxford.
18. Grabar, P. and Williams, C. A. (1953), *Biochim. biophys. Acta,* **10**, 193.
19. Ressler, N. (1960), *Clinica chim. Acta,* **5**, 359.
20. Laurell, C. -B. (1965), *Analyt. Biochem.,* **10**, 358.
21. Axelsen, N. H., Krϕll, J. and Weeke, B. (eds.) (1973), *A Manual of Quantitative Immunoelectrophoresis. Methods and Applications.* Universitetsforlaget, Oslo. alias *Scand. J. Immunol,* 2, Suppl. 1.
22. Weeke, B. (1973), *Scand. J. Immunol.,* **2**, Suppl. 1, 15.
23. Clarke, H. G. M. and Freeman, T. (1967), *Prot. Biol. Fluids,* (Peeters, H., ed.), **14**, 503, Elsevier, Amsterdam.
24. Laurell, C. -B. (ed.) (1972), *Scand. J. clin. Lab. Invest.,* **29**, Suppl. 124.
25. Bϕg-Hansen, T. C. (1973), *Analyt. Biochem.,* **56**, 480.
26. Axelsen, N. H. (ed.) (1975), *Quantitative Immunoelectrophoresis. New Developments and Applications,* Universitetsforlaget, Oslo, alias *Scand. J. Immunol.,* **4**, Suppl. 2.
27. Howe, C., Avremeas, S., de Vaux St. Cyr, C., Grabar, P. and Lee, L. T. (1963), *J. Immun.,* **91**, 683.
28. Vulpis, N. (1963), *Acta haemat.,* **30**, 229.
29. Furthmayr, M. and Timpl, R. (1970), *Eur. J. Biochem.,* **15**, 30.

30. Hamaguchi, H. and Cleve, H. (1971), *Biochim. biophys. Acta,* **233**, 320.
31. Lundahl, P. and Liljas, L. (1975), *Analyt. Biochem.*, **65**, 50
32. Bjerrum, O. J., Bhakdi, S., Knüfermann, H. and Bøg-Hansen, T. C. (1974), *Biochim. biophys. Acta,* **373**, 44.
33. Tillack, T. W., Marchesi, S. L., Marchesi, V. T. and Steers, E. (1970), *Biochim. biophys. Acta,* **200**, 125.
34. Bøg-Hansen, T. C. and Bjerrum, O. J. (1973), *Prot. Biol. Fluids,* (Peeters, H., ed.), **21**, 39, Pergamon Press, Oxford.
35. Adachi, H. and Furisawa, M. (1968), *Expl. Cell Res.,* **50**, 490.
36. Howe, C. and Lee, L. T. (1969), *J. Immunn.,* **102**, 573.
37. Poulik, M. D. and Bron, C. (1969), in *The Red Cell Membrane*, (Jamieson, G. A. and Greenwald, T. J., eds.), pp. 131 , Lippincott, Philadelphia.
38. Poulik, M. D. and Bron, C. (1970), in *Blood and Tissue Antigens*, (Aminoff, D., ed.), pp. 343, Academic Press, New York and London.
39. Howe, C., Blumenfeld, O. O., Lee, L. T. and Copeland, P. C. (1971), *J. Immun.,* **106**, 1035.
40. Fletcher, M. A. and Woolfolk, B. J. (1972), *Biochim. biophys. Acta,* **278**, 163.
41. Bron, C. and Poulik, M. D. (1972), *Proc. Soc. exp. Biol. Med.,* **139**, 187.
42. Taketomi, T. and Kawamura, N. (1972), *J. Biochem.,* **72**, 799.
43. Mitchell, T. G., Lee, L. T. and Howe, C. (1973), *Immun. Commun.,* **2**, 35.
44. Argaman, M. and Razin, S. (1969), *J. gen. Microbiol.,* **55**, 45.
45. Corry, G. and Stone, S. S. (1969), *Immunochemistry,* **6**, 627.
46. Fukui, Y., Nachbar, M. S. and Salton, M. R. J., (1971), *J. Bact.,* **105**, 86.
47. Berti, G., Ventrelli, I. and Menini, C. (1973), *Haematologica,* **58**, 184.
48. Blomberg, F. (1974), *Enzyme-active Membrane Antigens in Rat Liver*, Thesis, University of Stockholm.
49. Langdon, R. G. (1974), *Biochim. biophys. Acta,* **342**, 213.
50. Raftell, M., Blomberg, F. and Perlman, P. (1974), *Cancer Res.,* **34**, 2300.
51. Smith, R. P.P. and Shapiro, B. M. (1974), *Biochim. biophys. Acta,* **356**, 331.
52. Nielsen, C. S. and Bjerrum, O. J. (1975), *Scand. J. Immunol.,* **4**, Suppl. 2, 73.
53. Bjerrum, O. J., Bhakdi, S., Knüfermann, H., Bøg-Hansen, T. C. and Wallach, D. F. H. (1975), *Biochem. biophys. Acta,* **406**, 489.
54. Blomberg, F. and Raftell, M. (1973), *Biochim. biophys. Acta,* **291**, 431.
55. Maretzki, D., Groth, J., Tsamaloukas, A. G., Grundel, M., Krüger, S. and Rapoport, S. (1973), *FEBS Letters,* **39**, 83.
56. Bjerrum, O. J. and Lundahl, P. (1974), *Biochim. biophys. Acta,* **324**, 69.
57. Blomberg, F. and Raftell, M. (1974), *Eur. J. Biochem.,* **49**, 21.
58. Bock, E., Jørgensen, O. S. and Morris, S. J. (1974), *J. Neurochem.,* **22**, 1013.
59. Johansson, K. -E. and Hjertén, S. (1974), *J. molec. Biol.,* **86**, 341.
60. Raftell, M. and Blomberg, F. (1974), *Eur. J. Biochem.,* **49**, 31.
61. Bron, C. and Poulik, M. D. (1971), *Immunochemistry,* **8**, 447.
62. Tillack, T. W., Rosai, J. and Vervynck, D. J. (1974), *J. natn. Cancer Inst.,* **5**, 1059.
63. Bjerrum, O. J. (1974), *Experientia,* **30**, 831.
64. Bøg-Hansen, T. C., Svendsen, P. J., Bjerrum, O. J., Nielsen, C. S. and Ramlau, J. (1975), *Prot. Biol. Fluids,* **22**, 679, (Peeters, H., ed.), Pergamon Press, Oxford.
65. Schmidt-Ullrich, R., Wallach, D. F. H. and Hendricks, J. (1975), *Biochim. biophys. Acta,* **382**, 295.

66. Jϕrgensen, O. S. and Bock, E. (1974), *J. Neurochem.*, **23**, 879.
67. Vestergaard, B. F. (1973), *Acta path. microbiol. scand.*, **81**, 808.
68. Vestergaard, B. F. (1975), *Scand. J. Immunol.*, **4**, Suppl. 2, 203.
69. Vestergaard, B. F. and Bϕg-Hansen, T. C. (1975), *Scand. J. Immunol.*, **4**, Suppl. 2, 211.
70. Kirkpatrick, F. H. and LaCelle, P.L. (1974), *Experientia*, **30**, 140.
71. Maddy, A. H. (1970), *Seminars in Hematology*, **7**, 275
72. Razin, S. (1972) *Biochim. biophys. Acta*, **265**, 241.
73. Steck, T. L. and Fox, C. F. (1973), in *Membrane Molecular Biology*, pp. 43, (Fox, C. F. and Keith, A. D., eds.), Sinauer, Stamford.
74. Miller, D. M. (1970), *Biochem. biophys. Res. Commun.*, **40**, 716.
75. Crumpton, M. J. and Parkhouse, R. M. E. (1972), *FEBS Letters*, **22**, 210.
76. Kirkpatrick, F. H. and Sandberg, H. E. (1973), *Biochim. biophys. Acta*, **298**, 209.
77. Vestergaard, B. F. and Bjerrum, O. J. (1975). Unpublished work.
78. Yu, J., Fishman, D. A. and Steck, T. L. (1973), *J. superamol. Struct.*, **1**, 233.
79. Liljas, L., Lundahl, P. and Hjertén, S. (1974), *Biochim. biophys. Acta*, **352**, 327.
80. Kamat, V. B., Wyatt,A. J. and Davis, M. A. F. (1972), *Chem. Phys. Lipids* **8**, 341.
81. Nielsen, C. S. (1975). Unpublished work.
82. Hjertén, S. and Johansson, K. -E. (1972), *Biochim. biophys. Acta*, **288**, 312.
83. Svendsen, P. J. (1973), *Scand. J. Immunol.*, **2**, Suppl. 1, 69.
84. Lowry, O. H., Rosebrough, N. J., Farr, A. L. and Randall, R. J. (1951), *J. biol. Chem.*, **193**, 265.
85. Bonsall, R. W. and Hunt, S. (1971), *Biochim. biophys. Acta*, **249**, 266.
86. Futterman, S. and Rollins, M. H. (1973), *Analyt. Biochem.*, **51**, 443.
87. Lundahl, P. (1974), *Biochim. biophys. Acta.*, **379**, 304.
88. Bjerrum, O. J. and Bϕg-Hansen, T. C. (1975), *Scand. J. Immunol.*, **4**, Suppl. 2.
89. Bjerrum, O. J., Lundahl, P., Brogren, C. -H. and Hjertén, S. (1975), *Biochim. biophys. Acta*, **394**, 173
90. Helenius, A. and Simons, K. (1972), *J. biol. Chem.*, **247**, 3656.
91. Tanford, C. (1972), *J. molec. Biol.*, **67**, 59.
92. Makino, S., Reynolds, J. A. and Tanford, C. (1973), *J. biol. Chem.*, **248**, 4926.
93. Bjerrum, O. J. (1975). Unpublished work.
94. Lundahl, P. and Hjertén, S. (1974), *Prot. Biol. Fluids*, (Peeters, H. ed.) **21**, 27, Pergamon Press, Oxford.
95. Weeke, B. (1970), *Scand. J. clin. Lab. Invest.*, **25**, 161.
96. Gill, T. J., Kunz, H. W. and Papermaster, D. S. (1967), *J. biol. Chem.*, **242**, 3308.
97. Harboe, N. and Ingild, A. (1973), *Scand. J. Immunol.*, **2**, Suppl. 1, 161.
98. Bjerrum, O. J., Ramlau, J., Clemmensen, I., Ingild, A. and Bϕg-Hansen, T. C. (1975), *Scand. J. Immunol.*, **4**, Suppl. 2.
99. Axelsen, N. H. (1973), *Infect. Immun.*, **7**, 949.
100. Krϕll, J. (1973), *Scand. J. Immunol.*, **2**, Suppl. 1, 61.
101. Weeke, B. (1973), *Scand. J. Immunol.*, **2**, Suppl. 1, 47.
102. Johansson, B. G. (1972), *Scand. J. clin. Lab. Invest.*, **29**, Suppl. 124, 7.
103. Weeke, B. (1973), *Scand. J. Immunol.*, **2**, Suppl. 1, 37.

104. Bjerrum, O. J., Ingild, A., Løwenstein, H. and Weeke, B. (1973), *Clinica chim. Acta,* **46**, 337.
105. Grubb, A. (1974), *J. Immun.,* **112**, 1420.
106. Bjerrum, O. J., Ingild, A. and Løwenstein, H. (1974), *Immunochemistry,* **11**, 797.
107. Kostner, G. and Molasek, A. (1972), *Analyt. Biochem.,* **46**, 680.
108. Laurell, C. -B. (1972), *Scand. J. clin. Lab. Invest.,* **29**, Suppl. 124, 21.
109. Kindmark, C. -O. and Thorell, J. (1972), *Scand. J. clin. Lab. Invest.,* **29**, Suppl. 124, 49.
110. Nørgaard-Pedersen, B. (1973), *Clinica chim. Acta,* **48**, 345.
111. Ramlau, J. and Bjerrum, O. J. (1975),
112. Moore, G. L., Kocholaty, W. F., Cooper, D. A., Gray, J. L. and Robinson S. L. (1970), *Biochim. biophys. Acta,* **212**, 126.
113. Fairbanks, G., Steck, G. and Wallach, D. F. H. (1971), *Biochemistry,* **10**, 2606.
114. Bernacki, R. J. and Bosman, H. B. (1972), *J. Memb. Biol.,* **7**, 1.
115. Bosmann, H. B. (1971), *J. Memb. Biol.,* **4**, 114.
116. Bosman, H. B. (1974), *Vox. Sang.,* **26**, 497.
117. Brogren, C. -H., Bjerrum, O. J. and Bøg-Hansen, T. C. (1975). Unpublished work.
118. Clarke, H. G. M. and Freeman, T. (1968), *Clin. Sci.,* **35**, 403.
119. Bhakdi, S., Bjerrum, O. J. and Knüfermann, H. (1975), *Scand. J. Immunol.,* **4**, Suppl. 2, 67.
120. McDaniel, C. F., Kirtley, M. E. and Tanner, M. J. A. (1974), *J. biol. Chem.,* **249**, 6478.
121. Krøll, J. (1973), *Scand. J. Immunol.,* **2**, Suppl.1, 79.
122. Axelsen, N. H., Bock, E. and Krøll, J. (1973), *Scand. J. Immunol.,* **2**, Suppl. 1, 91.
123. Krøll, J. (1973), *Scand. J. Immunol.,* **2**, Suppl. 1, 57.
124. Axelsen, N. H. (1973), *Scand. J. Immunol.,* **2**, Suppl. 1, 71.
125. Knüfermann, H., Bhakdi, S. and Wallach, D. F. H. (1975), *Biochim. biophys. Acta,* **389**, 464.
126. Zimmermann, B. (1974), *J. Immun.,* **112**, 249.
127. Bron, C. and Gallagher, G. (1974), *Immunochemistry,* **11**, 409.
128. Ruoslahti, E. and Vaheri, A. (1974), *Nature,* **248**, 789.
129. Ruoslahti, E., Vaheri, A., Kuusela, P. and Linder, F. (1973), *Biochim. biophys. Acta,* **322**, 352.
130. Uriel, J. (1971), in *Methods in Immunology and Immunochemistry,* **3**, 294, (Williams, C. A. and Chase, M. W. eds.), Academic Press, New York and London.
131. Brogren, C.-H. and Bøg-Hansen, T. C. (1975), *Scand. J. Immunol.,* **4**, Suppl. 2, 37.
132. Gurd, J. W., Evans, W. H. and Perkins, H. R. (1973), *Biochem. J.,* **135**, 827.
133. Bjerrum, O. J. and Bøg-Hansen, T. C. (1974), *Dansk Kemi,* **4**, 13.
134. Augustinsson, K. -B., Axenfors, B., Andersson, I. and Eriksson, H. (1973), *Biochim. biophys. Acta,* **293**, 424.
135. Sharon, N. and Lis, H. (1972), *Science,* **177**, 949.
136. Lis, H. and Sharon, N. (1973), *A. Rev. Biochem.,* **42**, 54.
137. Bøg-Hansen, T. C., Brogren, C. -H. and McMurrough, I. (1974), *J. Inst. Brew.,* **61**, 522.
138. Bøg-Hansen, T. C., Bjerrum, O. J. and Ramlau, J. (1975), *Scand. J. Immunol.,* **4**, Suppl. 2, 141.

139. Bøg-Hansen, T. C., and Brogren, C. -H. (1975), *Scand. J. Immunol.*, **4**, Suppl. 2, 135.
140. Fukuda, M., Osawa, T. (1973), *J. biol. Chem.*, **248**, 5100.
141. Schultze, H. E. (1962), *Archs Biochem. Biophys.*, **98**, Suppl. 1, 298.
142. Brogren, C. -H., Bøg-Hansen, T. C., Svendsen, P. J. and Bjerrum, O. J. (1975), *Scand. J. Immunol.*, **4**, Suppl. 2, 59.
143. Brogren, C. -H., Svendsen, P. J. and Bøg-Hansen, T. C. in *Progress in isoelectrofocusing and Isotachophoreis 1975*, 359, (Righetti, P. G. ed.), Elsevier, Amsterdam.
144. Eylar, E. H., Madaff, M. A., Brody, O. V. and Oncley, J. L. (1962), *J. biol. Chem.*, **237**, 1992.
145. Wallach, D. F. H. (1972), *Biochim. biophys. Acta*, **265**, 61.
146. Zwall, R. F. A., Roelofsen, B. and Calley, C. M. (1973), *Biochims biophys. Acta*, **300**, 159.
147. Bøg-Hansen, T. C., Bjerrum, O. J. and Svendsen, P. J. (1974), *Sci. Tools*, **21**, 33.
148. Berthoud, J. D., Gallagher, G. and Bron, C. (1973), *Experientia*, **29**, 750.
149. Bjerrum, O. J. (1975), *Int. J. Biochem.*, **6**, 513.
150. Lennard, J. (1970), *Biochemistry*, **9**, 5037.
151. Hamaguchi, H. and Cleve, H. (1972), *Biochem. biophys. Res. Commun.*, **47**, 459.
152. Knüfermann, H., Schmidt-Ullrich, R., Ferber, E., Rischer, H. and Wallach, D. F. H. (1973), in *Erythrocytes, Thrombocytes, Leucocytes*, (Gerlach, E., Moser, K., Deutch, E. and Wilmanns, W. eds.), Georg Thieme, Stuttgart.
153. Platt, H. S., Sewell, B. M., Feldman, T. and Souhami, R. L. (1973), *Clinica chim. Acta*, **46**, 419.
154. Bock, E. and Axelsen, NH. (1973), *Scand. J. Immunol.*, 2, Suppl. 1, 95.
155. Axelsen, N. H. and Svendsen, P. J. (1973), *Scand. J. Immunol.*, **2**, Suppl. 1, 155.
156. Skude, G. and Jeppsson, J. -O. (1972), *Scand. J. clin. Lab. Invest.*, **29**, Suppl. 124, 55.
157. Nielsen, C. S. (1975), *Scand. J. Immunol.*, **4**, Suppl. 2.
158. Dodge, J. T., Mitchell, C. and Hanahan, D. J. (1963), *Archs. Biochem. Biophys.*, **100**, 119.
159. Marchesi, S. L., Steers, E., Marchesi, V. J. and Tillach, J. W. (1970), *Biochemistry*, **9**, 50.
160. Nørgaard-Pedersen, B. and Bjerrum, O. J. (1975). Unpublished work.
161. Weeke, B. and Thomsen, J. P. (1968), *Scand. J. clin. Lab. Invest.*, **22**, 165.
162. Laurell, C. -B. (1966), *Analyt. Biochem.*, **15**, 45.
163. Svendsen, P. J. and Axelsen, N. H. (1972), *J. immunol., Methods*, **1**, 169.
164. Stern, J. and Lewis, W. P. H. (1962), *J. ment. Defic. Res.*, **6**, 13.
165. Reichstein, E. and Blostein, R. (1973), *Biochem. biophys. Res. Commun.*, **54**, 494.
166. Green, F. R., Dunn, M. J., Spooner, R. L., Maddy, A. H. (1974), *Biochim. biophys. Acta*, **373**, 51.
167. Verbruggen, R. (1975), *Clin. Chem.*, **21**, 5.
168. Schick, M. J. and Manning, D. J. (1966), *J. Am. Oil Chem. Soc.*, **43**, 133.
169. Schick, M. J. (ed.) (1968), *Nonionic Surfactants* Vol. I., Marcel Dekker, New York.
170. Demus, H. (1973), *Biochim. biophys. Acta*, **291**, 93.
171. Ouchterlony, Ö. (1948), *Acta path. microbiol. scand.*, **25**, 186.
172. Grabar, P. and Burtin, P. (eds.) (1964), *Immuno-electrophoretic analysis*, Elsevier, Amsterdam.

# 12 Specific labels for cell surfaces

A. L. HUBBARD and Z. A. COHN



Abbreviations used. ATP, adenosine triphosphate; Con A, concanavalin A; DABS, diazobenzene sulphonate; DFP, diisopropylfluorophosphate; DIDS, 4,4′-diisothiocyano-stilbene-disulphonate; dansyl chloride, dimethylamino naphthalene-5-sulphonylchloride; EAI, ethyl acetamidate; FDNB, 1-fluoro-2,4-dinitrobenzene; FMMP, formyl methionyl-sulphone methyl phosphate; GO, glucose oxidase; G3PD, glyceraldehyde-3-phosphate dehydrogenase; IAI, isethionyl acetamidate; LPO, lactoperoxidase; NaK-ATPase, $Na^+$,$K^+$-stimulated adenosine triphosphatase; NEM, N-ethylmaleimide; NAP-taurine, N-(4-nitreno-2-nitrophenyl)-2 aminoethyl sulphonate; PCMBS, para-chloromercuribenzene sulphonate; PBS, phosphate-buffered saline; SITS, 4-acetamido-4′-isothiocyano-2,2′-stilbene disulphonate; SDS, sodium dodecyl sulphate; TCA, trichloroacetic acid; TNBS, 2,4,6-trinitrobenzene sulphonate; WGA, wheat germ agglutinin.

## 12.1 Introduction

There are numerous uses for chemical reagents which, when added to the external environment of a cell, will specifically react with membrane consitiuents: as probes of protein and/or lipid disposition in the membrane; in the study of membrane flow during endocytosis; in membrane turnover; as membrane markers in subcellular fractionation; as probes of membrane changes through the cell cycle, of membrane differences between normal and transformed cells, and of surface changes following perturbation by hormones, drugs or enzymes. Interest in these and other areas of membrane research has led to the introduction of several different types of cell surface-specific labels. The characteristics of these membrane reagents have varied according to their proposed use. Most reagents have been designed to be impermeant and hence to label exclusively the external surface of the plasma membrane. In some cases permeant labels have been used to probe a membrane function or to compare results with those obtained using impermeant reagents. The spectrum of cell surface labels includes reagents that interact non-covalently with membrane constituents as well as those that form covalent bonds. Certain reagents interact (covalently or non-covalently) with a specific subset of membrane constituents (e.g., an active site of an enzyme or an oligosaccharide residue) while others are designed to react generally (e.g., with all accessible amino groups). The size of the label incorporated into or interacting with the cell surface also varies from a single tritium atom to a multimeric protein of greater than 40 nm. An awareness of the membrane reagents available, the characteristics of each, and the advantages as well as limitations of each, is necessary for the intelligent selection and effective use of a cell surface-specific label in any membrane study.

In the first part of this chapter we will review and evaluate the cell surface-specific reagents currently in use. We have classified the labels into four broad categories: impermeant, chemical (Section 12.2.1 and 12.2.2); permeant, chemical (Section 12.2.2–12.2.4); site-specific (Section 12.3); and enzyme-mediated (Section 12.4). In some cases the classification of a specific reagent has been rather arbitrary (e.g., a label could be both site-specific *and* impermeant) and in others perhaps wishful (e.g., a reagent historically accepted as impermeant but not clearly demonstrated to be so). Albeit, we will first present the general characteristics of a given class of cell surface label and then discuss each reagent in that class. Several general questions which we will ask of each type of label and which should be considered by anyone choosing or using a particular membrane reagent are: (1) to what extent will the properties of the label (size, charge, reaction specificity, etc.) determine which membrane constituents are labelled? (2) to what extent will the properties of the membrane being labelled (hydrophobicity, electronegativity, fluidity, secretory activity, etc.) influence the reaction of the label? (3) to what extent will the reaction conditions (pH, time, temperature, buffer, etc.) influence the constituents labelled, the permeance of the reagent, and the permeability or fragility of the membrane?

In the second part of this chapter we will review in detail the lactoperoxidase iodination method of cell surface labelling (Section 12.4.2) with particular emphasis on our own approach to this technique (Section 12.5). We have chosen to concentrate on enzymatic iodination because it has proven to be overwhelmingly the most popular of the labelling techniques currently in use, and so the detailed methodology, that is the conditions, controls, cautions, etc., deserves special emphasis in a chapter on cell surface labels.

In the final section of this chapter (Section 12.6) we will present examples from the literature and our own experience of the uses for cell surface labels in membrane research. Several recent reviews on membranes deal peripherally with surface labels [1–6].

## 12.2 Impermeant and permeant chemical labels

### *12.2.1 Covalent chemical labels*

Numerous molecules designed as non-permeant cell surface-specific reagents have been used to probe protein and lipid dispositions in the plasma membrane. Maddy was the first to outline several requirements of labels in this class [7]. They should (a) be non-permeant and yet small, for accessibility; (b) react under physiological conditions of pH, tonicity and temperature; (c) form a stable, covalent linkage with membrane constituents; and (d) be detectable in small amounts. An additional requirement (e), that introduction of the label should cause little or no perturbation of membrane structure or function, is relevant when probing general protein or lipid disposition in the membrane or studying the metabolic fate of labelled membrane constituents. As we review the various chemical labels used, it will become apparent that certain criteria (e.g., (a)) have been difficult to fulfill in practice; others may not be desirable if total label exclusion (b) or localization and purification of a membrane function (e.g., an enzyme) is a goal (e); and several criteria have often been inadequately assessed ((a) and (c)) or ignored (e).

A list of compounds classified as covalent chemical labels, together with information on structure, group specificity, size and reaction conditions is given in Table 12.1. Most of these membrane reagents contain ionic groups, usually sulphonyl residues, to convey some hydrophilic character to otherwise hydrophobic molecules. The net negative charge increases the reagent's solubility in an aqueous environment and reduces (hopefully) its penetration of the membrane lipid bilayer. However, the amphipathic nature of the molecule may affect its reactivity with different membrane constituents. The anionic group on the molecule may promote electrostatic interactions with positively-charged regions of the membrane (or reduce interactions with negatively-charged regions) and result in unintended site-specific labelling (or non-labelling). Conversely, the hydrophobic portions of the molecule may promote association with non-polar regions of the membrane. The molecular size of these compounds ranges from 0.4–2 nm, which is fairly large, considering that they displace a proton of radius 0.12 nm on an amino acid or lipid in the modification reaction. Another general characteristic of the chemical reagents in this class is their reactivity toward nucleophiles. Modification of an unprotonated amino group is the most probable

TABLE 12.1
*Impermeant and permeant chemical labels*

| Reagent | Class | Structure* |
|---|---|---|
| (a) 4-Acetamido-4′-*iso*-thiocyano-2,2′-stilbene disulphonate (SITS) | Impermeant, non-covalent | 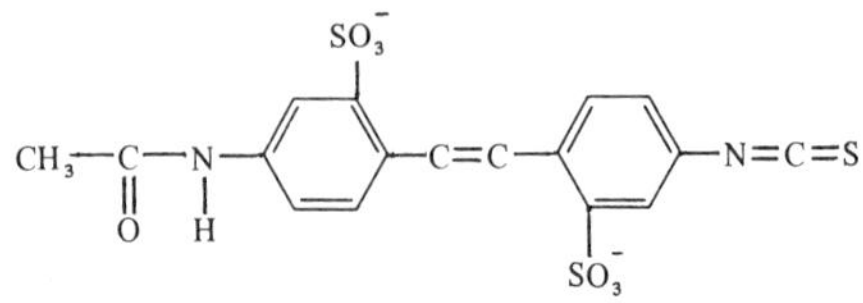 |
| (b) 4, 4′-Diisothiocyano stilbene disulphonate (DIDS) | Impermeant, covalent | 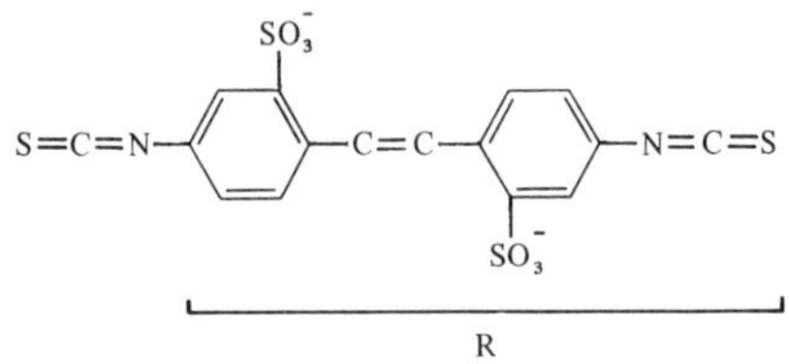 |
| (c) 4, 4′-Diisothiocyano ditritio-stilbene disulphonate ($^3$H-DIDS) | Impermeant, covalent | 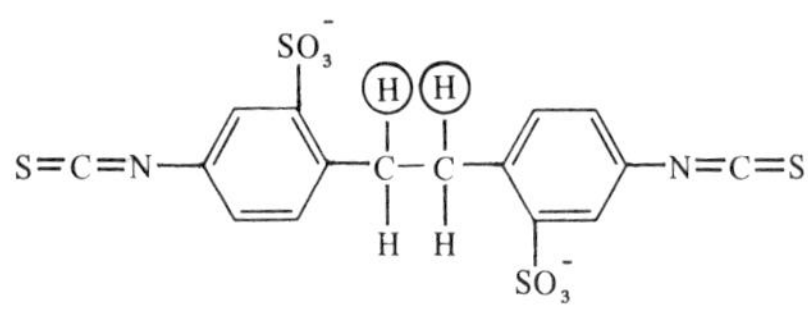 |
| (d) *para*-Chloromercuri-benzene sulphonate (PCMBS) | Slowly permeant, non-covalent | 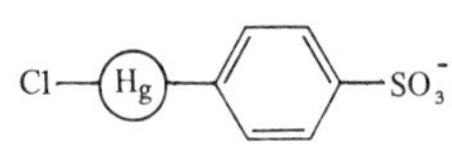 |
| (e) Diazobenzene sulphonate (DABS) | Impermeant, covalent | 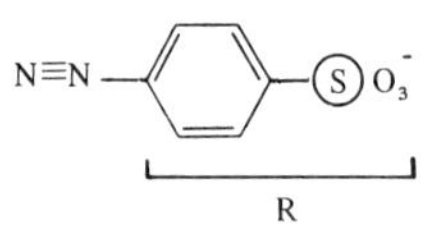 |
| (f) *N*-formyl-methionyl sulphone methyl phosphate ($^{35}$S-FMMP) | Impermeant(?) covalent | 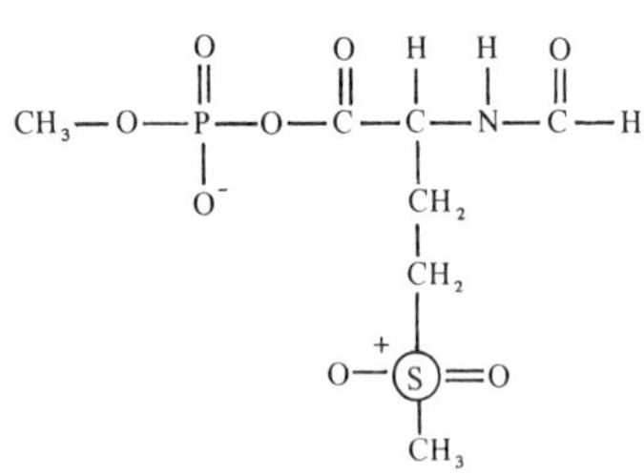 |

*A circled symbol denotes a radioactive isotope in the structure.
†M denotes membrane. These symbols are italicized to distinguish them from those of the chemical lablel.
‡Approximate molecular dimensions of the *incorporated* label, based on bond angles and lengths from Pauling [8].

| Reactive groups in membrane† | Product | Size‡ nm | Reaction conditions | References |
|---|---|---|---|---|
| *M–NH*$_2$ | – | 2.0 x 1.2 | 20°C, 10 min, PBS(pH 7.4), 0.1 mM SITS | 7 and 11, 12 13 |
| *M–NH*$_2$ | 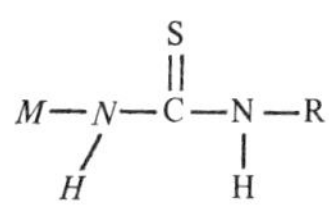 | 2.0 x 1.2 | 4°C, 20 min, PBS (pH 7.4), 0.1–1.0 mM DIDS | 13 |
| *M–NH*$_2$ | same as b. | 2.0 x 1.2 | 4°C, 15 to 30 min in the dark, PBS(pH 7.4), 0.1–0.5$\mu$M [$^3$H]–DIDS | 14, 16, 17 |
| *M–SH* | 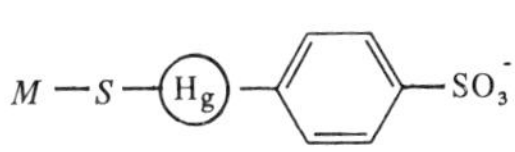 | 1.2 x 0.6 | 24°C, 0 to 180 min, saline (pH 6.8–7), 0.1mM PCMBS | 18, 21 |
| *M–NH*$_2$ | 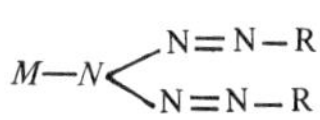 | | | |
| *M-tyrosine* | 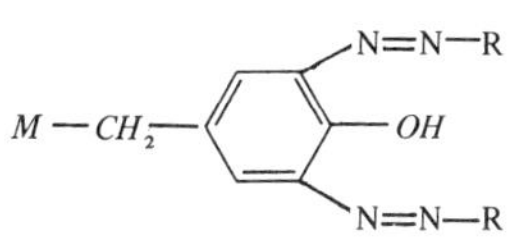 | 1.1 x 0.6 | 20°C, 20 min, isotonic $PO_4$ (pH 7.5), 3mM DABS | 26, 28, 29 |
| *M-histidine* | 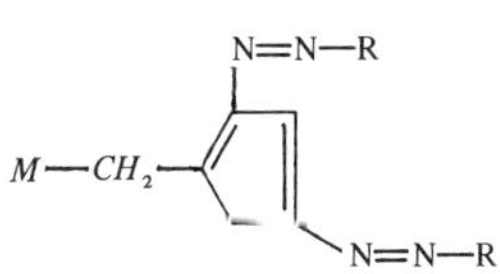 | | | |
| *M–NH*$_2$ | 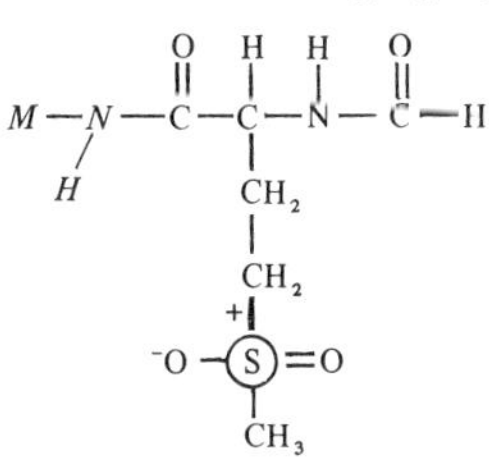 | 0.6 x 0.8 | 23°C, 10 min, NaCl–$NaHCO_3$ (pH 10), ~30$\mu$M FMMP | 30 |

TABLE 12.1 (*Continued*)

| Reagent | Class | Structure* |
|---|---|---|
| (g) Pyridoxal phosphate plus $NaB^3H_4$ | Impermeant(?), covalent | 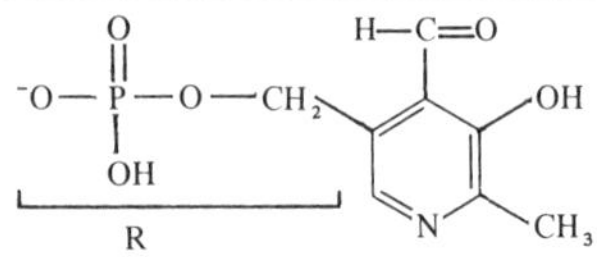 |
| (h) *N*-(4-nitreno-2-nitrophenyl)-2-aminoethyl sulphonate (NAP-taurine) | Impermeant (at 0°C), covalent | 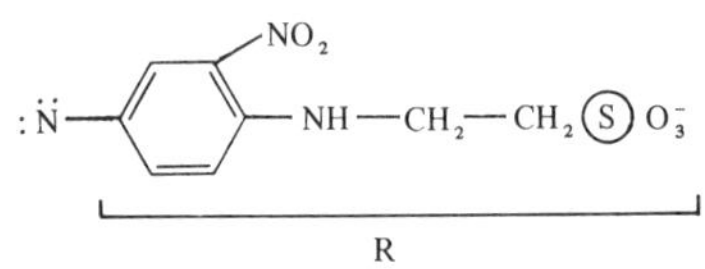 |
| (i) 1-Dimethylnaphthalene 5-sulphonyl chloride (dansyl chloride) | Slowly permeant, covalent | 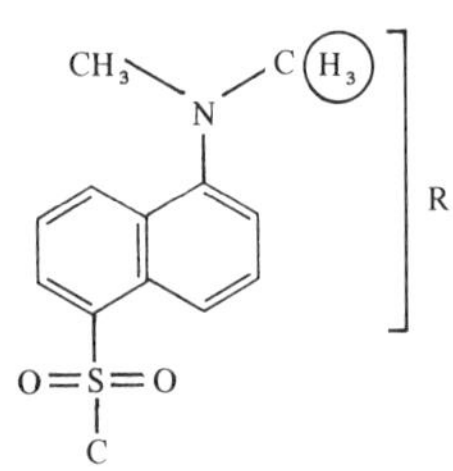 |
| (j) Isethionyl acetamidate (IAI) | Impermeant, covalent | 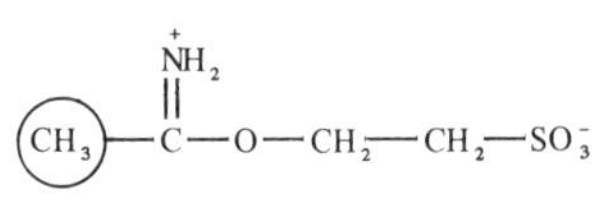 |
| (k) Ethyl acetamidate (EAI) | Permeant, covalent | 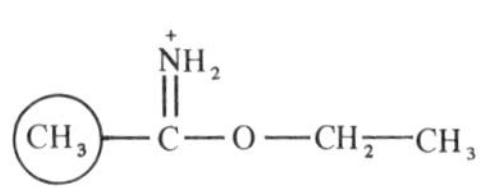 |
| (l) Acetic anhydride | Permeant, covalent | 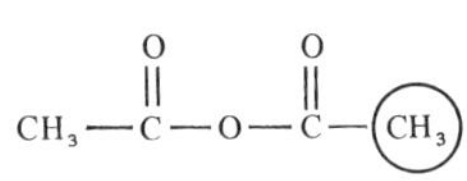 |
| (m) 1-Fluoro-2, 4-dinitro benzene (FDNB) | Permeant, covalent | 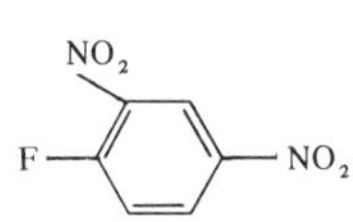 |
| (n) N-Ethylmaleimide (NEM) | Permeant, covalent | 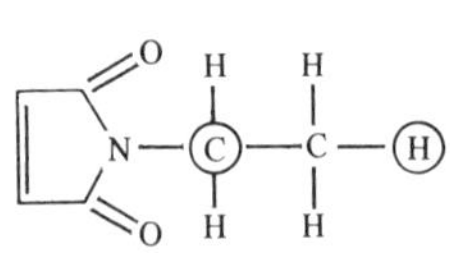 |
| (o) 2, 4, 6-Trinitrobenzene sulphonate (TNBS) | Permeant (at 37°C), covalent | 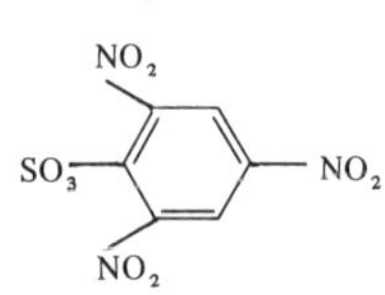 |

| Reactive groups in membrane† | Product | Size‡ nm | Reaction conditions | References |
|---|---|---|---|---|
| *M–NH*$_2$ | 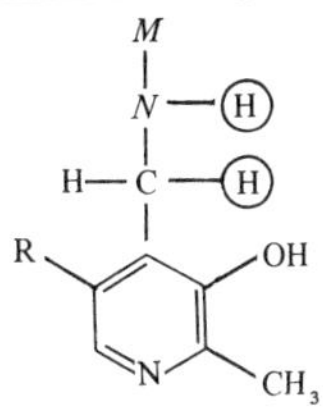 | 1.2 x 0.6 | 37°C, 30 min, Hepes (pH 7.5) | 36–40 |
| *M–C–H* | 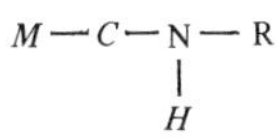 | 1.4 x 0.7 | 0°C, 20 min, PBS (pH 7.4), 0.1–0.5 mM NAP-taurine | 41, 43 |
| *M–NH*$_2$ | 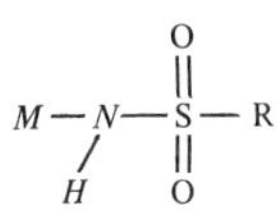 | 0.7 x 1.2 | 4°C, 180 min, sucrose-$PO_4$ (pH 9.0) | 45 |
| *M–NH*$_2$ | 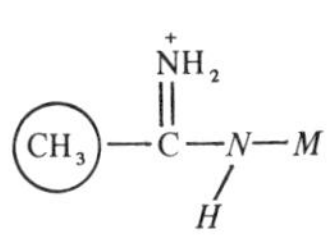 | 0.4 x 0.4 | 23°C, 30 to 120 min, "total buffer" (pH 8.0), ∿ 5 mM IAI | 35 |
| *M–NH*$_2$ | | 0.4 x 0.4 | same as (j) | 35 |
| *M–NH*$_2$ | 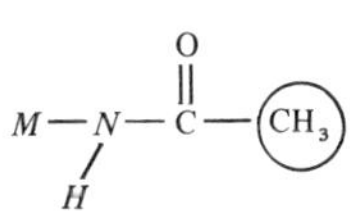 | 0.4 x 0.4 | 23°C, 15 min, isotonic $PO_4$ (pH 7.4), 10 mM acetic anhydride | 51 |
| *M–tyrosine*<br>*M–NH*$_2$<br>*M SH* | 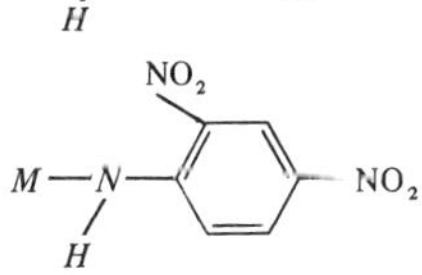 | 1.0 x 0.7 | 23°C, 60 to 180 min, NaCl-$NaHCO_3$ (pH 8,6), 1–5 mM FDNB | 52, 53, 60–63 |
| *M–SH* | 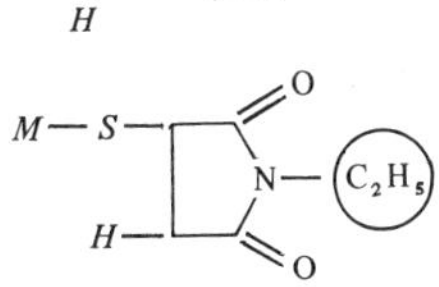 | 0.9 x 0.7 | 37°C, 30 to 45 min, pH 7.4, 0.1 mM NEM | 57 |
| *M–NH*$_2$ | 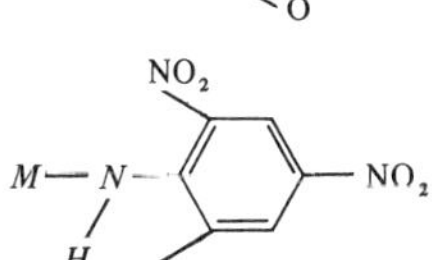 | 1.0 x 0.9 | 23–37°C, to 24 hours, NaCl-$NaHCO_3$ (pH 8.6), 1.5 mM TNBS | 70, 67–69 |

TABLE 12.1 *(Continued)*

| Reagent | Class | Structure* |
|---|---|---|
| (p) Sodium periodate plus $NaB^3H_4$ | Permeant, covalent | $NaIO_4$ as an oxidant |

reaction and constitutes the most stable covalent bond. Since the p$K_a$ range of the ϵ-amine of lysine in proteins is 9.4–10.6 and of most α-amines 7.5–8.5, many investigators label at elevated pH's, such as 8–10, to increase the number of reactive amines. Reactions with sulphydryls, imidazoles or tyrosine hydroxyls may also occur, but the bonds formed are less stable and may hydrolyse, even at pH 7.4 [9].

The mammalian erythrocyte has been the cell chosen for most studies utilizing chemical surface labels to probe membrane structure. Cells are available in large quantities, the isolation of plasma membrane is uncomplicated by intracellular organelles or membranes, and the release of haemoglobin, an intracellular protein present in high concentrations, allows a quick and accurate measure of cell damage. In addition the reagent's penetration of the membrane can be assessed by measuring the extent to which haemoglobin is labelled when intact erythrocytes are exposed to the molecule. An ideal non-permeant reagent is one that reacts *exclusively* with groups at the external surface, with no intracellular labelling whatsoever. In practice this has not been achieved and the reason is not clear. The assumption has been that the labelling reagent is permeant to all cells in the population. There are other explanations for the occurrence of labelled intracellular molecules in a preparation: (1) the presence of a small sub-population of cells which is very permeable to the label; or (2) adsorption of intracellular molecules from a small number of lysed cells to intact cells. Until further investigation establishes the exact cause of internal labelling in each system, a reasonable criterion of non-penetration would be that greater than 95% of the incorporated label is associated with the membrane. In the case of the erythrocyte this means that the specific activity (cpm $mg^{-1}$ protein) of the ghost, which constitutes approximately 1% of the cellular protein [10], should be 2000 times greater than the specific activity of haemoglobin, or

| Reactive groups in membrane† | Product | Size‡ nm | Reaction conditions | References |
|---|---|---|---|---|
| | 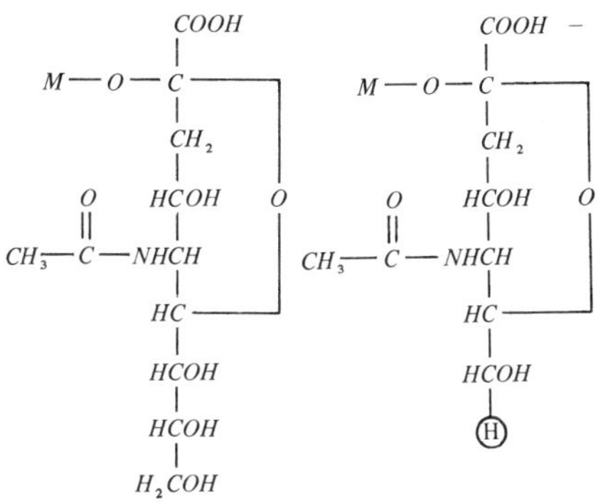 | – | 23°C, 10 min, isotonic $PO_4$ (pH 7.4), 2mM periodate | 72–74 |

conversely, haemoglobin specific activity 0.05% that of the ghost. (If 95% of the label and 1% of the cellular protein is associated with the ghost, and 5% of the label and 99% of the protein with haemoglobin, then the relative specific activity of ghosts over haemoglobin is $95/1 \div 5/99 = 2000$. The relative specific activity of haemoglobin ghosts is $1/2000 \times 100 = 0.05\%$.) As we review the studies using small chemical labels, this criterion of impermeance should be kept in mind.

In 1964 Maddy [7] designed a fluorescent compound, 4-acetamido-4′-isothiocyano-2,2′-stilbene disulphonate (SITS, Table 12.1, a) which appeared to satisfy the requirements of a surface label. A maximum of $5 \times 10^5$ molecules per ox erythrocyte were bound within five minutes at a 20-fold excess free SITS concentration. Haemoglobin was unlabelled under these conditions, lipid labelling was negligible and no cell lysis was observed. Similar results were reported using human erythrocytes ($4 \times 10^6$ SITS bound per cell). In addition, anion permeability was found to be completely inhibited by SITS [11]. SITS was thought to react covalently with amino, imidazole or guanidyl groups, but the nature of the linkage was not investigated in either study and the binding was measured indirectly by the reduction of SITS fluorescence in the medium. In a study on the plasma membrane of isolated rat liver cells [12], partial inhibition of SITS binding by sulphydryl reagents suggested reactive SH groups were involved, but again, there was no chemical analysis of the amino acids labelled. Although fluorescence is a sensitive method of detection, direct measurements are difficult due to possible autofluorescence and/or quenching by endogenous proteins in the sample. Extreme pH shifts also affect fluorescence.

Utilizing ($^{35}S$)-SITS, which made the direct measurement of label bound to human red cell membranes possible, Cabantchik and Rothstein [13] found that the SITS-membrane interaction was largely non-covalent and reversible. 85–90%

of SITS could be removed, and anion permeability restored, if a suitable acceptor, such as albumin, was in the rinse medium. This finding emphasizes the danger of using an operational characterization of a surface label's interaction with the membrane (such as rinsing cells till no additional label is removed). Chemical analysis of the label-membrane linkage is necessary for the adequate characterization of any labelling system. In addition, the use of extracellular proteins as scavengers to remove non-covalently-bound molecules simplifies subsequent fractionation and analysis of the label.

Extending the initial studies of SITS [11], Rothstein and colleagues designed a number of sulphonated stilbene compounds with high affinity and specificity for the substrate binding site of the anion transport system in the erythrocyte [13,14]. Although anion fluxes were inhibited by these reagents when they were added to intact cells, several other membrane functions were *not* perturbed [11]. Consequently, these reagents were used to probe the environment of the permeability site [13], to quantitate the number of these sites and to identify the membrane components involved [14]. Two impermeant, covalent labels, 4,4′-diisothiocyano stilbene disulphonate (DIDS, Table 12.1, b) and a radioactively-labelled derivative (($^3$H)DIDS, Table 12.1, c), were obtained by replacing the acetamido on SITS with a second isothiocyano group. When intact erythrocytes were exposed to ($^3$H)DIDS, primarily one class of membrane polypeptide was labelled (Band III in the nomenclature of Fairbanks *et al.* [15]). The assumption that the transport function also resided in this macromolecule was supported by indirect evidence in recent studies [16]. However, the number of stilbene disulphonate molecules bound per cell, hence the number of anion transport sites, has been variously reported at $1.8 \times 10^6$ [11] and $4 \times 10^6$ [13] in early studies, to $3 \times 10^5$ [14] and $1.2 \times 10^5$ [16] recently, and $3 \times 10^5$ after proteolytic treatment of intact human erythrocytes [16]. The basis for concluding that the interaction of ($^3$H)DIDS with the membrane was covalent was the stability of the label-protein complex in the presence of albumin and sodium dodecyl sulphate (SDS). Although direct proof was not presented, the covalent linkage was assumed to involve an amine, because ($^3$H)DIDS binding increased with pH and earlier studies indicated that membrane sulphydryl groups (which could react with isothiocyanate) were not involved in anion permeability [11]. Presumably the second isothiocyanate in DIDS increases the rate of the covalent reaction with a reactive amino group, since SITS already contains one reactive isothiocyanate but its binding is converted from reversible to irreversible very slowly (over 90 min) [13]. Although DIDS is bifunctional, it did not produce cross-linking artifacts, at least when intact cells were labelled. Interestingly, ($^3$H)DIDS reacted extensively with many membrane proteins and lipids of leaky ghosts, thus acting as a general membrane probe under certain conditions [14]. However, its specificity for one class of polypeptides at the external cell surface, where other proteins have been reported to be exposed, limits its usefulness as a general cell surface probe and cautions against assigning protein asymmetry in membranes based solely on the use of one label. Conversely, the successful use of these sulphonated stilbene molecules as site-specific labels (Section 12.3.1) should emphasize the importance and potential value in assessing the effect of a cell surface label on membrane function. Others have labelled tissue culture cells with ($^3$H)DIDS and found no adverse effect on viability but have not measured anion transport [17].

Several organic mercurials have been utilized to probe the role of sulphydryl groups in the membrane [18]. A presumed non-permeant reagent, *p*-chloromercuribenzene sulphonate (PCMBS, Table 12.1, d) inhibited glucose transport in intact erythrocytes, suggesting sulphhydryl involvement in this function. Approximately 1.4 x $10^6$ molecules were bound per cell, only 1.5% of the total SH groups in the ghost membrane. (The total number of sulphydryls in the human erythrocyte membrane, as measured by $HgCl_2$ titration, is 1 x $10^8$ [19], which is less than 5% of the total cellular sulphydryl groups [20]). A more detailed study [21] of the kinetics of PCMBS binding and effects showed that inhibition of glucose transport was related to an initial rapid phase of PCMBS binding to the cell surface but that PCMBS slowly penetrated the membrane (over 5 h a maximum of ~7 x $10^6$ molecules were bound per cell) with accompanying cation leakage. The surface-bound molecules were readily desorbed and glucose transport restored by the addition of small amounts of protein. By maintaining low PCMBS concentrations, using short incubations and low temperatures, one could presumably use this reagent to probe the surface disposition of different membrane functions. This has been done in bacteria [22] and eucaryotes [23]. However, the unstable interaction of PCMBS with surface SH groups certainly precludes its general use as a covalent membrane marker. In addition, PCMBS has been reported to extract a large fraction of the protein from erythrocyte ghosts [24,25], but intact erythrocytes were not studied.

Berg synthesized a small diazonium compound, ($^{35}$S)diazobenzene sulphonate (DABS, Table 12.1, e) for use as a probe of the human erythrocyte membrane [26]. At the level of reagent used, 3 mM, approximately 7 x $10^6$ molecules were bound per cell. (Using 2.5 mM DABS, Berg achieved 3.5% incorporation, or 0.0875 mM (87.5 nmol ml$^{-1}$). Assuming there were 7 x $10^9$ red blood cells ml$^{-1}$, then ~7 x $10^6$ DABS molecules were bound per cell.) Diazonium salts react with amino, imidazole and tyrosyl residues (Table 12.1, e), but the exact nature of DABS linkage to the membrane was not investigated extensively in this study. 20% of the bound DABS was soluble in organic solvents and 50% of this was presumed to be labelled phospholipid, but very little label had an electrophoretic mobility in SDS gels corresponding to that of lipids (just behind the dye front [27]). Haemoglobin was labelled to less than 0.2% of the level it was labelled in broken cells, suggesting that the membrane of intact erythrocyte was excluding extracellular DABS from reacting with intracellular haemoglobin. However, this is much higher than the 0.05% value which represents 5% internal labelling (as calculated above) and indicates

~17–20% reagent penetration (assuming the specific radioactivity of haemoglobin from broken labelled cells is the same as that of membrane proteins from intact labelled cells). A more accurate indication of the extent of DABS penetration would have been the percentage distribution of incorporated DABS in both haemoglobin and ghosts isolated from labelled intact cells, but this was not reported. In addition to the apparent permeant nature of this reagent a number of other difficulties has precluded its general use: DABS-modified (but not saturated) cells eventually lysed due to the rapid loss of $K^+$ [26]; facilitated glucose diffusion was inhibited at 1,3 and 10 mM DABS, as was acetylcholinesterase activity [28]. Furthermore, DABS was reported to degrade proteins in an undefined manner [28]. Others reduced the concentration of DABS to 0.3 mM but did not present data on cell stability [29]. Information on membrane protein dispostion obtained with this reagent should be viewed with caution considering the geneal deleterious effects it has on several membrane functions and thus, possibly, on membrane structure.

An amino-acylating reagent, ($^{35}$S)-formyl-methionyl sulphone methyl phosphate (FMMP, Table 12.1, f) was synthesized at very high specific activity (10–20 Ci mmol$^{-1}$) and used to study the transmembrane disposition of red cell proteins [30–32]. The labelling reaction was conducted at pH 10, presumably to optimize the number of accessible unprotonated amines, but tyrosine residues could also be acylated under these conditions [9]. The nature of the linkage with membrane constituents was not investigated. While only ~$1.8 \times 10^4$ FMMP molecules were incorporated per cell out of a maximum of ~$2.3 \times 10^8$ protein and lipid amino groups per ghost, 16% of that label was found in haemoglobin, suggesting membrane permeability to the label. (The total number of amino groups in the human erythrocyte membrane ($2.4 \times 10^8$) is the sum of protein $\alpha$-amino residues ($2.4 \times 10^6$) plus protein $\epsilon$-amino residues ($1.4 \times 10^8$), as reported by Rosenberg and Guidotti [33], and the lipid amino groups ($9 \times 10^7$) as reported by Reed *et al.* [34].) Others have achieved substantial labelling of membrane amino groups between pH 7.4 and 8.0 [35]. Perhaps incubation at this more physiological range would reduce the apparent penetration of the chemical label into the cell.

Despite the questions raised about the impermeant nature of FMMP (or the effect of the labelling conditions on the cells), the chemical evidence presented in these studies (peptide maps of FMMP-labelled components) for the existence of two classes of transmembrane proteins has been largely substantiated by others (Section 12.6.1).

Rifkin *et al.* [36] labelled the external proteins (glycoproteins) of influenza virus by reducing the Schiff's base formed by pyridoxal phosphate and protein amino groups with tritiated borohydride (Table 12.1, g). The conditions for labelling were physiological. Virus infectivity and haemagglutinin activity were unaffected by pyridoxalation of an estimated 3% of the external viral protein. However, the limitations of this surface labelling technique were indicated by (1) the labelling of internal proteins as demonstrated by SDS gels, (2) a rather high background of label, ascribed to adsorption of serum proteins to viral membranes prior to labelling, and (3) 6% non-specific reduction of endogenous Schiff's bases in the absence of pyridoxal phosphate. Others have utilized this technique to probe the structure of isolated rat liver microsomes but found that pyridoxal phosphate penetrated the microsomal membrane quite easily [37]. The plasma membrane of tissue culture cells is evidently impermeable to it under certain incubation conditions [17,38,39]. Given long incubations (12 h) and high reagent concentration, pyridoxal phosphate was shown to permeate the intact erythrocyte, although at shorter times it was reportedly impermeant [40]. This finding emphasizes the importance of establishing the optimal conditions for label exclusion in each new system, especially if the primary goal is to probe the external cell surface.

Another approach to the design of impermeant membrane-specific labels was the synthesis of a photoactivatable reagent with low group specificity and high reactivity [41]. In this study an aryl azide (the photoactivatable segment, see [42]) was derivatized to a sulphonated amine (the hydrophilic segment). ($^{35}$S)N-(4-nitreno-2-nitrophenyl)-2-aminoethylsulphonate (NAP-taurine, Table 12.1, h), a highly reactive aryl nitrene photogenerated from the inactive aryl azide precursor, inserts into any C–H bond at close proximity to the nitrene. Such low specificity should maximize the probability of labelling all externally accessible membrane components without regard to the nature of the amino acid residues available (i.e., lysine, cysteine, tyrosine, etc.). The half-life of the reactive chemical species is approximately 1 ms. Calculations from the data presented in this study indicate approximately $3 \times 10^5$ molecules were incorporated per erythrocyte, certainly less than saturation. Using this small probe, at least nine distinct proteins and glycoproteins in the membrane were found to be accessible to the external environment. Approximately 84% of the label was found in bands III, PAS-1 and PAS-2, the proteins others have reported to be externally disposed using different surface probes. The increased number was apparently due to the more general nature of the label, its smaller size and high reactivity, although the possibility of proteolysis or leukocyte contamination was not considered. The extent of intracellular labelling was not dealt with in detail; rather, the absence of label in the two bands of spectrin (I and II [15]), a fibrous protein present on the inner face of the erythrocyte membrane [43], was taken as evidence of the impermeant character of the label. The temperature of the reaction (0°C) was reported to be important since at 37°C the reagent slowly leaked into the cell. Although the degree to which lipids were labelled was not discussed, a peak of radioactivity is present in the region of SDS gels where lipids migrate and might be expected, given the non-specific nature of the label. A more recent study using NAP-taurine reported that membrane polypeptides inaccessible in the intact cell could be labelled in resealed ghosts. The extent of membrane reorganization (i.e., exposure) was apparently dependent on the presence of residual haemoglobin and/or ionic constituents in the environment during ghost formation [44].

These initial studies indicate that photoactivatable molecules have potential as general probes of membrane structure. Nitrene and carbene derivatives are also being used as affinity labels of membrane proteins and will be discussed in more detail below (Section 12.3.1).

### *12.2.2 Non-penetration versus low reactivity*

In many of the studies employing impermeant chemical reagents as probes of erythrocyte membrane structure and 'sidedness', the general approach has been to expose intact cells to the radioactive reagent, remove any nonreacted label, prepare ghosts, dissolve them in SDS, electrophorese in SDS-polyacrylamide gels and analyse the gels for radioactivity and protein. The protein bands coinciding with the labelled peaks are then presumably the only proteins exposed at the outer surface. A control experiment involving 'leaky' ghosts is required to demonstrate that all proteins can be labelled when they are made accessible to the chemical reagent. Schmidt-Ullrich *et al.* [45] have questioned the validity of the conclusions derived from experiments designed in this manner. They contend that the processes of haemolysis and ghost isolation change the chemical reactivity of some of the membrane proteins. In addition, the removal of intracellular haemoglobin as a 'sink' for the chemical label will affect membrane reactivity. To test this hypothesis they prepared 1-($^3$H)-dimethyl amino naphthalene-5-sulphonyl chloride dispersed in lecithin-cholesterol micelles (dansyl chloride, Table 12.1, i). Dansyl chloride is a lipophilic (i.e., membrane 'permeant') molecule which is readily hydrolysed in aqueous medium [46] but apparently can be protected when in a micelle (whose permeant characteristics are not known). In this study, the dansyl chloride-micelles were incubated with intact human erythrocytes for up to 3 h at 4°C and pH 9.0 in iso-osmolar phosphate-buffered sucrose without cell lysis. The use of sucrose presumably reduced osmotic lysis or cell fragility in the presence of the micelles. 36% of the label incorporated was present in haemoglobin. This indicated that dansyl-chloride had penetrated the membrane. However, the specific activity of this haemoglobin label was 0.7% that of the membrane protein, indicating preferential labelling of the membrane components despite the permeant nature of the reagent. (Although these authors state 0.07% to be the relative specific activity of haemoglobin to ghost labelling (Table 3 of [45]), ghost protein was apparently only 0.12% of the total. Others [10] report 1.2% as more accurate, and so the relative specific activity is 0.7%.) SDS gel analysis of ghost protein demonstrated that dansyl chloride labelled mainly the same two proteins when added to intact cells as the putative non-permeant surface labels discussed above (Table 12.1, a–h). Dansyl chloride reacted with all membrane components when isolated ghosts were used. The interpretation of these results was that the extent of any labelling reaction (by a permeant or impermeant molecule) will depend upon the accessibility and number of the reactive membrane sites, the intrinsic reactivity of the groups involved, and the effective concentration of the label at the reactive sites. The differing locales of membrane proteins, at the external surface, in the hydrophobic interior and at the cytoplasmic face near the large 'sink' of haemoglobin, will definitely affect the extent of labelling by even a permeating reagent, as dansyl chloride is suggested to be. Therefore, these investigators stressed caution in any interpretation of membrane 'sidedness' based on differential labelling of intact cells and isolated ghosts by small reagents.

Controversy [47,48] has arisen over interpretation of the results of Schmidt-Ullrich *et al.* [45]. Two questions are involved: (1) do membrane components reorganize during membrane isolation; (2) are the putative impermeant reagents actually impermeant? The dansyl chloride study addresses only the second question. Using a 'permeant' reagent, these investigators obtained results similar to those using 'non-permeant' reagents. Unfortunately, dansyl chloride is *not* freely permeant but rather slowly penetrating and thus the conclusions derived from its use are not clear-cut. However, the fact that intracellular haemoglobin *was* labelled when FMMP [30] and DABS [26] were exposed to intact erythrocytes does suggest membrane penetration by these molecules. The low specific activity (cpm label $mg^{-1}$ protein) of haemoglobin relative to that of ghosts [30] or to that of haemoglobin from lysed cells [26] which was reported in these studies (and discussed as criterion above, Section 12.2.1) indicates a low degree of penetration and a concentration of label in the membrane, but is somewhat misleading since it implies label exclusion. The lack of (or low) labelling of cytoplasmically oriented membrane proteins, such as spectrin, observed with FMMP or DABS could be the result of the low concentration of these proteins relative to that of intracellular haemoglobin. In isolated ghosts, their relative concentration is increased and thus the extent of their modification increased. Altered reactivity does *not* imply membrane constituents have become reorganized (i.e. exposure of sites previously unavailable even to a permeant label). This question will be discussed below (Section 12.2.3).

The experiments necessary to resolve the question of reagent permeation and low reactivity of membrane proteins on the cytoplasmic surface are: (1) a more rigorous assessment (e.g., purification) of the extent to which intracellular haemoglobin is *covalently* labelled with dansyl chloride, because these data are the basis for defining dansyl chloride as permeant (this was done, R. Schmidt-Ullrich, personal communication); (2) comparison of the specific activities of such cytoplasmically-oriented membrane proteins as spectrin [43] or glyceraldehyde-3-phosphate dehydrogenase (G3PD [49]) to that of haemoglobin isolated from cells labelled with FMMP (or DABS) and dansyl chloride. Since none of these membrane reagents is freely permeant, the extent to which an inner surface protein is labelled will certainly be less than that of the external surface proteins but should be comparable to haemoglobin. Spectrin and G3PD can be specifically eluted from the membrane [50] and their specific radioactivity measured directly, rather than relying on SDS gel analysis of total ghost material to assess their labelling, as was done by Schmidt-Ullrich *et al.* [45]. Polypeptides labelled to only 0.7% the relative specific activity of externally-disposed polypeptides would certainly be missed in the latter type of analysis.

### *12.2.3 Permeant labels with high reactivity*

The questions of erythrocyte membrane reorganization and altered protein reactivity after ghost isolation have been investigated using two chemicals of freely permeant nature. Labelling of intact cells with ($^3$H)-acetic anhydride (Table 12.1,1), then again with ($^{14}$C)-acetic anhydride after ghost isolation and SDS solubilization, revealed that the same proteins were labelled to the same extent by both procedures [51]. Evidently the presence of haemoglobin did not alter the *relative* reactivity of cytoplasmic surface proteins before and after ghost isolation. The disadvantage of acetic anhydride is that saturation labelling cannot be achieved without cell lysis so only *relative* differences in exposure of new sites on different membrane constituents can be assessed. However, lipid labelling was enhanced after ghost isolation, indicating some degree of sensitivity to the technique. Tissue culture cells have also been double-labelled with acetic anhydride in an effort to study membrane protein turnover [39].

The question of erythrocyte membrane reorganization during isolation was also investigated using another permeant chemical reagent, ethyl acetamidate (EAI, Table 12.1, k), together with a homologous imidoester, isethionyl acetamide (IAI, Table 12.1, j), which is non-permeant [35]. Both reagents convert primary amines to amidines without alteration of charge and without a significant increment in the size of the modified amine. In this study, both reagents exhibited saturation binding of cells or ghosts, only 5% more label being incorporated in a second amidination of the same sample. EAI was demonstrated to be freely permeant, because it labelled haemoglobin in intact cells to the same specific activity as that in lysed cell preparations. No significant reorganization of membrane components (protein or lipid) as a result of ghost isolation was found using double-labelling techniques with EAI. That is, after intact cells were saturated with the permeant molecule, EAI, leaky ghosts were isolated and again labelled with ($^3$H)-EAI. Only 5% more label was incorporated. Exposure of previously unavailable sites (i.e., reorganization) could be achieved when the membrane was solubilized by SDS, Triton X-100 or Non-Idet P40.

Amidination with IAI appears advantageous for the chemical modification of intact cells in that it has little effect on acetylcholinesterase activity or glucose diffusion and no effect on $K^+$ levels in labelled red blood cells. (Acetylcholinesterase activity was reduced by 30% after cells were labelled with EAI). Unfortunately, the percent distribution of IAI between haemoglobin and ghost protein after amidination of intact cells was not presented and is necessary in order to critically evaluate the contention that this reagent is non-permanent. In intact cells IAI labelled predominantly the two protein classes (Band III and PAS-1 and 2) others had reported to be present on the external surface of human erythrocytes. When EAI was exposed to intact cells, every protein was labelled. IAI labelled lipids more extensively in ghosts than in intact cells as have several other amino-reactive reagents (e.g. FMMP [30]).

The use of these two imidoesters, especially the permeant molecule, EAI, has provided the most convincing evidence to date that proteins are not drastically reorganized during membrane isolation. In addition, since exposure of previously-unreactive amine-containing lipids (phosphatidylethanolamine or phosphatidylserine) was not observed with EAI, membrane lipids were apparently not reorganized. This finding is in disagreement with two studies where permeant nucleophile-reactive chemicals, 1-fluoro-2,4-dinitrobenzene (FDNB, Table 12.1, m, [52,53]) and acetic anhydride (discussed above, [51]) labelled a higher percentage of membrane lipids in ghosts than in intact erythrocytes. The discrepancy might be due to differences in the methods of ghost preparation (leading to possible loss of spectrin in the FDNB study) or the lack of saturation labelling of intact cells prior to ghost isolation (in the acetic anhydride study). Saturation is necessary in order to distinguish between altered reactivity and exposure of new sites. Using phospholipases, others have presented data suggesting phospholipids are accessible on the external surface of resealed erythrocyte ghosts but not of intact cells [54]. The altered disposition (i.e., from the internal to external membrane surface) of previously-reacted (or reactive) sites has not yet been investigated using IAI and EAI, but could perhaps explain the apparent discrepancy. Clearly these membrane labels have great potential as probes of membrane structure and 'sidedness' in the intact cell. Furthermore, IAI appears to satisfy many of the requirements of the ideal non-permeant surface label (outlined in the introduction of Section 12.2.1).

### 12.2.4 *Additional permeant molecule labels*

Insight into the structural relationships of the membrane can be gained by the judicious use of small, permeant, group-specific reagents, as was shown with EAI above. In addition, permeant molecules which perturb a specific membrane function can provide information about the amino acid residues required for that function. Almost every chemical modifier used by protein chemists to probe enzyme structure and function has also been used in the study of membranes and an exhaustive review of this field is not within our scope. The reader is directed to relevant reviews on protein modification [9,55].

In studies using mercurials such as *p*-chloromercuribenzoate, chloromerodrin 1-bromercuri-2-hydroxypropane and other sulphydryl reagents, N-ethylamaleimide, 2-mercaptoethylguanidine and reduced glutathione, the erythrocyte membrane was found to contain multiple classes of sulphydryl groups, the alteration of which had varying effects on cation permeability and glucose transport [11,18,56,57]. Red cell ghosts from four different species were labelled with ($^{14}$C)N-ethylmaleimide (NEM) and most proteins were found to have reactive sulphydryl groups [58]. Amino-reactive reagents were used to probe the potential role (since established with impermeant labels [13]) of positively-charged amino groups in ion selectivity in the erythrocyte membrane [59]. FDNB (discussed above as a probe of membrane lipid reorganization) inhibited both anion and cation permeability [11,53,60–62] but was of limited value in determining the number of membrane-amino sites affecting permeability. FDNB lacks specificity, reacting with sulphydryl, imidazole and phenolic hydroxyl functions as well as amino groups, and it is extremely lipid soluble, penetrating the cell to react with numerous intracellular sites [62,63]. 2, 4, 6-Trinitrobenzene sulphonate (TNBS, Table 12.1, o) was chosen as an alternative probe of ion permeability because it is quite specific for amino groups [64,65], forming an unstable bond with sulphydryls at physiological pH [66]. Unfortunately the impermeant character of TNBS has not been conclusively demonstrated. Conflicting results on the extent of TNBS' penetration of the erythrocyte membrane [11,67–69] have been clarified to some extent by a recent report which showed that the use of a chloride-bicarbonate rather than phosphate buffer, lower temperatures and lower pH during incubation of cells with TNBS reduced membrane penetration by the reagent [70]. These interesting results suggest that physiological labelling conditions may not be compatible with the goal of label exclusion, particularly when small anionic molecules, which could be transported to the interior via the anion transport system, are utilized as putative labels of the external cell surface. In order to obtain complete label exclusion, and thus a truly impermeant membrane reagent, the effect of variables such as temperature, pH, ionic conditions, etc., on label distribution in haemoglobin and ghosts should be studied systematically.

A chemical labelling technique for soluble glycoproteins [71] utilizing mild (2 mM) periodate oxidation at physiological pH followed by tritiated borohydride reduction (Table 12.1, p) has been applied to a study of the human erythrocyte membrane [72–74]. Only sialo-glycoproteins were labelled, indicating a possible use for the method in the identification of these components (and sialo-lipids) in isolated preparations of different intracellular membranes. Current chemical methods [75] are not as sensitive as a radiochemical assay or as specific for sialic acid residues [76]. However, there are several limitations to the use of the periodate-borohydride method as a cell surface-specific labelling system. First, both reagents are permeant and could react intracellularly. Second, although periodate preferentially oxidizes sialic acid [77], it can also oxidize other moieties [78]. The specificity for sialic acid can be established by measuring the molar ratio of periodate utilized to formaldehyde produced, which is two to one only for sialic acid oxidation [71]. This ratio has not been established in the membrane studies to date. Also, cellular function can be perturbed by periodate oxidation. Very low concentrations of periodate (0.4–2 mM) were found to stimulate blast transformation in lymphocytes from different sources [79–81]. The cellular response was shown indirectly to require oxidation of surface sialic acid groups by reduction of the periodate effect following exposure of cells to neuraminidase, which removed the potentially reactive membrane sites [82]. Borohydride reduction of oxidized cells reversed the periodate stimulation [82,83]. ($^3$H)-Borohydride, the tritium of which constitutes the actual label incorporated in the membrane labelling system of Blumenfeld *et al.* [70], can also cause problems. It can reduce not only the C7 aldehyde of oxidized sialic acid to an alcohol (Table 12.1,p) but also endogenous aldehydes and Schiff's bases. This would produce non-specific labelling as was found by Rifkin *et al.* [36] with pyridoxalation (Table 12.1, g). Use of unlabelled borohydride prior to oxidation could reduce this background. Borohydride can also promote β-elimination of seryl or threonyl-linked oligosaccharides [84] and change the nature of the glycoproteins being studied. Clearly, the use of low concentrations of each reagent and careful analysis of the two reactions to establish their specificity for sialic acid in each membrane type studied are required for the effective use of the method.

## 12.3 Site-specific labels

Covalent and non-covalent membrane labels which interact only with certain constituents in the cell surface, have been used to probe membrane structure and to identify the components of several membrane enzymes. Labels in this class are rarely neutral probes. That is, they often (intentionally) perturb membrane or cellular functions. However, insight into membrane structure-function relationships has been gained by this approach.

### *12.3.1 Covalent, active-site labels*

Several labelling systems have been published in which a covalent label was introduced into a specific membrane site, usually the active site of an enzyme or transport system. Active-site labelling techniques have been used extensively in protein chemistry [85] and provide perhaps the only approach to the isolation and purification of specific membrane receptors and enzymes when retention of biological activity is difficult or the amount of material very small. Conversely, a chief disadvantage is the irreversible inactivation by active-site labels. However, once the component is purified, other techniques (e.g., immunological) exist for the subsequent isolation of the native protein. A few examples are presented below and listed by classes in Table 12.2.

One class of active-site label involves the use of substrate analogues which protect the site of interest during exposure of the cells to an unlabelled group-specific reagent, followed by removal of the analogue and exposure of the site to the same group-specific reagent, now radioactively-labelled. Fox and Kennedy [86,87] found that a reactive sulphydryl group was involved in the *E.*

*coli* β-galactoside transport system and utilized a protective agent, thiodigalactoside, together with the group-specific reagent, $(^{14}C)$N-ethylmaleimide, to differentially label the specific membrane protein and effect its partial purification (Table 12.2,A,1). The irreversible serine esterase inhibitor, diisopropylfluorophosphate (Table 12.2,A,2), was used with a non-hydrolysable substrate analogue to differentially label and identify the acetylcholinesterase of the human erythrocyte membrane [88]. This enzyme constitutes less than 0.2% of the total ghost protein and would have been overlooked with a more general surface probe.

A second class of active-site labels extended to membrane biology has been that of specific affinity labels. In theory, the labelling reagent first interacts

TABLE 12.2
*Site-Specific labels*

| Class and examples | Structures |
|---|---|
| A. Group-specific labels with site-protective agents | |
| 1. N-ethylmaleimide + β-thiodigalactoside | 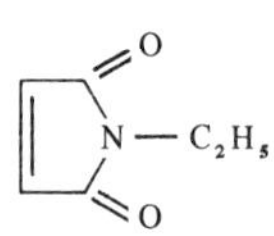 |
| 2. Diisopropylfluorophosphate + thioacetylcholine | 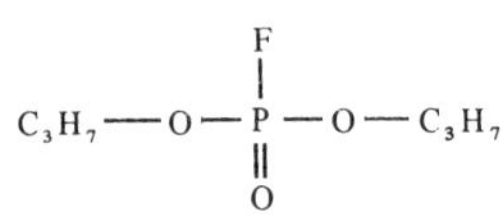 |
| B. Affinity labels | |
| 1. 3-haloacetate-strophanthidin | 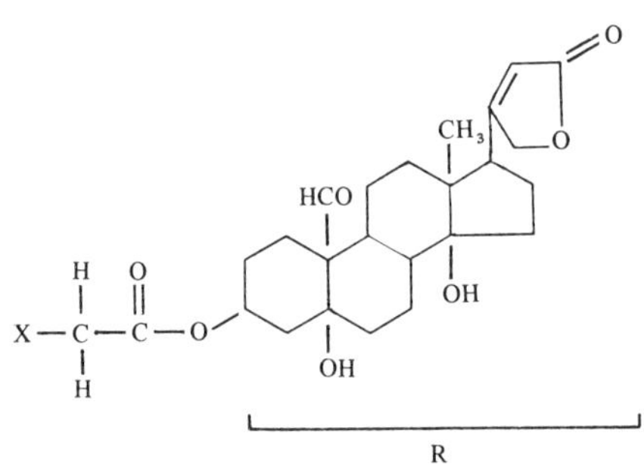 |
| 2. 4-(N-maleimido)-α-benzyltrimethyl ammonium | 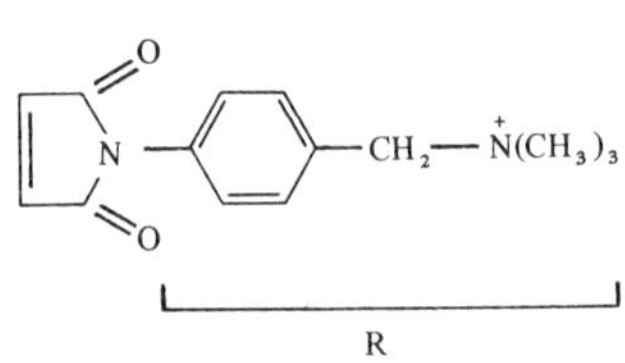 |

specifically and reversibly with the active site to which it is directed. By virtue of a reactive group on the labelling reagent, it then irreversibly and covalently reacts with an amino acid residue within or near the binding site and thus labels the protein of interest. Haloacetate derivatives of strophanthidin, a cardiotonic steroid [89], (Table 12.2,B,1), were originally reported to specifically alkylate the regulatory site of an isolated brain $Na^+$, $K^+$-stimulated adenosine triphosphatase (Na, K-ATPase) and irreversibly inactivate the enzyme. Later studies (cited in [90]) indicated that a rather high level of non-specific binding was occurring under the conditions used. Another alkylating agent with active-site directed properties, (methyl-$^3$H)-4-(N-maleimido)-benzyltrimethyl ammonium iodide (Table 12.2,B,2), was used to label the *Electrophorus electricus*

| Reactive site in Membrane | Modified site | Reference |
|---|---|---|
| –SH Groups near the β-galactoside binding site | 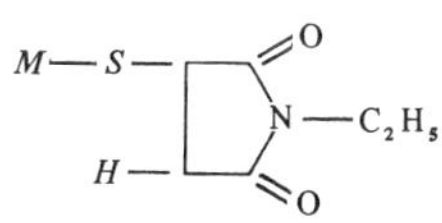 | 86, 87 |
| Reactive serine at the active site of acetylcholinesterase | 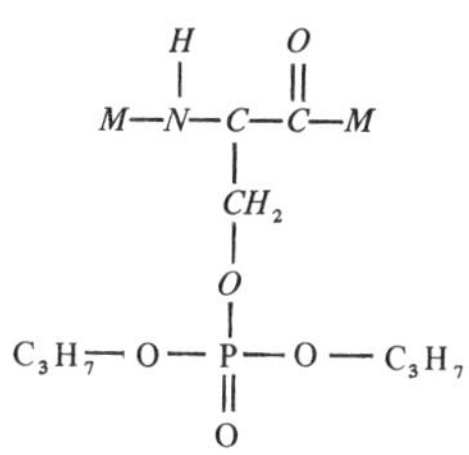 | 88 |
| –SH Groups near the cardiac glycoside binding site of Na, K–ATPase | 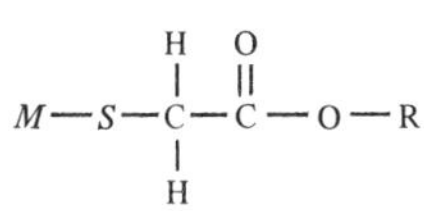 | 89 |
| –SH Groups near the binding site for acetylcholine | 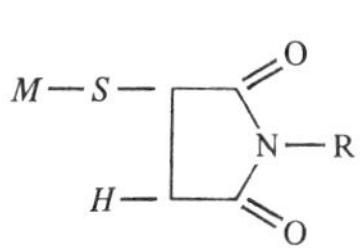 | 91–94 |

TABLE 12.2 (*Continued*)

| Class and examples | Structure |
|---|---|
| C. Photoaffinity labels | |
| 1. Carbenes | |
| a. 4-(ethyl-diazomalonyl) cyamarin | 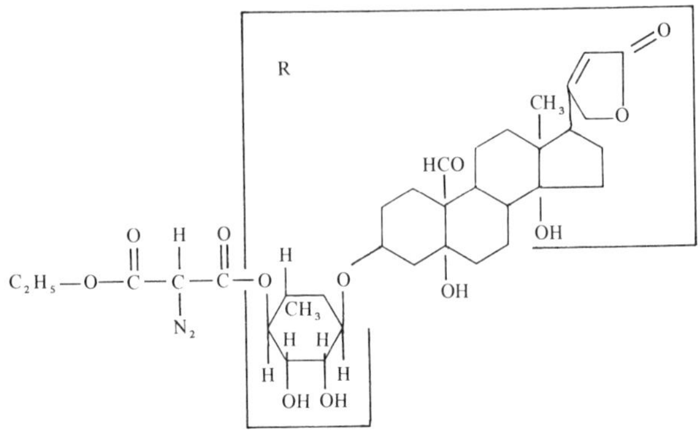 |
| b. N[6]-(ethyl-2-diazo-malonyl)-adenosine 3′-5′-cyclic monophosphate | 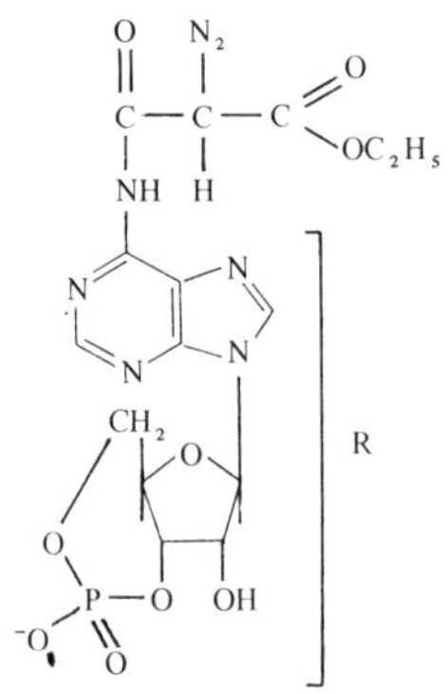 |
| 2. Nitrenes | |
| a. HK-83 | 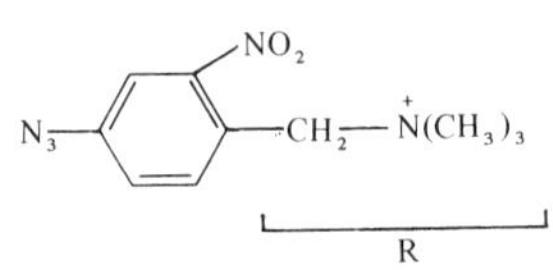 |
| b. 8-azido-adenosine triphosphate | 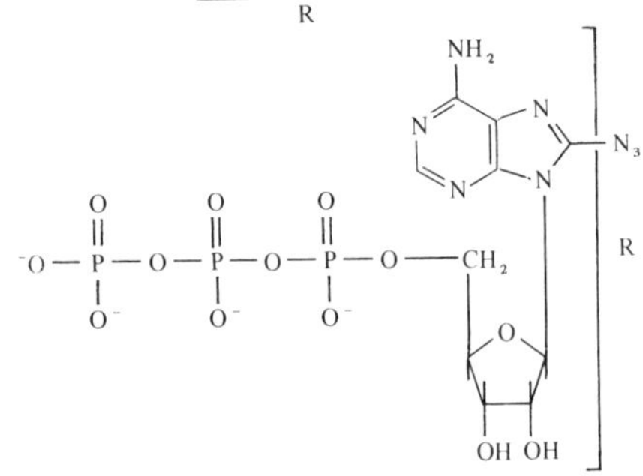 |

| Reactive site in membrane | Modified site | Reference |
|---|---|---|
| Any –CH groups in the cardiac glycoside binding site of Na, K–ATPase | 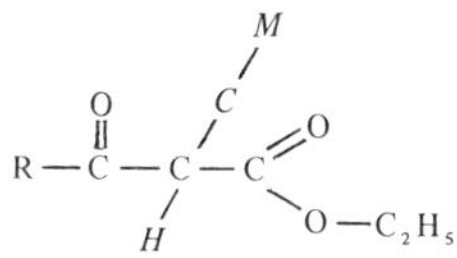 | 90 |
| Any –CH groups in cyclic AMP binding sites | 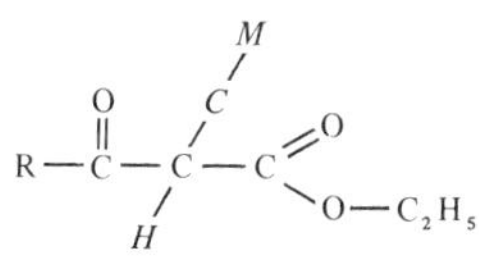 | 96 |
| Any –CH groups in the binding sites of acetyl-cholinesterase or acetyl-choline receptor | 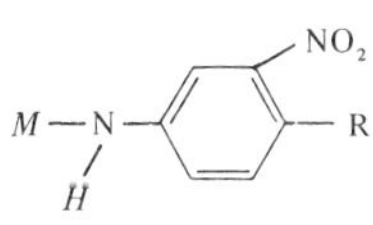 | 97, 98 |
| Any –CH groups in the ATP binding site |  | 99 |

acetylcholine receptor, in both isolated single cells [91,92] and subcellular membrane fragments [93,94]. Careful choice of conditions and concentrations of the affinity label, a non-specific alkylating agent (NEM) and dithiothreitol (to reduce a critical disulphide near the active site) was necessary in order to decrease non-specific sulphydryl labelling (discussed in [95]). The problems of non-specific labelling encountered in these studies indicate that affinity labels may *preferentially* interact with the active site to which they are directed but that their general reactivity (e.g., alkylating property) can be greater than their affinity for the protein of interest. These molecules then become quasi-general chemical labels with all the problems attendant to this class of membrane probes (discussed in Section 12.2.1–12.2.3). However, since protein purification is usually the goal of such studies, affinity labels and reaction conditions must be chosen which avoid non-specific labelling.

The stilbene disulphonates discussed in Section 12.2.1 were originally employed as general membrane probes but were found to be site-specific labels of the anion transport system in the erythrocyte membrane [13]. They have since been used to isolate and characterize this membrane component [14,16].

Affinity labels which contain photoactivable residues [42] have become increasingly popular as active-site membrane labels for the following reasons: (1) the parent compound is chemically unreactive, but active biologically so the parameters of the biological reaction can be measured (e.g., binding constants, inhibitory or stimulatory effects, reversibility) prior to photolysis and generation of the chemically-reactive species; (2) unbound, unreactive molecules can potentially be removed prior to photolysis, thus reducing the possibility of non-specific labelling; (3) the short half-life of some photogenerated species should also reduce the probability of non-specific labelling by free molecules; and finally, (4) the reactivity is directed to any C–H bond and thus does not require a specific reactive group (e.g. sulphydryl) in the active-site for covalent attachment of the label. This last characteristic of photoactivatable labels has prompted their introduction as general membrane probes ([41], Section 12.2.1).

Two classes of photoactivatable compounds have been used in membrane research: ethyl-2-diazomalonyl derivatives which generate carbenes upon photolysis (reaction **1** and Table 12.2,C,1) and azides (usually aryl) which generate nitrenes (reaction **2** and Table 12.2,C,2).

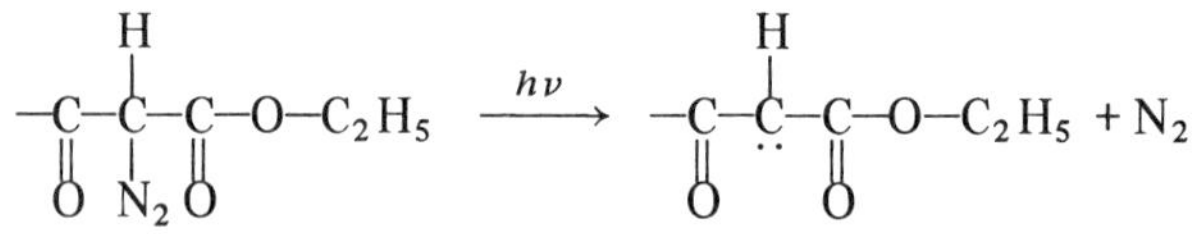


1

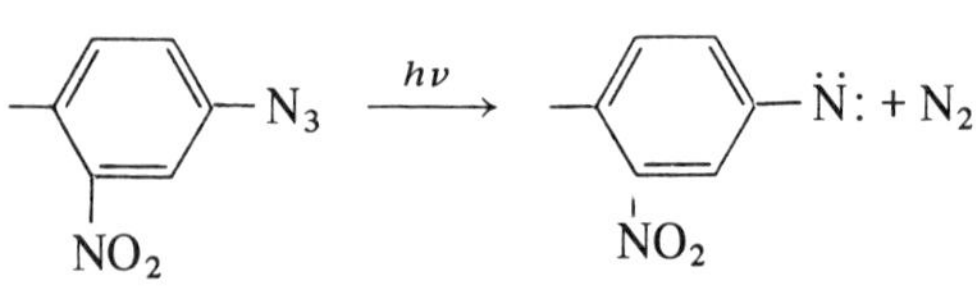


2

The advantages each class offers as a photoaffinity label, as well as the limitations, have been discussed by Knowles [42]. Briefly, carbenes have a much shorter lifetime than nitrenes ($10^{-6}$ s versus $10^{-3}$–$10^{-4}$ s), so non-specific labelling is less of a problem; but they also have less selectivity for C–H bonds and will react with water readily, thus reducing specific, active-site labelling. In addition, carbenes can undergo several rearrangements which generate chemically-reactive species. Both classes (with the exception of acyl nitrenes) can be photolytically generated at wavelengths long enough to avoid radiation damage to the system being labelled.

4-(Ethyl-diazomalonyl) cymarin, a photoactivatable derivative of a cardiac glycoside (Table 12.2,C,1-a), was used to label the ouabain-sensitive site on a 130 000 dalton subunit of Na, K-ATPase from canine renal medulla [90]. A second subunit of the enzyme was not significantly labelled. Another label of this class, $N^6$-(ethyl-2-diazomalonyl)-adenosine-3:5′ cyclic monophosphate (Table 12.2,C,1-b) was exposed to leaky human erythrocyte ghosts, photolysed, and upon analysis found to label predominantly a minor membrane protein at ~75 000 daltons with a rather high background of presumably non-specific labelling [96]. In both these studies, photolytic events evidently unrelated to formation of the carbene resulted in covalent attachment of label to the specific proteins. The mechanisms involved have not been elucidated. Photolytic cross-linking of proteins was also observed and presumably caused by free radical formation by the carbene [42]. This could lead to misinterpretation of results where more general photoaffinity probes are used.

Kiefer *et al.* [97] explored the feasibility of using nitrenes to photoaffinity label the erythrocyte membrane acetylcholinesterase. Aryl azide analogues of acetylcholine (Table 12.2, C, 2-a) were first bound reversibly to the active-site, with accompanying inhibition of the enzyme, and then the system was photolyzed. The aryl nitrene generated was presumed to react near the site of binding, because the enzyme was irreversibly inactivated. Later attempts to quantitate and characterize the membrane components using radiolabelled compounds revealed extensive non-specific labelling [98]. Scavenger molecules were added to the medium in 1000-fold molar excess of the nitrene in order to react with free nitrenes and reduce the non-specific labelling. However, the specific labelling of acetylcholinesterase was completely eliminated and the enzyme activity fully restored when scavengers of small or large size, having no affinity for the active-site, were included during photolysis. The interpretation that the dissociation rate of the enzyme-substrate complex was faster than the rate of aryl nitrene insertion into a C–H bond at or near the active-site, cautions against the use of long-lived nitrenes (which aryl nitrenes are) in systems where dissociation rates are known to be rapid. In addition, the use of scavengers which could disproportionately reduce the non-specific labelling by photoaffinity probes seems warranted.

Specific labelling of only three erythrocyte membrane polypeptides was obtained [99] using the photogenerated nitrene of ($\beta$, $\gamma$-$^{32}$P)-adenosine triphosphate (ATP, Table 12.2, C, 2-b). Both the Ca-ATPase and Na, K-ATPase were irreversibly inhibited after being photoaffinity labelled. The apparent molecular weights of two of the labelled species were similar to red cell membrane components known to be phosphorylated with ($\gamma$ $^{32}$P)ATP [100–103]. Leaky ghosts were used in all these studies, because the ATP-binding sites of the various red cell membrane ATPases are located on the inner surface [104].

Finally, three aryl azide derivatives of iodinated insulin have been prepared and found to be biologically active [104a], however photolysis and subsequent characterization of the labelled components have not yet been reported.

Photoactivatable reagents definitely have potential in membrane research as both specific and general probes of membrane structure, but the following problems must be resolved before the usefulness of these surface labels can be fully realized: (1) photolytic cross-linking; (2) non-specific labelling which may be a problem primarily with aryl nitrene derivatives due to the hydrophobicity of the benzyl group; (3) the uncertainty of covalently binding the actual receptor molecule or a very near neighbour. This last problem exists for affinity labels as well.

### *12.3.2 Lectins as cell surface labels*

Proteins, termed lectins [105], which agglutinate cells through surface carbohydrate determinants, have been used extensively to investigate membrane structure and function and to isolate specific membrane constituents. Rather than review the vast literature on lectins, which has been done recently by Sharon and Lis [106,107] and by Nicolson [108], we will discuss the properties of lectins in general and how these properties can be advantageous or disadvantageous in the use of lectins as cell surface labels.

Lectins are a class of site-specific labels, not directed to an active site but to specific oligosaccharide conformations. The use of a wide variety of lectins with different sugar specificities has facilitated the localization of different specific oligosaccharide residues on the cell surface and has contributed much to the current view of membrane structure [109]. Concanavalin A (Con A), a plant agglutinin, has been most commonly used, because it is commercially available in relatively pure form and its protein structure has been well-studied [108,110–115]. However, other lectins such as agglutinins from *Triticum vulgaris* (wheat germ, WGA, 116–118), *Glycine max* (soybean, SBA, 107, 119–121), *Phaseolus vulgaris* (red kidney bean, PHA, 122–125) and *Ricinus communis* (RCA, 126–128), have been purified and are being increasingly used, alone and in combinations, to probe the nature and relative localizations of their respective membrane binding sites. The large size of lectins, ranging from 23 000 to 200 000 daltons, certainly precludes their penetration through the membrane and thus restricts their binding exclusively to the external cell surface.

The lectin-ligand interaction is a non-covalent and reversible one, with a dissociation constant of $10^{-3}$–$10^{-4}$ M for simple sugars [106,108] and apparent dissociation constants of $10^{-6}$–$10^{-7}$ M for membrane binding sites [129], but these are likely to be heterogeneous. Provided conditions are carefully chosen, the non-covalent binding can be advantageous in the isolation of plasma membrane or specific membrane constituents (i.e., the 'receptors'), because at the end, the lectin can be dissociated from its binding site by competition with the appropriate simple sugar. However, noncovalent binding is disadvantageous should dissociation occur before desired, with possible binding of the free lectin to different, previously unavailable sites (in membrane isolations) or loss of low affinity membrane binding sites (in lectin receptor purification). This latter problem has often not been adequately appreciated nor quantitatively assessed in studies employing affinity chromatography ([130–132], but see [133]) chemical [134–137] or enzymatic methods ([138–140], but see [141]) to isolate and purify the putative membrane receptors of lectins.

Chang *et al.* [142] have recently reported the use of radiolabelled WGA complexed to its membrane receptor(s) as a marker in plasma membrane isolation. Their success presumably depended on the proper choice of lectin and on the use of lectin labelled to a high specific radioactivity (~500 Ci $mmol^{-1}$) in very low concentrations (~$10^{-9}$–$10^{-11}$ M) so that binding occurred principally to a few high affinity sites in the membrane (~$10^3$ per cell). The maintenance of low temperatures throughout the isolation was also critical in order to inhibit dissociation of the membrane ligand-lectin complex. This method was applied to both single cells and tissue, but unfortunately no assessment of the *in situ* localization of lectin was made. Despite this shortcoming, the use of lectins as membrane markers seems promising and certainly warrants further study.

By virtue of their large, polyvalent, protein character, lectins can be labelled in several ways for quantitative binding studies or for visualizaton at the ultrastructural level. Lectins radiolabelled with $^{125}$I [143,144] or ($^3$H)-acetic anhydride [145,146] have been used to quantitate the number of binding sites on normal versus transformed cells [143–146], on different subpopulations of lymphocytes [147,148] and a variety of other cell types [129,149]. $^{125}$I-labelled Con A and RCA were used to detect membrane changes in leukocytes following phagocytosis [150], in tissue culture cells during viral infection [151] and upon cell contact [152]. These are only a few examples where radiolabelled lectins have been used as cell surface probes ([108] for extensive review). Fluorescein-labelled Con A was used to probe oligosaccharide accessibility in a study on the disposition of rhodopsin (a glycoprotein) in retinal rod membranes [153]. The relative spatial distribution of different lectin receptors on the membranes of several cell types has been probed using radiolabelled lectins, competitive binding and agglutination inhibition assays [136,140,143,154]. The binding sites for Con A (specific for α-D-mannopyranosyl and α-D-glucopyranosyl groups) appear to be distinct oligosaccharide residues from those binding WGA (specific for β-*N*-acetyl-glucosaminyl groups), but it is not yet known whether these residues occur on the same or different membrane proteins.

The specificity of lectin binding in these biochemical studies has generally been assessed by simultaneous addition of the lectin and the appropriate saccharide inhibitor to the system. There should be no lectin bound to the cell surface under these conditions. Any lectin binding found is considered non-specific and is usually subtracted in quantitative studies. However, this non-specific binding can account for as much as 25% of the total [143], and thus, constitutes a major problem, particularly with a lectin such as WGA, which may bind to more than one saccharide [108]. Other sources of error in lectin binding measurements are: (1) the use of long incubations and high temperatures which could lead to endocytosis [155,156] and perhaps degradation of the bound, radiolabelled lectins [157]; (2) sugar-specific binding of lectins to components other than cell surfaces (i.e., material adsorbed to petri dishes [149]).

A very important use of lectins has been as probes of cell surface topography at the ultrastructural level. Lectins (and by extension, their membrane binding sites) have been visualized by noncovalent binding of the glycoproteins, haemocyanin [158] and horseradish peroxidase [159], to unoccupied ligand binding sites on polyvalent lectins which were already bound to cell surface glycoproteins. The distinctive shape and large size of the haemocyanin oligomer (~50 nm on the diagonal) allows its visualization using surface replica staining techniques. The topographical distribution of Con A-binding sites on erythrocyte, granulocyte and tissue culture cell surfaces has been studied using this method [158,160]. The presence of bound horseradish peroxidase is revealed by the diaminobenzidine reaction [161]. This localization method has been used to follow membrane movement in locomoting fibroblasts [162] and to map topographical differences in normal and transformed cell membranes [159,163,164]. The disadvantages of these two tracer methods are that they require numerous manipulations, that the tracer is not covalently attached to the lectin so release and non-specific reattachment could occur, and finally, that a divalent lectin already bound to two sites on the cell surface would not be detected since it has no unoccupied site for interaction with the tracer.

These problems have been avoided by the use of the electron-dense marker, ferritin, which has been covalently derivatized to numerous lectins for more direct visualization of surface receptors. Both freeze-etch [165–167] and conventional thin-section [168–170] techniques have been used to visualize lectin-ferritin complexes in studies probing the surface topography of numerous cell types. Nicolson and Singer introduced another method of specimen preparation which allows direct observation of the two dimensional topographical distribution of ferritin-conjugated lectins on the inner or outer membrane surface [171]. The technique involves cell lysis at an air-water interface and has proven applicable to nucleated cells [172–174] as well as to erythrocytes [171].

Some problems encountered in the use of lectins derivatized to tracers such as ferritin or haemocyanin have been: (1) non-specific binding of the tracer to the cell surface [171] or surfaces other than the cell [158] which apparently can be reduced if a protein such as albumin is included in the medium; (2) cross-linking and topographical redistribution of binding sites induced by the lectin (discussed below); and (3) limitation of the method to 10–50 nm resolution due to the large size of the tracers (ferritin at 10 nm and haemocyanin at 50 nm). In addition many investigators have not characterized the stoichiometry of the lectin-ferritin conjugate in order to establish whether one ferritin molecule might represent binding to one or more membrane sites, nor have they correlated lectin-tracer binding at the ultrastructural level with quantitative biochemical measurements in order to establish whether *all* membrane sites are being visualized.

Although lectins have proven to be useful probes of cell surface architecture and changes, it is important to recognize that they are not neutral membrane labels. The interaction of some lectins, especially Con A, with the plasma membrane of certain cells induces not only drastic cellular responses such as lymphocyte blast transformation ([175–178] and [108] for extensive reference list), cell-mediated cytoxicity [179], platelet release reaction [129], inhibition of phagocytosis by granulocytes [180], insulin-like stimulation of fat cells [181,182], and rapid internalization of membrane receptor-lectin complexes by macrophages with subsequent inhibition of phagolysosome formation [183,184], but also inhibition of membrane enzymes [185] and membrane transport systems [186,187]. A recent report of more subtle lectin-induced perturbations, involving clustering of membrane binding sites on unfixed cells at 37°C [160], has necessitated re-evaluation of the results obtained in several studies on the topographical distribution of lectin-binding sites. Many of the biological effects of lectin binding are hypothesized to be related to the capacity of these polyvalent molecules to cross-link membrane constituents ([108], but see [188], where succinylated Con A was shown to be monovalent, but still induced blastogenesis in lymphocytes). Certainly these lectin-induced cellular perturbations are of interest in themselves and constitute a major use for lectins in the study of the role of the membrane and its externally-disposed glycoproteins in transmission of environmental signals to the interior. However, when lectins are used strictly as probes of membrane architecture, their possible perturbing effect on it must be considered.

## 12.4 Enzyme-mediated labelling systems

### *12.4.1 Exogenous and endogenous enzymes*

Due to the possible problems of penetration by chemical labelling reagents and of membrane perturbation by lectins, a number of investigators have utilized exogenously-added enzymes, or even endogenous cell surface enzymes to catalyse the labelling of specific groups in proteins and lipids in the plasma membrane. The large size of an enzyme precludes its penetration through the membrane and limits its reaction to the exterior. However, the label actually incorporated (covalent) is usually small, in some cases merely a tritium atom, so protein conformation should not be perturbed drastically. The major disadvantage of enzyme-mediated labelling derives from the large size of the catalyst: only those reactive amino acid residues accessible to a 30–40 nm diameter enzyme will be able to form enzyme-substrate complexes and hence become labelled. Therefore the extent of labelling will depend on the relative exposure of a membrane protein (or target groups on a membrane protein) on the external cell surface, with less accessible proteins being missed although still located at the surface. A list of the enzyme-mediated labels in current use is given in Table 12.3.

The lactoperoxidase iodination method for cell surface labelling (Table 12.3,a) was the first reported enzyme-mediated system [189], and will be discussed in detail later (Section 12.4.2).

Kinzel amd Mueller (Table 12.3,b) explored the use of a partially purified phosphoprotein kinase from rat skeletal muscle and ($\gamma$-$^{32}$P)-ATP to label surface proteins of HeLa cells [190]. The incorporation of isotope was 85–90% enzyme-specific, only the $\gamma$-phosphate of ATP was incorporated and preliminary data on the linkage suggested that serine and/or threonine residues were phosphorylated. The external localization of 86–95% of the label was suggested by its removal with 500 $\mu$g ml$^{-1}$ trypsin. However, the cell surface localization and cell-derived nature of all the incorporated label has yet to be proven, because monolayer cells were used. It has been shown that there is a protein coat adherent to culture dishes [191], which is neither membranous nor necessarily cell-derived, but which could be phosphorylated by the protein kinase. In addition, the release of a portion of the label during membrane isolation suggests that adsorbed components may be labelled. A control experiment in which labelled culture medium (or washes) was mixed with unlabelled cells and the extent of adsorption measured would address this question. A major problem in the use of ($\gamma^{32}$P)-ATP in this labelling study was the high activity of surface ATPases which hydrolysed approximately 25% of the added isotope while the protein kinase catalysed incorporation of only 0.3%. A phosphate concentration of 12.5 mM had to be maintained in the external medium, rather than the physiological level of 1.2 mM, to minimize transport and utilization of the free radiolabelled phosphate by the cells.

Galactose oxidase (E.C.1.1.3.9) has been used by two groups of investigators to oxidize the C6 hydroxyl of D-galactose and *N*-acetyl-D-galactosamine residues found in cell surface glycoproteins and glycolipids ([74,192], Table 12.3,c). The resultant aldehydes were labelled by reduction with tritiated borohydride; hence, the modification was simply substitution of a proton by a tritium atom. Although only a subset of the externally-disposed membrane components, the sugar-bearing macromolecules, are labelled by the substitution, this class has been implicated in numerous intercellular and immunological interactions [193,194] and is being studied using lectins ([108] and Section 12.3.2). The conditions of labelling and results obtained using red cells differ between the two groups which have reported the galactose oxidase labelling technique. Steck and Dawson [74,195] used galactose oxidase from *Polyporus circinatus* (distinguished by its low protease contamination) at 3.5 $\mu$g ml$^{-1}$ and simultaneous addition of 0.1–0.5 mM ($^3$H)-borohydride, all at pH 8.5. They chose to label ghost vesicles, both outside and inside out. On the other hand, Gahmberg and Hakomori [192] used galactose oxidase (*Dactylium dendroides*) at 20 $\mu$g ml$^{-1}$, and ($^3$H)-borohydride at 2.5 mM followed by unlabelled borohydride at 4 mM, all at pH 7.0, to label intact erythrocytes. In both studies the principal lipid labelled was globoside while lactosyl ceramide, which has a potentially-reactive galactose, was unlabelled. However, Steck and Dawson found 18% of the total bound label to be in lipids of ghosts while Gahmberg and Hakomori reported 46%. In further contrast, 30% non-specific labelling of ghosts by ($^3$H)-borohydride in the absence of galactose oxidase was found by Gahmberg and Hakomori with only 3.5% in the intact erythrocyte, whereas Steck and Dawson reported 3% non-specific labelling of ghost vesicles. The extraordinarily high background of 30% suggests possible oxidation of unknown components during ghost isolation and cautions against any interpretation of labelling found in the presence of galactose oxidase. Steck and Dawson found no label in non-glycosylated proteins such as spectrin. The glycoproteins and glycolipids which were labelled were oriented exclusively toward the external environment, confirming ultrastructural studies where lectin-ferritin complexes were used to localize the carbohydrate components of the erythrocyte membrane [171,172].

The galactose oxidase technique has recently been employed to label the surface glycoprotein of several different cell types [17,38,39,196–198], both normal and transformed. However, several questions remain to be resolved before the galactose oxidase technique can be reliably extended to more complex eukaryotic cells. (1) Are *all* the galactosyl (or *N*-acetyl-D-glucosaminyl) residues which are oxidized subsequently reduced? This is important to establish because there is evidence that aldehyde groups generated by periodate oxidation or galactose oxidation can alter cellular activity (e.g., stimulate a mitogenic response in lymphocytes [199]), perhaps by cross-linking membrane proteins. Apparently significant delays between galactose oxidation and subsequent reduction *do* increase intermolecular cross-linking (Fig. 14, [192]). These perturbations could alter the dispositions of the very membrane components being studied and lead to erroneous interpretations of the results obtained. (2) Do the levels of peroxide generated by the galactose oxidase reaction peroxidize lipids or otherwise damage membranes? (3) Why is there increased efficiency of borohydride reduction (i.e., more label incorporated) in the absence of the erythrocyte cytoplasm [70,74]? Does the permeant $NaB^3H_4$ reduce intracellular constituents and transiently label them with tritium or is

TABLE 12.3
*Enzyme-mediated labelling systems*

| Enzyme (Source) | Label | Reactive group | Modified product* | Cells tested (Reference) |
|---|---|---|---|---|
| a Lactoperoxidase (bovine milk) | $^{125}I^-$, $H_2O_2$ | tyrosine | 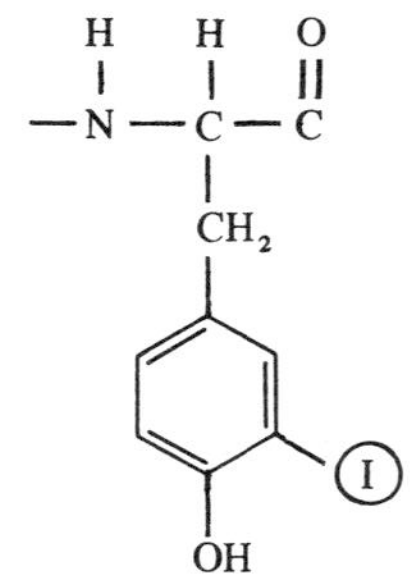 | Human erythrocytes (189,208) |
| b Phosphoprotein kinase (rat skeletal muscle) | $\{\gamma^{32}P-\}$ATP | serine, threonine | 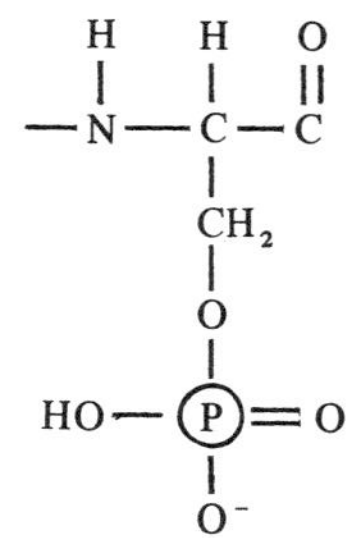 | HeLa cells (190) |

| | | | | | |
|---|---|---|---|---|---|
| c | Galactose oxidase<br>(1) (*Polyporus circinatus*)<br>(2) (*Dactylium dendroides*) | $NaB^3H_4$ | terminal<br>–galactose<br>–N-acetyl- glucosamine | 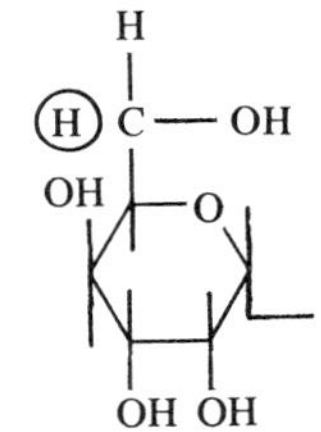 | Human erythrocytes (74,192) |
| d | Transglutaminase | $\{^{14}C\}$-Glycine methyl ester | Glutamine | 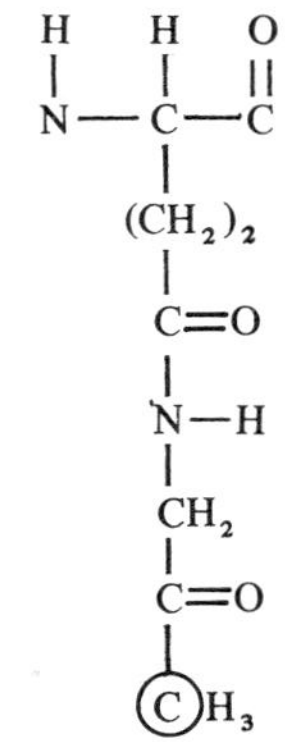 | PM-2 Bacteriophage (200) |
| e | Sialyl transferase (endogenous) | CMP-$\{^3H\}$N-Acetyl-Neuraminic acid | Suitable acceptors (usually terminal galactose residues) | Terminal sialyl residue | Mouse and hamster fibroblasts (201) |

*Circles denote position of radioisotope.

$NaB^3H_4$ hydrolysed intracellularly? These questions, as well as additional problems discussed by Steck and Dawson [74], should be considered prior to the use of the galactose oxidase technique.

Brewer and Singer [200] have reported use of the enzyme transglutaminase to catalyze substitution of ($^{14}$C)-glycine methyl ester for the amide group of glutaminyl residues in membrane proteins (Table 12.3,d). The enzyme is calcium-requiring, and at 5 mM $CaCl_2$ and 5 mM dithiothreitol, saturation labelling of 290 molecules per bacteriophage PM-2 was obtained (0.5% of the added isotope). Infectivity was not diminished by labelling. One membrane protein, considered to be external and peripheral [109] by other criteria, was not labelled by transglutaminase. Others, using a transglutaminase from guinea pig liver and radiolabelled putrescine, have been unable to label the external membrane proteins in intact erythrocytes or cultured fibroblasts, indicating the apparent paucity of glutamine residues exposed and/or available for reaction on cell surfaces. (V. Iwanij, personal communication.) These negative results emphasize the caution and collateral evidence required before assigning protein disposition in a membrane, especially when large enzymes are used as surface probes or the target amino acid is not abundant.

In a recent study [201] exogenously-added CMP-N-($^3$H)-acetylneuraminic acid and a sialyl transferase activity, endogenous to the cell (hamster and mouse fibroblasts) and presumed to be located at the surface, were utilized to label cell surface acceptors (Table 12.3,e). This technique was used earlier by others [202,203] to label isolated, desialylated rat liver plasma membranes and desialylated plasma glycoproteins with radioactive CMP-sialic acid. The problems encountered in these studies have been low incorporation efficiency of the radiolabel (0.05%), its low specific activity (200 Ci $mol^{-1}$), and its large expense. Consequently, long incubations (1–6 h) at 37°C are generally required for sufficient incorporation [201], but these conditions could lead to internalization of the nucleotide sugar and intracellular incorporation. In the study using tissue culture cells, at least 80% of the label was shown to be neuraminidase-sensitive and thus, externally-located sialic acid. However, monolayer cells were used, so the membrane localization of the incorporated label has yet to be demonstrated (see discussion of protein kinase-ATP labeling study [190] for reasons). Prior neuraminidase treatment of the tissue culture cells [201] increased labelling two- to three-fold. This pretreatment would seem preferable, since the ($^3$H)-sialic acid residues incorporated after neuraminidase would merely be replacing those previously removed enzymatically. However, failure to replace all the residues may affect cellular activity. Cell surface sialic acid groups have been implicated as receptors for plasma glycoproteins [203] and several types of viruses [204], and their removal has been shown to perturb the membrane [205]. Therefore, this labeling technique could be profitably used to study the effect on cellular function of altering the number and location of sialic acid residues on the external surface (with exogenous sialyl transferases) or to study differences in endogenous sialyl transferase and/or acceptor activity in normal and transformed cells [201]. However, it should not be considered a non-perturbing cell surface label.

### *12.4.2 Lactoperoxidase-mediated iodination*

The technique of iodinating cell surfaces with lactoperoxidase was first introduced by Phillips and Morrison [189,206]. Earlier, Marchalonis [207] had reported labelling soluble proteins using this enzyme or horseradish peroxidase. We later introduced the glucose oxidase-lactoperoxidase method [208]. Since 1971 there have been numerous reports on the use of enzymatic iodination, to probe the erythrocyte membrane [189,206,208–218], the plasma membranes of other eucaryotic cells [17,38,39,219–282] as well as bacteria and viruses [235,283–286], the membranes of subcellular organelles [287–292] and the topography of non-membranous organelles [293–295]. In addition, solid-state lactoperoxidase has been introduced for the iodination of soluble proteins [296,297]. The popularity of the method derives from the following: (1) it was the first enzyme-mediated labelling system published and it appeared to possess all the advantages of this class of cell surface reagents (i.e., covalent, impermeant and small label); (2) lactoperoxidase was commercially available, thus foregoing extensive synthesis of an unstable chemical reagent or enzyme purification; and (3) two iodine radionuclides, $^{125}$I and $^{131}$I, were available inexpensively and in carrier-free form, thus allowing high specific activity labelling with minimal membrane perturbation.

Lactoperoxidase (LPO) was first detected in milk in 1881 [298] and isolated in a number of laboratories [299–303] before Morrison and his colleagues published the purification method now used [304] with slight modifications [305]. The chemical and catalytic characteristics of the enzyme have been investigated primarily by the same laboratory [302,306–310]. The physiological function of LPO has been postulated to be bacterial killing [311–313].

LPO from bovine milk is a protein of 78 000 daltons with high carbohydrate content [305] and a haeme of the protoporphyrin IX structure linked via an ester bond to the apoprotein [306]. The ratio of absorbance at 412 nm to 280 nm (A 412/280 nm) has been the most common criterion for comparing various LPO preparations, with 0.93 the highest obtained [307]. The enzyme in this study was considered 99% pure and had a molar extinction coefficient of 114 000 at 412 nm.

Although all true peroxidases catalyse the oxidation of various electron donors with similar specific activitiies, only a small number of peroxidases will catalyze the formation of a carbon-halogen bond in the presence of hydrogen peroxide, a halide and a nucleophilic acceptor. These are chloroperoxidase from *Caldariomyces fumago* [314–316], myeloperoxidase from granulocytes [317], thyroid peroxidase [318] horseradish peroxidase [207] and lactoperoxidase. LPO has several advantages in that it is still very active at pH 7.4, whereas the other halogenating peroxidases are not, and it has greater specificity for iodide than chloride ions, while myeloperoxidase does not [317].

Using spectrophotometric methods, Bayse and Morrison [308] found at pH 3.6 that LPO catalysed two reactions: the oxidation of iodide to molecular iodine, followed by the iodination of tyrosine by iodine. Both reactions required $H_2O_2$, so a further oxidation of $I_2$ was postulated in the second reaction. The stoichiometry was as follows:

pH 3.6

$$H_2O_2 + 2I^- + 2H^+ \longrightarrow I_2 + 2H_2O \qquad 3$$

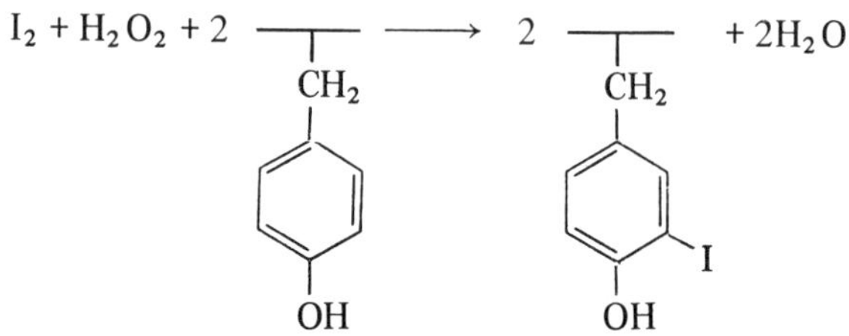

4

It is well-established that molecular iodine reacts spontaneously and rapidly with phenols at neutral and higher pH's, however, kinetic arguments were used to rule out such a non-enzymatic mechanism for reaction **4** at pH 3.6. At neutral pH there was no detectable formation of $I_2$ prior to the iodination of tyrosine by LPO. This data, together with the evidence that chemical and enzymatic iodination exhibited different rates of reaction with tyrosine derivatives [309], led Morrison to postulate that $I_2$ was not an obligatory intermediate in the LPO-catalysed iodination of tyrosine at pH 7.4. The stoichiometry of the reaction was thus

pH 7.4

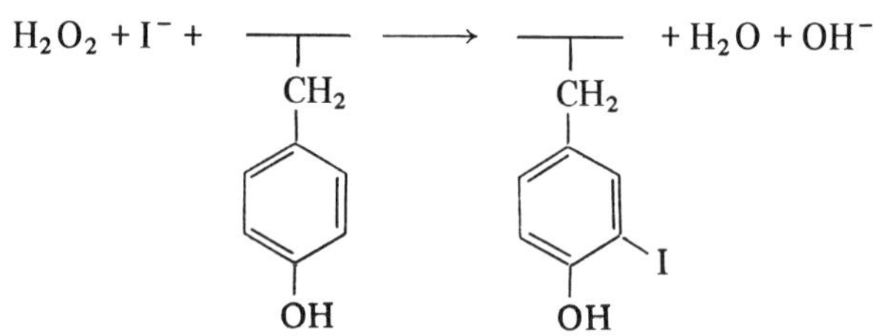

5

Apparently LPO will accept a second iodide if sufficient tyrosine is not available and $I_2$ can be formed at neutral pH [319]. Considering the chemical reactivity of $I_2$ discussed above, this statement has profound implications for the iodination of membranes and will be discussed below (Section 12.5). It is well-established for other halogenating peroxidases, and substantiated for LPO, that $I^-$ and $H_2O_2$ complex with the haeme prosthetic group of the enzyme (not

TABLE 12.4
*Enzyme and substrate concentrations for the iodination reaction*

| Reagent | Free tyrosine [307] | Iodination system (Reference) Membrane-bound tyrosine [189] | Membrane-bound tyrosine [208] |
|---|---|---|---|
| | | Concentrations ($10^{-6}$ M) | |
| LPO | 0.0074 | 0.3 | 0.05* |
| $H_2O_2$ | 100 | 8.0‡ | 5.0† |
| $I^-$ | 90 | 10‡ | 0.004–0.020 |
| Tyrosine | 810 | (0.000015–50)§ | (0.000015–50)§ |

*Equivalent to $H_2O_2$ utilization of 4 $\mu$M $min^{-1}$
†Equivalent to $H_2O_2$ production of 5 $\mu$M $min^{-1}$ by glucose oxidase (0.0026 x $10^{-6}$ M in protein).
‡Consumption of $H_2O_2$ indirectly recorded by iodide sensitive electrode and titrator and aliquots added to maintain 8 $\mu$M $H_2O_2$ and 10 $\mu$M $I^-$.
§Approximate concentration range of reactive tyrosines in 3 x $10^9$ human erythrocytes $ml^{-1}$ [208].

necessarily simultaneously) [306,309,315]. Evidence for the binding of the third reactant, tyrosine, to LPO was inferred from the stereospecificity of the enzyme for D- over L-tyrosine and the competitive inhibition of tyrosine iodination by di-iodotyrosine, an analogue which cannot be further iodinated. The finding that tyrosine must be bound by the enzyme to be iodinated suggests that LPO is not merely oxidizing iodide to a reactive form which diffuses away and reacts non-enzymatically, but is catalyzing the electrophilic substitution of an $H^+$ for an $I^+$ on the phenyl residue of an enzyme-bound tyrosine.

When free tyrosine was iodinated by LPO at pH 7.4, the optimal reagent concentrations were reported to be those listed in Table 12.4 [307]. The conditions for enzymatic iodination of the tyrosines in membrane proteins of intact cells, as first published by Phillips and Morrison [189], are also given (Table 12.4). The reagent concentrations in the membrane labelling system were generally lower with the exception of that of LPO, which was increased a relative five-fold. These conditions for iodination of intact human erythrocytes resulted in no haemoglobin labelling, no lipid labelling and a 10% loss of $K^+$ for iodinated cells over a four-hour period. However, the rather high levels of iodide ($^{125}$I and $^{127}$I), maintained at 10 $\mu$M throughout the reaction by an $I^-$-sensitive electrode and titrator, resulted in low incorporation of $^{125}$I (incorporation efficiencies were not reported). In more recent publications [212,219,229], these investigators have utilized carrier-free $^{125}$I. Also recently, $H_2O_2$ has been added at intervals of approximately two aliquots per minute to give a final $H_2O_2$ concentration of 8 $\mu$M, rather than in response to $I^-$ additions by the titrator. The conditions used by many investigators in later LPO studies were similar to those originally employed by Phillips and Morrison in that 10 $\mu$M $K^{127}I$ was present in addition to $Na^{125}I$: however, others (particularly Uhr and his colleagues [247]) have used carrier-free $^{125}$I to increase the specific activity of labelling. The conditions first reported by Marchalonis *et al.* [262] for the iodination of lymphoid cells have continued to be used by this group and others. Apparently much lower $H_2O_2$ concentrations are used, but the iodide incorporation efficiency is reportedly 10%. We introduced the use of glucose-glucose oxidase to generate continuously micromolar quantities of $H_2O_2$ (Table 12.4), the use of lower LPO concentrations and the use of carrier-free radiolabelled iodide for high specific activity labelling, or 'tagging' of the membrane [208]. These modifications improved the enzymatic iodination method in our laboratory, and others have reported successful use of the glucose oxidase-lactoperoxidase labelling method on a variety of membrane systems. We will discuss specific details of our technique below (Section 12.5).

Many of the publications in which LPO iodination was used, the cell type (or system) labelled, and the general iodination conditions chosen (i.e., those of Phillips and Morrison [189], Vitetta *et al.* [246], Marchalonis *et al.* [262] or Hubbard and Cohn [208]) are listed in Table 12.5 for reference. As yet there has been no experimental study comparing critically the four iodination

TABLE 12.5
*Lactoperoxidase-mediated iodination*

| Object labelled | Iodination system used* | Reference and method of surface localization (where performed)† |
|---|---|---|
| A. Blood cells | | |
| Human erythrocytes | LPO + $H_2O_2$ (I) | 189,206(A),209, 210(B),211(B), 212–215,227 |
| Human erythrocytes | LPO + GO | 208(A,B),216–218 |
| Human platelets | LPO + $H_2O_2$ (II) | 219(B,D),220 |
| Human platelets | LPO + GO | 221(B,C,D),222,223 |
| B. Tissue culture cells | | |
| Mouse fibroblasts (L) | LPO + GO | 224(C,D),225,240 |
| Mouse fibroblasts (L) | LPO + $H_2O_2$ (II) | 226,38(A, also used pyridoxal phosphate (36) and galactose oxidase (192)) |
| Mouse neuroblastoma | LPO + $H_2O_2$ (II) | 233 |
| Human epitheloid carcinoma (HeLa) | LPO + $H_2O_2$ (I) | 227(B),228 |
| Ehrlich ascites tumor | LPO + $H_2O_2$ (II) | 229 |
| Sarcoma 180 ascites tumor | LPO + $H_2O_2$ (I) | 230(A,D) |
| Baby hamster kidney | LPO + GO | 231(A) |
| Baby hamster kidney | LPO + $H_2O_2$ (II) | 232 |
| Chinese hamster ovary | LPO + $H_2O_2$ (III) | 39(C,D, also used pyridoxal phosphate (36), galactose oxidase (192) and DIDS (14)) |
| Secondary chick embryo fibroblasts | LPO + GO | 234,244 |
| Secondary chick embryo fibroblasts | LPO + $H_2O_2$ (I) | 235 |

TABLE 12.5 (*Continued*)

| Object labelled | Iodination system used* | Reference and method of surface localization (where performed)† |
|---|---|---|
| C. Normal versus virally-transformed tissue culture cells | | |
| Baby hamster kidney (BHK) vs. polyoma-transformed (Py-BHK) | LPO + GO | 236(B) |
| BHK vs. Rous sarcoma-transformed ($C_{13}/B_4$-BHK) | LPO + $H_2O_2$ (II) | 226 |
| Hamster cells (NIL) vs. Py-NIL and hamster sarcoma-transformed (HSV–NIL) | LPO + GO | 237–239,241 |
| Mouse fibroblasts (3T3) vs. Py-3T3 and simian virus-transformed (SV-3T3) | LPO + GO | 240(D) |
| $SV_{40}$-3T3 | LPO + $H_2O_2$ (II) | 242(D, also used TNBS) |
| Chick embryo fibroblasts (CEF) vs. RSV-CEF | LPO + $H_2O_2$ (I) | 243(B) |
| D. Primary lymphoid cells | | |
| Mouse spleen | LPO + $H_2O_2$ (II) | 246,247,249–259, 273 |
| Mouse spleen | LPO + $H_2O_2$ (III) | 245,263–265 |
| Mouse thymus | LPO + $H_2O_2$ (II) | 249,255–257,259, 274 |
| Mouse thymus | LPO + $H_2O_2$ (III) | 245,263–267 |
| Other mouse lymphoid cells | | |
| –bone marrow | LPO + $H_2O_2$ (II) | 249,256 |
| –lymph nodes | LPO + $H_2O_2$ (II) | 249,252,256,257, 259 |
| –thoracic duct | LPO + $H_2O_2$ (II) | 252 |
| –thoracic duct | LPO + $H_2O_2$ (III) | 265 |
| –other | LPO + $H_2O_2$ (II) | 245 |
| –other | LPO + $H_2O_2$ (III) | 264 |
| Human peripheral blood | LPO + $H_2O_2$ (II) | 272 |
| Human neonatal thymus | LPO + $H_2O_2$ (III) | 266 |
| Calf and rabbit thymus | LPO + $H_2O_2$ (II) | 270 |
| E. Lymphoid lines | | |
| Daudi (human B cells) | LPO + $H_2O_2$ (II) | 246,248(D) |
| $P_3K$ (mouse plasmacytoma) | LPO + GO | 271 |
| $WIL_2$-A3 (human diploid) | LPO + $H_2O_2$ (II) | 269 |
| WEHI-22T (mouse lymphoma) | LPO + $H_2O_2$ (III) | 262,268 |
| S1AT.4 (mouse lymphoma) | LPO + $H_2O_2$ (III) | 262(D),263 |
| MOPC460 (mouse myeloma) | LPO + $H_2O_2$ (III) | 262 |
| P815 (mouse mastocytoma) | LPO + $H_2O_2$ (III) | 262 |
| S-49.1 (mouse lymphoma) | LPO + $H_2O_2$ (II) | 249 |
| F. Primary cells, isolated and *in situ* | | |
| Rat adipocytes | LPO + $H_2O_2$ (I) | 275 |
| Rat adipocytes | LPO + $H_2O_2$ (II) | 276 |
| Toad bladder epithelium | LPO + GO | 277(C) |
| Cat spinal cord | LPO + $H_2O_2$ (II) | 278 |

TABLE 12.5 (*Continued*)

| Object labelled | Iodination system used* | Reference and method of surface localization (where performed)† |
|---|---|---|
| G. Intracellular organelles | | |
| Rat liver microsomes | LPO + $H_2O_2$ (I) | 287 |
| Rat liver microsomes | LPO + GO | 288(A,B) |
| Rat liver mitochondria | LPO + $H_2O_2$ (I) | 289 |
| Rat liver mitochondria | LPO + $H_2O_2$ (II) | 290 |
| Chloroplasts | LPO + $H_2O_2$ (I) | 291 |
| Sarcoplasmic reticulum | LPO + $H_2O_2$ (II) | 292 |
| Ribosomes | LPO + $H_2O_2$ (II) | 293–295 |
| H. Viruses | | |
| Influenza | LPO + $H_2O_2$ (I) | 285 |
| Sindbis | LPO + $H_2O_2$ (I) | 235 |
| Mouse RNA tumor viruses | LPO + $H_2O_2$ (II) | 286 |
| I. Other cellular organisms | | |
| Bacteria | LPO + $H_2O_2$ (II) | 280,281,282 |
| Other | LPO + GO | 279 |
| Other | LPO + $H_2O_2$ | 283,284 |
| J. Other systems | | |
| Soluble proteins | Soluble LPO + $H_2O_2$ | 207 |
| Soluble proteins | Solid-state LPO + $H_2O_2$ | 296,297 |

*LPO + $H_2O_2$ (I) refers to iodination conditions similar to those initially used by Phillips and Morrison [189] where carrier $^{127}$I was added; LPO + $H_2O_2$ (II) refers to the modified conditions (carrier-free $^{125}$I and periodic $H_2O_2$ additions) of Uhr and colleagues ([247], LPO reduced in [251]) and Morrison and colleagues [219,229]; LPO + $H_2O_2$ (III) refers to the conditions of Marchalonis and colleagues [262], LPO + GO refers to conditions similar to those of Hubbard and Cohn [208].

†See Section 12.5.5 for discussion of four methods for localization of label. Briefly, A = analysis of high-speed supernatant and pellet radioactivities; B = use of proteases to remove surface label; C = autoradiographic analysis; D = biochemical and SDS-electrophoretic analysis of membrane-enriched fractions.

methods in one cell system. Such a study is needed in order to establish whether the different details of each method affect the results obtained. In a brief report, Schenkein *et al.* [271], did demonstrate that the use of glucose oxidase rather than chemical $H_2O_2$ significantly improved lymphoid cell survival and increased incorporation efficiency.

## 12.5 The glucose oxidase-lactoperoxidase iodination method for labelling cell surfaces

Although the glucose oxidase-lactoperoxidase (GO-LPO) iodination method has been published [208] and utilized by ourselves [224,225], it seemed useful to provide details from our experience which might aid others attempting to use the technique and to review the controls, procedures and precautions we have used in an effort to adequately characterize the labelling method.

### *12.5.1 The enzymes and other reagents*

In our studies, we have chosen to specify the activities of lactoperoxidase and glucose oxidase, as well as the quantities of enzyme protein used in the iodination, because we consider it important to have equivalent $H_2O_2$ production (GO) and consumption (LPO) in the reaction mixture, to minimize possible peroxide excess and resultant cellular damage. Different enzyme preparations vary in purity so specification of the protein quantities used is not helpful in comparing different iodination methods. Admittedly the assays we use to specify LPO and GO activities involve oxidation of an artificial substrate, not iodination of tyrosine. However, the assay for glucose oxidase is valid since the ratio of glucose to glucose oxidase concentrations used in the spectrophotometric assay is lower than in the iodination reaction, insuring substrate saturation for GO and comparable $H_2O_2$ production in both systems. The concentration of LPO, based on its molar extinction coefficient of 114 000 at 412 nm, might be a better designation of the amount of enzyme being used, but some knowledge of its rate of $H_2O_2$ utilization is important.

The activity of glucose oxidase (D-glucose $O_2$:oxidoreductase, E.C.1.1.3.4) is measured using a coupled peroxidase-*o*-dianisidine system with the following final concentrations of reagents: 0.005 M phosphate buffer, pH 7.2; 0.005% ($1.9 \times 10^{-6}$ M) horseradish peroxidase; 0.5% (0.027 M) glucose; and 0.01% (0.0004 M) *o*-dianisidine. The change in absorbance with time at 460 nm is recorded and a molar extinction coefficient of 11 300 used to establish the activity. 1 unit is that amount of enzyme producing 1 $\mu$mol peroxide min$^{-1}$ at 25°C. The six lots of glucose oxidase (Type V, aqueous, Sigma Chemical Co., St. Louis, Mo.) we used exhibited activities ranging from 200–350 units ml$^{-1}$ (~40–70 units mg$^{-1}$) under the above specified conditions. It should be noted that the units designated and assay conditions used by the supplier (Sigma) are different from ours and cannot be readily converted. Preparations of glucose oxidase other than the Sigma Type V were found to have lower specific activities. Although GO is stable indefinitely at 4°C at the concentration supplied, at the concentrations used in the iodination procedure (~5 milliunits ml$^{-1}$ or 0.1–0.2 $\mu$g protein ml$^{-1}$) it is not stable for longer than ~8 h and therefore should be diluted immediately before use.

The activity of lactoperoxidase (donor: $H_2O_2$ oxidoreductase E.C. 1.11.1.7) is measured using *o*-dianisidine as substrate, with the following final concentrations of reagents: 0.01 M phosphate buffer, pH 7.2; 0.003% (0.00088 M) $H_2O_2$; 0.008% (0.00033 M) *o*-dianisidine. Again, a molar extinction coefficient of 11 300 for *o*-dianisidine at 460 nm was used to calculate the enzyme activity and the units are similar to those of GO except that 1 unit of LPO utilizes 1 $\mu$mol peroxide min$^{-1}$ at 25°C. The commercial preparations we used (Calbiochem.) ranged in specific activities from 1–2 units $mg^{-1}$ protein and A 412/280 of > 0.6. This enzyme is stable at ~6 mg $ml^{-1}$, pH 7.2, at 4°C for up to one week (longest used) or –20°C for six months. We routinely freeze LPO in solution in small aliquots, thaw and dilute it to a final activity of ~5 milliunits $ml^{-1}$ (~5 μg $ml^{-1}$) immediately before use.

LPO activity has been assayed in the presence of two detergents, deoxycholate and Triton X-100. Deoxycholate, from 0.5–5%, enhanced activity approximately 2.5 fold as did Triton at low concentrations (0.05–0.1%). At 1% Triton the activity was equal to control values. It is important to assess the effect of varying ionic, detergent and pH conditions on enzyme activity before enhanced iodination is interpreted as the availability of more tyrosines [37].

$Na^{125}I$ (carrier-free, New England Nuclear, Boston, Mass.) is obtained at 20 mCi $ml^{-1}$ in 0.1 N NaOH, stored at 4°C and used within two weeks. It is particularly important to use fresh isotope that has not been exposed to air for long periods of time, since in our experience oxidized iodide results in a high level of incorporation even in the absence of LPO (see Section 12.5.3). When $K^{127}I$ is used, it too should be prepared fresh. The addition of sulphite to iodide preparations has been reported to reduce any oxidized iodine without inhibiting subsequent LPO iodination [319].

### 12.5.2 Iodination conditions

The effect on iodination of such variables as time, temperature, iodide and chloride concentrations (all discussed below), cell number, and LPO and GO activities was examined first using human erythrocytes [208] and then mouse fibroblasts [224]. We will present here the conditions for handling and iodinating mouse fibroblasts of the L strain, since nucleated cells are generally more fastidious in their requirements than red blood cells and thus the iodination conditions more restrictive. The following conditions were chosen for routine iodinations so as to achieve reasonable incorporation efficiency in a short time but also to minimize membrane activity such as pinocytosis and diffusion of proteins in the membrane plane. L cells which had been cultured in suspension [224] were rinsed twice in cold phosphate-buffered saline (PBS) at ~2 x $10^5$ cells $ml^{-1}$, pelleted and resuspended in cold PBS containing 20 mM glucose. The cell concentration was 4–6 x $10^6$ cells $ml^{-1}$. 4–8 milliunits LPO (~0.08 nmol) and slightly less GO (3–6 milliunits or ~0.1 μg) were added, together with 100 μCi $Na^{125}I$, to one ml of cells, and the reaction mixture rotated at 5 revs/min for 10 min at 4°C. Violent mixing or agitation of the cells in this protein-free environment (e.g. use of a Vortex) led to cell damage. The reaction was stopped by the addition of 20 vol. cold Minimum Essential Medium (Schwarz/Mann Division, Becton, Dickinson and Co., Orangeburg, New York) and the cells were centrifuged away from the soluble LPO and GO at 1000 revs/min for 10 min. The cells were then rinsed at least twice in large volumes of ice-cold PBS. The entire procedure took ~1.5 hours and cell viability was maintained at 95% throughout as assessed by exclusion of trypan blue (0.16%). Enhanced cell viability has been reported when Earle's balanced salt solution is used.

Incorporated radioactivity was measured by trichloroacetic acid (TCA) precipitation and centrifugation or by the 'disc-batch' method [320] when many samples were handled (see [224] for procedural details). It has been our experience that protein precipitation and thorough rinsing in the presence of $^{127}I$ (50 mM) is necessary to remove adsorbed but unincorporated $^{125}I$.

The effect of iodide concentration and temperature on the extent of iodination was graphically demonstrated by two experiments. In the first (Fig. 12.1), a time course of iodination of L cells was performed at 4°C and 37°C. Carrier-free $^{125}I$ was used. The amount of isotope incorporated under these conditions was temperature-independent. That is, the reactions at 4°C and 37°C proceeded in parallel over 30 min and were most rapid in the first five minutes. After 30 min the reaction at 4°C appeared to plateau but not at 37°C. Additional LPO did not significantly enhance incorporation at either

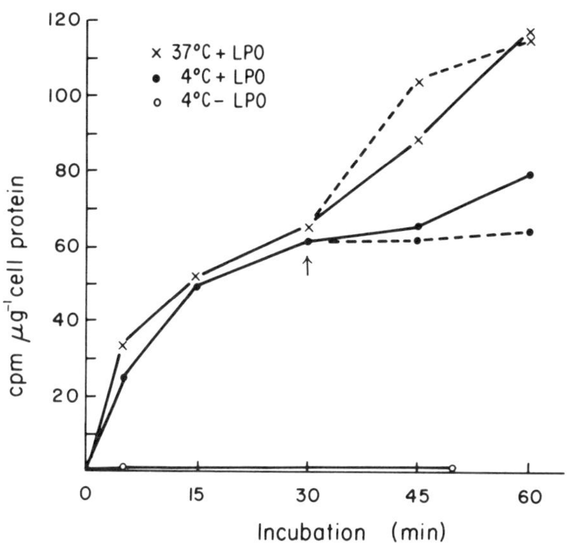


**Fig. 12.1** Effect of temperature on the time course of L cell iodination with carrier-free $^{125}I$. Cells and reagents were preincubated separately for 15 min at the appropriate temperature and then mixed. 1 ml of the reaction mixture contained 9 x $10^6$ cells in PBS, 5 $\mu$mol glucose, 6 milliunits GO, 3 milliunits LPO and 50 $\mu$Ci $Na^{125}I$. At indicated times 1 ml was removed, added to 10 ml $10^{-5}$ M $Na_2S_2O_3$ in PBS, and the cells were pelleted and rinsed five times, lysed in 0.5 ml $H_2O$ and aliquots analysed for TCA-insoluble radioactivity. Arrow indicates the addition of 3 milliunits more LPO: (– – – – –), additional LPO added; (———) no additional LPO added. 1 $\mu$g L cell protein $\simeq$ 6 x $10^3$ cells. (Figure from [224], courtesy of the Journal of Cell Biology).

temperature. Cells incubated at 4°C (or 37°C) without LPO incorporated less than 1% the radioactivity of those which were exposed to the complete system (discussed in Section 12.5.3).

The lack of a temperature effect on iodination was reproducible when carrier-free concentrations of iodide were used (50 $\mu$Ci or 2 x $10^{-8}$ M). However, in the second experiment, increasing amounts of $K^{127}I$ were added to the system and the effect of temperature on the reaction became apparent (Fig. 12.2). A $Q_{10}$ of >3 was obtained, in agreement with other enzymatic reactions. At the highest iodide concentration used, 12.5 x $10^{-6}$ M and at 37°C, approximately 5–10 x $10^6$ iodide atoms were incorporated per cell. At 4°C and 12.5 x $10^{-6}$ M iodide, 1.0 x $10^5$ atoms per cell were incorporated (Fig. 12.2). Only 2 x $10^3$ iodide atoms were incorporated per cell at 4°C (or 37°C) when carrier-free isotope (2 x $10^{-8}$ M) was used. Incorporation at 4°C could be increased to 10 000–50 000 atoms per cell by increasing the $^{125}I$ concentration to 100–150 $\mu$Ci ml$^{-1}$ (6 x $10^{-8}$ M iodide) and doubling the LPO concentration.

The chemical concentration of iodide in a carrier-free iodide preparation can be calculated from the following facts: (1) $t$ 1/2 of $^{125}I$ = 60 days; (2) $\lambda$, the

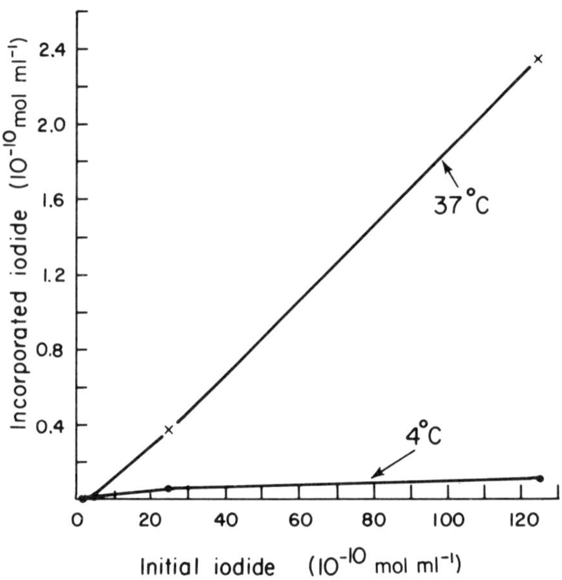


**Fig. 12.2** The effects of temperature and iodide ($^{125}I$ + $^{127}I$) concentration on the iodination of L cells. 1 ml of reaction mixture contained 4 x $10^6$ cells in SP, 5 $\mu$mol glucose, 5 milliunits GO, 3 milliunits LPO, 50 $\mu$Ci $Na^{125}I$, and varying amounts of $K^{127}I$. Cells were iodinated 30 min at either 4° or 37°C, rinsed five times in SP, lysed in 0.5 ml $H_2O$ and aliquots analysed for TCA-insoluble radioactivity. (Figure from [224], courtesy of the Journal of Cell Biology).

decay constant (probability of decay) for $^{125}I = 0.0115\ d^{-1}$ or $8 \times 10^{-6}\ min^{-1}$; (3) 1 $\mu$Ci = $2.2 \times 10^6$ disintegrations $min^{-1}$. Therefore, the occurrence of 1 disintegration in a minute for $^{125}I$ means $1/(8 \times 10^{-6})$ or $1.2 \times 10^5$ atoms of $^{125}I$ were present per dpm. Converting to moles (using $6.02 \times 10^{23}$ atoms $mol^{-1}$) one obtains $4 \times 10^{-13}$ moles of $^{125}I$ per $\mu$Ci of $^{125}I$. This value can also be calculated from the listed specific activity of $^{125}I$ (New England Nuclear), which is ~17 $\mu$Ci $mg^{-1}$ of $^{125}I$.

In our L cell studies, we chose to use a low concentration of iodide which is ~99% $^{125}I$ (carrier-free) rather than to introduce non-radioactive $^{127}I$ into the system. Of course, every tyrosine on the exposed membrane proteins is not labelled, but the extent to which the iodinated population is *representative* can easily be assessed by the addition of carrier $^{127}I$. Also, the use of carrier-free isotope avoids the possibility of denaturing proteins by excessive iodination since less than 0.1% of the available sites per cell are usually iodinated. We have also chosen to iodinate routinely at 4°C and would recommend this, because pinocytosis is inhibited at this temperature [321], but incorporation efficiency is not decreased when carrier-free isotope is used (Fig. 12.1).

To assess possible inhibition of the iodination reaction by chloride ions, erythrocytes were iodinated in iso-osmolar phosphate buffer, pH 7.4 or PBS (containing 0.154 M NaCl). A significant but variable inhibition was found in the presence of chloride, but the inhibition was less than 50% and yet the chloride to iodide ratio in the medium was $6 \times 10^6$ to 1. These results indicate that chloride is an extremely poor inhibitor of the iodination reaction as catalysed by LPO (see Fig. 1 of [208]), and suggest the use of physiological salt solutions which contain chloride rather than phosphate solutions [226] which may harm cells.

Although the viability of cells immediately after iodination may be greater than 95%, it is necessary to establish that iodinated cells remain viable over a longer period and behave like uniodinated control cells. Toward this end, the growth of $^{125}I$-labelled, $^{127}I$-labelled and unlabelled L cells was followed and each cell population was found to double within the 16–20 h generation time of the L cell line used (Fig. 12.3). The plating efficiency of all these cells was more than 95% at each time point. In an additional experiment cells were prelabelled with ($^3H$)-thymidine, then rinsed and iodinated with $^{127}I$, rinsed but not iodinated, or neither rinsed nor iodinated. Aliquots from the three treatments were analysed for acid-insoluble ($^3H$)-thymidine over 24 h. The curves were identical for each treatment, indicating that the iodination and rinsing procedures had no adverse effect on cell viability. Others have reported similar findings with different tissue culture cells [17,227,269].

The possible deleterious effects of hydrogen peroxide and of iodination on membrane proteins, lipids and even on cell viability should be considered in light of recent reports of lipid peroxidation [287], alteration of electro-physiological properties [277], enzyme inhibition [291], cell cytotoxicity [322] and loss of viral infectivity [200] during LPO iodination. As little as $5 \times 10^{-6}$ M $H_2O_2$

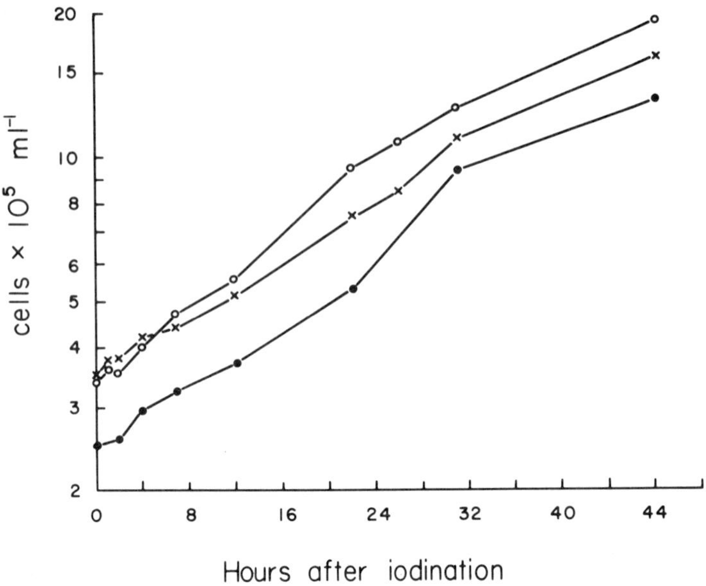


**Fig. 12.3** Increase in population density of L cells after iodination. (○———○), cells iodinated with $^{125}$I; (●———●), cells iodinated with $^{127}$I; (x———x), cells untreated. All cells resuspended in fresh growth medium at zero hour. See [225] for methodology. (Figure from [225], courtesy of the Journal of Cell Biology).

added to rat liver microsomes (at the very dilute concentration of 0.5 mg protein ml$^{-1}$) caused significant peroxidation of lipids, loss of cytochrome $P^{450}$ and inhibition of several enzymes [287]. The first two effects could be reversed by low concentrations (0.0001%) of an antioxidant, butylated hydroxytoluene. These results caution strongly against the high peroxide concentrations ($>50 \times 10^{-6}$ M) used in numerous iodination studies [e.g., 38,39,207,215, 230,232,245,246–259,292,293]. Even when the final peroxide levels used are low, periodic addition of small aliquots of concentrated $H_2O_2$ to the system creates both temporal and spatial gradients of peroxide which may be harmful. In contrast, the glucose-glucose oxidase couple generates a continuous, micromolar supply of hydrogen peroxide throughout the system and avoids these gradients. However, the use of glucose-glucose oxidase does not necessarily guarantee success. A methionine residue in adreno-corticotrophic hormone was reportedly oxidized by $H_2O_2$ generated by glucose-glucose oxidase [323], but the GO activities used were apparently five- to ten-fold greater than those we have routinely used. Dimethyl sulphoxide has been reported to protect sensitive methionine residues in peptide hormones from peroxide oxidation during chemical iodination using chloramine T without reduction in the specific activity of labelling [324]. Such a scavenger might also be used in the enzymatic iodination of membrane systems, provided the levels of dimethyl sulphoxide (~1000-fold molar excess) did not interfere with function.

In cases where inhibition of a membrane function [277,287,291] or cell cytoxicity [322] have been reported, carrier $^{127}$I was added to final concentrations of 10 – 63 x $10^{-6}$ M. In two of these studies, no perturbation was found when lower iodide [277] or lower GO or LPO levels [322] were used. These results indicate (but do not prove conclusively) that the extent of iodination does affect membrane function and integrity and that saturation labelling (with $\gtrsim$10 x $10^{-6}$ M iodide or very low cell concentrations [227,322] using red blood cells) should not be a goal. It should be noted that in none of the above studies was the nature of the iodinated species examined in order to establish that the perturbation was due only to the iodination of tyrosine residues and that side reactions were not occurring (see Section 12.5.4). This important consideration is often overlooked when interpreting results. Using two parameters for assessing membrane perturbation, cell viability and phagocytic capacity, we have shown that L cells were unaffected by our iodination conditions (4°C and 2–6 x $10^{-8}$ M iodide). Usually 1–5 x $10^4$ atoms per cell are incorporated, or approximately 30 atoms of 0.3 nm diameter per $10^6$ nm$^2$ of L-cell surface area. Others have also reported normal functional activity after iodination of fat cells [275] and ribosomes [293]. These variable effects of iodination on specific cellular functions emphasize the need to establish the optimal labelling conditions for each new membrane system in order not to perturb membrane structure and alter function. Of course, the sensitivity to the iodinated conditions of the particular membrane function being measured (whether phagocytosis, transport, etc.) will determine the amount of label which can be incorporated into a given membrane. The effect of iodination on the function of individual, isolated, labelled membrane proteins has not yet been investigated; however, soluble proteins have been enzymatically iodinated with no loss in activity [143,191]. To date, only peptide hormones with sensitive tyrosines cannot be iodinated without loss of biological activity [325].

### *12.5.3 Controls*

To ascertain the specificity of the iodination reaction, controls were performed which consisted of deleting each reagent or combinations from the complete iodination system and assessing the extent of incorporation (Table 12.6). The deletion of any one reagent reduced incorporation into ghosts by approximately 99%. Even the deletion of exogenous glucose (Table 12.6,e) inhibited iodination to 1.3% of the complete system. The relative amount of incorporated label in the first lysis supernatant was much larger in all controls than in the complete system. This may reflect the adsorption of unreacted isotope during the precipitation of large amounts of protein and trapping during centrifugation. Certainly, more extensive characterization of the labelled species would be

TABLE 12.6
*Controls of the iodination reaction*

| Reagent deleted | | Radioactivity (cpm per $10^9$ cells) Ghosts | First supernatant (Haemoglobin) |
|---|---|---|---|
| (a) | None | $390 \times 10^4$ | $20 \times 10^4$ |
| (b) | LPO (in phosphate buffer)* | $3 \times 10^4$ | $6 \times 10^4$ |
| (c) | LPO (in saline)* | $3 \times 10^4$ | $4 \times 10^4$ |
| (d) | GO | $2 \times 10^4$ | $5 \times 10^4$ |
| (e) | Glucose | $5 \times 10^4$ | $7 \times 10^4$ |
| (f) | GO, LPO | $3 \times 10^4$ | $5 \times 10^4$ |
| (g) | GO, LPO, glucose | $4 \times 10^4$ | $5 \times 10^4$ |

Iodination conditions for the complete system (a) per millilitre of final solution: $1 \times 10^9$ washed red blood cells, 5 $\mu$mol glucose, 7.2 milliunits glucose oxidase 3.6 milliunits lactoperoxidase, 50 $\mu$Ci Na $^{125}$I. Appropriate volumes of buffer replaced the specific reagent deleted. Incubation was 30 min at 37°C.
*Iodination performed in 310 milliosmols Na phosphate, pH 7.2 or 0.154 M NaCl, 0.01 M Na phosphate, pH 7.2. (Table from [208], courtesy of the Journal of Cell Biology.)

necessary before this radioactivity could be interpreted as 'intracellular labelling' (see discussion, Section 12.2.2).

Similar controls were performed with equivalent results in the L cell study ([224] and Fig. 12.1). In general, the extent of non-enzymatic iodination (i.e., iodination in the absence of LPO) has been found to be negligible (1–2%), where carefully studied [e.g., 247,263,277,288]. Several investigators have reported very high levels of iodination (20–72%) in the absence of LPO [17,226]. There are several possible explanations: (1) oxidized iodine ($I_2$) may be present in the isotope preparation and reacting chemically with membrane or intracellular constituents (see Section 12.5.4); (2) intracellular peroxidases may be present and catalysing the iodination of cytoplasmic tyrosines by added $I^-$ and $H_2O_2$, which are permeant; (3) unincorporated isotope may still be present, due to failure to precipitate the material and/or rinse sufficiently, prior to the radioactivity measurements [226]. A high background of non-enzymatic iodination should not occur if the technique is to be useful as a specific label of the external cell surface. The most direct methods for determining the source of the background are first to use fresh isotope; then, if the problem persists, analyse the species labelled under these conditions (Section 12.5.4) and localize the non-specific label using autoradiography (Section 12.5.5).

We performed two additional controls in order to assess the possible adsorption of LPO to L cells or its transport into cells during iodination. In the first, soluble LPO and GO were enzymatically iodinated with insolubilized enzymes (provided by Dr. J. P. Kraehenbuhl, Yale University) and then added for 10 min at 4°C to unlabelled L cells in the same concentrations used for routine iodination but in the absence of glucose and $^{125}$I. 96% of the radioactivity was recovered in the first supernatant and none remained cell-associated by the fourth rinse. The second control consisted of prior incubation of unlabelled LPO with L cells, or erythrocytes, followed by two washes and addition of the remaining reagents. Incorporation was less than 1% of the level obtained in the presence of LPO. These controls indicated that LPO, iodinated or uniodinated, did not adsorb to cells and cause localized, non-representative iodination or contribute to the membrane labelling pattern seen on SDS gels [224].

### *12.5.4 Characterization of the iodinated species*

It is critical to the subsequent use and interpretation of any labelling technique to first adequately characterize the labelled species: is the linkage covalent, what is the distribution of label between lipid and protein, and specifically, which amino acid (or lipid) residues are labelled? In our studies the initial evidence that the cell-associated iodide was bound covalently and largely in protein was its precipitation in TCA and subsequent complete recovery in SDS gels in the approximate molecular weight range of 10 000 to >200 0000 daltons [208,224]. To determine the amino acid(s) iodinated, we chose enzymatic hydrolysis since acid or alkaline hydrolysis cause extensive deiodination [326]. Iodinated cell material was precipitated with TCA onto discs [224], rinsed, dried and placed in glass scintillation vials. Crude trypsin or pancreatin at 2.5 mg ml$^{-1}$ in 0.05 M Tris-HCl, pH 8.0, 0.004 M $CaCl_2$, and one drop toluene were added to the vial, and the mixture was incubated for 24 h at 37°C with agitation. Additional trypsin was added for another 24 h. After 48 h of enzymatic digestion 91–95% of the initially acid-insoluble radioactivity was soluble in TCA. The success of crude trypsin in degrading proteins to the level of amino acids depends presumably upon the presence of other proteases and peptidases in the preparation.

Two methods have been used to analyse the trypsin digest of iodinated material [208]. The first is semi-quantitative and involves acidification, *n*-butanol extraction of the digest and chromatography on flexible thin-layer gel plates (MN, Sil G, Brinkman Instruments, Inc., Westbury, N.J.) in a solvent system of *n*-butanol/acetone/1 N $NH_4OH$ at 1/4/1 [327]. Monoiodotyrosine (MIT) of $R_F$ 0.30 and iodide, $R_F$ 0.85, were resolved in this solvent system, but mono- and di-iodotyrosine were not. Greater than 95% of the radioactivity recovered in the *n*-butanol extract (80% of initial counts from precipitated, $^{125}$I-ghosts) migrated with an $R_F$ identical to that of both chemical and radiolabelled MIT (Fig. 12.4, top). The identity of the unknown species with MIT was established by co-running a radioactive standard with the sample. The second method for analysing tryptic digests yields a more quantitative recovery of the protein-bound radioactivity and involves gel filtration of supernatants (from enzymatic digests that had been precipitated with TCA) on a column of Sephadex-G25 Fine in 1 M acetic acid. MIT ($K_d$ = 2.3) was well-separated from di-iodotyrosine ($K_d$ = 3.65) and iodide ($K_d$ = 1.3). As can be seen in Fig. 12.4 (bottom) greater than 85% of the soluble radioactivity was mono-iodotyrosine and the remainder either $^{125}I$ (~10%) or material eluting at the void volume (presumably small peptides). Di-iodotyrosine was never detected although columns were routinely run past the elution volume of this amino acid.

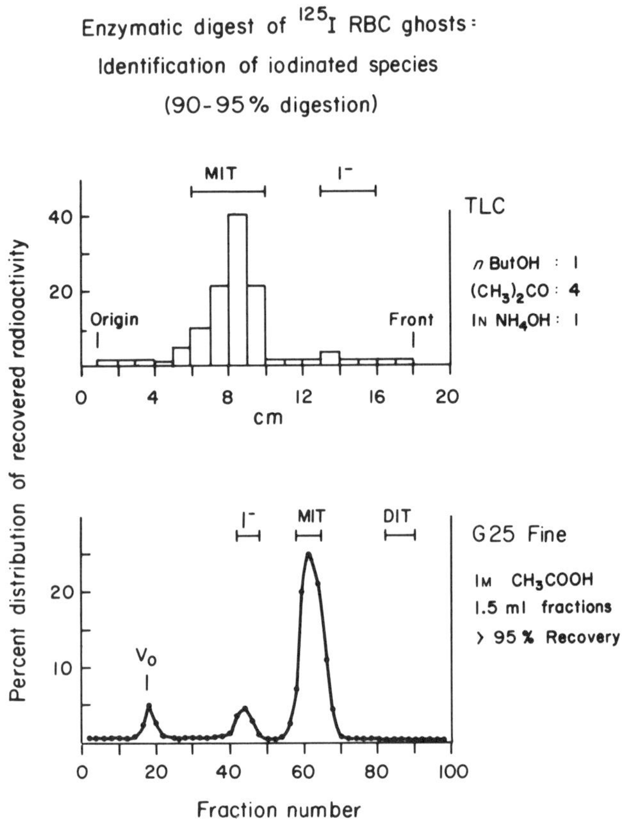


**Fig. 12.4** Characterization of the iodinated species: *Top panel*; thin layer chromatographic analysis; ghost protein precepitated with TCA, extracted with 3/1 ethanol/ether and then digested with trypsin as described in Section 12.5.4. 80% of the radioactivity was hydrolysed and extracted into acidified *n*-butanol. See [208] for methodology. MIT and di-iodotyrosine were not well-separated in the solvent system used. *Bottom panel*; gel filtration (G25 Fine) analysis; ghost protein precipitated with TCA and digested as described in Section 12.5.4. The digested material was treated with TCA and the supernatant (85% of the original radioactivity) applied to a column (1.2 x 56 cm) with 96% recovery. See [208] for methodology.

TABLE 12.7
*Organic solvent extraction of acid-insoluble label*

| Sample | Organic solvent | Soluble material cpm | Soluble material protein (μg) | Residue cpm | Percent change |
|---|---|---|---|---|---|
| 1* | None | – | – | 300 000† | 0 |
| | 10% $H_2O$ in acetone | nd | nd | 275 000 | –8% |
| | 3/1 Ethanol/ether | nd | nd | 278 000 | –7% |
| | 2/1 $CCl_3$/methanol | nd | nd | 272 000 | –9% |
| 2‡ | None | – | – | 309 000§ | 0 |
| | 10% $H_2O$ in acetone | 49 500 | 8.4 | 248 000 | –16%¶ |
| | 3/1 Ethanol/ether | 34 000 | 11.4 | 182 000 | –16%¶ |

*25 μl iodinated L cell samples (see [224] for methodology) spotted on 2.5 cm diameter Whatman GF/C glass fibre discs and washed in TCA. Solvent added at –20°C for 30 min.
†Average of 3 determinations, all within 3%.
‡Well-washed ghosts TCA precipitated, rinsed four times in 10% TCA, dried and the solvents added for 1 h at –20°C. (See [208] for methodology.)
§100 μg protein.
¶Percent of recovered radioactivity.

The filter disc method can also be used to assess the amount of lipid labelling of cells during iodination. After a suitable number of rinses in TCA, the filter discs containing iodinated material are air-dried immediately (to measure the total TCA-insoluble radioactivity incorporated) or rinsed in solvents that have been reported to extract neutral lipids as well as phospholipids [33,328]. 8% of the TCA-insoluble label could be extracted with 90% acetone, 3/1 ethanol/ether or 2/1 chloroform/methanol when $^{125}$I-labelled L cells were spotted onto filter discs and processed using the 'disc-batch' method (Table 12.7, 1). Approximately 15% of the recovered label could be extracted if the samples were first rinsed by repeated centrifugation in 10% TCA and the pellets then extracted with these organic solvents (Table 12.7, 2). The soluble radioactive material (from sample 2, Table 12.7) did not migrate with any known phospholipid or neutral lipid standard co-chromatographed in two different solvent systems. A variable percentage (10–30%) ran with $^{125}$I. When the lipid extract was digested with trypsin, 50% of the label chromatographed with an M$^{125}$IT standard. In addition, about 10–15% of the total protein could be extracted with either organic solvent mixture. This suggested that the majority of the labelled material extracted into lipid solvents was protein or $^{125}$I. The 'disc-batch' method resulted in less trapped $^{125}$I and therefore released less label in the presence of organic solvents.

The occurrence of lipid labelling has been reported in a number of enzymatic iodination studies [226,228,251,275]. Different criteria have been applied for defining label as being lipid-associated: (1) migration in SDS gels at the dye front [228,275]; (2) extraction with organic solvents [226,228,251]; (3) chemical characterization of the organic solvent extract. Only the third criterion is a valid test of the occurrence of lipid iodination and where this has been performed, less than 5% of the label has been identified as lipid-associated [224,251]. Other investigators have reported comparable levels [288], or lower [206,290]. The variability of results with regard to lipid labelling emphasizes the necessity of characterizing the components labelled *prior* to use of this method. Since LPO has not been found to catalyse the direct iodination of lipids [206], another mechanism must be involved. Morrison [319] has raised the possibility of $I_2$ formation catalysed by LPO when the iodide to tyrosine ratio is very high (i.e., carrier $^{127}I$ present). $I_2$ could diffuse into the membrane interior or cell and non-enzymatically iodinate unsaturated fatty acids and tyrosine residues. Thus, the occurrence of lipid labelling in a system certainly casts doubt on the putative external disposition of any proteins found to be labelled.

### *12.5.5 Methods for assessing the cell surface localization of the label*

Characterization of the iodination reaction, the controls and the species labelled must be accompanied by localization of the incorporated label to the cell surface prior to use of the method. A number of procedures are available for assessing localization and several should be performed since each method has certain limitations.

One of the simplest localization methods (but not an unequivocal one) is preparation of a high-speed supernatant and pellet (100 000 x *g* for 60 min to sediment all membrane vesicles) from a homogenate of iodinated cells and measurement of the label distribution and specific activity in the two fractions. Table 12.8 lists such a distribution for a typical erythrocyte experiment [208]. Ghosts, from labelled cells which were lysed, were rinsed four times and contained no measurable haemoglobin. 97% of the incorporated label was found in the ghosts and since ghost protein represented 1.3% of the total, this resulted in an 80-fold increase in specific activity of ghosts over the intact erythrocyte and a 2600-fold increase of ghosts over haemoglobin. When haemoglobin was directly exposed to the LPO system, it was heavily iodinated. The results clearly indicate that the membrane is serving as a barrier to the iodination of intracellular haemoglobin by LPO. This method for the localization of label has been used in iodination studies on more complex eucaryotic cells and lower specific activities were found in the sedimentable fraction [230,231,233,249]. Of course, the pellet from these cells, unlike that of the erythrocyte, contained unlabelled, intracellular membranous organelles as well as labelled plasma membrane vesicles, so a lower specific activity would be expected. However, the distribution of label should approximate that found in the erythrocyte. When greater than 3% of the label is found in a high-speed supernatant, it is possible that (1) adsorbed proteins were labelled, released during homogenization and did not sediment; (2) dead cells were labelled and account for some of the soluble label; (3) intracellular, non-enzymatic iodination occurred in addition to LPO-catalysed surface iodination (see Section 12.5.4). A recent report claims

TABLE 12.8
*Distribution of label after iodination of intact red blood cells*

| Fraction | Protein (mg) | Radioactivity (cpm) | Distribution of label (%) | Specific activity (cpm $mg^{-1}$ protein) | Relative enrichment |
|---|---|---|---|---|---|
| Intact cells | 422 | $5.1 \times 10^7$ | 100 | $0.12 \times 10^6$ | 1 |
| Ghosts | 5.2 | $4.94 \times 10^7$ | 97 | $9.6 \times 10^6$ | 80 |
| Haemoglobin | 420 | $0.14 \times 10^7$ | 3 | $0.0033 \times 10^6$ | 0.03 |

Iodination conditions per millilitre (total 10 ml): $1 \times 10^9$ washed cells in SP, 5 $\mu$mol glucose, 3.6 milliunits LPO, 7.2 milliunits GO, 50 $\mu$Ci carrier-free Na $^{125}$I. Incubation for 30 min at 37°C. Ghosts and haemoglobin were prepared as described in [208]. (Table from [208], courtesy of the Journal of Cell Biology.)

that LPO preferentially iodinates membranes (intracellular or plasma) rather than soluble cytoplasmic proteins when exposed to them [17]. However, the control in this study indicated a very high level of non-enzymatic iodination, hence the results with LPO are somewhat questionable. In addition, others have reported comparable iodination of the cytoplasm and membranes when cell homogenates were labelled [226], indicating that this method of localization is a valid one.

If the cell (or organelles) being studied contains a specialized protein(s) whose intracellular location is known, the amount of label in that purified component would indicate the extent to which iodination occurred intracellularly. This has been done with red blood cells [208], rat liver microsomes [288] and lymphoid cells [247].

Another method for localizing the label to the external surface is the use of proteolytic enzymes on intact viable cells to remove the label. A number of precautions on the use and limitations of this method are warranted. The viability of cells before and after protease treatment should be assessed, because the use of high concentrations of enzymes may lead to cell damage. The use of broad spectrum proteases, such as pronase or bromelain, would seem preferable for maximum hydrolysis. Finally, lack of complete removal does not necessarily mean the incorporated label is not on the surface, as illustrated by the action of trypsin and pronase on intact iodinated red blood cells [208]. Although 97% of the radioactivity was present in ghosts, when intact cells were exposed to 250–1000 $\mu$g $ml^{-1}$ trypsin, only 40–45% of the label was released. Preincubating the cells with neuraminidase had no effect on the amount of label subsequently released. Pronase also released 45% of the label. The combination of trypsin and pronase still gave 45% release. Analysis of the ghosts from trypsinized cells indicated that one class of labelled polypeptides (100 000 daltons, Band III) was insensitive to trypsin and was converted to a smaller membrane component after pronase but with retention of greater than 90% of its label. Similar results were presented by Tsai *et al.*, where a maximum of 70% of the label on HeLa cells could be removed in 2 h with 25 $\mu$g ml$^{-1}$ pronase [227]. Others have reported smaller percentages released [219,229,288]. Qualitative analysis of the effect of proteases on specific, labelled polypeptides, as seen by SDS-polyacrylamide gels, has also been used as evidence that the polypeptides lost after protease treatment were externally-located [219,223,236–238].

The subcellular localization of the label can be studied most unequivocably by autoradiography since all of the incorporated label in each cell can be visualized without resorting to fractionation procedures. Standard iodination, fixation and processing procedures are followed [224] except ~5–10 times more radioactivity must be incorporated per cell than for biochemical experiments and $K^{127}I$ might be included in an overnight rinse following aldehyde fixation to enhance removal of any free $^{125}I$ [277]. Various control experiments are necessary to establish that the grains seen in autoradiography represent specifically-incorporated $^{125}I$. In our study [224] L cells incubated with all reagents except LPO, fixed and processed for autoradiography were negative. Cells incubated with the complete iodination system in the presence of $10^{-5}$ M $Na_2S_2O_3$ and processed contained 5% of the label associated with cells iodinated in the usual manner. Finally, only 0.5% of the incorporated label was extracted in the dehydrations preparatory to epon embedding. It was concluded that more than 95% of the grains seen by autoradiography represented specifically incorporated $^{125}I$.

When electron microscopic autoradiographs of iodinated L cells were examined (Fig. 12.5), 85–90% of the grains were located at the cell periphery, 5–14% in the centrosphere region, and 0–4% over the nucleus. Grain distribution appeared to be random over the periphery, occurring at the tips and depressions of surface projections. A tangential section through the surface projections of an iodinated cell (Fig. 12.5d) demonstrated the random labelling obtained. Labelling over the centrosphere was somewhat variable between cells in the same preparation and grains appeared to be localized primarily over vesicles. It is possible that cells warmed to 22°C for iodination and rinsed with 10°C solutions might have interiorized membrane before fixation.

This sensitive method has been successfully used in several iodination studies to localize the incorporated label [221,262,269,277]. Others have reported extensive intracellular labelling [229] or a high degree of labelling of dead cells [17]. The latter investigators cautioned against the indiscriminate use of surface labelling. Dead cell labelling is a problem and could lead to erroneous results, but the fourth method (see below) for surface localization, together with autoradiography, in our opinion, establishes the localization of the label and all the labelled components to the plasma membrane and allows further use of the method.

The fourth method, preparation of membrane-enriched fractions from iodinated cells, does not conclusively localize all of the label to the plasma membrane because the yield of membranes is always less than 100%. However, comparison of SDS-polyacrylamide gel labelling patterns from appropriate fractions will reinforce other evidence of the membrane localization of label and provide information on the contribution of labelled dead cells. First, the enrichment of acid-insoluble label in the membrane should parallel that of a number of putative plasma membrane markers, such as cholesterol and several enzyme activities. Then, the SDS gel of solubilized membranes isolated from intact, iodinated cells should exhibit the same label distribution pattern as

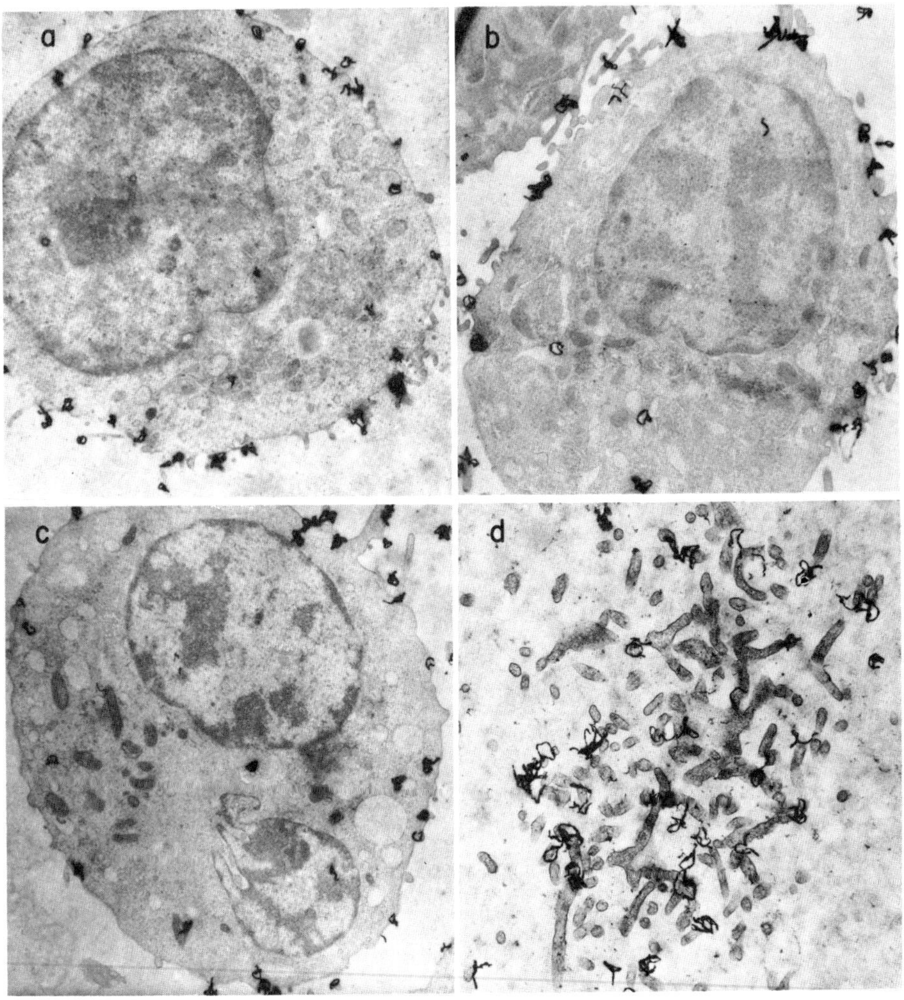


**Fig. 12.5** Electron microscopic autoradiographs of three L cells (a,b,c) iodinated on three different occasions. Grains represent $^{125}$I atoms covalently linked to tyrosine residues in proteins. Grains largely localized to cell periphery (x 7800). 5d: tangential section of an iodinated cell (x 27 000). (Figure from [224], courtesy of the Journal of Cell Biology).

that of the intact cell but with more radioactivity in each peak or band per unit protein applied. This increase reflects a relative enrichment of the labelled proteins in the membrane preparation over that of the intact cell. This is illustrated in Fig. 12.6 A. Membranes prepared from iodinated, intact L cells and the total homogenate from the same cells were solubilized and electrophoresed in two different tube gel systems (see [224] for details), a homogeneous 7.5% acrylamide gel and a discontinuous 5.0% over 7.5% gel. The homogenate and membrane coomassie blue staining patterns were quite different with large polypeptides predominating in the latter (Fig. 12.6 C). The parallel distribution of radioactivity of the homogenate and membrane fractions from iodinated L cells (Fig. 12.6 A and 12.6 B) demonstrates that all the peaks present in the

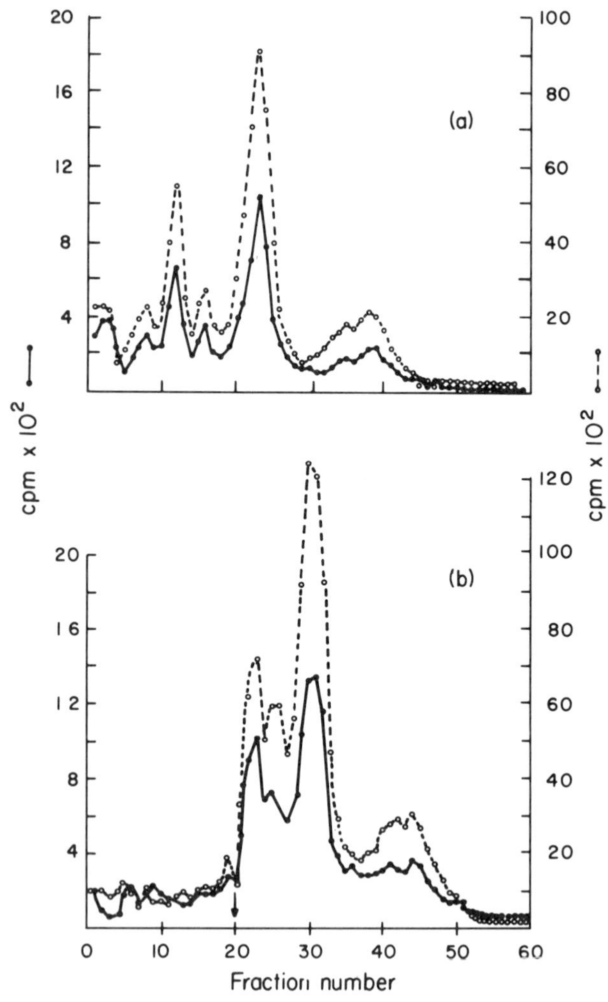

intact cell were also present in the isolated membrane fraction. However, the radioactivity in each peak of the membrane fraction was increased 14-fold over the homogenate (Fig. 12.6 A and 12.6 B). These results are further evidence that only the externally-disposed plasma membrane fractions were being iodinated in this system and also that these proteins were firmly anchored in the membrane since none were lost during the isolation procedure. A number of investigators have used this comparison method for localization of incorporated label [219,221,229,230] while others have prepared membrane fractions, analysed the label on SDS gels (usually <20% of the total cell label), but not compared it to that of the total cell [227,228,269,275,276].

A recent study [17] evaluated four surface labelling techniques using tissue

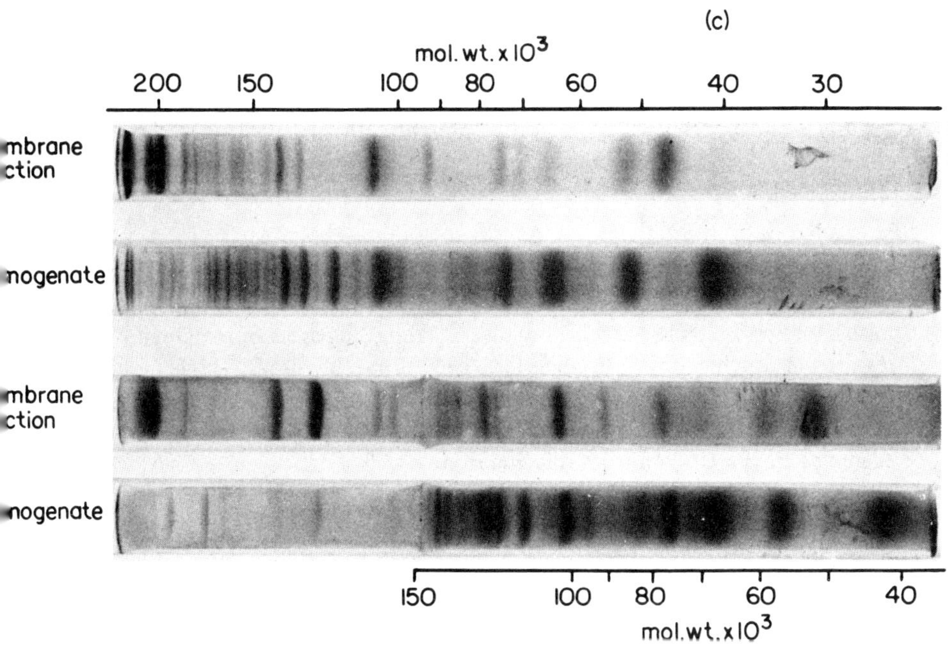


**Fig. 12.6** SDS-polyacrylamide tube gel electrophoresis patterns of iodinated L cells and membranes isolated from iodinated L cells. See Fig. 10 [224] for details of electrophoresis conditions. A, 7.5% acrylamide gels. Graph represents distribution of $^{125}$I in 80 μg total cell protein (●———●), 50 μg membrane protein (○ — — — — ○). B, Gels of 4 cm of 5% acrylamide over 6 cm 7.5% acrylamide. Gels cut and counted and the results plotted as in A. C, gels stained for protein with coomassie blue. Composition of gel I identical to 6A, gels II to 6B. (From [224], courtesy of the Journal of Cell Biology).

culture cells: galactose oxidase [192], pyridoxal phosphate [36], ($^3$H)-DIDS [14], and lactoperoxidase (according to Marchalonis *et al.* [262]). The high background of iodination obtained by these investigators has already been discussed. The methods for localization of label were light microscopic autoradiography, subcellular fractionation and SDS gel analysis of total cell and membrane label (only for galactose oxidase). The results suggested that labelled dead cells contribute significantly to the total incorporated label and could yield incorrect information if only the radioactivity profiles of total cells rather than of membranes were analysed on SDS gels. Our gel electrophoresis results of membrane fractions and total cells indicate that, under the iodination conditions we have used for L cells, this is not a problem. However, with lymphoid cell lines it has been necessary to remove dead cells on Ficoll-Hypaque gradients after iodination (G. Hoffman, personal communication). The question of whether the procedures of membrane isolation and dead cell removal also release loosely-attached (peripheral [109]) membrane proteins is difficult since the definition of this class of membrane component is largely operational and only slightly different from that of adsorbed proteins. Perhaps intrinsic isotope incorporation could establish whether the material removed was at least cell-derived.

## 12.6 Application of LPO-iodination to problems in membrane research

In this section we will review briefly several interesting areas of membrane research in which cell surface labels have been profitably used. A large number of the recent studies utilizing surface labels have chosen the LPO-iodination technique (see Table 12.5), and since we are most familiar with this system, the examples cited are taken from this literature.

### *12.6.1 Membrane Protein Disposition*

The transmembrane orientation of the two size classes of externally disposed glycoproteins in the erythrocyte was initially studied using non-permeant chemical labels ([30–32], Table 12.1, f). Many investigators have recently utilized LPO to explore the erythrocyte membrane in more detail (Table 12.5, A). The advantage of LPO is in the possibility of introducing the enzyme inside ghost vesicles during the process of resealing but not 'activating' the system until $H_2O_2$ (chemical or via glucose-glucose oxidase) and $I^-$ are added. Although the 100 000 dalton class of glycoproteins (Band III, [15]) has been shown to span the membrane using LPO [217], there is apparent disagreement as to the transmembrane character of the major glycoprotein (PAS-1 and 2). In the original study by Segrest *et al.* [218], intact cells, leaky ghosts and resealed ghosts were iodinated and label was found in a cyanogen bromide cleavage fragment from the C-terminal (presumed to be internal) end of the molecule only when leaky ghosts were iodinated. In all cases they reported label in a fragment presumed to be external due to its high carbohydrate content. Others have examined the label distribution of resealed ghosts iodinated from the inside only and reported no labelling of PAS–1, 2 [215,216]. Even though PAS-1 is highly accessible at the external surface, since it is only ~ 5% of total membrane protein and Band III at least 20% [15], the expectation for comparable labelling at the internal surface was probably unrealistic. In those studies in which the PAS-1 was reported to span the membrane, the glycoprotein was isolated and analysed free of other labelled components [213,217]. Examination of the gel from one study [213] suggests that PAS-1 is labelled to ~1.25% the extent inside as it is outside and thus would be missed if a total membrane label profile was analysed. Controls performed in this same study [213], consisting of the presence of catalase in the extracellular medium to scavenge $H_2O_2$ when inside iodination was the goal, suggested that the small amount of labelling was probably not due to PAS-1 labelled externally. Cyanogen bromide cleavage and peptide analysis would have been more conclusive.

The number and character of the externally-accessible membrane proteins in a variety of cell types has been studied using LPO-iodination (see Table 12.5,B–F for cells examined and references). The general finding is that there are more proteins of different size classes (by SDS gels) exposed on the surface of platelets [219,221], adipose cells [275,276], tissue culture cells [224,228–231,238,240] and lymphoid cells [269] than were found on the erythrocyte surface. The predominant species labelled have been glycoproteins of apparent large size (the uncertainty of glycoprotein molecular weight determinations on SDS gels is discussed in [329]). Although presence of incorporated label in various bands in SDS gels indicates an external localization of those polypeptides (provided sufficient controls (Section 12.5.3 and 12.5.4) and surface localization (Section 12.5.5) have been performed), the amount of label depends not only on the concentration of the polypeptide at the surface, but on the number of tyrosine residues in the polypeptide and the accessibility of those tyrosines to LPO during the reaction. Thus, a species which is prominently iodinated may or may not be a major membrane protein. Other labelling reagents which have low group specificity, such as the general photoaffinity probe discussed above (Section 12.2.1 and [41]), provided it is neutral in other regards, could be used to establish the relative concentrations of the exposed membrane polypeptides.

The disposition of proteins in membranes of subcellular organelles such as rat liver rough microsomes [288] and mitochondria [289,290], and rabbit skeletal muscle sarcoplasmic reticulum [292] has been studied by LPO-iodination. In addition, the external-orientation of membrane glycoproteins of several enveloped viruses has been confirmed using this surface label [235,285,286]

### 12.6.2 *The membrane in normal and transformed cells and membrane changes through the cell cycle*

Interest in the nature of the cell surface changes accompanying transformation has led a number of investigators to use LPO-iodination as a probe of the membranes of normal and transformed cells (Table 12.5, C). A large (> 200 000 daltons), external, transformation-sensitive glycoprotein has been reported to be labelled on normal cells from various lines but is absent or much reduced in their transformed counterparts [236–238,240,242]. This polypeptide accounts for 25–35% of the total label incorporated when confluent cells are exposed to LPO. Other membrane proteins are labelled in both normal and transformed cells. Trypsinization removes this protein from normal cells. The expression of the 200 000 dalton glycoprotein is apparently cell cycle-dependent as well as density dependent [230]. Normal cells in mitosis or in non-confluent, growing cultures have labelling profiles on SDS gels very similar to transformed cells, with only ~4.5% of the label in a 200 000 dalton band.

The characteristics of this protein are consistent with other findings on the differences between normal and transformed cells [330], but rigorous quantitation has yet to be performed. For instance, it would be of interest to measure the lectin-binding capacity of this protein(s), the percentage and nature of the label removed during trypsinization and of the labelled and unlabelled segments of the polypeptide which remain embedded in the membrane. Also, although a number of control experiments have suggested that this protein is not a fibroblast secretion product, such as hyaluronic acid or collagen [238], its localization to the membrane has not been firmly established since gel patterns are of total cells. There has been one report of the presence of the 200 000 dalton protein in isolated, uncharacterized membrane preparations [240]. Inititial attempts to purify the material have also suggested that a percentage of the intact protein (~15%) can be easily extracted from living cells without cell death [234]. Others, using immunological [331–334] and ultrastructural [335] analyses, have identified a fibroblast surface antigen of fibrillar morphology with characteristics very similar to those reported in the LPO studies. Although the antigen is synthesized by normal chicken and human cells, it is evidently released intact into the plasma and is readily detectable as an α-globulin [331]. It is not detectable in cultured transformed cells, but the medium from these cells has not been assayed for presence of the antigen. These data, together with another LPO study, indicate that the transformation-sensitive protein is not an integral component of the cell surface [241], but whether it is a peripheral membrane protein or a transiently-adsorbed secretion product has yet to be established.

A major problem in most of the LPO studies on normal and transformed cells is that cells iodinated are in monolayer culture. In initial autoradiographic experiments we found that a proteinaceous layer adherent to the culture dish was iodinated much more heavily than the plasma membrane of the cells spread on this layer. The extensive labelling of non-membrane elements in a system obscures analysis of those membrane proteins which are labelled and should be avoided or recognized and carefully controlled. Mastro *et al.* [231] found differences in the extent of iodination and in the polypeptides iodinated between monolayer BHK cells and the same cells suspended after mechanical detachment. When LPO-iodination is performed with carrier-free iodide usually less than 1% of the available sites are labelled. The labelling pattern is assumed to be representative of saturation labelling but iodination of monolayer cells might result in preferential, non-representative, iodination of one surface (the upper), in addition to the possibility of labelling non-membranous material.

### *12.6.3 The metabolic fate of membrane proteins*

Use of the enzymatic iodination technique to label only the externally-disposed membrane proteins provides an approach for studying membrane turnover, without the necessity of isolating membranes of varying purity and yield and without the problem of reutilization of label. Of course, it is necessary first to demonstrate that MIT, a non-reutilizable amino acid, is the species labelled and to localize the label to the surface before studying the fate of the labelled components. After establishing these two points with L cells [224], we followed the distribution of incorporated-$^{125}$I in L cells and medium over a period of 47 h after iodination (Table 12.9, see [225] for details). Analysis of constant volume aliquots excluded the necessity of correcting for label dilution due to cell division and allowed the distribution of the same *total* radioactivity in the population to be analysed at each time point. The appearance of 51% soluble M$^{125}$IT in the medium (not intracellular presumably because the MIT pool is small) by 47 h indicated that degradation was the predominant fate of the labelled membrane proteins (more precisely, the labelled peptide regions of these membrane proteins). Degradation was found to be cell-mediated and

TABLE 12.9
*Distribution of radioactivity after iodination*

| Hours after iodination | Acid-insoluble radioactivity | | {$^{125}$I} mono-iodotyrosine in medium* | Recovery of 0 hour acid-insoluble radioactivity |
|---|---|---|---|---|
| | Cells | Medium | | |
| | % | % | % | % |
| 0 | 97.4 | 2.6 | 0.5 | 100† |
| 3 | 82.4 | 5.8 | 8.1 | 96.3 |
| 20 | 46.5 | 13.0 | 40.5 | 100 |
| 47 | 26.4 | 16.5 | 51.0 | 93.9 |

*Percent of zero time, acid-insoluble radioactivity.
†Number of acid-insoluble counts at zero time set at 100%. (Table from [225], courtesy of the Journal of Cell Biology.)

temperature-dependent. A variable percentage of acid-insoluble label was also released from the cell population. The kinetics of label loss from the cells revealed that 80–95% was lost rather slowly, with a half-life of 25–33 h, and SDS gel analysis indicated a relatively synchronous loss of all the labelled components. The possible involvement of lysosomes in the degradation process was studied by allowing iodinated L cells to ingest non-digestible, polystyrene latex particles of low density, isolating latex-filled phagolysosomes over 47 h and comparing the intracellular fate of the labelled membrane proteins that had been internalized to that of the label remaining at the cell surface. First, SDS gel analysis demonstrated that the labelled proteins of the plasma membrane were present in the same relative proportions on the membranes surrounding interiorized latex particles (phagolysosomes). This suggested random internalization. However, the half-life of the majority of label inside phagolysosomes was found to be ~1–2 h, with some heterogeneity of degradation rates when individual bands on SDS gels were analysed. What, then, is the explanation for the apparent slow degradation of total cell (*i.e.*, cell surface) label in the non-phagocytic case, when forced internalization of plasma membrane does not occur? Our working hypothesis is that surface label enters the lysosomal pool at a very slow rate (~3% $h^{-1}$) but is degraded rapidly once it enters. A study of pinocytic rates in L cells [321] reported a much faster rate of interiorization of plasma membrane (25% $h^{-1}$) than the rate of degradation we found. (This value has been revised to ~50% $h^{-1}$ on the basis of morphometry (R. M. Steinman, personal communication).) Although the conditions of endocytosis differed (digestible versus non-digestible material; pinocytosis versus phagocytosis), these two studies suggest that membrane may be recycled before it enters the lysosomal compartment and is degraded, at least when digestible material is endocytosed. Another interesting problem as yet unsolved, is the fate of the unlabelled peptide segments of the externally-disposed proteins, which are presumably embedded in the lipid bilayer but non-functional after degradation of the exposed labelled segments.

Others have also studied the turnover of iodinated surface proteins in HeLa cells [228], baby hamster kidney (BHK) fibroblasts [231,236] and lymphoid cells [250–253,263]. Mouse splenic lymphocytes have been reported to shed intact surface immunoglobulin, as well as other uncharacterized, labelled material at a rapid rate ($t$ 1/2 ~ 6–8 h), but retain the surface histocompatibility antigens [251]. Lymphoid lines do not exhibit this extensive release [260]. Recent results suggest that the shedding phenomenon may not be physiological and that the actual rate of turnover of cell surface innunoglobulin is much slower [253]. BHK cells reportedly exhibit differential rates of loss and/or appearance of a 200 000 dalton surface component relative to the remainder of the labelled surface proteins [236]. Puromycin was found to inhibit this turnover. Since quantitation of label loss from cells, distribution of the total label during the experiment, or characterization of the labelled products has generally not been reported in these studies, comparison to our L-cell results is difficult.

### *12.6.4 Plasma membrane isolation*

To date there have been few reports of the use of enzymatic iodination as a plasma membrane marker in the development of new isolation procedures or in the evaluation of cross-contamination in other subcellular fractions. This use for surface labels should become more important in the future. One study on platelet subcellular fractionation reportedly established the identity of a membrane fraction as surface-derived using GO-LPO iodination [222]. In several studies [219–221,224,226,229,230,248,269,275,276], existing homogenization and isolation procedures were followed after intact cell iodination and the yield and relative enrichment of plasma membrane over homogenate assessed using incorporated-$^{125}$I as a membrane marker. Generally the purity of these preparations has been rather low (~6–20-fold purification).

### *12.6.5 Immunologically-important surface molecules in lymphoid cells*

Cellular immunologists have used enzymatic iodination extensively to label lymphocyte cell surfaces in order to characterize those molecules involved in phenomena such as histocompatability and antigen recognition and to study their fate. The principal techniques used have been iodination of intact cells, immunoprecipitation of a specific antigen(s) in the presence of non-ionic detergents and SDS-polyacrylamide gel electrophoretic analysis of the labelled molecules which were specifically precipitated. This approach to the identification of membrane constituents present in trace amounts, has general applicability to many areas of membrane research and will undoubtedly be used to a greater extent in the future.

The laboratories of Uhr ([246–259,271,274] and [260] for review) and Marchalonis ([262–268] and [261] for review) have been principally involved in LPO-iodination of a wide variety of lymphocytes from different lymphoid organs of mouse as well as lymphoid cell lines of both mouse and human origin (Table 12.5, D and E). Both these groups, and others, have reported the presence of monomers of the M class of immunoglobulin (8S IgM) on the surfaces of bone marrow-derived (B) mouse lymphocytes [247,249–255,258,264] and mouse and human B cell lines, both secretors and non-secretors of IgG [246,247,248, 268,269]. In addition, another immunoglobulin has been found to be externally-disposed and LPO-iodinated on mouse spleen cells and tentatively identified as IgD [254,256]. Histocompatibility antigens in mouse lymphocytes of both the B and thymus-derived (T) cell type [255,258], the thymus leukemia [255] and thy-1 ($\theta$) antigens on mouse thymocytes [257,259], and an oncorna virus glycoprotein on normal and malignant thymocytes [273,274] have also been localized to cell surfaces. In addition the fate of several of these surface molecules has been studied [250–253,258,259,263] and was discussed above (Section 12.6.4). To date, approximately 5–10% of the total surface label on B cells has been specifically identified. The remaining 90% has yet to be investigated.

In all of these studies utilizing immunoprecipitation, non-specific precipitation with control serum has usually accounted for ~10–50% of the specific immunoprecipitable radioactivity. This is a rather high background. Also, the SDS gel radioactivity profiles have generally revealed very broad peaks which suggests either molecular heterogeneity or overloading of gels due to the presence of unlabelled antibody protein as well as the labelled surface molecules. The use of insolubilized antibodies might allow a greater subsequent resolution of the proteins of interest.

## 12.7 Summary and prospects

The current list of cell surface labels is diverse and growing rapidly. In this chapter we have classified, reviewed and evaluated these membrane reagents in an effort to aid the investigator who must choose a labelling system suitable to his particular problem in membrane research. It should be apparent that there is no ideal, universal label. Rather, each membrane reagent has advantages which can be maximized as well as disadvantages which should be clearly recognized and minimized. First the reagent whose physical and chemical characteristics best satisfy the requirements of the specific membrane system and question being studied should be selected. Then, the parameters of the labelling reaction (e.g., reagent concentrations, time, temperature, pH, ions) should be explored in order to achieve maximal labelling without introducing undesirable side reactions (e.g., non-specific labelling, reagent permeation, membrane reorganization, cell lysis). There are no ideal conditions which are applicable to all membranes and all labelling reagents. Finally, one should recognize that the characteristics of both the reagent used and the membrane system studied will determine which available membrane constituents are actually labelled. In studies of membrane asymmetry this point is of particular importance and suggests the use of at least two labels of different reaction specificities in order to maximize the number of exposed membrane components labelled.

A large part of this chapter was devoted to the methodology of and problems with the enzymatic iodination technique as used by ourselves and many others. We hope our treatment provides some general guide-lines for use of this labelling method as well as the others reviewed here.

The prospects for cell surface labels are bright. We have just begun to explore the numerous uses for these reagents in membrane research. The erythrocyte has been the model cell for initial studies but cells having more complex membrane systems are now being studied with surface labels. In addition, isolation and functional characterization of the external membrane proteins can now be more easily achieved using surface labels and even a combination of labels. For instance, a general probe could be used to tag all external molecules, together with a site-specific probe to label a restricted number of exposed molecules. As yet there is no label which is specific only for the lipids in a membrane, which is understandable considering the paucity of reactive groups on phospholipid molecules. The most likely candidate for use as a lipid label would be a photoactivatable reagent sufficiently hydrophobic to insert into membranes without passing through. Certainly the next review on specific labels for cell surfaces will list such a reagent and many more still to be developed.

## Acknowledgements

We appreciate the advice and helpful criticisms of T. Brockmeyer, D. Castle, P. Edelson, G. Hoffman, J. P. Kraehenbuhl, G. E. Palade and S. Silverstein, who read the manuscript. In addition, we would like to thank V. Iwanij, G. Hoffman and R. Steinman for their personal communications, and M. Grossi, F. Hubbard, and G. Silvestri for the help in preparing the manuscript. A. L. H. was a Post-doctoral Fellow of the Leukemia Society of America, Inc.

## References

1. Guidotti, G. (1972), *Ann. Rev. Biochem.*, **41**, 731.
2. Wallach, D. F. H. (1972), *Biochim. biophys. Acta.*, **265**, 61.
3. Oseroff, A. R., Robbins, P. W. and Burger, M. M. (1973), *Ann. Rev. Biochem.*, **42**, 647.
4. Juliano, R. L. (1973), *Biochim. biophys. Acta.*, **300**, 341.
5. Steck, T. L. (1974), *J. Cell Biol.*, **62**, 1.
6. Zwall, R. F. A., Roelofsen, B. and Colley, C. M. (1973), *Biochim. biophys. Acta.*, **300**, 159.
7. Maddy. A. H. (1964), *Biochim. biophys. Acta.*, **88**, 390.
8. Pauling, L. (1960), *The Nature of the Chemical Bond*, Cornell University Press, Ithaca, N.Y.
9. Means, G. E. and Feeney, R. E. (1971), *Chemical Modification of Proteins*, Holden Day, San Francisco, California.
10. Deenen, L. L. M. van and Gier, J. de. (1964), in *The Red Blood Cell*, (Bishop, C. and Surgenor, D. M., eds.), p. 243, Academic Press, New York.
11. Knauf, P. A. and Rothstein, A. (1971), *J. gen. Physiol.*, **58**, 190.
12. Marinetti, G. V. and Gray, G. M. (1967), *Biochim. biophys. Acta.*, **135**, 580.
13. Cabantchik, Z. I. and Rothstein, A. (1972), *J. Memb. Biol.*, **10**, 311.
14. Cabantchik, Z. I. and Rothstein, A. (1974), *J. Memb. Biol.*, **15**, 207.
15. Fairbanks, G., Steck, T. L. and Wallach, D. F. H. (1971), *Biochemistry*, **10**, 2606.
16. Cabantchik, Z. I. and Rothstein, A. (1974), *J. Memb. Biol.*, **15**, 227.
17. Juliano, R. L. and Behar-Bannelier, M. (1975), *Biochim. biophys. Acta.*, **375**, 249.
18. Van Steveninck, J., Weed, R. I. and Rothstein, A. (1965), *J. gen. Physiol.*, **48**, 617.

19. Weed, R. I., Eber, J. and Rothstein, A. (1962), *J. gen. Physiol.*, **45**, 395.
20. Weed, R. I., Reed, C. F. and Berg, G. (1963), *J. clin. Invest.*, **42**, 581.
21. Sutherland, R. M., Rothstein, A. and Weed, R. I. (1967), *J. Cell Physiol.*, **69**, 185.
22. Pardee, A. B. and Watanabee, K. (1968), *J. Bact.*, **96**, 1049.
23. Tsan, M. and Berlin, R. D. (1971), *J. exp. Med.*, **134**, 1016.
24. Godin, D. V. and Schrier, S. L. (1972), *J. Memb. Biol.*, **7**, 285.
25. Carter, J. R., Jr. (1973), *Biochemistry*, **12**, 171.
26. Berg, H. C. (1969), *Biochim. biophys. Acta.*, **183**, 65.
27. Lenard, J. (1970), *Biochemistry*, **9**, 1129.
28. Bender, W. W., Garan, H. and Berg, H. C. (1971), *J. molec. Biol.*, **58**, 783.
29. Carraway, K. L., Kobylka, D. and Triplett, R. B. (1971), *Biochim. biophys. Acta.*, **241**, 934.
30. Bretscher, M. S. (1971), *J. molec. Biol.*, **58**, 775.
31. Bretscher, M. S. (1971), *J. molec. Biol.*, **59**, 351.
32. Bretscher, M. S. (1971), *Nature New Biology*, **231**, 229.
33. Rosenberg, S. A. and Guidotti, G. (1968), *J. biol. Chem.*, **243**, 1985.
34. Reed, C. F., Swisher, S. N., Marinetti, G. V. and Eden, E. G. (1960), *J. Lab. clin. Med.*, **56**, 281.
35. Whiteley, N. M. and Berg, H. C. (1974), *J. molec. Biol.*, **87**, 541.
36. Rifkin, D. B., Compans, R. W. and Reich, E. (1972), *J. biol. Chem.*, **247**, 6432.
37. Kreibich, G., Hubbard, A. L. and Sabatini, D. D. (1974), *J. Cell. Biol.*, **60**, 616.
38. Hunt, R. C. and Brown, J. C. (1974), *Biochemistry*, **13**, 22.
39. Roberts, R. M. and Yuan, B. O. (1974), *Biochemistry*, **13**, 4846.
40. Cabantchik, Z. I., Balshin, M., Breuer, W., Markus, H. and Rothstein, A. (1975), *Biochim. biophys. Acta.*, **382**, 621.
41. Staros, J. V. and Richards, F. M. (1974), *Biochemistry*, **13**, 2720.
42. Knowles, J. R. (1972), *Accounts of Chemical Research*, **5**, 155.
43. Marchesi, V. T., Steers, E. Jr., Tillack, T. W. and Marchesi, S. L. (1969), in *Red Cell Membrane Structure and Function*, (Jamieson, G. A. and Greenwalt, T. J., eds.), p. 117, J. B. Lippincott Co., Philadelphia.
44. Staros, J. V., Haley, B. E. and Richards, F. M. (1974), *J. biol. Chem.*, **249**, 5004.
45. Schmidt-Ullrich, R., Knüfermann, H. and Wallach, D. F. H. (1973), *Biochim. biophys. Acta.*, **307**, 353.
46. Gray, W. R. (1967), in *Meth. Enzym.*, (Colowick, S. P., Kaplan, N. O. and Hirs, G. H. W., eds.), Vol. XI, p. 139, Academic Press, New York.
47. Bretscher, M. S. (1973), *Nature New Biology*, **245**, 116.
48. Wallach, D. F. H., Schmidt-Ullrich, R. and Knüfermann, H. (1974), *Nature, Lond.*, **248**, 623.
49. Kant, J. A. and Steck, T. L. (1973), *J. biol. Chem.*, **248**, 8457.
50. Steck, T. L. and Yu, J. (1973), *J. supramolec. Struct.*, **1**, 220.
51. Carraway, K. L., Kobylka, D., Summers, J. and Carraway, C. A. (1972), *Chem. Phys. Lipids*, **8**, 65.
52. Gordesky, S. E., Marinetti, G. V. and Segel, G. B. (1972), *Biochem. biophys. Res. Comm.*, **47**, 1004.
53. Gordesky, S. E., Marinetti, G. V. and Segel, G. B. (1973), *J. Memb. Biol.*, **14**, 229.
54. Woodward, C. B. and Zwaal, R. F. A. (1972), *Biochim. biophys. Acta.*, **274**, 272.

55. Shaw, E. (1970), *Physiol. Rev.*, **50**, 244.
56. Shapiro, B., Kollman, G. and Martin, D. (1970), *J. Cell. Physiol.*, **75**, 281.
57. Knauf, P. A. and Rothstein, A. (1971), *J. gen. Physiol.*, **58**, 211.
58. Lenard, J. (1970), *Biochemistry*, **9**, 5037.
59. Passow, H. (1969), *Prog. Biophys. Mol. Biol.*, **19**, 424.
60. Berg, H. C., Diamond, J. M. and Marfey, P. S. (1965), *Science*, **150**, 64.
61. Passow, H. and Schnell, K. F. (1969), *Experientia*, **25**, 460.
62. Poensgen, J. and Passow, H. (1971), *J. Memb. Biol.*, **6**, 210.
63. Krupka, R. M. (1972), *Biochim. biophys. Acta.*, **282**, 326.
64. Okuyama, T. and Satake, K. (1960), *J. Biochem., Tokyo*, **47**, 454.
65. Means, G. E., Congdon, W. I. and Bender, M. L. (1972), *Biochemistry*, **11**, 3564.
66. Kotaki, A., Harada, M. and Yagi, K. (1964), *J. Biochem., Tokyo*, **55**, 553.
67. Bonsall, R. W. and Hunt, S. (1971), *Biochim. biophys. Acta.*, **249**, 281.
68. Arrotti, J. J. and Garvin, J. E. (1972), *Biochim. biophys. Acta.*, **255**, 79.
69. Gordesky, S. E. and Marinetti, G. V. (1973), *Biochem. biophys. Res. Comm.*, **50**, 1027.
70. Gordesky, S. E., Marinetti, G. V. and Love, R. (1975), *J. Memb. Biol.*, **20**, 111.
71. Van Lenten, L. and Ashwell, G. (1971), *J. biol. Chem.*, **246**, 1889.
72. Blumenfeld, O. O., Gallop, P. M. and Liao, T. H. (1972), *Biochem. biophys. Res. Comm.*, **48**, 242.
73. Liao, T., Gallop, P. M. and Blumenfeld, O. O. (1973), *J. biol. Chem.*, **248**, 8247.
74. Steck, T. L. and Dawson G. (1974), *J. biol. Chem.*, **249**, 2135.
75. Warren, L. (1959), *J. biol. Chem.*, **234**, 1971.
76. Zacharias, R. M., Zell, T. E., Morrison, J. H. and Woodlock, J. J. (1969), *Analyt. Biochem.*, **30**, 148.
77. Spiro, R. G. (1964), *J. biol. Chem.*, **239**, 567.
78. Rothfus, J. A. and Smith, E. L. (1963), *J. biol. Chem*, **238**, 1402.
79. Novogrodsky, A. and Katchalski, E. (1971), *F.E.B.S. Letters*, **12**, 297.
80. Novogrodsky, A. and Gery, I. (1972), *J. Immunol.*, **109**, 1278.
81. McClain, D. A., Wang, J. L. and Edelman, G. M. (1975), *Cell. Immunol.*, **15**, 287.
82. Novogrodsky, A. and Katchalski, E. (1972), *Proc. natn. Acad. Sci. U.S.A.*, **69**, 3207.
83. Zatz, M. M., Goldstein, A. L., Blumenfeld, O. O. and White, A. (1972), *Nature New Biology*, **240**, 252.
84. Paz, M. A., Henson, E., Rombauer, R., Abrash, L., Blumenfeld, O. O. and Gallop, P. M. (1970), *Biochemistry*, **9**, 2123.
85. Singer, S. J. (1967), *Adv. Protein Chem.*, **22**, 1.
86. Fox, C. F. and Kennedy, E. P. (1965), *Proc. natn. Acad. Sci. U.S.A.*, **54**, 891.
87. Fox, C. F., Carter, J. R. and Kennedy, E. P. (1967), *Proc. natn. Acad. Sci. U.S.A.*, **572**, 698.
88. Bellhorn, M. B., Blumenfeld, O. O. and Gallop, P. M. (1970), *Biochem. biophys. Res. Comm.*, **39**, 267.
89. Hokin, L. E., Mokotoff, M. and Kupchan, S. M. (1966), *Proc. natn. Acad. Sci. U.S.A.*, **55**, 797.
90. Ruoho, A. and Kyte, J. (1974), *Proc. natn. Acad. Sci. U.S.A.*, **71**, 2352.
91. Reiter, M. J., Cowburn, D. A., Prives, J. M. and Karlin, A. (1972), *Proc. natn. Acad. Sci. U.S.A.*, **69**, 1168.

92. Karlin, A., Prives, J., Deal, W. and Winnik, M. (1971), *J. molec. Biol.*, **61**, 175.
93. Karlin, A. and Cowburn, D. (1973), *Proc. natn. Acad. Sci. U.S.A.*, **70**, 3636.
94. Karlin, A., Cowburn, D. A. and Reiter, M. J. (1973), in *Drug Receptors, A Biological Council Symposium*, (Rang, H. P., ed.), p. 193, MacMillan, London.
95. Karlin, A. (1973), *Fedn. Proc., Fedn. Am. Socs. exp. Biol.*, **32**, 1847.
96. Guthrow, E. C., Rasmussen, H., Brunswick, D. J. and Cooperman, B. S. (1973), *Proc. natn. Acad. Sci. U.S.A.*, **67**, 1688.
97. Kiefer, H., Lindstrom, J., Lennox, E. S. and Singer, S. J. (1970), *Proc. natn. Acad. Sci. U.S.A.*, **67**, 1688.
98. Ruoho, A. E., Kiefer, H., Roeder, P. E. and Singer, S. J. (1973), *Proc. natn. Acad. Sci. U.S.A.*, **70**, 2567.
99. Haley, B. E. and Hoffman, J. F. (1974), *Proc. natn. Acad. Sci. U.S.A.*, **71**, 3367.
100. Knauf, P. A., Proverbio, F. and Hoffman, J. F. (1974), *J. gen. Physiol.*, **63**, 305.
101. Avruch, J. and Fairbanks, G. (1972), *Proc. natn. Acad. Sci. U.S.A.*, **69**, 1216.
102. Knauf, P. A., Proverbio, F. and Hoffman, J. F. (1974), *J. gen. Physiol.*, **63**, 305.
103. Knauf, P. A., Proverbio, F. and Hoffman, J. F. (1974), *J. gen. Physiol.*, **63**, 324.
104. Skou, J. C. (1965), *Physiol. Rev.*, **45**, 596.
104a. Levy, D. (1973), *Biochim. biophys. Acta.*, **322**, 329.
105. Boyd, W. C. and Shapleigh, E. (1954), *J. Lab. clin. Med.*, **44**, 235.
106. Sharon, N. and Lis, H. (1972), *Science*, **177**, 949.
107. Lis, H. and Sharon, N. (1973), *Ann. Rev. Biochem.*, **42**, 541.
108. Nicolson, G. L. (1974), *Int. Rev. Cytol.*, **39**, 89.
109. Singer, S. J. and Nicolson, G. L. (1972), *Science*, **175**, 720.
110. Agrawal, B. B. L. and Goldstein, I. J. (1967), *Biochim. biophys. Acta.*, **147**, 262.
111. Olson, M. O. J. and Liener, I. E. (1967), *Biochemistry*, **6**, 3801.
112. Kalb, A. J. and Lustig, A. (1968), *Biochim biophys. Acta.*, **168**, 366.
113. Becker, J. W., Reeke, G. N. and Edelman, G. M. (1971), *J. biol. Chem.*, **246**, 6123.
114. Hardman, K. D. and Ainsworth, C. F. (1972), *Nature New Biology*, **237**, 54.
115. Edelman, G. M., Cunningham, B. A., Reeke, G. N. Jr., Becker, J. W., Waxdal, M. J. and Wang, J. L. (1972), *Proc. natn. Acad. Sci. U.S.A.*, **69**, 2580.
116. Burger, M. M. and Goldberg, A. R. (1967), *Proc. natn. Acad. Sci. U.S.A.*, **57**, 359.
117. Allen, A. K., Neuberger, A. and Sharon, N. (1973), *Biochem. J.*, **131**, 155.
118. Wright, C. S., Sato, T., Nagata, Y., MacMillan, J., Langridge, R. and Burger, M. M. (1972), *Fed. Proc., Fedn. Am. Socs. exp. Biol.*, **31**, 923.
119. Liener, I. E. and Pallansch, M. J. (1952), *J. biol. Chem.*, **197**, 29.
120. Lis, H., Fridman, C., Sharon, N. and Katchalski, E. (1966), *Archs. Biophy. Biochem.*, **117**, 301.
121. Lis, H., Sharon, N. and Katchalski, E. (1969), *Biochim. biophys. Acta.*, **192**, 364.

122. Weber, T., Nordman, C. T. and Gräsbeck, R. (1967), *Scand. J. Haemat.*, **4**, 77.
123. Allen, L. W., Svenson, R. H. and Yachnin, S. (1969), *Proc. natn. Acad. Sci. U.S.A.*, **63**, 334.
124. Rigas, D. A. and Head, C. (1969), *Biochem. biophys. Res. Comm.*, **34**, 633.
125. Miller, J. B., Noyes, C., Heinrikson, R., Kingdon, H. S. and Yachnin, S. (1973), *J. exp. Med.*, **138**, 939.
126. Nicolson, G. L. and Blaustein, J. (1972), *Biochim. biophys. Acta.*, **266**, 543.
127. Tomita, M., Kurokawa, T., Onozaki, K., Ichiki, M., Osawa, T. and Ukita, T. (1972), *Experientia*, **28**, 84.
128. Nicolson, G. L., Blaustein, J. and Etzler, M. E. (1974), *Biochemistry*, **13**, 196.
129. Majerus, P.and Brodie, G. N. (1972), *J. biol. Chem.*, **247**, 4253.
130. Allan, D., Auger, J. and Crumpton, M. J. (1972), *Nature New Biology*, **236** 23.
131. Hayman, M. J. and Crumpton, J. J. (1972), *Biochem. biophys. Res. Comm.*, **49**, 923.
132. Hunt, R. C., Bullis, C. M. and Brown, J. C. (1975), *Biochemistry*, **14**, 109.
133. Findlay, J. B. C. (1974), *J. biol. Chem.*, **249**, 4398.
134. Burger, M. M. (1968), *Nature*, **219**, 499.
135. Janson, V. K. and Burger, M. M. (1973), *Biochim. biophys. Acta.*, **291**, 127.
136. Janson, V. K., Sakamoto, C. K. and Burger, M. M. (1973), *Biochim. biophys. Acta.*, **291**, 136.
137. Fukuda, M. and Osawa, T. (1973), *J. biol. Chem.*, **248**, 5100.
138. Wray, V. P. and Walborg, E. F. Jr. (1971), *Cancer Res.*, **31**, 2072.
139. Smith, D. F. and Walborg, E. F. Jr. (1972), *Cancer Res.*, **32**, 543.
140. Smith, D. F., Neri, G. and Walborg, E. F. Jr. (1973), *Biochemistry*, **12**, 2111.
141. Kornfeld, R., Keller, J., Baenziger, J. and Kornfeld, S. (1971), *J. biol. Chem.*, **246**, 3259.
142. Chang, K. J., Bennett, V. and Cuatrecasas, P. (1975), *J. biol. Chem.*, **250**, 488.
143. Ozanne, B. and Sambrook, J. (1971), *Nature New Biology*, **232**, 156.
144. Arndt-Jovin, D. J. and Berg, P. (1971), *J. Virol.*, **8**, 716.
145. Cline, M. J. and Livingston, D. C. (1971), *Nature New Biology*, **232**, 155.
146. Noonan, K. D. and Burger, M. M. (1973), *J. biol. Chem.*, **248**, 4286.
147. Greaves, M. F., Bauminger, S. and Janossy, G. (1972), *Clin. exp. Immunol.*, **10**, 537.
148. Boldt, D., Skinner, A. M. and Kornfeld, S. (1972), *J. clin. Invest.*, **51**, 3225.
149. Lutton, J. D. (1973), *J. Cell Biol.*, **56**, 611.
150. Oliver, J. M., Ukena, T. E. and Berlin, R. D. (1974), *Proc. natn. Acad. Sci. U.S.A.*, **71**, 394.
151. Penhoet, E., Olsen, C., Carlson, S., Lacorbiere, M. and Nicolson, G. L. (1974), *Biochemistry*, **13**, 3561.
152. Nicolson, G. L. and Lacorbiere, M. (1973), *Proc. natn. Acad. Sci. U.S.A.*, **70**, 1672.
153. Steinemann, A. and Stryer, L. (1973), *Biochemistry*, **12**, 1499.
154. Kornfeld, S. and Kornfeld, R. (1971), in *Glycoproteins of Blood Cells and*

*Plasma*, (Jamieson, G. A. and Greenwalt, T. J., eds.), p. 50, J. B. Lippincott Co., Philadelphia.
155. Taylor, R. B., Duffus, W. P. H., Raff, M. and dePetris, S. (1971), *Nature New Biology,* **233**, 225.
156. Unanue, E. R., Perkins, W. D. and Karnovsky, M. J. (1972), *J. exp. Med.,* **136**, 885.
157. Unanue, E. R., Karnovsky, M. J. and Engers, H. D. (1973), *J. exp. Med.,* **137**, 675.
158. Smith, S. B. and Revel, J. P. (1972), *Devl. Biol.,* **27**, 434.
159. Bernhard, W. and Avramaes, S. (1971), *Exp. Cell. Res.,* **64**, 232.
160. Rosenblith, J. Z., Ukena, T. E., Yin, H. H., Berlin, R. D. and Karnovsky, M. J. (1973), *Proc. natn. Acad. Sci. U.S.A.,* **70**, 1625.
161. Graham, R. C., Jr. and Karnovsky, M. J. (1966), *J. Histochem. Cytochem.,* **14**, 291.
162. Abercrombie, M., Heaysman, J. E. M. and Pegrum, S. M. (1972), *Exp. Cell Res.,* **73**, 536.
163. Martinez-Paloma, A., Wicker, R. and Bernhard, W. (1972), *Int. J. Cancer,* **9** 676.
164. Bretten, R., Wicker, R. and Bernhard, W. (1972), *Int. J. Cancer,* **10**, 397.
165. Pinto da Silva, P. and Nicolson, G. L. (1974), *Biochim. biophys. Acta.,* **363**, 311.
166. Tillack, T. W., Scott, R. E. and Marchesi, V. T. (1972), *J. exp. Med.,* **135**, 1209.
167. Marchesi, V. T., Tillack, T. W., Jackson, R. L., Segrest, J. P. and Scott, R. D. (1972), *Proc. natn. Acad. Sci. U.S.A.,* **69**, 1445.
168. Klein, P. A. and Adams, W. R. (1972), *J. Virol.,* **10**, 844.
169. Williams, M. A. and Voaker, D. (1972), *Br. J. Haemat.,* **23**, 427.
170. Stobo, J. D., Rosenthal, A. S. and Paul, W. E. (1972), *J. Immunol.,* **108**, 1.
171. Nicolson, G. L. and Singer, S. J. (1971), *Proc. natn. Acad. Sci. U.S.A.,* **68**, 942.
172. Nicolson, G. L. and Singer, S. J. (1974), *J. Cell Biol.,* **60**, 236.
173. Nicolson, G. L. (1971), *Nature New Biology,* **233**, 244.
174. Nicolson, G. L. (1972), *Nature New Biology,* **239**, 193.
175. Novogrodsky, A. and Katchalski, E. (1971), *Biochim. biophys. Acta.,* **228**, 579.
176. Andersson, J., Edelman, G. M., Möller, G. and Sjöberg, O. (1972), *Eur. J. Immunol.,* 2, 233.
177. Greaves, M. F. and Bauminger, S. (1972), *Nature New Biology,* **235**, 67.
178. Betel, I. and van den Berg, K. J. (1972), *Eur. J. Biochem.,* **30**, 571.
179. Holm, G., Perlmann, P. and Werner, B. (1964), *Nature,* **203**, 841.
180. Berlin, R. D. (1972), *Nature New Biology,* **235**, 44.
181. Cuatrecasas, P. and Tell, G. P. E. (1973), *Proc. natn. Acad. Sci. U.S.A.,* **70**, 485.
182. Czech, M. P. and Lynn, W. S. (1973), *Biochim. biophys. Acta.,* **297**, 368.
183. Edelson. P. J. and Cohn, Z. A. (1974), *J. exp. Med.,* **140**, 1364.
184. Edelson, P. J. and Cohn, Z. A. (1974), *J. exp. Med.,* **140**, 1387.
185. Riordan, J. R. and Slavik, M. (1974), *Biochim. biophys. Acta.,* **373**, 356.
186. Inbar, M., Ben-Bassat, H. and Sachs, L. (1971), *J. Memb. Biol.,* **6**, 195.
187. Isselbacher, K. J. (1972), *Proc. natn. Acad. Sci. U.S.A.,* **69**, 585.
188. Gunther, G. R., Wang, J. L., Yahara, I., Cunningham, B. A. and Edelman, G. M . (1973), *Proc. natn. Acad. Sci. U.S.A.,* **70**, 1012.
189. Phillips, D. R. and Morrison, M. (1971), *Biochemistry,* **10**, 1766.

190. Kinzel, V. and Mueller, G. C. (1973), *Biochim. biophys. Acta.*, **322**, 337.
191. Steinman, R. S. and Cohn, Z. A. (1972), *J. Cell Biol.*, **55**, 186.
192. Gahmberg, C. G. and Hakomori, S. (1973), *J. biol. Chem.*, **248**, 4311.
193. Kraemer, P. M. (1971), in *Biomembranes I*, (Manson, L. A., ed.), p. 67, Plenum Press, New York.
194. Ginsburg, V. and Kobata, A. (1971), in *Structure and Function of Biological Membranes*, (Rothfield, L. A., ed.), p. 439, Academic Press, New York.
195. Steck, T. L. (1972), in *Membrane Research*, (Fox, C. F., ed.), p. 71, Academic Press, New York.
196. Gahmberg, C. G. and Hakomori, S. (1973), *Proc. natn. Acad. Sci. U.S.A.*, **70**, 3329.
197. Gahmberg, C. G., Kiehn, D. and Hakomori, S. (1974), *Nature*, **248**, 413.
198. Critchley, D. R. (1974), *Cell*, **3**, 121.
199. Novogrodsky, A. and Katchalski, E. (1973), *Proc. natn. Acad. Sci. U.S.A.*, **70**, 1824.
200. Brewer, G. J. and Singer, S. J. (1974), *Biochemistry*, **13**, 3580.
201. Datta, P. (1974), *Biochemistry*, **13**, 3987.
202. Pricer, W. E., Jr. and Ashwell, G. (1971), *J. biol. Chem.*, **246**, 4825.
203. Hickman, J., Ashwell, G., Morell, A. G., van den Hamer, C. J. A. and Scheinberg, I. H. (1970), *J. biol. Chem.*, **245**, 759.
204. Kodza, H. and Jungeblut, C. W. (1958), *J. Immunol.*, **81**, 76.
205. Kraemer, P. M. (1967), *J. Cell Physiol.*, **69**, 199.
206. Phillips, D. R. and Morrison, M. (1970), *Biochem. biophys. Res. Comm.*, **40**, 284.
207. Marchalonis, J. J. (1969), *Biochem. J.*, **113**, 299.
208. Hubbard, A. L. and Cohn, Z. A. (1972), *J. Cell Biol.*, **55**, 390.
209. Phillips, D. R. and Morrison, M. (1971), *F.E.B.S Letters*, **18**, 95.
210. Phillips, D. R. and Morrison, M. (1971), *Biochem. biophys. Res. Comm.*, **45**, 1103.
211. Phillips, D. R. and Morrison, M. (1973), *Nature New Biology*, **242**, 213.
212. Teuch, J. K. and Morrison, M. (1974), *Biochem. biophys. Res. Comm.*, **59**, 352.
213. Morrison, M., Mueller, J. J. and Huber, C. T. (1974), *J. biol. Chem.*, **249**, 2658.
214. Atlas, S. J., Shapiro, B. and Green, J. W. (1973), *Biochim. biophys. Acta.*, **323**, 194.
215. Shin, B. C. and Carraway, K. L. (1974), *Biochim. biophys. Acta.*, **345**, 141.
216. Reichstein, E. and Blostein, R. (1973), *Biochem. biophys. Res. Comm.*, **54**, 494.
217. Boxer, D. H., Jenkins, R. E. and Tanner, M. J. A. (1974), *Biochem. J.*, **137**, 531.
218. Segrest, J., Kahane, I., Jackson, R. L. and Marchesi, V. T. (1973), *Archs. Biophys. Biochem.*, **155**, 167.
219. Phillips, D. R. (1972), *Biochemistry*, **11**, 4582.
220. Barber, A. J. and Jamieson, G. A. (1971), *Biochemistry*, **10**, 4711.
221. Nachman, R. L., Hubbard, A. and Ferris, B. (1973), *J. biol. Chem.*, **248**, 2928.
222. Taylor, D. G. and Crawford, M. (1974), *F.E.B.S Letters*, **41**, 317.
223. Steiner, M. (1973), *Biochim. biophys. Acta.*, **323**, 653.
224. Hubbard, A. L. and Cohn, Z. A. (1975), *J. Cell Biol.*, **64**, 438.

225. Hubbard, A. L. and Cohn, Z. A. (1975), *J. Cell Biol.,* **64**, 461.
226. Poduslo, J. F., Greenberg, G. S. and Glick, M. C. (1972), *Biochemistry,* **11**, 2616.
227. Tsai, C. M., Huang, C. C. and Canellakis, E. S. (1973), *Biochim. biophys. Acta.,* **332**, 47.
228. Huang, C. C., Tsai, C. M. and Canellakis, E. S. (1973), *Biochim. biophys. Acta.,* **332**, 59.
229. Gates, R. E., McClain, M. and Morrison, M. (1974), *Exp. Cell. Res.,* **83**, 344.
230. Shin, B. C. and Carraway, K. L. (1973), *Biochim. biophys. Acta.,* **330**, 254.
231. Mastro, A. M., Beer, C. T. and G. C. Mueller, (1974), *Biochim. biophys. Acta.,* **352**, 38.
232. Owen, E., Knight, V. A. and Thomas, H. W. (1973), *Exp. Cell Res.,* **82**, 461.
233. Truding, R., Shelanski, M. L., Daniels, M. P. and Morell, P. (1974), *J. biol. Chem.,* **249**, 3973.
234. Yamada, K. M. and Weston, J. A. (1974), *Proc. natn. Acad. Sci. U.S.A.,* **71**, 3492.
235. Sefton, B. M., Wickus, G. G. and Burge, B. W. (1973), *J. Virol.,* **11**, 730.
236. Pearlstein, E. and Waterfield, M. D. (1974), *Biochim. biophys. Acta.,* **326**, 1.
237. Hynes, R. O. (1973), *Proc. natn. Acad. Sci. U.S.A.,* **70**, 3170.
238. Hynes, R. O. and Humphryes, K. C. (1974), *J. Cell Biol.,* **62**, 438.
239. Hynes, R. O. and Bye, J. M. (1974), Cell, **3**, 113.
240. Hogg, N. M. (1974), *Proc. natn. Acad. Sci. U.S.A.,* **71**, 489.
241. Graham, J. M., Hynes, R. O. Davidson, E. A. and Bainton, D. F. (1975), *Cell,* **4**, 353.
242. Vidal, R., Tarone, G., Peroni, F. and Comoglio, P. M. (1974), *F.E.B.S Letters,* **47**, 107.
243. Wickus, G. G., Branton, P. E. and Robbins, P. W. (1974), in *Control of Cell Proliferation,* (Clarkson, B. and Baserga, R., eds.), p. 541, Cold Spring Harbor Laboratory.
244. Wray, V. P. and Perdue, J. F. (1974), *J. biol. Chem.,* **249**, 1189.
245. Lisowska-Bernstein, B., Rinuy, A. and Vassalli, P. (1973), *Proc. natn. Acad. Sci. U.S.A.,* **70**, 2879.
246. Baur, S., Vitetta, E. S., Sherr, C. J., Schenkein, I. and Uhr, J. W. (1971), *J. Immunol.,* **106**, 1133.
247. Vitetta, E. S., Baur, S. and Uhr, J. W. (1971), *J. exp. Med.,* **134**, 242.
248. Sherr, C. J., Baur, S., Grundke, I., Zeligs, J. Zeligs, B. and Uhr, J. W. (1972), *J. exp. Med.,* **135**, 1392.
249. Vitetta, E. S., Bianco, C., Nussenzweig, V. and Uhr, J. W. (1972), *J. exp. Med.,* **136**, 81.
250. Vitetta, E. S. and Uhr, J. W. (1972), *J. Immunol.,* **108**, 577.
251. Vitetta, E. S. and Uhr, J. W. (1972), *J. exp. Med.,* **136**, 676.
252. Vitetta, E. S., Grundke-Iqbal, I., Holmes, K. V. and Uhr, J. W. (1974), *J. exp. Med.,* **139**, 862.
253. Vitetta, E. S. and Uhr, J. W. (1974), *J. exp. Med.,* **139**, 1599.
254. Melcher, U., Vitetta, E. S., McWilliams, M., Lamm, M. E., Phillips-Quagliata, J. M. and Uhr, J. W. (1974), *J. exp. Med.,* **140**, 1427.
255. Vitetta, E. S., Uhr, J. W. and Boyse, E. A. (1972), *Cell Immunol.,* **4**, 187.
256. Vitetta, E. S., Melcher, U., McWilliams, M., Lamm, M. E., Phillips-Quagliata, J. M. and Uhr, J. W. (1974), *J. exp. Med.,* **141**, 206.

257. Vitetta, E. S., Boyse, E. A. and Uhr, J. W. (1972), *Eur. J. Immunol.*, **3**, 446.
258. Wernet, D., Vitetta, E. S., Uhr, J. W. and Boyse, E. A. (1973), *J. exp. Med.*, **138**, 847.
259. Vitetta, E. S., Uhr, J. W. and Boyse, E. A. (1974), *Eur. J. Immunol.*, **4**, 276.
260. Vitetta, E. S. and Uhr, J. W. (1973), *Transplant Rev.*, **14**, 50.
261. Marchalonis, J. J. and Cone, R. E. (1973), *Transplant Rev.*, **14**, 3.
262. Marchalonis, J. J., Cone, R. E. and Santer, V. (1971), *Biochem. J.*, **124**, 921.
263. Cone, R. E., Marchalonis, J. J. and Rolley, R. T. (1971), *J. exp. Med.*, **134**, 1373.
264. Marchalonis, J. J., Cone, R. E. and Atwell, J. L. (1972), *J. exp. Med.*, **135**, 956.
265. Cone, R. E., Sprent, J. and Marchalonis, J. J. (1972), *Proc. natn. Acad. Sci. U.S.A.*, **69**, 2556.
266. Marchalonis, J. J., Atwell, J. L. and Cone, R. E. (1972), *Nature New Biology*, **235**, 240.
267. Atwell, J. L., Cone, R. E. and Marchalonis, J. J. (1973), *Nature New Biology*, **241**, 251.
268. Haustein, D., Marchalonis, J. J. and Crumpton, M. J. (1974), *Nature*, **252**, 602.
269. Kennel, S. J. and Lerner, R. A. (1973), *J. molec. Biol.*, **76**, 485.
270. Schmidt-Ullrich, R., Ferber, E., Knüfermann, H., Fischer, H. and Wallach, D. F. H. (1974), *Biochim. biophys. Acta.*, **332**, 175.
271. Schenkein, I., Levy, M. and Uhr, J. W. (1972), *Cell Immunol.*, **5**, 490.
272. Wernet, P. and Kunkel, H. G. (1973), *J. exp. Med.*, **138**, 1021.
273. del Villano, B. C., Nave, B., Croker, B. P., Lerner, R. A. and Dixon, F. J. (1975), *J. exp. Med.*, **141**, 172.
274. Tung, J. S., Vitetta, E. S., Fleissner, E. and Boyse, E. A. (1975), *J. exp. Med.*, **141**, 198.
275. Trosper, T. and Levy, D. (1974), *Biochemistry*, **13**, 4284.
276. Czech, M. P. and Lynn, W. S. (1973), *Biochemistry*, **12**, 3597.
277. Strum, J. M. and Edelman, I. S. (1973), *J. Memb. Biol.*, **14**, 17.
278. Poduslo, J. F. and Braun, P. E. (1975), *J. biol. Chem.*, **250**, 1099.
279. Smart, J. E. and Hynes, R. O. (1974), *Nature*, **251**, 319.
280. Salton, M. R. J., Schor, M. T. and Ng, M. H. (1972), *Biochim. biophys. Acta.*, **290**, 408.
281. Rittenhouse, H. G., McFadden, B. A., Shumway, L. K. and Heptinst, J. (1973), *J. Bact.*, **115**, 338.
282. Lampe, H. H. and Drews, G. (1972), *Arch. Mikrobiol.*, **84**, 1.
283. Kusel, J. R. (1972), *Parasitology*, **65**, 55.
284. Amar, A., Rottem, S. and Razin, S. (1974), *Biochim. biophys. Acta.*, **352**, 228.
285. Stanley, P. and Haslam, E. A. (1971), *Virology*, **46**, 764.
286. Witte, O. N., Weissman, I. L. and Kaplan, H. S. (1973), *Proc. natn. Acad. Sci. U.S.A.*, **70**, 36.
287. Welton, A. F. and Aust, S. D. (1972), *Biochim. biophys. Res. Comm.*, **49**, 661.
288. Kreibich, G., Hubbard, A. L. and Sabatini, D. D. (1974), *J. Cell Biol.*, **60**, 616.
289. Astle, L. and Cooper, C. (1974), *Biochemistry*, **13**, 154.

290. Huber, C. T. and Morrison, M. (1973), *Biochemistry,* **12**, 4274.
291. Arntzen, J., Armond, P. A., Zettinger, C. S., Vernotte, C. and Briantais, J. M. (1974), *Biochim. biophys. Acta.,* **347**, 329.
292. Thorley-Lawson, D. A. and Green, N. M. (1973), *Eur. J. Biochem.,* **40**, 403.
293. Litman, D. J. and Cantor, C. R. (1974), *Biochemistry,* **13**, 512.
294. Michalski, C. J., Sells, B. H. and Morrison, M. (1973), *Eur. J. Biochem.,* **33**, 481.
295. Miller, R. V. and Sypherd, P. S. (1973), *J. molec. Biol.,* **78**, 527.
296. David, G. S. and Reisfeld, R. A. (1974), *Biochemistry,* **13**, 1014.
297. David, G. S. (1972), *Biochem. biophys. Res. Comm.,* **48**, 464.
298. Arnold, C. (1881), *Arch. Pharmac.,* **19**, 41.
299. Thurlow, S. (1925), *Biochem. J.,* **19**, 175.
300. Elliot, K. A. C. (1932), *Biochem. J.,* **26**, 10.
301. Polis, B. D. and Shmukler, H. W. (1953), *J. biol. Chem.,* **201**, 475.
302. Morrison, M., Hamilton, H. B. and Stotz, E. (1957), *J. biol. Chem.,* **228**, 767.
303. Paul, K. G. (1960), in *The Enzymes,* Vol. III, (Boyer, P. D., Lardy, H. and Myrbäck, K., eds.), Academic Press, New York.
304. Morrison, M. and Hultquist, D. E. (1963), *J. biol. Chem.,* **238**, 2847.
305. Rombauts, W. A., Schroeder, W. A. and Morrison, M. (1967), *Biochemistry,* **6**, 2965.
306. Hultquist, D. E. and Morrison, M. (1963), *J. biol. Chem.,* **238**, 2843.
307. Morrison, M. and Bayse, G. S. (1970), *Biochemistry,* **9**, 2995.
308. Bayse, G. S. and Morrison, M. (1971), *Archs. Biophys. Biochim.,* **145**, 143.
309. Morrison, M., Bayse, G. and Danner, D. J. (1970), in *Biochemistry of the Phagocytic Process,* (Schultz, J., ed.), p. 51, North-Holland, Netherlands.
310. Morrison, M. and Bayse, G. S. (1973), in *Oxidases and Related Redox Systems, Proc. 2nd Int. Symp.,* (King, T. E., Mason, H. S. and Morrison, M., eds.), p. 375, University Park Press, Baltimore.
311. Jago, G. R. and Morrison, M. (1962), *Proc. Soc. exp. Biol. Med.,* **111**, 585.
312. Clem, W. H. and Klebanoff, S. J. (1966), *J. Bact.,* **91**, 1848.
313. Hogg, D. McC. and Jago, G. R. (1970), *Biochem. J.,* **117**, 779.
314. Thomas, J. A., Morris D. R. and Hager, L. P. (1970), *J. biol. Chem.,* **245**, 3129.
315. Thomas, J. A., Morris, D. R. and Hager, L. P. (1970), *J. biol. Chem.,* **245**, 3135.
316. Hager, L. P., Thomas, J. A. and Morris, D. R. (1970), in *Biochemistry of the Phagocytic Process,* (Schultz, J., ed.), p. 67, North-Holland, Netherlands.
317. Klebanoff, S. J. (1967), *J. exp. Med.,* **126**, 1063.
318. Taurog, A., and Gamble, D. E. (1966), *Fedn. Proc., Fedn. Am. Socs. exp. Biol.,* **25**, 347.
319. Morrison, M. (1974), *Meth. Enzym.,* **32(B)**, 103.
320. Mans, R. J. and Novelli, G. M. (1961), *Archs. Biophys. Biochem.,* **94**, 48.
321. Steinman, R. M., Silver, J. M. and Cohn, Z. A. (1974), *J. Cell Biol.,* **63**, 949.
322. Edelson, P. J. and Cohn, Z. A. (1973), *J. exp. Med.,* **138**, 318.
323. Rae, P. A. and Schimmer, B. P. (1974), *J. biol. Chem.,* **249**, 5649.
324. Stagg, B. H., Temperley, J. M., Rochman, H. and Morley, J. S. (1970), *Nature,* **228**, 58.
325. Lambert, B., Sutter, B. C. J. and Jacquemain, C. (1972), *Horm. Met. Res.,* **4**, 149.

326. Roche, J., Lissitzky, S. and Michel, R. (1954), in *Methods of Biochemical Analysis*, (Glick, D. ed.), Vol. 1, p. 243, Interscience Div., John Wiley and Sons, New York.
327. Ouellette, R. P. and Balcius, J. F. (1966), *J. chromat.*, **24**, 465.
328. Fleischer, S., Fleischer, B. and Stoeckenius, W. (1967), *J. Cell Biol.*, **32**, 193.
329. Segrest, J. R., Jackson, F. L., Andrews, E. P. and Marchesi, V. T. (1971), *Biochem. biophys. Res. Comm.*, **44**, 390.
330. Burger, M. M. (1971), in *Current Topics in Regulation*, (Horecker, B., ed.), Vol. 3, p. 135, Academic Press, New York.
331. Ruoslahti, E., Vaheri, A., Kuusela, P. and Linder, E. (1973), *Biochim. biophys. Acta.*, **322**, 352.
332. Ruoslahti, E. and Vaheri, A. (1974), *Nature*, **248**, 789.
333. Vaheri, A. and Ruoslahti, E. (1974), *Int. J. Cancer*, **13**, 579.
334. Kuusela, P., Ruoslahti, E. and Vaheri, A. (1975), *Biochim. biophys. Acta.*, **379**, 295.
335. Wartiovaara, J., Linder, E. and Ruoslahti, E. (1974), *J. exp. Med.*, **140**, 1522.

# Index

## Index